Twenty Years of Kaniadakis Entropy: Current Trends and Future Perspectives

Twenty Years of Kaniadakis Entropy: Current Trends and Future Perspectives

Guest Editors

Dionissios T. Hristopulos
Sergio Luiz E. F. da Silva
Antonio M. Scarfone

MDPI

Basel • Beijing • Wuhan • Barcelona • Belgrade • Novi Sad • Cluj • Manchester

Guest Editors
Dionissios T. Hristopulos
School of Electrical &
Computer Engineering
Technical University of Crete
Chania
Greece

Sergio Luiz E. F. da Silva
Laboratory of Parallel
Architectures for Signal
Processing
Universidade Federal do Rio
Grande do Norte
Natal
Brazil

Antonio M. Scarfone
Complex Systems Institute
Consiglio Nazionale delle
Ricerche
Roma
Italy

Editorial Office
MDPI AG
Grosspeteranlage 5
4052 Basel, Switzerland

This is a reprint of the Special Issue, published open access by the journal *Entropy* (ISSN 1099-4300), freely accessible at: https://www.mdpi.com/journal/entropy/special_issues/kaniadakis_entropy.

For citation purposes, cite each article independently as indicated on the article page online and as indicated below:

Lastname, A.A.; Lastname, B.B. Article Title. *Journal Name* **Year**, *Volume Number*, Page Range.

ISBN 978-3-7258-3649-9 (Hbk)
ISBN 978-3-7258-3650-5 (PDF)
https://doi.org/10.3390/books978-3-7258-3650-5

Contents

About the Editors

Dionissios T. Hristopulos

Computer Engineering at the Technical University of Crete, where he directs the Master's program in Machine Learning and Data Science. He holds a Diploma in Electrical Engineering from the National Technical University of Athens and a PhD in Physics from Princeton University. He has held appointments at the Department of Environmental Sciences and Engineering (University of North Carolina at Chapel Hill) and the Pulp and Paper Research Institute of Canada (currently FPInnovations). Hristopulos serves on the Editorial Boards of *Computers* and *Geosciences* (Elsevier) and *Stochastic Environmental Research and Risk Assessment* (Springer). He has co-authored more than 100 peer-reviewed publications and is the sole author of "Random Fields for Spatial Data Modeling: A Primer for Scientists and Engineers" (Springer, 2020). His research interests include spatiotemporal statistics, statistical physics, and machine learning and applications. In 2003, he shared the Johannes A. Van den Akker International Prize for Advances in Paper Physics. In 2024, he was awarded the Georges Matheron lectureship by the International Association of Mathematical Geosciences.

Sergio Luiz E. F. da Silva

Signal Processing at the Federal University of Rio Grande do Norte (UFRN). He holds a PhD in Physics (2021), which he completed one year ahead of schedule, focusing on statistical physics and complex systems, and a Master's degree (2016) in Geophysics, both from the UFRN. He also obtained his degree in Physics from UFRN in 2013. His primary research interests are in the analysis of data-centered problems, especially in the theory of inverse problems. He is currently involved in projects aimed at improving seismic imaging techniques using wave equation-based approaches for complex geologic structures, overcoming challenges such as low resolution and non-Gaussian noise. These projects are conducted in collaboration with multinational companies and universities from the United States, Norway, France and Brazil. Over the years, Dr. da Silva has collaborated with international institutions, including Politecnico di Torino, Italy, and Mines ParisTech, France, where he expanded his research horizons in geophysical methodologies. He was a member of the Organizing Committee of the SigmaPhi International Conference series in 2023. Dr. da Silva has published papers in reputable journals, contributing to the advancement of knowledge in inverse problems and related fields.

Antonio M. Scarfone

Antonio M. Scarfone, working mainly in the field of statistical mechanics, is a senior researcher in mathematical physics at the Complex System Institute of the National Research Council, Italy. Born in 1968, he studied physics at the University of Torino, earning his degree in 1996, and later completed his Ph.D. in Physics at Polytechnic of Torino in 2000. To date, he has authored more than 100 scientific papers on statistical mechanics, kinetic theory, geometry information, nonlinear classical and quantum dynamics, and noncommutative algebras, all published in many international peer-review journals. He is Editor-in-Chief of the Section titled *Statistical Physics* in *Entropy* and a member of the Editorial Board of several journals in mathematical physics, such as *Advances in Mathematical Physics, Bulletin of Mathematical Analysis and Applications*, etc., and belongs to the advisory panel of *Journal of Physics A*. Lastly, he is one of the main organizers of the *SigmaPhi International Conference* series that has been held every three years starting from 2005.

Preface

Entropy is publishing a Special Issue in honor of the twentieth anniversary of Kaniadakis entropy's introduction. This then-new entropy is one of the most viable candidates for explaining the experimentally observed power-law-tailed statistical distributions in various physical, natural, and artificial complex systems. In this reprint, readers can find information about the recent applications of Kaniadakis entropy in mechanical statistics, high-energy physics, cosmology, information geometry, biology, and economic sciences. We hope that the papers gathered in this Special Issue may inspire the reader and encourage them to continue the development of this expanding field of statistical physics.

Entropy, introduced in statistical physics at the end of the 19th century, remains an important but enigmatic concept. Over the past century, several particularly intriguing and powerful extensions of the classical Shannon–Boltzmann–Gibbs entropic form have been proposed. Among these, Kaniadakis entropy offers a new perspective on systems with complex dynamics and non-extensive behavior exhibiting power-law statistical distributions. Its applications span cover varied fields, ranging from statistical mechanics to complex systems, including cosmology and quantum mechanics.

This Special Issue of *Entropy* commemorates two decades of developments related to Kaniadakis entropy. In the editorial, we discuss the introduction of entropy, beginning with the Kaniadakis deformation of Napier's number compared to the well-known Euler deformation. We then introduce Kaniadakis' exponential and logarithmic functions and emphasize the axiomatic structure of Kaniadakis' statistical mechanics. The Editorial also includes a list of over 250 papers written in the past twenty-five years up to January 2025 by various authors from around the world, which focus on the foundations, mathematical formalism, and applications of this theory. This compilation should not be considered a complete bibliography on Kaniadakis statistics since it includes works that explicitly refer to the subject in the title or abstract. This list will be accessible through the SCOPUS and Web of Science-Clarivate databases.

Dionissios T. Hristopulos, Sergio Luiz E. F. da Silva, and Antonio M. Scarfone
Guest Editors

entropy

Editorial

Twenty Years of Kaniadakis Entropy: Current Trends and Future Perspectives

Dionissios T. Hristopulos [1,2], Sérgio Luiz E. F. da Silva [3,4,*] and Antonio M. Scarfone [4]

1 School of Electrical and Computer Engineering, Technical University of Crete, 73100 Chania, Greece; dchristopoulos@tuc.gr
2 Southern Marine Science and Engineering, 2 University Street, Zhuhai 519082, China
3 Laboratory of Parallel Architectures for Signal Processing, Federal University of Rio Grande do Norte, Natal 59078-970, RN, Brazil
4 Istituto dei Sistemi Complessi—Consiglio Nazionale delle Ricerche (ISC-CNR), c/o Dipartimento di Scienza Applicata e Tecnologia del Politecnico di Torino, 10129 Torino, Italy; antoniomaria.scarfone@cnr.it
* Correspondence: sergio.silva.highres@imd.ufrn.br

Napier's number $e = 2.7182818284\ldots$ can be introduced as the limit of a numerical sequence $e = \lim_{n\to\infty} e_n$. There are infinite sequences that lead to e, the simplest of which was proposed by Euler of the form $e_n^E = (1+1/n)^n$. Euler's sequence converges very slowly and only reaches Napier's number to two decimal places for $n = 500$. Among the infinite sequences that converge to e, there is only one that has the property $e_{-n} = e_n$. This new numerical sequence converging toward Napier's number e emerges naturally within Kaniadakis formalism, proposed in a seminal paper from 2001 [1], and has the form

$$e_n^K = \left(\sqrt{1+\frac{1}{n^2}}+\frac{1}{n}\right)^n . \tag{1}$$

The Kaniadakis sequence converges very quickly to the Napier number, which is already calculated correctly for $n = 10$ to the first two decimal places, while for $n = 500$, the first five decimal places have been obtained correctly.

Starting from the two numerical sequences that lead to the Napier number, it is possible to construct the corresponding sequences of functions that converge to the exponential function. The Euler sequence of functions leading to the exponential function is very simple and is given by $\exp_n(x) = (1+x/n)^n$. This sequence represents a one-parameter deformation of the ordinary exponential function, with n being the deformation parameter. This deformed exponential of Euler's function $exp_n(x)$ has been used in statistical mathematics to construct the Student distribution, which has been employed to study astrophysical plasmas since the end of the 1960s. This distribution is also known as the κ distribution, and the plasmas described by this distribution are called κ plasmas. The name is derived from the fact that instead of the deformation parameter n, parameter κ is historically used so that in the $\kappa \to \infty$ limit, the κ distribution reduces to the corresponding ordinary exponential distribution. Since the late 1980s, Euler deformation with the new parameter $q = (1+n)/n$ has been used instead of n in the development of non-extensive statistical mechanics.

Given the impact of the Euler deformation in plasma physics, in statistical physics, and more generally in statistical sciences, in 2001, Kaniadakis introduced a deformation of the exponential function, which now bears his name. The new deformed function captures properties of the ordinary exponential function. The starting point is the property of the self-duality of the ordinary exponential function, that is, $\exp(x)\exp(-x) = 1$. The deformation parameter is denoted by κ, and the deformed exponential is defined by

Received: 13 February 2025
Accepted: 18 February 2025
Published: 27 February 2025

Citation: Hristopulos, D.T.; da Silva, S.L.E.F.; Scarfone, A.M. Twenty Years of Kaniadakis Entropy: Current Trends and Future Perspectives. *Entropy* **2025**, *27*, 247. https://doi.org/10.3390/e27030247

$$\exp_\kappa(x) = \left(\sqrt{1+\kappa^2x^2}+\kappa x\right)^{1/\kappa} . \tag{2}$$

In contrast with the homonymous parameter of κ plasmas, for the new deformed exponential $\exp_\kappa(x)$, the range of κ is $0<\kappa<1$. The ordinary exponential is obtained in the limit $\lim_{\kappa\to 0}\exp_\kappa(x)=\exp(x)$. Like the ordinary exponential function, the Kaniadakis exponential is self-dual, that is, $\exp_\kappa(x)\exp_\kappa(-x)=1$.

All properties of the ordinary exponential function $\exp(x)$ are transferred in an identical form to the function $\exp_\kappa(x)$, accordingly generalized by the parameter κ. This makes the function $\exp_\kappa(x)$ particularly interesting for the construction of a statistical theory in analogy to the standard statistical theory of Boltzmann–Gibbs–Maxwell, which is based on the function $\exp(x)$. Kaniadakis showed in 2001 [1] that κ-statistical mechanics preserve important features of classical statistical mechanics. The most interesting difference between the two theories is the behavior of the tails of the associated statistical distributions: while the ordinary theory leads to exponential tails, in κ-statistical mechanics, such tails are described by Pareto power laws.

It is noteworthy that power-law tails are also present in Euler's deformed exponential function and in the related statistical theory. The main difference between the two statistical theories is that Kaniadakis theory arises in the framework of special relativity, as shown in papers from 2002 and 2005 [2,3]. The Kaniadakis deformation is therefore a relativistic effect, which thus provides a robust physical basis for κ-statistical theory.

In the three seminal papers mentioned above, Kaniadakis introduced influential concepts that have transformed our understanding of the physics of complex systems. His pioneering work laid the foundation for what is now known as Kaniadakis entropy or κ-entropy. This concept emerged as a relativistic generalization of the classical Boltzmann–Shannon entropy, marking a significant advance in the framework of special relativity and statistical mechanics. Kaniadakis entropy is defined according to the following:

$$S_\kappa = -\sum_i f_i \ln_\kappa(f_i) = \sum_i f_i \ln_\kappa(1/f_i) , \tag{3}$$

where f_i represents the probability distribution function (probability mass function), while $\ln_\kappa(f)$ is the Kaniadakis logarithm, defined as the inverse of the $\exp_\kappa(x)$ function. The summation is over all the states of the system. In the continuum case, the summation is replaced by integration, and f_i is replaced by the probability density function $f(x)$.

The κ-deformed logarithm satisfies the *scaling property* $\frac{d}{df}[f\ln_\kappa(f)] = \frac{1}{\gamma_\kappa}\ln_\kappa(e_\kappa f)$, where γ_κ and e_κ are two scaling constants. The maximization of the entropy S_κ, subject to available moment constraints, leads to a maximum entropy probability distribution expressed univocally in terms of the $\exp_\kappa(x)$ function. This extends the exponential maximum entropy distributions obtained by maximizing Shannon's entropy.

Interestingly, the above scaling property is common to the following three functions: the ordinary logarithm, the Euler-deformed logarithm, and the κ-deformed logarithm. Of course, different scaling constants apply to the three logarithms. However, the self-duality property is only present for the ordinary and κ-deformed functions. The property of self-duality for $\exp(x)$ and $\exp_\kappa(x)$, i.e., $\exp_\kappa(x)=1/\exp_\kappa(-x)$, can also be expressed for the inverse functions. If we define $f=\exp_\kappa(x)$ and operate with $\ln_\kappa$ on both sides of the self-duality equation, we obtain $\ln_\kappa(1/f)=-\ln_\kappa(f)$ and, in the $\kappa\to 0$ limit, $\ln(1/f)=-\ln f$. The above *self-duality* property allows us to interpret the entropy S_κ as the mean value of the function $-\ln_\kappa(f_i)$ or equivalently that of $\ln_\kappa(1/f_i)$; thus, κ-entropy extends the definition of Boltzmann entropy while maintaining the self-duality property.

Ordinary statistical mechanics are based on the Boltzmann–Gibbs–Shannon (BGS) entropy or logarithmic entropy. The uniqueness of BGS entropy is derived from the four Khinchin–Shannon axioms, namely continuity (the entropy changes continuously with changes in the probability distribution), maximality (it is maximized by the uniform distribution), expansibility (adding a zero-probability event does not change the entropy), and strong additivity (the entropy of the sum of two independent variables is the sum of the individual entropies). In this Special Issue, Kaniadakis proposes that the Boltzmann–Gibbs–Shannon entropy be introduced by retaining the first three Khinchin–Shannon axioms while replacing the fourth axiom with two new axioms: the so-called *scaling* and *self-duality* axioms that arise naturally in special relativity. The new set of axioms allows for the introduction of ordinary entropy without using the concepts of additivity or statistical independence. Kaniadakis also shows that the new set of axioms is uniquely satisfied by κ-entropy in addition to ordinary entropy; the latter is obtained from κ-entropy in the classical limit $\kappa \to 0$. The emergence of Kaniadakis entropy in special relativity allows us to revisit relativistic thermodynamics problems, providing evidence for the Einstein–Planck proposal of relativistic temperature reduction in a moving frame (see Kaniadakis' contribution in this Special Issue).

The scientific community has embraced and expanded on the innovative framework of Kaniadakis entropy since its introduction. Hence, the mathematical and practical implications of κ-statistics have been the subject of over 260 papers published by more than 200 scientists. For a comprehensive list of these works, please refer to the References section. The applications of Kaniadakis entropy extend over a wide range of disciplines [1–48], including kinetic theory [49–64], thermodynamics [65–71], astrophysics [72–78], cosmology [79–104], dark energy models [105–127], quantum gravity [128–139], information geometry [140–147], information theory and optimization [148–150], classical statistics [151–169], quantum statistics [170–175], artificial intelligence [176–178], dynamical systems [179–189], econophysics, sociophysics and network theory [190–212], geophysics and geomechanics [213–223], genomic analysis [224–227], nuclear physics [228–236], and plasma physics [237–263], among other topics. The versatility and applicability of κ-entropy highlight its profound impact on theoretical foundations and practical applications across scientific fields.

This Special Issue aims to capture the current trends and future perspectives of Kaniadakis entropy and its applications:

- Tan et al. (2022) [253] investigate the dispersion and Landau damping of Langmuir and ion acoustic waves in a κ-deformed Kaniadakis distributed plasma system. Their analysis shows that dispersion increased with increasing κ, while Landau damping was suppressed. These results could provide insights into plasma particle trapping and energy transport.
- Biró (2022) [42] explores the use of generalized exponentials in particle–hole symmetry, demonstrating that the Kaniadakis κ-approach is compatible with the Kubo–Martin–Schwinger (KMS) relation, and discusses potential further generalizations.
- De Lima et al. (2022) [226] study the distribution of plant DNA lengths in three Cucurbitaceae species (gourd family). Using Bayesian analysis, they find that the sum of two κ-exponential distributions provides the best fit (among other models) to the empirical distribution curves.
- Hristopulos and Baxevani (2022) [220] introduce a nonlinear normalizing transformation and its inverse based on deformed logarithmic and exponential functions; the transformed pair is useful in the analysis of skewed data, and its inverse is stable, unlike the commonly used Box-Cox transform pair. They also discuss the connection between the heavy-tailed κ-Weibull distribution and the weakest-link scaling theory, showing that the former is suitable for modeling mechanical strength distributions of materials. The paper also introduces the κ-lognormal probability distribution,

which has a flexible right tail and can be used to model fluid permeability in random porous media.

- De Abreu et al. (2022) [234] review the application of Kaniadakis entropy to nuclear reactor physics. In particular, they focus on the Doppler broadening effect on neutron cross-sections if the Maxwell–Boltzmann distribution is replaced with the Kaniadakis distribution (which uses the deformed exponential). The authors claim more accurate radiative capture cross-section calculations using the Kaniadakis distribution.
- Guha (2022) [61] uses the Calogero–Leyvraz Lagrangian framework to construct 2D Lotka–Volterra replicator equations and $N = 2$ relativistic Toda lattice systems. The kinetic energy term is deformed using the κ-deformed logarithm, resulting in new formulations of the above systems.
- Luciano (2022) [89] explores recent advances and future challenges of Kaniadakis statistics in gravity and cosmology. He focuses on implications of κ-entropy on cosmological theories and Big Bang nucleosynthesis and uses observational evidence to constrain the κ-parameter.
- Wada and Scarfone (2023) [44] present applications of Kaniadakis distributions to various topics in statistical physics and natural science such as Gompertz functions, the Bloch equation for thermal states, contact density dynamics, and the law of large numbers under κ-addition.
- Martinez and de Abreu (2023) [236] present a review on the use of Kaniadakis entropy in nuclear reactor physics. They generate simulated nuclear data under non-thermal equilibrium conditions using the Kaniadakis distribution to address the limitations of the traditional Maxwell–Boltzmann statistics.
- da Silva et al. (2023) [223] introduce a novel objective function for full-waveform inversion problems, incorporating the Kaniadakis κ-Gaussian distribution and optimal transport theory, to mitigate non-Gaussian noise and phase ambiguity in seismic wave analysis.
- Pistone and Shoaib (2023) [147] propose using a specific case of the Kaniadakis logarithm for the exploratory analysis of compositional data. They show that the affine information geometry derived from Kaniadakis' algorithm provides a consistent framework for the geometric analysis of compositional data. Moreover, they propose a particular functional form of the κ-divergence.
- Clementi (2023) [211] discusses the application of the κ-generalized distribution in the statistical analysis of income data, highlighting the distribution's analytical properties, and relationships with other distributions. The paper also comments on the very good agreement of the κ-generalized distribution with empirical data.
- Evangelista and Lenzi (2023) [64] use the H-theorem to examine the dynamics of a system composed of two subsystems that obey nonlinear Fokker–Planck equations. They focus on the behavior of the entropy of the entire system considering subsystems that have (i) identical and (ii) different dynamics.
- Scarfone and Wada (2024) [45] demonstrate that Kaniadakis entropy can become multi-additive under a suitably defined constraint, that is, under the composition of two identically distributed probability distributions.
- Kaniadakis (2024) [46] highlights the importance of κ-entropy in statistical theory, identifying five axioms in κ-statistical theory that provide a solid foundation for understanding entropy in complex systems. This study sheds light on the physical origins of κ-entropy, emphasizing the self-duality and scaling axioms as fundamental elements. Kaniadakis also emphasizes the emergence of κ-entropy in Einstein's special theory of relativity and introduces relativistic statistical mechanics based on the new entropy. Finally, Kaniadakis shows that the new formalism allows us to re-discuss

in a modern way some remaining open problems of relativistic thermodynamics introduced by Planck concerning the transformation law of temperature and entropy.

The interdisciplinary collaboration of experts in different fields has substantiated the theoretical foundations of Kaniadakis entropy and established practical applications of the κ-entropy as well as other κ-deformed functions in various scientific fields. This Special Issue celebrates the insightful contributions of Giorgio Kaniadakis and the work of scholars who have significantly advanced the field of κ-statistics.

In addition to presenting a tribute to past achievements and accomplishments, this Special Issue aims to provide motivation and directions for future research. In this spirit, this introduction includes an extensive list of published research on κ-statistics and their applications. The list of references is organized in sections according to the scientific field of application. We believe that this information will be useful for researchers entering the field of κ-statistics.

Papers published in this Special Issue

- Biró, T.S. *Kaniadakis Entropy Leads to Particle–Hole Symmetric Distribution*, Entropy **2022**, 24 (9), 1217. (Ref. [42])
- Clementi, F. *The Kaniadakis Distribution for the Analysis of Income and Wealth Data*, Entropy **2023**, 25(8), 1141. (Ref. [211])
- da Silva, S.L.E.F.; de Araújo, J.M.; de la Barra, E.; Corso, G. *A Graph-Space Optimal Transport Approach Based on Kaniadakis κ-Gaussian Distribution for Inverse Problems Related to Wave Propagation*, Entropy **2023**, 25, 990. (Ref. [223])
- de Abreu, W.V.; Maciel, J.M.; Martinez, A.S.; Gonçalves, A.D.C.; Schmidt L. *Doppler Broadening of Neutron Cross-Sections Using Kaniadakis Entropy*, Entropy **2022**, 24(10), 1437. (Ref. [234])
- de Lima, M.M.F.; Anselmo, D.H.A.L.; Silva, R.; Nunes, G.H.S.; Fulco, U.L.; Vasconcelos, M.S.; Mello, V.D *A Bayesian Analysis of Plant DNA Length Distribution via κ-Statistics*, Entropy **2022**, 24, 1225. (Ref. [226])
- Evangelista, L.R.; Lenzi, E.K. *Nonlinear Fokker–Planck Equations, H-Theorem and Generalized Entropy of a Composed System*, Entropy **2023**, 25, 1357. (Ref. [64])
- Guha, P. *The κ-Deformed Calogero-Leyvraz Lagrangians and Applications to Integrable Dynamical Systems*, Entropy 2023, 24, 1673. (Ref. [61])
- Hristopulos, D.T.; Baxevani A. *Kaniadakis Functions beyond Statistical Mechanics: Weakest-Link Scaling, Power-Law Tails, and Modified Lognormal Distribution*, Entropy **2022**, 24(10), 1362. (Ref. [220])
- Kaniadakis, G. Relativistic Roots of κ-Entropy. *Entropy*. **26**, 406 (2024). (Ref. [46])
- Luciano G.G. *Gravity and Cosmology in Kaniadakis Statistics: Current Status and Future Challenges*, Entropy **2022**, 24(12), 1712. (Ref. [89])
- Martinez A.S.; de Abreu W.V. *The Scientific Contribution of the Kaniadakis Entropy to Nuclear Reactor Physics: A Brief Review*, Entropy **2023**, 25(3), 478. (Ref. [236])
- Pistone G.; Shoaib M. *Kaniadakis's Information Geometry of Compositional Data, Entropy* **2023**, 25(7), 1107. (Ref. [147])
- Scarfone A.M., Wada T. *Multi-Additivity in Kaniadakis Entropy*, Entropy **2024**, 26(1), 77. (Ref. [45])
- Tan, L.; Yang, Q.; Chen, H.; Liu S. *The Longitudinal Plasma Modes of κ-Deformed Kaniadakis Distributed Plasmas Carrying Orbital Angular Momentum*, Entropy **2022**, 24 (9), 1211. (Ref. [253])
- Wada T., Scarfone A.M. On the Kaniadakis Distributions Applied in Statistical Physics and Natural Sciences, Entropy **2023**, 25(2), 292. (Ref. [44])

Author Contributions: Conceptualization, D.T.H., S.L.E.F.d.S. and A.M.S.; writing—original draft preparation, D.T.H., S.L.E.F.d.S. and A.M.S.; writing—review and editing, D.T.H., S.L.E.F.d.S. and A.M.S.; supervision, D.T.H., S.L.E.F.d.S. and A.M.S. All authors have read and agreed to the published version of the manuscript.

Conflicts of Interest: The authors declare no conflicts of interest.

References

1. Kaniadakis, G. Non-linear kinetics underlying generalized statistics. *Phys. A* **2001**, *296*, 405–425. [CrossRef]
2. Kaniadakis, G. Statistical mechanics in the context of special relativity. *Phys. Rev. E* **2002**, *66*, 056125. [CrossRef]
3. Kaniadakis, G. Statistical mechanics in the context of special relativity II. *Phys. Rev. E* **2005**, *72*, 036108. [CrossRef]
4. Kaniadakis, G. H-theorem and generalized entropies within the framework of nonlinear kinetics. *Phys. Lett. A* **2001**, *288*, 283–291. [CrossRef]
5. Kaniadakis, G.; Scarfone, A.M. A new one-parameter deformation of the exponential function. *Phys. A* **2002**, *305*, 69–75.
6. Naudts, J. Deformed exponentials and logarithms in generalized thermostatistics. *Phys. A* **2002**, *316*, 323–334. [CrossRef]
7. Naudts, J. Continuity of a class of entropies and relative entropies. *Rev. Math. Phys.* **2004**, *16*, 809–822. [CrossRef]
8. Kaniadakis, G.; Scarfone, A.M. Lesche stability of κ-entropy. *Phys. A* **2004**, *340*, 102–109. [CrossRef]
9. Abe, S.; Kaniadakis, G.; Scarfone, A.M. Stabilities of generalized entropy. *J. Phys. A Math. Gen.* **2004**, *37*, 10513. [CrossRef]
10. Kaniadakis, G.; Lissia, M.; Scarfone, A.M. Deformed logarithms and entropies. *Phys. A* **2004**, *40*, 41–49. [CrossRef]
11. Kaniadakis, G.; Lissia, M.; Scarfone, A.M. Two-parameter deformations of logarithm, exponential, and entropy: A consistent framework for generalized statistical mechanics. *Phys. Rev. E* **2005**, *71*, 046128. [CrossRef]
12. Scarfone, A.M. Canonical quantization of nonlinear many-body systems. *Phys. Rev. E* **2005**, *71*, 051103-15. [CrossRef]
13. Kaniadakis, G. Towards a relativistic statistical theory. *Phys. A* **2006**, *365*, 17–23. [CrossRef]
14. Silva, R. The relativistic statistical theory and Kaniadakis entropy: An approach through a molecular chaos hypothesis. *Eur. Phys. J. B* **2006**, *54*, 499–502. [CrossRef]
15. Silva, R. The H-theorem in κ-statistics: Influence on the molecular chaos hypothesis. *Phys. Lett. A* **2006**, *352*, 17–20. [CrossRef]
16. Kaniadakis, G. Relativistic Entropy and related Boltzmann kinetics. *Eur. Phys. J. A* **2009**, *40*, 275–287. [CrossRef]
17. Kaniadakis, G. Maximum Entropy Principle and power-law tailed distributions. *Eur. Phys. J. B* **2009**, *70*, 3–13. [CrossRef]
18. Kaniadakis, G. Relativistic kinetics and power-law-tailed distributions. *Europhys. Lett.* **2010**, *92*, 35002. [CrossRef]
19. Oikonomou, T.; Bagci, G.B. A completeness criterion for Kaniadakis, Abe, and two-parameter generalized statistical theories. *Rep. Math. Phys.* **2010**, *66*, 137–146. [CrossRef]
20. Stankovic, M.S.; Marinkovic, S.D.; Rajkovic, P.M. The deformed exponential functions of two variables in the context of various statistical mechanics. *Appl. Math. Comput.* **2011**, *218*, 2439–2448. [CrossRef]
21. Tempesta, P. Group entropies, correlation laws, and zeta functions. *Phys. Rev. E* **2011**, *84*, 021121. [CrossRef]
22. Kaniadakis, G. Power-law tailed statistical distributions and Lorentz transformations. *Phys. Lett. A* **2011**, *375*, 356–359. [CrossRef]
23. Santos, A.P.; Silva, R.; Alcaniz, J.S.; Anselmo, D.H.A.L. Kaniadakis statistics and the quantum H-theorem. *Phys. Lett. A* **2011**, *375*, 352–355. [CrossRef]
24. Kaniadakis, G. Physical origin of the power-law tailed statistical distribution. *Mod. Phys. Lett. B* **2012**, *26*, 1250061. [CrossRef]
25. Kaniadakis, G. Theoretical Foundations and Mathematical Formalism of the Power-Law Tailed Statistical Distributions. *Entropy* **2013**, *15*, 3983–4010. [CrossRef]
26. Deossa Casas, D.E. Sobre Funciones Exponenciales y Logaritmicas Deformadas Segun Kaniadakis. Master's Thesis, Universidad EAFIT, Medellin, Colombia, 2011. Available online: http://hdl.handle.net/10784/156 (accessed on 21 February 2025).
27. Vigelis, R.F.; Cavalcante, C.C. On φ-Families of probability distributions. *J. Theor. Probab.* **2013**, *26*, 870–884. [CrossRef]
28. Wada, T.; Suyari, H. The κ-generalizations of stirling approximation and multinominal coefficients. *Entropy* **2013**, *15*, 5144–5153. [CrossRef]
29. Scarfone, A.M. Entropic Forms and Related Algebras. *Entropy* **2013**, *15*, 624–649. [CrossRef]
30. Ourabah, K.; Tribeche, M. Planck radiation law and Einstein coefficients reexamined in Kaniadakis κ statistics. *Phys. Rev. E* **2014**, *89*, 062130. [CrossRef]
31. Souza, N.T.C.M.; Anselmo, D.H.A.L.; Mello, V.D.; Silva, R. Analysis of fractal groups of the type $d-(m,r)$—Cantor within the framework of Kaniadakis statistics. *Phys. Lett. A* **2014**, *378*, 1691–1694. [CrossRef]
32. Bento, E.P.; Viswanathan, G.M.; Da Luz, M.G.E.; Silva, R. Third law of thermodynamics as a key test of generalized entropies. *Phys. Rev. E* **2015**, *91*, 022105. [CrossRef] [PubMed]
33. Kalogeropoulos, N. Entropies from coarse-graining: Convex polytopes vs. ellipsoids. *Entropy* **2015**, *17*, 6329–6378. [CrossRef]
34. Scarfone, A.M. On the κ-deformed cyclic functions and the generalized Fourier series in the framework of the κ-algebra. *Entropy* **2015**, *17*, 2812–2833. [CrossRef]

35. Scarfone A.M. κ-deformed Fourier transform. *Phys. A* **2017**, *480*, 63–78. [CrossRef]
36. Kaniadakis, G.; Scarfone, A.M.; Sparavigna, A.; Wada, T. Composition law of κ-entropy for statistically independent systems. *Phys. Rev. E* **2017**, *95*, 052112. [CrossRef] [PubMed]
37. da Costa, B.G.; Gomez, I.S.; Portesi, M. κ-Deformed quantum and classical mechanics for a system with position-dependent effective mass. *J. Math. Phys.* **2020**, *61*, 082105. [CrossRef]
38. da Silva, J.L.E.; da Silva, G.B.; Ramos, R.V. The Lambert-Kaniadakis W_κ function. *Phys. Lett.* **2020**, *384*, 126175. [CrossRef]
39. Kaniadakis, G. New power-law tailed distributions emerging in κ-statistics(a). *Europhys. Lett.* **2021**, *133*, 10002. [CrossRef]
40. Scarfone, A.M. Boltzmann configurational entropy revisited in the framework of generalized statistical mechanics. *Entropy* **2022**, *24*, 140. [CrossRef]
41. Guha, P. The κ-deformed entropic Lagrangians, Hamiltonian dynamics and their applications. *Eur. Phys. J. Plus* **2022**, *137*, 932. [CrossRef]
42. Biró, T.S. Kaniadakis Entropy Leads to Particle–Hole Symmetric Distribution. *Entropy* **2022**, *24*, 1217. [CrossRef] [PubMed]
43. Alves, T.F.A.; Neto, J.F.D.S.; Lima, F.W.S.; Alves, G.A.; Carvalho, P.R.S. Is Kaniadakis κ-generalized statistical mechanics general? *Phys. Lett. B* **2023**, *843*, 138005. [CrossRef]
44. Wada, T.; Scarfone, A.M. On the Kaniadakis Distributions Applied in Statistical Physics and Natural Sciences. *Entropy* **2023**, *25*, 292. [CrossRef]
45. Scarfone, A.M.; Wada, T. Multi-Additivity in Kaniadakis Entropy. *Entropy* **2024**, *26*, 77. [CrossRef]
46. Kaniadakis, G. Relativistic Roots of κ-Entropy. *Entropy* **2024**, *26*, 406. [CrossRef] [PubMed]
47. Nascimento, H.J.B.; Soura, P.R.M.; da Silva, J.L.E. Radial basis function network using Lambert–Kaniadakis W_κ function. *Commun. Nonlinear Sci. Numer. Simul.* **2025**, *142*, 108539. [CrossRef]
48. da Costa, B.G.; Gomez, I.S.; Portesi, M. κ–Deformed quantum mechanics: Information entropies for the Mathews–Lakshmanan oscillator. *Phys. A* **2025**, *661*, 130407. [CrossRef]
49. Kaniadakis, G.; Quarati, P.; Scarfone, A.M. Kinetical foundations of non-conventional statistics. *Phys. A* **2002**, *305*, 76–83. [CrossRef]
50. Rossani, A.; Scarfone, A.M. Generalized kinetic equations for a system of interacting atoms and photons: Theory and Simulations. *J. Phys. A* **2004**, *37*, 4955–4975. [CrossRef]
51. Biró, T.S.; Kaniadakis, G. Two generalizations of the Boltzmann equation. *Eur. Phys. J. B* **2006**, *50*, 3–6. [CrossRef]
52. Silva, J.M.; Silva, R.; Lima, J.A.S. Conservative force fields in non-Gaussian statistics. *Phys. Lett. A* **2008**, *372*, 5754–5757. [CrossRef]
53. Wada, T.; Scarfone, A.M. Asymptotic solutions of a nonlinear diffusive equation in the framework of κ-generalized statistical mechanics. *Eur. Phys. J. B* **2009**, *70*, 65–71. [CrossRef]
54. Wada, T. A nonlinear drift which leads to kappa-generalized distributions. *Eur. Phys. J. B* **2010**, *73*, 287–291.
55. Guo, L.N.; Du, J.L. The two parameters (κ, r) in the generalized statistics. *Phys. A* **2010**, *389*, 47–51. [CrossRef]
56. Guo, L.N. Physical meaning of the parametres in the two-parameter (κ, ζ) generalized theory. *Modern Phys. Lett. B* **2012**, *26*, 1250064. [CrossRef]
57. Casas, G.A.; Nobre, F.D.; Curado, E.M.F. Entropy production and nonlinear Fokker-Planck equations. *Phys. Rev. E* **2012**, *86*, 061136. [CrossRef] [PubMed]
58. Kaniadakis, G.; Hristopulos, D. Nonlinear Kinetics on Lattices Based on the Kinetic Interaction Principle. *Entropy* **2018**, *20*, 426. [CrossRef]
59. Soares, B.B.; Barboza, E.M., Jr.; Abreu, E.M.C.; Neto, J.A. Non-Gaussian thermostatistical considerations upon the Saha equation. *Phys. A* **2019**, *532*, 121590. [CrossRef]
60. Silva, M.V.D.; Martinez, A.S.; Gonçalves, A.C. Effective medium temperature for calculating the Doppler broadening function using Kaniadakis distribution. *Ann. Nucl. Energy* **2021**, *161*, 108500. [CrossRef]
61. Guha, P. The κ-Deformed Calogero-Leyvraz Lagrangians and Applications to Integrable Dynamical Systems. *Entropy* **2022**, *24*, 1673. [CrossRef]
62. Hirica, I.-E.; Pripoae, C.-L.; Pripoae G.-T.; Preda, V. Lie Symmetries of the Nonlinear Fokker-Planck Equation Based on Weighted Kaniadakis Entropy. *Mathematics* **2022**, *10*, 2776. [CrossRef]
63. Gomez I.S.; da Costa B.G.; dos Santos M.A.F. Inhomogeneous Fokker–Planck equation from framework of Kaniadakis statistics. *Commun. Nonlin. Sci. Num. Sim.* **2023**, *119*, 107131. [CrossRef]
64. Evangelista, L.R.; Lenzi, E.K. Nonlinear Fokker–Planck Equations, H-Theorem and Generalized Entropy of a Composed System. *Entropy* **2023**, *25*, 1357. [CrossRef]
65. Yamano, T. On the laws of thermodynamics from the escort average and on the uniqueness of statistical factors. *Phys. Lett. A* **2003**, *308*, 364–368. [CrossRef]
66. Kaniadakis, G.; Lissia, M. Editorial on News and expectations in thermostatistics. *Phys. A* **2004**, *340*, 15–19.
67. Wada, T. Thermodynamic stabilities of the generalized Boltzmann entropies. *Phys. A* **2004**, *340*, 126–130. [CrossRef]

68. Wada, T. Thermodynamic stability conditions for nonadditive composable entropies. *Contin. Mechan. Thermod.* **2004**, *16*, 263–267. [CrossRef]
69. Scarfone, A.M.; Wada, T. Canonical partition function for anomalous systems described by the κ-entropy. *Prog. Theor. Phys. Suppl.* **2006**, *162*, 45–52. [CrossRef]
70. Lucia, U. Maximum entropy generation and kappa-exponential model. *Phys. A* **2010**, *389*, 4558–4563. [CrossRef]
71. Lourek, I.; Tribeche, M. Thermodynamic properties of the blackbody radiation: A Kaniadakis approach. *Phys. Lett. A* **2017**, *381*, 452–456. [CrossRef]
72. Carvalho, J.C.; Silva, R.; do Nascimento, J.D., Jr.; De Medeiros, J.R. Power law statistics and stellar rotational velocities in the Pleiades. *Europhys. Lett.* **2008**, *84*, 59001. [CrossRef]
73. Carvalho, J.C.; do Nascimento, J.D., Jr.; Silva, R.; De Medeiros, J.R. Non-gaussian statistics and stellar rotational velocities of main sequence field stars. *Astrophys. J. Lett.* **2009**, *696*, L48–L51. [CrossRef]
74. Carvalho, J.C.; Silva, R.; do Nascimento jr, J.D.; Soares, B.B.; De Medeiros, J.R. Observational measurement of open stellar clusters: A test of Kaniadakis and Tsallis statistics. *Europhys. Lett.* **2010**, *91*, 69002. [CrossRef]
75. Bento, E.P.; Silva, J.R.P.; Silva, R. Non-Gaussian statistics, Maxwellian derivation, and stellar polytropes. *Phys. A* **2013**, *392*, 666–672. [CrossRef]
76. Curé, M.; Rial, D.F.; Christen, A.; Cassetti, J. A method to deconvolve stellar rotational velocities. *Astron. Astrophys.* **2014**, *565*, A85–A87. [CrossRef]
77. Kolesnichenko, A.V. Chandrasekhar's integral equilibrium theorems modified in the context of non-Gaussian kappa statistics for a spherically symmetric protostar cloud. *Sol. Syst. Res.* **2022**, *56*, 43–53. [CrossRef]
78. Shu C.-H.; Zhang K.-X.; He K.-R.; Chen H.; Liu S.-Q. Gravitational instability of dark-baryonic matter systems in $f(R)$ gravity. *Phys. Scr.* **2023**, *98*, 105213. [CrossRef]
79. Abreu, E.M.C.; Neto, J.A.; Barboza, E.M., Jr.; Nunes, R.C. Holographic considerations on non-gaussian statistics and gravothermal catastrophe. *Phys. A* **2016**, *441*, 141–150. [CrossRef]
80. Nunes, R.C.; Barboza, E.M., Jr.; Abreu, E.M.C.; Neto, J.A. Probing the cosmological viability of non-Gaussian statistics. *J. Cosmol. Astropart. Phys.* **2016**, *2016*, 051. [CrossRef]
81. Yang, W.-H.; Xiong, Y.-Z.; Chen, H.; Liu, S.-Q. Jeans gravitational instability with κ-deformed Kaniadakis distribution in Eddington-inspired Born-Infield gravity. *Chin. Phys. B* **2020**, *29*, 110401. [CrossRef]
82. Lymperis, A.; Basilakos, S.; Saridakis, E.N. Modified cosmology through Kaniadakis horizon entropy. *Eur. Phys. J. C* **2021**, *81*, 1037. [CrossRef]
83. Abreu, E.M.C.; Ananias Neto, J. Black holes thermodynamics from a dual Kaniadakis entropy. *EPL* **2021**, *133*, 49001. [CrossRef]
84. Nojiri, S.; Odintsov, S.D.; Paul, T. Modified cosmology from the thermodynamics of apparent horizon. *Phys. Lett. B* **2022**, *835*, 137553. [CrossRef]
85. Moradpour, H.; Javaherian, M.; Namvar, E.; Ziaie, A.H. Gamow Temperature in Tsallis and Kaniadakis Statistics. *Entropy* **2022**, *24*, 797. [CrossRef] [PubMed]
86. Luciano, G.G. Modified Friedmann equations from Kaniadakis entropy and cosmological implications on baryogenesis and 7Li -abundance. *Eur. Phys. J. C* **2022**, *82*, 314. [CrossRef]
87. Abreu, E.M.C.; Neto, J.A. Statistical approaches on the apparent horizon entropy and the generalized second law of thermodynamics. *Phys. Lett. B* **2022**, *824*, 136803. [CrossRef]
88. Abreu, E.M.C.; Neto, J.A. Statistical approaches and the Bekenstein bound conjecture in Schwarzschild black holes. *Phys. Lett. B* **2022**, *835*, 137565. [CrossRef]
89. Luciano, G.G. Gravity and Cosmology in Kaniadakis Statistics: Current Status and Future Challenges. *Entropy* **2022**, *24*, 1712. [CrossRef] [PubMed]
90. Sadeghnezhad, N. Entropic gravity and cosmology in Kaniadakis statistics. *Int. J. Mod. Phys. D* **2023**, *32*, 2350002. [CrossRef]
91. Luciano, G.G.; Saridakis, E.N. P-v criticalities, phase transitions and geometrothermodynamics of charged AdS black holes from Kaniadakis statistics. *J. High Energy Phys.* **2023**, *2023*, 114. [CrossRef]
92. Lambiase, G.; Luciano, G.G.; Sheykhi, A. Slow-roll inflation and growth of perturbations in Kaniadakis modification of Friedmann cosmology. *Eur. Phys. J. C* **2023**, *83*, 936. [CrossRef]
93. Sheykhi, A. Corrections to Friedmann equations inspired by Kaniadakis entropy. *Phys. Lett. B* **2024**, *850*, 138495. [CrossRef]
94. Yarahmadi, M.; Salehi, A. Using the Kaniadakis horizon entropy in the presence of neutrinos to alleviate the Hubble and S_8 tensions. *Eur. Phys. J.* **2024**, *84*, 443. [CrossRef]
95. Raoa, B.; Mohantyb, D.; Adityac, Y.; Prasanthid, U. Cosmological Evolution Of Bianchi Type-V I0 Kaniadakis Holographic Dark Energy Model. *East Eur. J. Phys.* **2024**, *2024*, 43–54. [CrossRef]
96. Xia, J.; Ong, Y. Upper Bound of Barrow Entropy Index from Black Hole Fragmentation. *Universe* **2024**, *10*, 177. [CrossRef]
97. Prasanthan, P.; Nelleri, S.; Poonthottathil, N.; Sreejith, E.K. Emergence of Cosmic Space and Horizon Thermodynamics from Kaniadakis Entropy. *arXiv* **2024**, arXiv:2405.03592. [CrossRef]

98. Santos, F.F.; Boschi-Filho, H. Black branes in asymptotically Lifshitz spacetimes with arbitrary exponents in κ-Horndeski gravity. *Phys. Rev. D* **2024**, *109*, 064035. [CrossRef]
99. Kumar, N. Relativistic correction to black hole entropy. *Gen. Relativ. Gravit.* **2024**, 56, 47. [CrossRef]
100. Vijaya Prasanthi, A.; Suryanarayana, G.; Aditya, Y.; Divya Prasanthi, U.Y. Cosmological Dynamics of Anisotropic Kaniadakis Holographic Dark Energy Model in Brans-Dicke Gravity. *East Eur. J. Phys.* **2024**, *2*, 10–20. [CrossRef]
101. Alsaedi, R.H.; Azizi, T.; Sadeghi, J. Nonextensive Statistical Mechanics and Black Hole Thermodynamics: Tsallis and Kaniadakis Entropies. *J. Hologr. Appl. Phys.* **2024**, *4*, 3.
102. Ribeiro, B.W.; Macêdo, I.M.; Carvalho, F.C. Generalized inflation in the context of κ-deformed theories. *arXiv* **2024**, arXiv:2409.07678.
103. Shokri, M. Bekenstein bound on black hole entropy in non-Gaussian statistics. *Phys. Lett. B* **2025**, *860*, 139193. [CrossRef]
104. Bhattacharjee, A.; Phukon, P. BTZ Black Hole In The Non-Extensive Generalizations of Gibbs Entropy. *Prog. Theor. Exp. Phys.* **2025**, *2025*, 023E01. [CrossRef]
105. Moradpour, H.; Ziaie, A.H.; Zangeneh, M.K. Generalized entropies and corresponding holographic dark energy models. *Eur. Phys. J. C* **2020**, *80*, 732–737. [CrossRef]
106. Jawad, A.; Sultan, A.M. Cosmic Consequences of Kaniadakis and Generalized Tsallis Holographic Dark Energy Models in the Fractal Universe. *Adv. High Energy Phys.* **2021**, *2021*, 5519028. [CrossRef]
107. Rani, S.; Jawad, A.; Sultan, A.M.; Shad, M. Cosmographic and thermodynamic analysis of Kaniadakis holographic dark energy. *Int. J. Mod. Phys. D* **2022**, *31*, 2250078. [CrossRef]
108. Ghaffari, S. Kaniadakis holographic dark energy in Brans-Dicke cosmology. *Mod. Phys. Lett. A* **2022**, *37*, 2250152. [CrossRef]
109. Sharma, U.K.; Dubey, V.C.; Ziaie, A.H.; Moradpour, H. Kaniadakis holographic dark energy in nonflat universe. *Int. J. Mod. Phys. D* **2022**, *31*, 2250013. [CrossRef]
110. Nojiri, S.; Odintsov, S.D.; Faraoni, V. From nonextensive statistics and black hole entropy to the holographic dark universe. *Phys. Rev. D* **2022**, *105*, 044042. [CrossRef]
111. Nojiri, S.; Odintsov, S.D.; Paul, T. Early and late universe holographic cosmology from a new generalized entropy. *Phys. Lett. B* **2022**, *831*, 137189. [CrossRef]
112. Drepanou, N.; Lymperis, A.; Saridakis, E.N.; Yesmakhanova, K. Kaniadakis holographic dark energy and cosmology. *Eur. Phys. J. C* **2022**, *82*, 449. [CrossRef]
113. Korunur, S. Kaniadakis holographic dark energy with scalar field in Bianchi type-V universe. *Int. J. Mod. Phys. A* **2022**, *37*, 2250214. [CrossRef]
114. Hernandez-Almada, A.; Leon, G.; Magana, J.; Garcia-Aspeitia, M.A.; Motta, V.; Saridakis, E.N.; Yesmakhanova, K.; Millano, A.D. Observational constraints and dynamical analysis of Kaniadakis horizon-entropy cosmology. *Mon. Not. R. Astron. Soc.* **2022**, *512*, 5122–5134. [CrossRef]
115. Hernandez-Almada, A.; Leon, G.; Magana, J.; Garcia-Aspeitia, M.A.; Motta, V.; Saridakis, E.N.; Yesmakhanova, K. Kaniadakis-holographic dark energy: Observational constraints and global dynamics. *Mon. Not. R. Astron. Soc.* **2022**, *511*, 4147–4158. [CrossRef]
116. Blasone, M.; Lambiase, G.; Luciano, G.G. Kaniadakis entropy-based characterization of IceCube PeV neutrino signals. *Phys. Dark Universe* **2023**, *42*, 101342. [CrossRef]
117. Sania, A.N.; Rani, S.; Jawad, A. Cosmic and Thermodynamic Consequences of Kaniadakis Holographic Dark Energy in Brans-Dicke Gravity. *Entropy* **2023**, *25*, 576. [CrossRef] [PubMed]
118. Dubey, V.C.; Kumar, M.; Sharma, L.K.; Sharma, U.K. Some features of Kaniadakis holographic dark energy model. *Int. J. Geom. Meth. Mod. Phys.* **2023**, *20*, 2350036. [CrossRef]
119. Jawad, A.; Ul Abideen, Z.; Rani, S. Study of cosmic acceleration in modified theories of gravity through Kaniadakis holographic dark energy. *Mod. Phys. Lett. A* **2023**, *38*, 2350037 [CrossRef]
120. Singh, B.K.; Sharma, U.K.; Sharma, L.K.; Dubey, V.C. Statefinder hierarchy of Kaniadakis holographic dark energy with composite null diagnostic. *Int. J. Geom. Meth. Mod. Phys.* **2023**, *20*, 2350074. [CrossRef]
121. Kumar, P.S.; Pandey, B.D.; Sharma, U.K.; Pankaj. Holographic dark energy through Kaniadakis entropy in non flat universe. *Eur. Phys. J. C* **2023**, *83*, 143. [CrossRef]
122. Sharma, U.K.; Kumar, P.S.; Pankaj. Quintessence scalar field of Kaniadakis holographic dark energy model with statefinder analysis. *Int. J. Geom. Meth. Mod. Phys.* **2023**, *20*, 2450004. [CrossRef]
123. Sadeghi, J.; Gashti, S.N.; Azizi, T. Complex quintessence theory, Tsallis and Kaniadakis holographic dark energy and Brans-Dicke cosmology. *Mod. Phys. Lett. A* **2023**, *38*, 2350076. [CrossRef]
124. Kumar, P.S.; Pandey, B.D.; Pankaj; Sharma, U.K. Kaniadakis agegraphic dark energy. *New Astr.* **2024**, *105*, 102085. [CrossRef]
125. Sultana, S.; Chattopadhyay, S. Intermediate inflation through Nojiri–Odintsov holographic dark fluid with the cosmological settings of Kaniadakis. *Int. J. Geom. Meth. Mod. Phys.* **2024**, *21*, 2450133. [CrossRef]

126. Chokyi, K.K.; Chattopadhyay, S. Cosmology of Tsallis and Kaniadakis holographic dark energy in Saez-Ballester theory and consideration of viscous van der Waals fluid. *Ann. Phys.* **2024**, *463*, 169611. [CrossRef]
127. Sadeghi, J.; Afshar, M.A.S.; Alipour, M.R.; Gashti, S.N. Phase transition dynamics of black holes influenced by Kaniadakis and Barrow statistics. *Phys. Dark Universe* **2025**, *47*, 101780. [CrossRef]
128. Abreu, E.M.C.; Neto, J.A.; Barboza, E.M.; Nunes, R.C. Jeans instability criterion from the viewpoint of Kaniadakis statistics. *EPL* **2016**, *114*, 55001. [CrossRef]
129. Abreu, E.M.C.; Neto, J.A.; Barboza, E.M.; Nunes, R.C. Tsallis and Kaniadakis statistics from the viewpoint of entropic gravity formalism. *Int. J. Mod. Phys.* **2017**, *32*, 1750028. [CrossRef]
130. Chen, H.; Zhang, S.X.; Liu, S.Q Jeans gravitational instability with kappa-deformed Kaniadakis distribution. *Chin. Phys. Lett.* **2017**, *34*, 075101. [CrossRef]
131. Abreu, E.M.C.; Neto, J.A.; Mendes, A.C.R.; Bonilla, A. Tsallis and Kaniadakis statistics from a point of view of the holographic equipartition law. *EPL* **2018**, *121*, 45002. [CrossRef]
132. Abreu, E.M.C.; Neto, J.A.; Mendes, A.C.R.; Bonilla, A.; de Paula, R.M. Cosmological considerations in Kaniadakis statistics. *EPL* **2018**, *124*, 30003. [CrossRef]
133. Abreu, E.M.C.; Neto, J.A.; Barboza, E.M.; Soares, B.B. On incomplete statistics and the loop quantum gravity Immirzi parameter. *EPL* **2019**, *127*, 10006. [CrossRef]
134. Abreu, E.M.C.; Neto, J.A.; Mendes, A.C.R.; de Paula, R.M. Loop quantum gravity Immirzi parameter and the Kaniadakis statistics. *Chaos Sol. Fractals* **2019**, *118*, 307–310. [CrossRef]
135. Yang, W.; Xiong, Y.; Chen, H.; Liu, S. Jeans instability of dark-baryonic matter model in the context of Kaniadakis' statistic distribution. *J. Taibah Univ. Sci.* **2022**, *16*, 337–343. [CrossRef]
136. He, K.-R. Jeans analysis with κ-deformed Kaniadakis distribution in f (R) gravity. *Phys. Scr.* **2022**, *97*, 025601. [CrossRef]
137. Çimdiker, I.; Dąbrowski, M.P.; Gohar, H. Generalized uncertainty principle impact on nonextensive black hole thermodynamics. *Class. Quantum Gravity* **2023**, *40*, 145001. [CrossRef]
138. Salehi, A. The analytical approach in testing the Kaniadakis cosmology. *Class. Quantum Gravity* 2024, *to appear*. [CrossRef]
139. Prasanthan, P.; Nelleri, S.; Pradeepan, A.K.; Poonthottathil, N.; Tom, E. On the interrelation of the generalized holographic equipartition and entropy maximization in Kaniadakis paradigm. *Gen. Relativ. Gravit.* **2025**, *57*, 6. [CrossRef]
140. Pistone, G. $\kappa-$exponential models from the geometrical point of view. *Eur. Phys. J. B* **2009**, *70*, 29–37. [CrossRef]
141. Scarfone, A.M.; Wada, T. Legendre structure of kappa-thermostatistics revisited in the framework of information geometry. *J. Phys. A Math. Theor.* **2014**, *47*, 275002. [CrossRef]
142. Wada, T.; Matsuzoe, H.; Scarfone, A.M. Dualistic Hessian structures among the thermodynamic potentials in the κ-thermostatistics. *Entropy* **2015**, *17*, 7213–7229. [CrossRef]
143. Wada, T.; Scarfone, A.M.; Information geometry on the κ-thermostatistics. *Entropy* **2015**, *17*, 1204–1217. [CrossRef]
144. Quiceno Echavarría, H.R.; Arango Parra, J.C. A statistical manifold modeled on Orlicz spaces using Kaniadakis κ-exponential models. *J. Math. Anal.* **2015**, *431*, 1080–1098. [CrossRef]
145. Scarfone, A.M.; Matsuzoe, H.; Wada, T. Information geometry of κ-exponential families: Dually-flat, Hessian and Legendre structures. *Entropy* **2018**, *20*, 436. [CrossRef] [PubMed]
146. Mehri-Dehnavi, H.; Mohammadzadeh, H. Thermodynamic geometry of Kaniadakis statistics. *J. Phys. A Math. Gen.* **2020**, *53*, 375009. [CrossRef]
147. Pistone G.; Shoaib M. Kaniadakis's Information Geometry of Compositional Data. *Entropy* **2023**, *25*, 1107. [CrossRef] [PubMed]
148. Wada, T.; Suyari, H. A two-parameter generalization of Shannon-Khinchin axioms and the uniqueness teorem. *Phys. Lett. A* **2007**, *368*, 199–205. [CrossRef]
149. Lei, B.; Fan, J.-L. Adaptive Kaniadakis entropy thresholding segmentation algorithm based on particle swarm optimization. *Soft Comput.* **2020**, *24*, 7305–7318. [CrossRef]
150. Huo, X.; Zhang, F.; Shao, K.; Tan, J.-Q. Improved meta-heuristic optimization algorithm and its application in image segmentation. *J. Softw.* **2021**, *32*, 3452–3467.
151. Topsoe, F. Entropy and equilibrium via games of complexity. *Phys. A* **2004**, *340*, 11–31. [CrossRef]
152. Wada, T.; Suyari, H. κ-generalization of Gauss' law of error. *Phys. Lett. A* **2006**, *348*, 89–93. [CrossRef]
153. Abul-Magd, A.Y. Nonextensive random-matrix theory based on Kaniadakis entropy. *Phys. Lett. A* **2007**, *361*, 450–454. [CrossRef]
154. Abul-Magd, A.Y. Nonextensive and superstatistical generalizations of random-matrix theory. *Eur. Phys. J. B* **2009**, *70*, 39–48. [CrossRef]
155. Mattone, M.; Miraldi E.; Scarfone, A.M. Statistical analysis of the time base distortion of a digital oscilloscope. *Mod. Phys. Lett. B* **2009**, *23*, 3179. [CrossRef]
156. Abul-Magd, A.Y.; Abdel-Mageed, M. Kappa-deformed random-matrix theory based on Kaniadakis statistics. *Mod. Phys. Lett. B* **2012**, *26*, 1250059. [CrossRef]

157. Tapiero, O.J. A maximum (non-extensive) entropy approach to equity options bid-ask spread. *Phys. A* **2013**, *392*, 3051–3060. [CrossRef]
158. Biró, T.S.; Schram, Z. Non-extensive entropic distance based on diffusion: Restrictions on parameters in entropy formulae. *Entropy* **2016**, *18*, 42–49. [CrossRef]
159. Eck, D.J.; McKeague, I.W. Central Limit Theorems under additive deformations. *Stat. Prob. Lett.* **2016**, *118*, 156–162. [CrossRef]
160. da Silva, S.L.E.F.; dos Santos Lima, G.Z.; Volpe, E.V.; de Araújo, J.M.; Corso, G. Robust approaches for inverse problems based on Tsallis and Kaniadakis generalised statistics. *Eur. Phys. J. Plus* **2021**, *136*, 518. [CrossRef]
161. da Silva, S.L.E.F.; Silva, R.; dos Santos Lima, G.Z.; de Araújo, J.M.; Corso, G. An outlier-resistant κ-generalized approach for robust physical parameter estimation. *Phys. A* **2022**, *600*, 127554. [CrossRef]
162. Sfetcu, R.-C.; Sfetcu, S.-C.; Preda, V. Some Properties of Weighted Tsallis and Kaniadakis Divergences. *Entropy* **2022**, *24*, 1616. [CrossRef] [PubMed]
163. Sfetcu, R.-C.; Sfetcu, S.-C.; Preda, V. On Tsallis and Kaniadakis Divergences. *Math. Phys. An. Geom.* **2022**, *25*, 7. [CrossRef]
164. Dănilă-Cernat, I. Nearest neighbor estimates of Kaniadakis entropy. *An. St. Univ. Ovidius Constanta* **2022**, *30*, 171–189. [CrossRef]
165. dos Santos Lima, G.Z.; de Lima, J.V.T.; de Araújo, J.M.; Corso, G.; da Silva, S.L.E.F. Generalized statistics: Applications to data inverse problems with outlier-resistance. *PLoS ONE* **2023**, *18*, e0282578. [CrossRef]
166. Yoshioka, H.; Yoshioka, Y. Generalized divergences for statistical evaluation of uncertainty in long-memory processes. *Chaos Solitons Fract.* **2024**, *182*, 114627. [CrossRef]
167. Trifonov, M. A note about set of (κ, μ)-deformed Gaussian distributions. *Researchgate* **2024**. [CrossRef]
168. Sfetcu, R.; Sfetcu, S.; Preda, V. Discrete Entropies of Chebyshev Polynomials. *Mathematics* **2024**, *12*, 1046. [CrossRef]
169. Newton, N. Global differentiable structures for the Fisher-Rao and Kantorovich-Wasserstein-Otto metrics. *J. Math. Anal. Appl.* **2024**, *538*, 128435. [CrossRef]
170. Aliano, A.; Kaniadakis, G.; Miraldi, E. Bose-Einstein condensation in the framework of kappa-statistics. *Phys. B* **2003**, *325*, 35–40. [CrossRef]
171. Pereira, F.I.M.; Silva, R.; Alcaniz, J.S. Non-gaussian statistics and the relativistic nuclear equation of state. *Nucl. Phys. A* **2009**, *828*, 136–148. [CrossRef]
172. Santos, A.P.; Silva, R.; Alcaniz, J.S.; Anselmo, D.H.A.L. Generalized quantum entropies. *Phys. Lett. A* **2011**, *375*, 3119–3123. [CrossRef]
173. Santos, A.P.; Silva, R.; Alcaniz, J.S.; Anselmo, D.H.A.L. Non-Gaussian effects on quantum entropies. *Phys. A* **2012**, *391*, 2182–2192. [CrossRef]
174. Ourabah, K.; Hamici-Bendimerad, A.H.; Tribeche, M. Quantum Kaniadakis entropy under projective measurement. *Phys. Rev. E* **2015**, *92*, 032114. [CrossRef]
175. Ourabah, K.; Hamici-Bendimerad, A.H.; Tribeche, M. Quantum entanglement and Kaniadakis entropy. *Phys. Scr.* **2015**, *90*, 045101. [CrossRef]
176. Passos, L.A.; Cleison Santana, M.; Moreira, T.; Papa, J.P. κ-Entropy based restricted Boltzmann machines. In Proceedings of the 2019 International Joint Conference on Neural Networks (IJCNN), Budapest, Hungary, 14–19 July 2019; p. 8851714.
177. Lan, R.; Gao, X. Threshold segmentation based on Fuzzy Kaniadakis entropy for criminal investigation images. In Proceedings of the 2021 4th International Conference on Artificial Intelligence and Pattern Recognition, Xiamen, China, 24–26 September 2021; pp. 102–108.
178. Lei, B.; Fan, J. Infrared pedestrian segmentation algorithm based on the two-dimensional Kaniadakis entropy thresholding. *Knowl. Based Syst.* **2021**, *225*, 107089. [CrossRef]
179. Cravero, M.; Iabichino, G.; Kaniadakis, G.; Miraldi, E.; Scarfone, A.M. A κ-entropic approach to the analysis of the fracture problem. *Phys. A* **2004**, *340*, 410–417. [CrossRef]
180. Coraddu, M.; Lissia, M.; Tonelli, R. Statistical descriptions of nonlinear systems at the onset of chaos. *Phys. A* **2006**, *365*, 252–257. [CrossRef]
181. Tonelli, R.; Mezzorani, G.; Meloni, F.; Lissia, M.; Coraddu, M. Entropy production and Pesin identity at the onset of chaos. *Prog. Theor. Phys.* **2006**, *115*, 23–29. [CrossRef]
182. Celikoglu, A.; Tirnakli, U. Sensitivity function and entropy increase rates for z-logistic map family at the edge of chaos. *Phys. A* **2006**, *372*, 238–242. [CrossRef]
183. Olemskoi, A.I.; Kharchenko, V.O.; Borisyuk, V.N. Multifractal spectrum of phase space related to generalized thermostatistics. *Phys. A* **2008**, *387*, 1895–1906. [CrossRef]
184. Olemskoi, A.I.; Borisyuk, V.N.; Shuda, I.A. Statistical field theories deformed within different calculi. *Eur. Phys. J. B* **2010**, *77*, 219–231. [CrossRef]
185. Kaniadakis, G.; Baldi, M.M.; Deisboeck, T.S.; Grisolia, G.; Hristopulos, D.T.; Scarfone, A.M.; Sparavigna, A.; Wada, T.; Lucia, U. The κ-statistics approach to epidemiology. *Sci. Rep.* **2020**, *10*, 19949. [CrossRef]
186. da Silva, S.L.E.F. Newton's cooling law in generalised statistical mechanics. *Phys. A* **2021**, *565*, 125539. [CrossRef]

187. Bushinskaya, A.; Timashev, S. Application of Kaniadakis κ-statistics to extreme wind speed load distributions. *Reliab. Theory Appl.* **2022**, *17*, 188–199.
188. da Silva, J.L.E.; Mendes, F.V.; Ramos, R.V. The $R_{\kappa\kappa}$ function and its applications: Disentropy, image processing and the κ-diode. *Eur. Phys. J. Plus* **2022**, *137*, 320–329. [CrossRef]
189. Bushinskaya, A.; Timashev, S. Application of Kaniadakis κ-statistics to load and impact distributions. *Lect. Notes Civ. Eng.* **2023**, *308*, 489–499.
190. Rajaonarison, D.; Bolduc, D.; Jayet, H. The K-deformed multinomial logit model. *Econ. Lett.* **2005**, *86*, 13–20. [CrossRef]
191. Scarfone, A.M. A mechanism to derive multi-power law functions: An application in the econophysics framework. *Phys. A* **2007**, *382*, 271–277. [CrossRef]
192. Clementi, F.; Gallegati, M.; Kaniadakis, G. κ-generalized statistics in personal income distribution. *Eur. Phys. J. B* **2007**, *57*, 187–193. [CrossRef]
193. Rajaonarison, D. Deterministic heterogeneity in tastes and product differentiation in the K-logit model. *Econ. Lett.* **2008**, *100*, 396–399. [CrossRef]
194. Clementi, F.; Di Matteo, T.; Gallegati, M.; Kaniadakis, G. The κ-generalized distribution: A new descriptive model for the size distribution of incomes. *Phys. A* **2008**, *387*, 3201–3208. [CrossRef]
195. Clementi, F.; Gallegati, M.; Kaniadakis, G. A κ-generalized statistical mechanics approach to income analysis. *J. Stat. Mech.* **2009**, P02037. [CrossRef]
196. Clementi, F.; Gallegati, M.; Kaniadakis, G. A model of personal income distribution with application to Italian data. *Empir. Econ.* **2010**, *39*, 559–591. [CrossRef]
197. Clementi, F.; Gallegati, M.; Kaniadakis, G. A new model of income distribution: The kappa-generalized distribution. *J. Econ.* **2012**, *105*, 63–91. [CrossRef]
198. Clementi, F.; Gallegati, M.; Kaniadakis, G. A generalized statistical model for the size distribution of wealth. *J. Stat. Mech.* **2012**, *2012*, P12006.
199. Bertotti, M.L.; Modenese, G. Exploiting the flexibility of a family of models for taxation and redistribution. *Eur. Phys. J. B* **2012**, *85*, 261–270. [CrossRef]
200. Trivellato, B. The minimal κ-entropy martingale measure. *Int. J. Theor. Appl. Financ.* **2012**, *15*, 1250038. [CrossRef]
201. Trivellato, B. Deformed exponentials and applications to finance. *Entropy* **2013**, *15*, 3471–3489. [CrossRef]
202. Macedo-Filho, A.; Moreira, D.A.; Silva, R.; da Silva, L.R. Maximum entropy principle for Kaniadakis statistics and networks. *Phys. Lett. A* **2013**, *377*, 842–846. [CrossRef]
203. Stella, M.; Brede, M. A κ-deformed model of growing complex networks with fitness. *Phys. A* **2014**, *407*, 360–368. [CrossRef]
204. Preda, V.; Dedu, S.; Sheraz, M. New measure selection for Hunt-Devolder semi-Markov regime switching interest rate models. *Phys. A* **2014**, *407*, 350–359. [CrossRef]
205. Khordad, R.; Rastegar Sedehi, H.R. Application of different entropy formalisms in a neural network for novel word learning. *Eur. Phys. J. Plus* **2015**, *130*, 246. [CrossRef]
206. Clementi, F.; Gallegati, M.; Kaniadakis, G.; Landini, S. κ-generalized models of income and wealth distributions: A survey. *Eur. Phys. J. Spec. Top.* **2016**, *225*, 1959–1984. [CrossRef]
207. Bertotti, M.L.; Modanese, G. Statistics of binary exchange of energy or money. *Entropy* **2017**, *19*, 465. [CrossRef]
208. Moretto, E.; Pasquali, S.; Trivellato, B. A non-Gaussian option pricing model based on Kaniadakis exponential deformation. *Eur. Phys. J. B* **2017**, *90*, 179. [CrossRef]
209. Sheraz, M.; Preda, V.; Dedu, S. Tsallis and Kaniadakis entropy measures for risk neutral densities. In Proceedings of the Computer Aided Systems Theory—EUROCAST 2017, 16th International Conference, Las Palmas de Gran Canaria, Spain, 19–24 February 2017; Lecture Notes in Computer Science (LNTCS); Springer: Cham, Switzerland, 2018; Volume 10672, pp. 55–63.
210. Vallejos, A.; Ormazábal, I.; Borotto, F.A.; Astudillo, H.F. A new κ-deformed parametric model for the size distribution of wealth. *Phys. A* **2019**, *514*, 819–829. [CrossRef]
211. Clementi, F. The Kaniadakis Distribution for the Analysis of Income and Wealth Data. *Entropy* **2023**, *25*, 1141. [CrossRef] [PubMed]
212. Baldi, M.M.; Mammana, C.; Michetti, E. The κ-logistic growth model. Qualitative and quantitative dynamics. *Math. Comput. Simul.* **2024**, *225*, 350–369. [CrossRef]
213. Hristopulos, D.T.; Petrakis, M.P.; Kaniadakis, G. Finite-size effects on return interval distributions for weakest-link-scaling systems. *Phys. Rev. E* **2014**, *89*, 052142. [CrossRef]
214. Hristopulos, D.T.; Petrakis, M.P.; Kaniadakis, G. Weakest-Link Scaling and Extreme Events in Finite-Sized Systems. *Entropy* **2015**, *17*, 1103–1122. [CrossRef]
215. Oreste, P.; Spagnoli, G. Statistical analysis of some main geomechanical formulations evaluated with the Kaniadakis exponential law. *Geomech. Geoengin.* **2018**, *13*, 139–145. [CrossRef]
216. Spagnoli, G.; Oreste, P. Relation water content ratio-to-liquidity index versus the Atterberg limits ratio evaluated with the Kaniadakis exponential law. *Geomech. Geoengin.* **2019**, *14*, 148–153. [CrossRef]

217. da Silva, S.L.E.F.; Carvalho, P.T.; de Araújo, J.M.; Corso, G. Full-waveform inversion based on Kaniadakis statistics. *Phys. Rev. E* **2020**, *101*, 053311. [CrossRef]
218. da Silva, S.L.E.F. κ-generalised Gutenberg–Richter law and the self-similarity of earthquakes. *Chaos Solitons Fractals* **2021**, *143*, 110622. [CrossRef]
219. da Silva, S.L.E.F.; dos Santos Lima, G.Z.; de Araújo, J.M.; Corso, G. Extensive and nonextensive statistics in seismic inversion. *Phys. A* **2021**, *563*, 125496. [CrossRef]
220. Hristopulos, D.T.; Baxevani, A. Kaniadakis Functions beyond Statistical Mechanics: Weakest-Link Scaling, Power-Law Tails, and Modified Lognormal Distribution. *Entropy* **2022**, *24*, 1362. [CrossRef] [PubMed]
221. da Silva, S.L.E.F.; Kaniadakis, G. κ–statistics approach to optimal transport waveform inversion. *Phys. Rev. E* **2022**, *106*, 034113. [CrossRef] [PubMed]
222. da Silva, S.L.E.F.; Kaniadakis, G. A graph-space optimal transport FWI approach based on κ-generalized Gaussian distribution. In Proceedings of the Third International Meeting for Applied Geoscience & Energy Technical Program Expanded Abstracts, Houston, TX, USA, 28 August–1 September 2023; pp. 670–674. [CrossRef]
223. da Silva, S.L.E.F.; de Araújo, J.M.; de la Barra, E.; Corso, G. A Graph-Space Optimal Transport Approach Based on Kaniadakis κ-Gaussian Distribution for Inverse Problems Related to Wave Propagation. *Entropy* **2023**, *25*, 990. [CrossRef]
224. Souza, N.T.C.M.; Anselmo, D.H.A.L.; Silva, R.; Vasconcelos, M.S.; Mello, V.D. A kappa-statistical analysis of the Y-chromosome. *EPL* **2014**, *108*, 38004. [CrossRef]
225. Costa, M.O.; Silva, R.; Anselmo, D.H.A.L.; Silva, J.R.P. Analysis of human DNA through power-law statistics. *Phys. Rev. E* **2019**, *99*, 022112. [CrossRef] [PubMed]
226. de Lima, M.M.F.; Anselmo, D.H.A.L.; Silva, R.; Nunes, G.H.S.; Fulco, U.L.; Vasconcelos, M.S.; Mello, V.D. A Bayesian Analysis of Plant DNA Length Distribution via κ-Statistics. *Entropy* **2022**, *24*, 1225. [CrossRef] [PubMed]
227. Lima, M.; Costa, M.; Silva, R.; Fulco, U.; Oliveira, J.; Vasconcelos, M.; Anselmo, D. Viral proteins length distributions: A comparative analysis. *Phys. A.* **2024**, *633*, 129367. [CrossRef]
228. Guedes, G.; Gonçalves, A.C.; Palma, D.A.P. The Doppler Broadening Function using the Kaniadakis distribution. *Ann. Nucl. Energy* **2017**, *110*, 453–458. [CrossRef]
229. Shen, K.-M.; Biro, T.S.; Wang, E.-K. Different non-extensive models for heavy-ion collisions. *Phys. A* **2018**, *492*, 2353–2360. [CrossRef]
230. de Abreu, W.V.; Goncalves, A.C.; Martinez, A.S. Analytical solution for the Doppler broadening function using the Kaniadakis distribution. *Ann. Nucl. Energy* **2019**, *126*, 262–268. [CrossRef]
231. Shen, K.-M. Analysis on hadron spectra in heavy-ion collisions with a new non-extensive approach. *J. Phys. G* **2019**, *46*, 105101. [CrossRef]
232. de Abreu, W.V.; Martinez, A.S. New analytical formulations for the Doppler broadening function and interference term based on Kaniadakis distributions. *Ann. Nucl. Energy* **2020**, *135*, 106960. [CrossRef]
233. Guedes, G.; Palma, D.A.P. Quasi-Maxwellian interference term functions. *Ann. Nucl. Energy* **2021**, *151*, 107914. [CrossRef]
234. de Abreu, W.V.; Maciel, J.M.; Martinez, A.S.; Gonçalves, A.D.C.; Schmidt, L. Doppler Broadening of Neutron Cross-Sections Using Kaniadakis Entropy. *Entropy* **2022**, *24*, 1437. [CrossRef] [PubMed]
235. de Abreu, W.V.; Martinez, A.S.; do Carmo, E.D.; Gonçalves, A.C. A novel analytical solution of the deformed Doppler broadening function using the Kaniadakis distribution and the comparison of computational efficiencies with the numerical solution. *Nucl. Eng. Technol.* **2022**, *54*, 1471–1481. [CrossRef]
236. Martinez, A.S.; de Abreu, W.V. The Scientific Contribution of the Kaniadakis Entropy to Nuclear Reactor Physics: A Brief Review. *Entropy* **2023**, *25*, 478. [CrossRef] [PubMed]
237. Teweldeberhan, A.M.; Miller, H.G.; Tegen, R. κ-deformed Statistics and the formation of a quark-gluon plasma. *Int. J. Mod. Phys. E* **2003**, *12*, 669–673. [CrossRef]
238. Lapenta, G.; Markidis, S.; Marocchino, A.; Kaniadakis, G. Relaxation of relativistic plasmas under the effect of wave-particle interactions. *Astrophys. J.* **2007**, *666*, 949–954. [CrossRef]
239. Guo, L.; Du, J.; Liu, Z. The property of κ-deformed statistics for a relativistic gas in an electromagnetic field: κ parameter and κ-distribution. *Phys. Lett. A* **2007**, *367*, 431–435. [CrossRef]
240. Guo, L.; Du, J. The κ parameter and κ-distribution in κ-deformed statistics for the sysstems in an external field. *Phys. Lett. A* **2007**, *362*, 368–370. [CrossRef]
241. Lapenta, G.; Markidis, S.; Kaniadakis, G. Computer experiments on the relaxation of collisionless plasmas. *J. Stat. Mech.* **2009**, *2009*, P02024. [CrossRef]
242. Gougam, L.A.; Tribeche, M. Electron-acoustic waves in a plasma with a kappa-deformed Kaniadakis electron distribution. *Phys. Plasmas* **2016**, *23*, 014501. [CrossRef]
243. Lourek, I.; Tribeche, M. On the role of the κ-deformed Kaniadakis distribution in nonlinear plasma waves. *Phys. A* **2016**, *441*, 215–220. [CrossRef]

244. Lopez, R.A.; Navarro, R.E.; Pons, S.I.; Araneda, J.A. Landau damping in Kaniadakis and Tsallis distributed electron plasmas. *Phys. Plasmas* **2017**, *24*, 102119. [CrossRef]
245. Chen, H.; Zhang, S.X.; Liu, S.Q. Te longitudinal plasmas modes of κ-deformed Kaniadakis distributed plasmas. *Phys. Plasmas* **2017**, *24*, 022125. [CrossRef]
246. Saha, A.; Tamang, J. Qualitative analysis of the positron-acoustic waves in electron-positron-ion plasmas with kappa deformed Kaniadakis distributed electrons and hot positrons. *Phys. Plasmas* **2017**, *24*, 082101. [CrossRef]
247. Lourek, I.; Tribeche, M. Dust charging current in non equilibrium dusty plasma in the context of Kaniadakis generalization. *Phys. A* **2019**, *517*, 522–529. [CrossRef]
248. Khalid, M.; Rahman, A.-U. Oblique ion acoustic excitations in a magnetoplasma having κ-deformed Kaniadakis distributed electrons. *Astr. Space Sc.* **2020**, *365*, 75. [CrossRef]
249. Naeem, S.N.; Qamar, A.; Khalid, M.; Rahman, A. Coexistence of positive and negative polarity dust ion acoustic excitations with κ-deformed Kaniadakis distribution. *EPJ Plus* **2021**, *136*, 1205–1211. [CrossRef]
250. Khalid, M.; Khan, A.; Khan, M.; Hadi, F.; Ata-ur-Rahman. Dust ion acoustic solitary waves in unmagnetized plasma with Kaniadakis distributed electrons. *Braz. J. Phys.* **2021**, *51*, 60–65. [CrossRef]
251. Khalid, M.; Khan, M.; Muddusir; Ata-Ur-Rahman; Irshad, M. Periodic and localized structures in dusty plasma with Kaniadakis distribution. *Z. Naturforsch. A* **2021**, *76*, 891–897. [CrossRef]
252. Bala, P.; Kaur, A. Quantum electron acoustic solitons and double layers with κ-deformed Kaniadakis distributed electrons. *Indian J. Pure Appl. Phys.* **2021**, *59*, 577–585.
253. Tan, L.; Yang, Q.; Chen, H.; Liu, S. The Longitudinal Plasma Modes of κ-Deformed Kaniadakis Distributed Plasmas Carrying Orbital Angular Momentum. *Entropy* **2022**, *24*, 1211. [CrossRef]
254. Irshad, M.; Khalid, M.; Ata-ur-Rahman. Modulational instability of ion acoustic excitations in a plasma with a κ-deformed Kaniadakis electron distribution. *Eur. Phys. J. Plus* **2022**, *137*, 893. [CrossRef]
255. Bellahsene, Z.; Bacha, M.; Zerguini, T.H. The role of κ-deformed Kaniadakis distributed electrons on the dust ion-acoustic waves in charge-varying dusty plasma. *Contrib. Plasma Phys.* **2023**, *64*, e202300155. [CrossRef]
256. Dubinov, A.E. Gas-dynamic approach to the theory of non-linear ion-acoustic waves in plasma with Kaniadakis' distributed species. *Adv. Space Res.* **2023**, *71*, 1108–1115. [CrossRef]
257. Raut, S.; Mondal, K.K.; Chatterjee, P.; Roy, S. Dust ion acoustic bi-soliton, soliton, and shock waves in unmagnetized plasma with Kaniadakis-distributed electrons in planar and nonplanar geometry. *Euro. Phys. J. D* **2023**, *77*, 100. [CrossRef]
258. Irshad, M.; Ata-Ur-Rahman; Khalid, M.; Khan, S.; Alotaibi, B.M.; El-Sherif, L.S.; El-Tantawy, S.A. Effect of κ-deformed Kaniadakis distribution on the modulational instability of electron-acoustic waves in a non-Maxwellian plasma. *Phys. Fluids* **2023**, *35*, 105116. [CrossRef]
259. Khalid, M.; Kabir, A.; Jan, S.U.; Eldin, S.M. Coexistence of Compressive and Rarefactive Positron-Acoustic Electrostatic Excitations in Unmagnetized Plasma with Kaniadakis Distributed Electrons and Hot Positrons. *Braz. J. Phys.* **2023**, *53*, 66. [CrossRef]
260. Bala, P.; Kaur, G. Modulational instability of ion-acoustic waves in multicomponent plasma using κ-deformed Kaniadakis distribution. *Pramana J. Phys.* **2024**, *98*, 7. [CrossRef]
261. Housset, J.; Saavedra, J.; Tello-Ortiz, F. Cosmological FLRW phase transitions and micro-structure under Kaniadakis statistics. *Phys. Lett. B.* **2024**, *853*, 138686. [CrossRef]
262. Kalita, J.; Das, R.; Hosseini, K.; Salahshour, S.; Baleanu, D. Some Models in Unmagnetized Plasma Involving Kaniadakis Distributed Electrons and Temperature Ratio: Dust Ion Acoustic Solitary Waves. *J. Appl. Comp. Mech.* **2024**, *10*, 792–800.
263. Imon, U.; Bhuyan, M.S.; Alam, M.S. Phase Shifts Due to Head-on Collision of Electron Acoustic Waves and Production of Rogue Waves: Kappa-Deformed Kaniadakis Electrons. *Brazilian J. Phys.* **2024**, *55*, 61. [CrossRef]

entropy

MDPI

Article

The Longitudinal Plasma Modes of κ-Deformed Kaniadakis Distributed Plasmas Carrying Orbital Angular Momentum

Ling Tan, Qiaoyun Yang, Hui Chen * and Sanqiu Liu *

Jiangxi Province Key Laboratory of Fusion and Information Control, Department of Physics, Nanchang University, Nanchang 330031, China
* Correspondence: hchen61@ncu.edu.cn (H.C.); sqlgroup@ncu.edu.cn (S.L.)

Abstract: Based on plasma kinetic theory, the dispersion and Landau damping of Langmuir and ion-acoustic waves carrying finite orbital angular momentum (OAM) were investigated in the κ-deformed Kaniadakis distributed plasma system. The results showed that the peculiarities of the investigated subjects relied on the deformation parameter κ and OAM parameter η. For both Langmuir and ion-acoustic waves, dispersion was enhanced with increased κ, while the Landau damping was suppressed. Conversely, both the dispersion and Landau damping were depressed by OAM. Moreover, the results coincided with the straight propagating plane waves in a Maxwellian plasma system when $\kappa = 0$ and $\eta \to \infty$. It was expected that the present results would give more insight into the trapping and transportation of plasma particles and energy.

Keywords: orbital angular momentum; Langmuir waves; ion-acoustic waves; κ-deformed Kaniadakis distribution function

PACS: 52.20.-j; 52.25.Dg; 52.35.Fp

Citation: Tan, L.; Yang, Q.; Chen, H.; Liu, S. The Longitudinal Plasma Modes of κ-Deformed Kaniadakis Distributed Plasmas Carrying Orbital Angular Momentum. *Entropy* **2022**, *24*, 1211. https://doi.org/10.3390/e24091211

Academic Editors: Dionissios T. Hristopulos, Sergio Luiz E. F. da Silva and Antonio M. Scarfone

Received: 7 August 2022
Accepted: 25 August 2022
Published: 29 August 2022

1. Introduction

In 1990, Tamm et al. produced a Laguerre–Gaussian (LG)-mode laser beam with helical wave fronts that can drive molecules and neutral atoms [1]. Allen et al. subsequently demonstrated that the angular momentum carried by a laser beam with azimuth phase distribution was unrelated to the state of the polarized photons, and that the missing parts of the angular momentum of photons are to be found in twisted electromagnetic beams or optical vortices [2]. It is well recognized that the angular momentum of electromagnetic radiation has two distinct components. The first is the intrinsic part associated with wave polarization, or spin; the second is an extrinsic part related to the orbital angular momentum (OAM) that depends on the spatial radiation distribution [3]. Hence, the laser beams, as depicted by the LG function that satisfies the basic orthogonal condition, possess spin and angular momentum as well as OAM, which lays the foundation for numerous important scientific applications because of its inherent orthogonality and production techniques that have matured in the laboratory [4–8]. For example, owing to the special helical phase and hollow light field of the OAM beam, OAM can be applied to micro-control technologies in the microscopic world, such as optical tweezers and micromotors [9]. As an independent degree of freedom for wireless and quantum communications, OAM can be used to achieve a higher communication capacity through the simultaneous transmission of multiple orthogonal OAM mode vortex beams [10–13]. It has even been suggested that telescopes equipped with OAM diagnostic instruments can be made to detect rotating black holes [14,15].

The energy, momentum and angular momentum of laser beams can be transferred to matter by interacting with it. Moreover, the propagation of an OAM beam in plasma is associated with the excitation of a plasma wave, which may likewise carry OAM. In recent years, related research carried out on OAM in plasma has attracted much attention, and the significance of collective plasma oscillations with OAM has been recognized in different contexts. For instance, Mendonça et al. first employed the concepts of photon OAM states in plasma systems to investigate the stimulated Raman and Brillouin backscattering of collimated beams [16]. It was shown that the exchange of OAM between electromagnetic and electrostatic waves occurred in stimulated Raman and Brillouin backscattering, which implied that plasmon and phonon states carried OAM. The idea opened the door for various studies on OAM in plasma. By using two fluid models and Ampere's law in a magnetized plasma, Shukla indicated that three-dimensional modified-kinetic Alfvén waves can propagate in the shape of Alfvénic tornadoes featuring plasma density whirls or magnetic flux ropes with OAM [17]. Vieira et al. confirmed the existence of LG modes in particle-in-cell (PIC) simulations of intense laser–plasma interactions, and revealed the mechanism of high OAM harmonic generation and amplification through stimulated Raman backscattering in plasma [18,19]. Ali et al. calculated the quasistatic axial magnetic field generated during laser propagation in plasma by taking into account both the spin and OAM of the laser pulse [20,21]. The fluid theory of electron-acoustic waves in a two-temperature electron plasma was considered by Shahzad et al., who derived the OAM density of electron-acoustic waves [22]. Ali et al. subsequently presented dust oscillons with distinct OAM states in a collisionless unmagnetized self-gravitating dusty plasma [23].

The OAM modes in plasma also introduced other significant effects such as single-electron level twisted photon emission [24], toroidal shaped plasma turbulence in radio-pumping [25], and a helical plasma accelerator [26].

Moreover, the quasistatic axial magnetic field generated by OAM beams in plasma also has prospective applications in deep resource exploration, atmospheric science, and underwater communications [27].

In the study of the interplay between light beams and plasma, wave-particle interaction plays a pivotal role in particle acceleration, wave mixing, and the nonlinear decay of laser beams with OAM, for which a kinetic framework is necessary. Relying on the plasma kinetic description, Mendonça derived the dispersion relation and Landau damping of helical electron plasma waves with OAM in cylindrical geometry under the paraxial approximation. It was shown that the vertical velocity component of a helical Langmuir wave also contributes to Landau resonance [28,29]. Following the work of Mendonça, Khan et al. extended the theory to ion-acoustic plasma vortices with OAM and indicated that the azimuthal component of an electric field produces optical torque on the medium, which results in increased OAM of the plasma vortex [30]. Rehman et al. studied the propagation characteristics of an electronic acoustic wave in a two-electron component plasma. The results showed that the electronic acoustic wave carrying OAM was strongly damped at large and intermediate wavelengths, whereas it was weakly damped at small wavelengths [31]. Recently, Khan et al. described the helical structure of electrostatic plasma waves carrying OAM by introducing a variable transformation. The proposed idea improved the method of accessing wave damping [32]. The kinetic theory for these OAM-carrying plasma waves were investigated in a Maxwellian distributed plasma system. However, during solar wind or flares, pulsars and other complex environments, plasma systems with superheated electrons exhibit energetic tails in particle velocity or energy distribution; therefore, the Maxwellian distribution is not applicable [33].

To process some emerging physical problems in complex environments, attempts were made to generalize statistical mechanics on conventional Boltzmann–Gibbs (BG) entropy. In this context, Rényi proposed the non-extensive generalization of BG entropy [34], which was later also suggested by Tsallis [35], whose non-extensive entropy was in excellent agreement [36] with experimental data [37,38]. Afterwards, Kaniadakis put forward a new so-called κ-deformed distribution in 2001 [39], which has been widely applied to the kinetics of interaction atoms and photons [40], nonlinear kinetics [41–43], cosmic rays [33], blackbody radiation [44], quantum entanglement [45], quark–gluon plasma formation [46], and even financial systems [47,48] and epidemiology [49]. The κ-deformed distribution arising from Kaniadakis entropy covers both nonextensive and the classical Maxwell–

Boltzman distributions [39]. In subsequent studies, Beck and Cohen proposed that this κ-deformed distribution can be regarded as the result of more generalized statistics known as superstatistics [50]. Ourabah et al. also verified that the nonthermal and suprathermal empirical distributions can be recovered from Beck–Cohen superstatistics [51]. The κ-deformed distribution can be represented as a more universal form of the distribution functions mentioned in the above work. Consequently, many studies have been revisited under the κ-deformed distribution, such as longitudinal plasma modes [52] and Jeans gravitational instability [53–55]. The κ-deformed distribution as a generalized statistic may be able to characterize plasma waves in a fusion device. The power density of the OAM beam highly localized away from the propagation axis could be an efficient device for transferring concentrated heating power, which could be used to heat the fusion plasma at certain locations. The OAM state may also act as a potential plasma diagnostic technique since it can be modulated by various anisotropic and nonuniform structures in plasma [56].

To explore the effects of OAM and the κ-deformed Kaniadakis distribution function on the dispersion and Landau damping of longitudinal plasma waves in this paper, longitudinal plasma modes carrying OAM were considered in κ-deformed Kaniadakis distributed plasmas based on kinetic theory. This manuscript is organized in the following fashion. In Section 2, linear kinetic theory is employed to derive a dielectric function for the longitudinal plasma waves in κ-deformed Kaniadakis distributed plasmas. Sections 3 and 4 describe the linear dispersion relation and the damping rate of Langmuir and ion-acoustic waves, respectively. The numerical results and a brief summary are given in Section 5.

2. The Longitudinal Dielectric Function

According to plasma kinetic theory, the dispersion relation and Landau damping of electrostatic waves in an unmagnetized collisionless isotropic plasma are determined by

$$\mathrm{Re}\,\varepsilon_l(\omega,k)=0, \tag{1}$$

and

$$\gamma(\omega,k)=-\frac{\mathrm{Im}\,\varepsilon_l(\omega,k)}{\partial/\partial\omega\,\mathrm{Re}\,\varepsilon_l(\omega,k)}, \tag{2}$$

respectively, where $\mathrm{Re}\,\varepsilon_l(\omega,k)$ and $\mathrm{Im}\,\varepsilon_l(\omega,k)$ are real and imaginary parts of the plasma longitudinal dielectric function $\varepsilon_l(\omega,k)$. To investigate the novel properties of finite OAM carried by electrostatic waves in a κ-deformed Kaniadakis distributed plasma system, the LG function is used to describe the perturbed electrostatic potential and distribution function, which followed the same method as mentioned in Refs. [29–31]. Then the dielectric function was obtained by the linearized Vlasov–Poisson equation [28] given by

$$\varepsilon_l\left(\omega,\mathbf{q}_{eff}\right)=1+\sum_\alpha\frac{\omega_{p\alpha}^2}{k^2}\int\frac{\mathbf{q}_{eff}\cdot\partial f_\alpha/\partial\mathbf{v}}{\omega-\mathbf{q}_{eff}\cdot\mathbf{v}}d\mathbf{v}. \tag{3}$$

Here, $\omega_{p\alpha}$ is the plasma frequency; α represents plasma species ($\alpha=i$ for ion, $\alpha=e$ for electron, respectively); and $\mathbf{q}_{eff}=-iq_r\hat{\mathbf{e}}_r+lq_\theta\hat{\mathbf{e}}_\theta+(k-iq_z)\hat{\mathbf{e}}_z$. More details about the relevant parameters are given in Refs. [29–31]). The κ-deformed Kaniadakis distribution function is written as [39]

$$f_\alpha(\mathbf{v})-A_\kappa\exp_\kappa\left(-\frac{\mathbf{v}^4}{4\tau^4}\right), \tag{4}$$

with

$$\exp_\kappa(x)=\left(\sqrt{1+\kappa^2x^2}+\kappa x\right)^{\frac{1}{\kappa}}, \tag{5}$$

where $\tau=k_BT_\alpha/m_\alpha$ in relation to the thermal velocity $v_{T\alpha}=\sqrt{k_BT_\alpha/m_\alpha}$; and A_κ is the normalized constant given by

$$A_k = \frac{n_{\alpha 0}}{(2\pi)^{\frac{3}{2}}\tau^3}|2\kappa|^{\frac{3}{2}}\left(1+\frac{3}{4}|2\kappa|\right)\frac{\Gamma\left(\frac{1}{|2\kappa|}+\frac{3}{4}\right)}{\Gamma\left(\frac{1}{|2\kappa|}-\frac{3}{4}\right)}. \tag{6}$$

Compared to the Maxwellian case, the effective temperature of the superstatistics depends on the deformation parameter κ, which was explicitly given in Ref. [53], as

$$T_{\alpha(eff)} = \frac{1}{|2\kappa|}\frac{1+\frac{3}{4}|2\kappa|}{1+\frac{5}{4}|2\kappa|}\frac{\Gamma\left(\frac{1}{|2\kappa|}-\frac{5}{4}\right)}{\Gamma\left(\frac{1}{|2\kappa|}+\frac{5}{4}\right)}\frac{\Gamma\left(\frac{1}{|2\kappa|}+\frac{3}{4}\right)}{\Gamma\left(\frac{1}{|2\kappa|}-\frac{3}{4}\right)}, \tag{7}$$

In Equation (6), κ represents the strength of the deformation and the symbol Γ pinpoints the gamma function. To obtain A_k, the following integral was used [33,57]

$$\int_0^\infty x^{r-1}\exp_\kappa(-x)dx = \frac{|2\kappa|^{-r}}{1+r|\kappa|}\frac{\Gamma\left(\frac{1}{2|\kappa|}-\frac{2}{r}\right)}{\Gamma\left(\frac{1}{2|\kappa|}+\frac{2}{r}\right)}\Gamma(r). \tag{8}$$

When $\kappa \to 0$, it is important to note that the Kaniadakis distribution function is reduced to the Maxwellian distribution with $\lim_{\kappa\to 0}\exp_\kappa(x) \equiv \exp(x)$ [33].

The use of the κ-deformed Kaniadakis distribution results in the dielectric function of longitudinal waves carrying OAM which correlated strongly with the azimuthal velocity contribution, as

$$\varepsilon_l(\omega,k) = 1+\sum_\alpha \frac{\omega_{p\alpha}^2}{C_k k^2 v_{T\alpha}^2}[B_k - Z(\xi_z^\alpha) - Z(\xi_\theta^\alpha)], \tag{9}$$

where

$$B_k = \frac{|2\kappa|^{-\frac{1}{2}}\Gamma\left(\frac{1}{|2\kappa|}-\frac{1}{4}\right)\sqrt{\pi}}{1+\frac{1}{4}|2\kappa|\Gamma\left(\frac{1}{|2\kappa|}+\frac{1}{4}\right)},$$

$$C_k = \frac{|2\kappa|^{-\frac{3}{2}}\Gamma\left(\frac{1}{|2\kappa|}-\frac{3}{4}\right)\sqrt{\pi}}{1+\frac{3}{4}|2\kappa|\Gamma\left(\frac{1}{|2\kappa|}+\frac{3}{4}\right)},$$

$Z(\xi_s^\alpha)$ is the modified plasma dispersion function that includes both axial and azimuthal contributions with $\xi_z^\alpha = \omega/\sqrt{2}kv_{T\alpha}$ and $\xi_\theta^\alpha = \omega/\sqrt{2}lq_\theta v_{T\alpha}$, respectively, in the presence of the κ-deformed Kaniadakis distribution function, which can be written as

$$Z(\xi_s^\alpha) = \int_{-\infty}^\infty \frac{\xi_s^\alpha}{x-\xi_s^\alpha}\left(\sqrt{1+\kappa^2x^4}-\kappa x^2\right)^{\frac{1}{\kappa}}dx. \tag{10}$$

By making use of the Plemelj formula [58] and integrating for Equation (10), one can obtain the modified dispersion functions under the limitation $\xi_{z,\theta}^\alpha \gg 1$ and $\xi_{z,\theta}^\alpha \ll 1$, as

$$\begin{aligned} Z(\xi_z^\alpha)+Z(\xi_\theta^\alpha) = {} & B_k + \frac{1}{2(\xi_z^\alpha)^2}C_k + \frac{3}{4(\xi_z^\alpha)^4}D_k + \frac{1}{2(\xi_\theta^\alpha)^2}C_k + \frac{3}{4(\xi_\theta^\alpha)^4}D_k \\ & - i\pi\left[\xi_z^\alpha\exp_\kappa\left(-(\xi_z^\alpha)^2\right) + \xi_\theta^\alpha\exp_\kappa\left(-(\xi_\theta^\alpha)^2\right)\right], \end{aligned} \tag{11}$$

and

$$Z(\xi_z^\alpha)+Z(\xi_\theta^\alpha) = -i\sqrt{\pi}\xi_z^\alpha - i\sqrt{\pi}\xi_\theta^\alpha, \tag{12}$$

respectively, with

$$D_k = \frac{|2\kappa|^{-\frac{5}{2}}\Gamma\left(\frac{1}{|2\kappa|}-\frac{5}{4}\right)\sqrt{\pi}}{1+\frac{5}{4}|2\kappa|\Gamma\left(\frac{1}{|2\kappa|}+\frac{5}{4}\right)},$$

which were given by Chen [53].

3. The Disperation and Landau Damping of Langmuir Waves

It is generally acknowledged that the existence of a Langmuir wave requires that $\frac{\omega}{k} \gg v_{Te}$, namely, $\xi^e_{z,\theta} \gg 1$, and then from Equations (9) and (11) the real part of the longitudinal dielectric function for an electron can be given by

$$\operatorname{Re}\varepsilon^e_l(\omega,k) = 1 - \frac{1}{k^2\lambda^2_{De}}\frac{k^2v^2_{Te} + (lq_\theta)^2v^2_{Te}}{\omega^2} - \frac{3D_k}{C_k}\frac{\omega^2_{pe}}{k^2}\frac{k^4v^2_{Te} + (lq_\theta)^4v^2_{Te}}{\omega^4}, \tag{13}$$

which, in combination with Equation (1), gives rise to the dispersion relation for Langmuir waves with OAM as

$$\omega^2 = \omega^2_{pe}\left(1 + \frac{1}{\eta^2}\right) + \frac{3D_k}{C_k}k^2v^2_{Te}\frac{\left(1 + \frac{1}{\eta^4}\right)}{\left(1 + \frac{1}{\eta^2}\right)}. \tag{14}$$

In the above expression, $\eta = \frac{k}{lq_\theta}$ is the dimensionless parameter showing the helical phase structure involving the plasma oscillations which are directly associated with OAM. Here, it was obvious that the kinetic dispersion relation was similar to ordinary plane waves in form. Nevertheless, it should be pointed out that the dispersion of a Langmuir wave in an OAM state relies on the deformation parameter κ and OAM parameter η. Moreover, when the azimuthal wave number approached zero ($\eta \to \infty$), Equation (14) ultimately recovered the following plane wave dispersion relation

$$\omega^2 = \omega^2_{pe} + \frac{3D_k}{C_k}k^2v^2_{Te}, \tag{15}$$

which was consistent with the results of Langmuir waves in the κ-deformed Kaniadakis distributed plasma system studied by Chen [53].

In the limit $\kappa \to 0$, one has $D_k/C_k \to 1$, and then Equation (14) readily reduces to the result of the Langmuir waves with OAM derived in Maxwellian plasma [29],

$$\omega^2 = \omega^2_{pe}\left(1 + \frac{1}{\eta^2}\right) + 3k^2v^2_{Te}\frac{\left(1 + \frac{1}{\eta^4}\right)}{\left(1 + \frac{1}{\eta^2}\right)}. \tag{16}$$

The expression for the imaginary part of the dielectric function was obtained from Equations (9) and (11)

$$\operatorname{Im}\varepsilon^e_l(\omega,k) = \frac{\pi}{C_k k^2\lambda^2_{De}}\left[\xi^e_z \exp_\kappa\left((\xi^e_z)^2\right) + \xi^e_\theta \exp_\kappa\left((\xi^e_\theta)^2\right)\right]. \tag{17}$$

Then, the Landau damping for Langmuir waves carrying OAM with Equation (2) yielded

$$\gamma = -\frac{\omega^4\pi}{2\sqrt{2}C_k k^3 v^3_{Te}\left(1 + \frac{1}{\eta^2}\right)}\left[\exp_\kappa\left(-\frac{\omega^2}{2k^2v^2_{Te}}\right) + \eta\exp_\kappa\left(-\frac{\omega^2\eta^2}{2k^2v^2_{Te}}\right)\right]. \tag{18}$$

Here λ_{De} is the electron Debye length. Note that the inclusion of the azimuthal velocity component led to the existence of an OAM parameterv η. Equation (18) shows that the Landau damping of Langmuir waves in a κ-deformed Kaniadakis distributed plasma system was significantly modified by the OAM parameter η and the parameter κ of the distribution function. Again by setting the OAM parameter $\eta \to \infty$, the Landau damping eventually arrived at the simplified expression

$$\gamma = -\frac{\omega^4\pi}{2\sqrt{2}C_k k^3 v_{Te}^3}\exp_\kappa\left(-\frac{\omega^2}{2k^2v_{Te}^2}\right), \tag{19}$$

which was obtained in the absence of the Landau damping OAM in conventional κ-deformed Kaniadakis distributed electron plasma [53].

In Equation (18), when $\kappa = 0$, the coefficient $C_k \to \pi$, and the Landau damping for Langmuir waves with OAM was reduced to the form of a Maxwellian case [29]:

$$\gamma = -\sqrt{\frac{\pi}{8}}\omega_{pe}\frac{\omega_{pe}^3}{k^3v_{Te}^3}\left(1+\frac{1}{\eta^2}\right)\left[\exp\left(-\omega^2/2k^2v_{Te}^2\right)+\eta\exp\left(-\eta^2\omega^2/2k^2v_{Te}^2\right)\right]. \tag{20}$$

4. The Disperation and Landau Damping of Ion-Acoustic Waves

Under the ion-acoustic time scale $\xi_{z,\theta}^e \ll 1$ and $\xi_{z,\theta}^i \gg 1$, one has the ion-dielectric function for low-frequency longitudinal modes from Equations (9) and (11),

$$\begin{aligned}\varepsilon_l^i(\omega,k) = 1 - \frac{\omega_{pi}^2}{C_k k^2 v_{Ti}^2}\left[\frac{C_k}{2}\left(\frac{1}{(\xi_z^i)^2}+\frac{1}{(\xi_\theta^i)^2}\right)+\frac{3D_k}{4}\left(\frac{1}{(\xi_z^i)^4}+\frac{1}{(\xi_\theta^i)^4}\right)\right]+ \\ i\frac{\pi}{C_k k^2\lambda_{Di}^2}\left[\xi_z^i\exp_\kappa\left(-\left(\xi_z^i\right)^2\right)+\xi_\theta^i\exp_\kappa\left(-\left(\xi_\theta^i\right)^2\right)\right].\end{aligned} \tag{21}$$

Substituting Equations (13) and (21) into the relational expression $\varepsilon_l(\omega,k) = 1 + \left[\varepsilon_l^e - 1\right] + \left[\varepsilon_l^i - 1\right]$ along with Equation (1), the dispersion equation yielded

$$\mathrm{Re}\,\varepsilon_l = 1 + \frac{\omega_{pe}^2}{C_k k^2 v_{Te}^2}B_k - \frac{\omega_{pi}^2}{C_k k^2 v_{Ti}^2}\left[\frac{C_k}{2}\left(\frac{1}{(\xi_z^i)^2}+\frac{1}{(\xi_\theta^i)^2}\right)+\frac{3D_k}{4}\left(\frac{1}{(\xi_z^i)^4}+\frac{1}{(\xi_\theta^i)^4}\right)\right] = 0. \tag{22}$$

Then we arrived at the dispersion relation for ion-acoustic waves carrying OAM, as

$$\omega^2 = \omega_{pi}^2\left(1+\frac{1}{\eta^2}\right)\left[\frac{k^2\lambda_{De}^2\frac{C_k}{B_k}}{k^2\lambda_{De}^2\frac{C_k}{B_k}+1}+\frac{3D_k}{C_k}k^2\lambda_{Di}^2\frac{1+\frac{1}{\eta^4}}{\left(1+\frac{1}{\eta^2}\right)^2}\right], \tag{23}$$

where λ_{Di} is the ion Debye length. In the limit $\eta \to \infty$, $C_k/B_k \to 1$, and the Equation (23) was reduced to the same dispersion relation for planar ion-acoustic waves in the κ-deformed Kaniadakis distributed plasma system [53], as

$$\omega^2 = \omega_{pi}^2\frac{k^2\lambda_{De}^2 C_k/B_k}{k^2\lambda_{De}^2 C_k/B_k + 1}. \tag{24}$$

In addition, the Maxwellian limit of Equation (23) can essentially be recovered by setting $\kappa = 0$:

$$\omega^2 = \omega_{pi}^2\left(1+\frac{1}{\eta^2}\right)\frac{k^2\lambda_{De}^2}{k^2\lambda_{De}^2+1} + 3k^2\lambda_{Di}^2\frac{1+\frac{1}{\eta^4}}{1+\frac{1}{\eta^2}}. \tag{25}$$

Likewise, according to Equations (2) and (21), the Landau damping for ion-acoustic waves in an OAM state can easily be derived as follows:

$$\gamma = \frac{\omega^4\pi}{2\sqrt{2}C_k k^3 v_{Ti}^3\left(1+\frac{1}{\eta^2}\right)}\left[\exp_\kappa\left(-\frac{\omega^2}{2k^2v_{Ti}^2}\right)+\eta\exp_\kappa\left(-\frac{\omega^2\eta^2}{2k^2v_{Ti}^2}\right)\right]. \tag{26}$$

Similarly, in Equation (26), by taking $\eta \to \infty$ and $\kappa = 0$, the Landau damping for ion-acoustic waves carrying OAM was then reduced to the standard result in a Maxwellian plasma system:

$$\gamma = -\sqrt{\frac{\pi}{8}}\omega\frac{\omega^3}{k^3 v_{Te}^3}\exp\left(-\omega^2/2k^2 v_{Ti}^2\right). \tag{27}$$

5. Discussion and Conclusions

For graphical illustration, we analyzed the dispersion relation and Landau damping of Langmuir waves and ion-acoustic waves with OAM in a κ-deformed Kaniadakis distributed plasma. How the dispersion and the damping rate of Langmuir waves varied with wave number, for various values of the deformation parameter κ while retaining the OAM parameter $\eta = 1$ are graphically displayed in Figures 1 and 2.

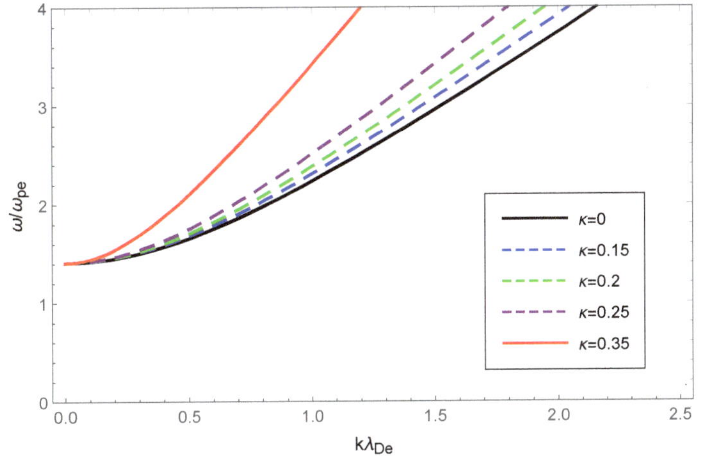

Figure 1. Variation in the normalized wave frequency ω/ω_{pe} of Langmuir waves with the normalized wave number $k\lambda_{De}$ for the deformation parameter $\kappa = 0$ (black), $\kappa = 0.15$ (dashed blue), $\kappa = 0.20$ (dashed green), $\kappa = 0.25$ (dashed purple) and $\kappa = 0.35$ (solid red), respectively, with the OAM parameter $\eta = 1$.

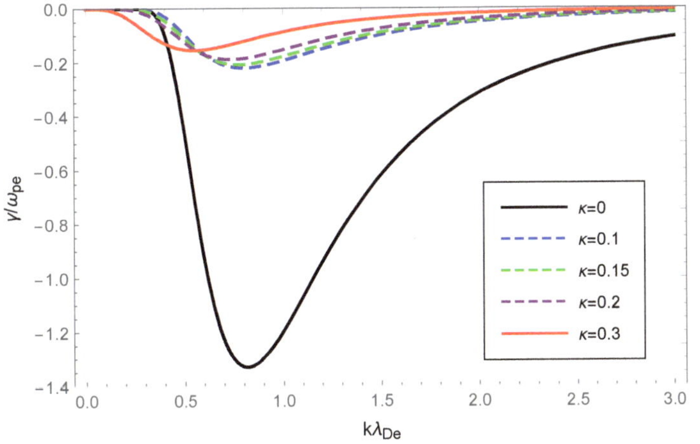

Figure 2. Variation of the Landau damping rate γ/ω_{pe} of Langmuir waves with the normalized wave number $k\lambda_{De}$ for deformation parameter $\kappa = 0$ (black), $\kappa = 0.15$ (dashed blue), $\kappa = 0.2$ (dashed green), $\kappa = 0.25$ (dashed purple) and$\kappa = 0.35$ (solid red), respectively, with the OAM parameter $\eta = 1$.

In Figure 1, the dashed blue, green, purple and solid red curves correspond to $\kappa = 0.15$, $\kappa = 0.20$, $\kappa = 0.25$ and $\kappa = 0.35$, respectively, and were acquired through solving Equation (14) numerically, whereas the solid black curve corresponding to $\kappa = 0$ was obtained by solving the analytically derived Equation (16). It is clear that as the κ increased, so did the phase velocity, which indicated that increasing the deformation parameter κ enhanced Langmuir wave dispersion.

In Figure 2, the dashed blue, green, purple and solid red curves, corresponding to $\kappa = 0.15$, $\kappa = 0.20$, $\kappa = 0.25$ and $\kappa = 0.35$, respectively, were obtained by solving Equation (20), whereas the solid black curve corresponding to $\kappa = 0$ was acquired through the numerical solution of Equation (18). Interestingly, the amplitude of the Landau damping declined drastically as κ increased, which demonstrated that the Landau damping was intensely depressed in the κ-deformed Kaniadakis distributed plasma system compared to the Maxwellian.

Next, to illustrate the effect of OAM on the Langmuir waves, we showed the dispersion variation and Landau damping with different values of the OAM parameter η while keeping κ fixed. The results are shown in Figures 3 and 4. From Figure 3, it cqn be seen that the Langmuir wave dispersion shrank as the curves moved away from each other for higher wave number values. In addition, Figure 3 shows that increasing values of η resulted in a narrowing of Langmuir wave dispersion. In particular, the dispersion was very sensitive to the OAM parameter values that varied in the range $0.4 \leq \eta \leq 1$ while there was little change beyond 1. This behavior was realized by looking at the term $1 + 1/\eta^2$ in Equation (14), which increased sharply in the range $0 \leq \eta \leq 1$. In Figure 4, as η increased within the range $0.4 \leq \eta \leq 1$, the Landau damping diminished. Conversely, when η increased within the range $1 \leq \eta \leq \infty$, the Landau damping was enhanced at small and diminished at large wave numbers. Similarly, it was enhanced with the increased value of $1 + 1/\eta^2$ as the wave number gradually increased, which can be understood more naturally from Equation (18).

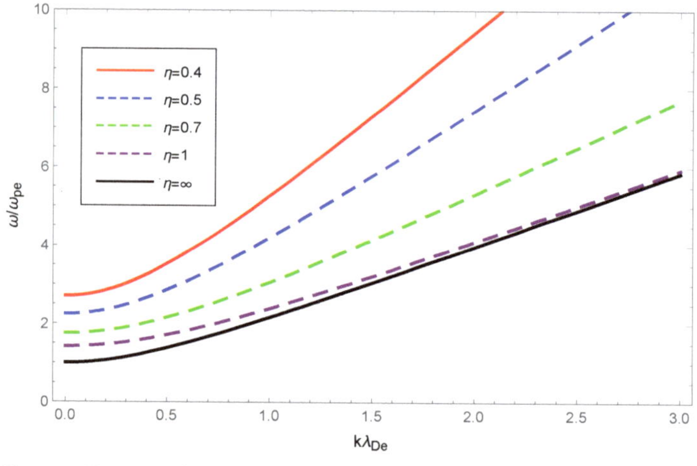

Figure 3. Variation of the normalized wave frequency $\omega/\omega_{\text{pe}}$ of Langmuir waves with the normalized wave number $k\lambda_{De}$ for OAM parameter $\eta = 0.4$ (solid red), $\eta = 0.5$ (dashed blue), $\eta = 0.7$ (dashed green), $\eta = 1$ (dashed purple), and $\eta = \infty$ (solid black), with the deformation parameter $\kappa = 0.2$.

With the value of the OAM parameter η fixed, the dispersion variation and Landau damping of ion-acoustic waves with varying values of the deformation parameter κ are illustrated in Figures 5 and 6. We solved Equation (23) numerically and plotted the Figure 5 with dashed blue, green, purple, and solid red curves corresponding to $\kappa = 0.15$, $\kappa = 0.20$, $\kappa = 0.25$ and $\kappa = 0.35$, respectively, while the solid black line corresponded to $\kappa = 0$ according to the numerical solution to Equation (25).

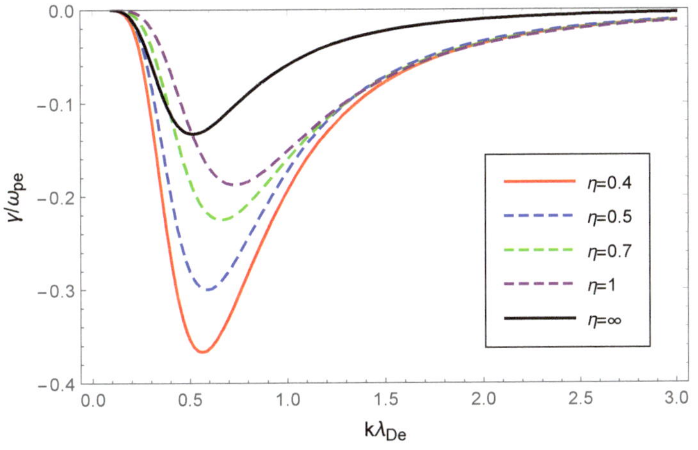

Figure 4. Variation of the Landau damping rate $\gamma/\omega_{\rm pe}$ of Langmuir waves with the normalized wave number $k\lambda_{De}$ for OAM parameter $\eta = 0.4$ (solid red), $\eta = 0.5$ (dashed blue), $\eta = 0.7$ (dashed green), $\eta = 1$ (dashed purple) and $\eta = \infty$ (solid black) with the deformation parameter $\kappa = 0.2$.

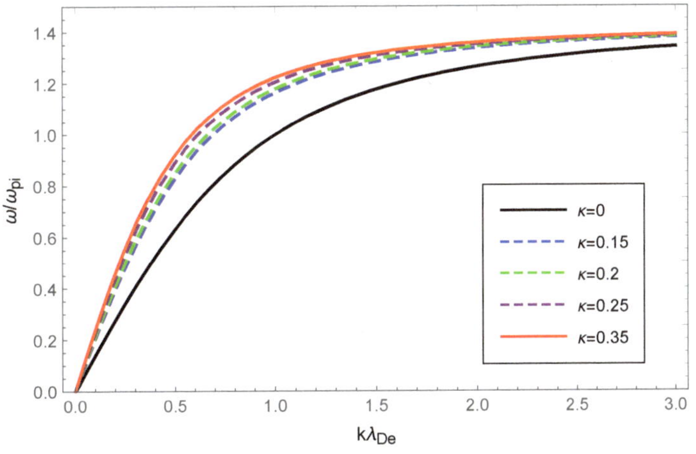

Figure 5. Variation of the normalized wave frequency $\omega/\omega_{\rm pi}$ of ion-acoustic waves with the normalized wave number $k\lambda_{De}$ for deformation parameter $\kappa = 0$ (soild black), $\kappa = 0.15$ (dashed blue), $\kappa = 0.2$ (dashed green), $\kappa = 0.25$ (dashed purple) and $\kappa = 0.35$ (soild red) with OAM parameter $\eta = 1$.

From Figure 5, a very slight change was seen in the dispersion of ion-acoustic waves carrying OAM with increased κ, but the dispersion increased overall. The Landau damping of ion-acoustic waves with OAM was calculated primarily from Equation (26). In Figure 6, the dashed blue curve for $\kappa = 0.15$ is clearly above the solid black line where $\kappa = 0$ (Maxwellian case). Therefore, it was evident from the figure that the Landau damping rate of the non-Maxwellian ion-acoustic waves was smaller compared to the Maxwellian. The Figures 7 and 8 show how the dispersion relation and the damping rate of ion-acoustic waves were affected by the presence of the OAM parameter η while keeping κ fixed. In Figure 7, the dispersion of ion-acoustic waves diminished as the OAM parameter η increased. From Figure 8, the Landau damping rate for ion-acoustic waves shifted to smaller negative values with increased η. In the case of the fixed deformation parameter κ, both dispersion and damping for ion-acoustic waves were suppressed by the OAM effect.

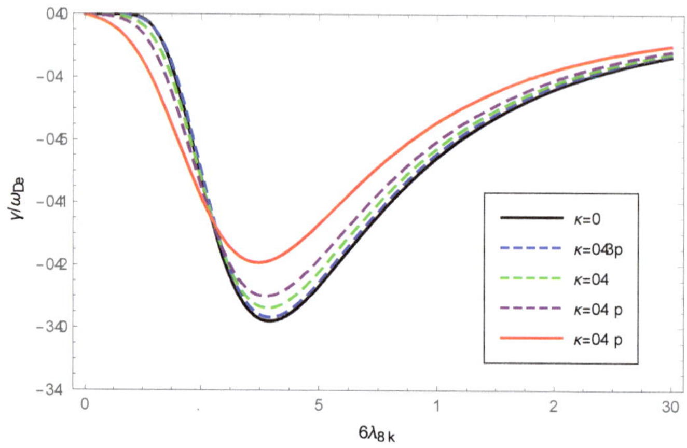

Figure 6. Variation of the Landau damping rate $\gamma/\omega_{\mathrm{pi}}$ of ion-acoustic waves with the normalized wave number $k\lambda_{De}$ for deformation parameter $\kappa = 0$ (solid black), $\kappa = 0.15$ (dashed blue), $\kappa = 0.2$ (dashed green), $\kappa = 0.25$ (dashed purple) and $\kappa = 0.35$ (solid red) with OAM parameter $\eta = 1$.

Physically, the thermal motion of particles gave rise to the dispersion of the longitudinal plasmon collective mode; the Landau damping was attributed to the resonance interaction between the plasma wave and the particles that had a velocity close to the phase velocity; therefore, the damping rate depended on the number of resonant particles. When the value of η was fixed, owing to Equations (14) and (23), the dispersion of Langmuir waves and ion-acoustic waves was viewed as being approximately proportional to $\sqrt{T_{eff}/m_e}$ and $\sqrt{T_{eff}/m_i}$, respectively. From Figure 1 of Ref. [53], as the κ increased, so did the effective temperature of the κ-deformed Kaniadakis distributed plasma system. Since $m_e \ll m_i$, the significant enhancement of Langmuir wave dispersion and the slight enhancement of ion-acoustic wave dispersion with increasing κ in Figures 1 and 5 can be explained. Gougam showed that in the κ-deformed Kaniadakis distributed plasma system, the presence of high-energy states became more plausible as κ increased [57], which indicated an increase in the number of fast particles or a decrease in the number of slow particles with respect to the Maxwellian case. As a result, with increasing κ for both Langmuir and ion-acoustic waves, the phase velocity was skewed toward the trailing part of the distribution function; thus, the Landau damping amplitude diminished. When the κ value was fixed, the decline in η, namely, the rise in $1 + 1/\eta^2$, provoked the dissipation of Langmuir and ion-acoustic waves, the intensity of which was related to the magnitude of the OAM. This was due to the participation of partially resonant particles in the resonance of the wave OAM, which allowed a relative increase in the number of resonances obtaining energy from the wave, thereby leading to enhanced wave damping.

To summarize, electrostatic waves carrying OAM were first considered in a κ-deformed Kaniadakis plasma system by keeping kinetic theory in view. Statistically, the κ-deformed Kaniadakis distribution function was the product of superstatistics, which are perceived as a more generalized statistic in contrast to non-generalized and the traditional Boltzman–Gibbs statistics. Because it served the same purpose as the κ of the suprathermal distribution and the parameter q of the non-extensive distribution, the parameter κ of the κ-deformed Kaniadakis distribution function also measured the derivation from the Maxwellian distribution function, so the collective mode features of the associated plasma system were modified. The analytical expressions were derived for the dispersion relation and the Landau damping of both the Langmuir and ion-acoustic waves. It was shown that the presence of the κ-deformed Kaniadakis distribution function and OAM modified the propagation properties of the Langmuir and ion-acoustic waves. For both, the dispersion was enhanced with increased κ, while Landau damping was suppressed. Conversely, both the dispersion and Landau damping were depressed by the OAM effect. As expected, when

$\kappa = 0$ and $\eta \to \infty$, the results coincided with the straight propagating plane waves in a Maxwellian plasma system. It was also expected that the results of this study would offer more insight into the trapping and transportation of plasma particles and energy in a κ-deformed Kaniadakis distributed plasma system. In addition, the results may also provide a reference for studying its nonlinearity.

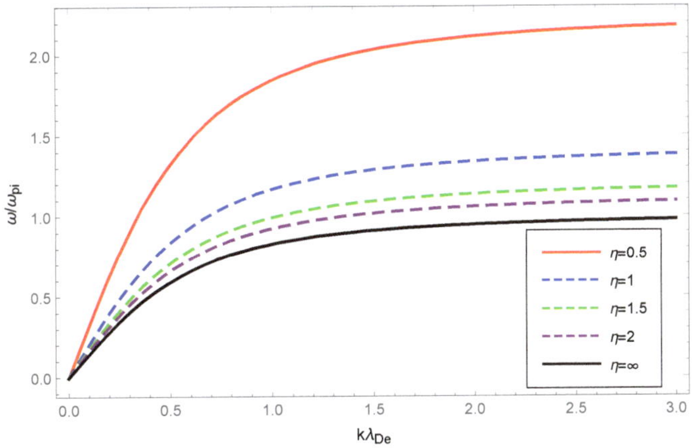

Figure 7. Variation of the normalized wave frequency $\omega/\omega_{\rm pi}$ of ion-acoustic waves with the normalized wave number $k\lambda_{De}$ for OAM parameter $\eta = 0.5$ (solid red), $\eta = 1$ (dashed blue), $\eta = 1.5$ (dashed green), $\eta = 2$ (dashed purple) and $\eta = \infty$ (solid black) with deformation parameter $\kappa = 0.2$.

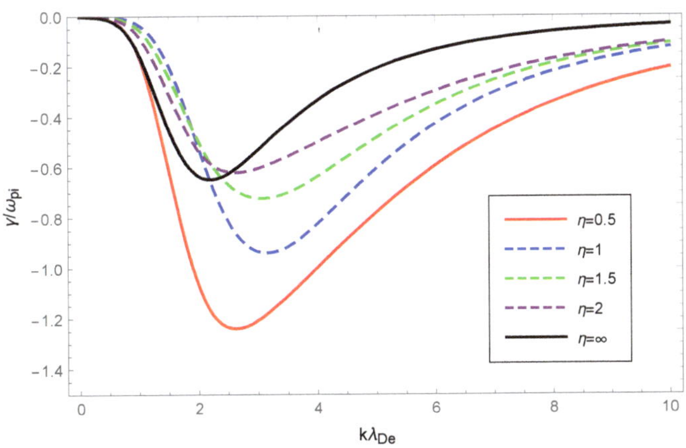

Figure 8. Variation of the Landau damping rate $\gamma/\omega_{\rm pi}$ of ion-acoustic waves with the normalized wave number $k\lambda_{De}$ for OAM parameter $\eta = 0.5$ (solid red), $\eta = 1$ (dashed blue), $\eta = 1.5$ (dashed green), $\eta = 2$ (dashd purple) and $\eta = \infty$ (solid black) with deformation parameter $\kappa = 0.2$.

Author Contributions: Funding acquisition, S.L.; Investigation, Q.Y.; Writing—original draft, L.T.; Writing—review & editing, H.C. All authors have read and agreed to the published version of the manuscript.

Funding: This work was supported by the National Natural Science Foundation of China (Grant Nos. 11763006 and 11863004), the fund from the Jiangxi Provincial Key Laboratory of Fusion and Information Control (Grant No. 20171BCD40005), and the Project of Scientific and Technological Innovation Base of Jiangxi Province (Grant No. 20203CCD46008).

Conflicts of Interest: The authors declare no conflict of interest.

References

1. Tamm, C.C.O.; Weiss, B. Instability and optical switching of spatial patterns in a laser. *J. Opt. Soc. Am. B* **1990**, *7*, 1034–1038. [CrossRef]
2. Allen, L.; Beijersbergen, M.W.; Spreeuw, R.J.C.; Woerdman, J.P. Orbital angular momentum of light and the transformation of Laguerre-Gaussian laser modes. *Phys. Rev. A* **1992**, *45*, 8185. [CrossRef] [PubMed]
3. Jackson, J.D. *Classical Electrodynamics*, 2nd ed.; Wiley: New York, NY, USA, 1962.
4. Ohtake, Y.; Ando, T.; Fukuchi, N.; Matsumoto, N. Universal generation of higher-order multiringed Laguerre-Gaussian beams by using a spatial light modulator. *Opt. Lett.* **2007**, *32*, 1411–1413. [CrossRef] [PubMed]
5. Matsumoto, N.; Ando, T.; Inoue, T.; Ohtake, Y.J. Generation of high-quality higher-order Laguerre-Gaussian beams using liquid-crystal-on-silicon spatial light modulator. *Opt. Soc. Am. A* **2008**, *25*, 1642–1651. [CrossRef]
6. Verbeeck, J.; Tian, H.; Schattschneider, P. Production and application of electron vortex beams. *Nature* **2010**, *467*, 301–304. [CrossRef]
7. Uchida, M.; Tonomura, A. Generation of electron beams carrying orbital angular momentum. *Nature* **2010**, *464*, 737–739. [CrossRef]
8. Taira, Y.; Hayakawa, T.; Katoh, M. Gamma-ray vortices from nonlinear inverse Thomson scattering of circularly polarized light. *Sci. Rep.* **2017**, *7*, 5018. [CrossRef]
9. Zhang, Z.C.; Hai, L.; Fu, S.Y. Advances on solid-state vortex laser. *Photonics* **2022**, *9*, 215. [CrossRef]
10. Ruan, X.C.; Shi, W.H.; Chen, G.J.; Zhao, W. High-rate continuous-variable quantum key distribution with orbital angular momentum multiplexing. *Entropy* **2021**, *23*, 1187. [CrossRef]
11. Willner, A.E.; Ren, Y.; Xie, G.; Yan, Y.L. Recent advances in high-capacity free-space optical and radiofrequency communications using orbital angular momentum multiplexing. *Philos. Trans. R. Soc. A Math. Phys. Eng. Sci.* **2017**, *375*, 20150439. [CrossRef]
12. Chen, R.; Zhou, H.; Moretti, M.; Wang, X. Orbital angular momentum waves: Generation, detection, and emerging applications. *IEEE Commun. Surv. Tutor.* **2020**, *22*, 840–868. [CrossRef]
13. Willner, A.E.; Liu, C. Perspective on using multiple orbital-angular-momentum beams for enhanced capacity in free-space optical communication links. *J. Nanophotonics* **2020**, *10*, 225. [CrossRef]
14. Tamburini, F.; Thidé, B.; Molina, T.G. Twisting of light around rotating black holes. *Nat. Phys.* **2011**, *7*, 195–197. [CrossRef]
15. Gwak, B. Coalescence of Kerr black holes—Binary systems from GW150914 to GW170814. *Entropy* **2019**, *21*, 1017. [CrossRef]
16. Mendonça, J.T.; Thidé, B.; Then, U. Stimulated Raman and Brillouin backscattering of collimated beams carrying orbital angular momentum. *Phys. Rev. Lett.* **2009**, *102*, 185005. [CrossRef]
17. Shukla, P.K. Alfvénic tornadoes in a magnetized plasma. *J. Geophys. Res. Space Phys.* **2013**, *118*, 1–4. [CrossRef]
18. Vieira, J.; Mendonça, J.T. Nonlinear laser driven donut wakefields for positron and electron acceleration. *Phys. Rev. Lett.* **2014**, *112*, 215001. [CrossRef]
19. Vieira, J. Amplification and generation of ultra-intense twisted laser pulses via stimulated Raman scattering. *Nat. Commun.* **2016**, *7*, 10371. [CrossRef]
20. Ali, S.; Davies, J.R.; Mendonça, J.T. Inverse Faraday effect with linearly polarized laser Pulses. *Phys. Rev. Lett.* **2010**, *105*, 035001. [CrossRef]
21. Ali, S.; Mendonça, J.T. Inverse Faraday effect with plasmon beams. *Plasma Phys. Control. Fusion* **2011**, *53*, 045007. [CrossRef]
22. Shahzad, K.; Ali, S. Finite orbital angular momentum states and Laguerre-Gaussian potential in two-temperature electron plasmas. *Space Sci.* **2014**, *353*, 3–8. [CrossRef]
23. Ali, S.; Bukhari, S.; Ikram, M.; Mendonça, J.T. Dust oscillons with finite OAM and dust self-gravity effects. *Phys. Scr.* **2019**, *105*, 055602. [CrossRef]
24. Katoh, M.; Fujimoto, M.; Kawaguchi, H.; Tsuchiya, K. Angular momentum of twisted radiation from an electron in spiral motion. *Phys. Rev. Lett.* **2017**, *118*, 094801. [CrossRef]
25. Leyser, T.B.; Norin, L.; McCarrick, M.; Pedersen, T.R. Radio pumping of ionospheric plasma with orbital angular momentum. *Phys. Rev. Lett.* **2009**, *102*, 065004. [CrossRef] [PubMed]
26. Vieira, J.; Mendonça, J.T.; Quere, F. Optical control of the topology of laser-plasma accelerators. *Phys. Rev. Lett.* **2018**, *121*, 054801. [CrossRef] [PubMed]
27. Shi, Y.; Vieira, J.; Trines, R.M.G.M.; Bingham, R. Magnetic field generation in plasma waves driven by copropagating intense twisted lasers. *Phys. Rev. Lett.* **2018**, *121*, 145002. [CrossRef]
28. Mendonça, J.T.; Ali, S.; Thidé, B. Plasmons with orbital angular momentum. *Phys. Plasmas* **2009**, *16*, 112103. [CrossRef]
29. Mendonça, J.T. Kinetic description of electron plasma waves with orbital angular momentum. *Phys. Plasmas* **2012**, *9*, 112113. [CrossRef]
30. Khan, S.A.; Rehman, A.; Mendonça, J.T. Kinetic study of ion-acoustic plasma vortices. *Phys. Plasmas* **2014**, *21*, 092109. [CrossRef]
31. Rehman, A.; Ali, S.; Khan, S.A. Twisted electron-acoustic waves in plasmas. *Phys. Plasmas* **2016**, *23*, 082122. [CrossRef]
32. Khan, S.A.; Fukuyama, A. On kinetic electrostatic plasma waves carrying orbital angular momentum. *Phys. Plasmas* **2021**, *28*, 047507. [CrossRef]
33. Kaniadakis, G. Statistical mechanics in the context of special relativity. *Phys. Rev. E* **2002**, *66*, 056125. [CrossRef]
34. Rényi, A. On a new axiomatic theory of probability. *Acta Math. Hung.* **1955**, *6*, 285–335. [CrossRef]
35. Tsallis, C. Possible generalization of Boltzmann-Gibbs statistics. *J. Stat. Phys.* **1988**, *52*, 479–487. [CrossRef]

36. Tsallis, C.; Cirto, L.J.L. Black hole thermodynamical entropy. *Eur. Phys. J. C* **2013**, *73*, 2487. [CrossRef]
37. Douglas, P.; Bergamini, S.; Renzoni, F. Tunable tsallis distributions in dissipative optical lattices. *Phys. Rev. Lett.* **2006**, *96*, 110601. [CrossRef]
38. Liu, B.; Goree, J. Superdiffusion and non-Gaussian statistics in a driven-dissipative 2D dusty plasma. *Phys. Rev. Lett.* **2008**, *100*, 055003. [CrossRef]
39. Kaniadakis, G. Non-linear kinetics underlying generalized statistics. *Physica A* **2001**, *296*, 405–425. [CrossRef]
40. Rossani, A.; Scarfone, A.M. Generalized kinetic equations for a system of interacting atoms and photons: Theory and simulations. *J. Phys. A* **2004**, *37*, 4955. [CrossRef]
41. Biro, T.S.; Kaniadakis, G. Two generalizations of the Boltzmann equation. *Eur. Phys. J. B* **2006**, *50*, 3–6. [CrossRef]
42. Casas, G.A.; Nobre, F.D.E.; Curado, M.F. Entropy production and nonlinear Fokker-Planck equations. *Phys. Rev. E* **2012**, *86*, 061136. [CrossRef]
43. Kaniadakis, G.; Hristopulos, D.T. Nonlinear kinetics on lattices based on the kinetic interaction principle. *Entropy* **2018**, *20*, 426. [CrossRef]
44. Ourabah, K.; Tribeche, M. Planck radiation law and Einstein coefficients reexamined in Kaniadakis κ statistics. *Phys. Rev. E* **2014**, *89*, 062130. [CrossRef]
45. Ourabah, K.; Hamici-Bendimerad, A.H.; Tribeche, M. Quantum entanglement and Kaniadakis entropy. *Phys. Scr.* **2015**, *90*, 045101. [CrossRef]
46. Teweldeberhan, A.M.; Miller, H.G.; Tegen, G. κ-deformed statistics and the formation of a quark-gluon plasma. *Int. J. Mod. Phys. E* **2003**, *12*, 669. [CrossRef]
47. Moretto, E.; Pasquali, S.; Trivellato, B. A non-Gaussian option pricing model based on Kaniadakis exponential deformation. *Eur. Phys. J. B.* **2017**, *90*, 179. [CrossRef]
48. Trivellato, B. Deformed exponentials and applications to finance. *Entropy* **2013**, *15*, 3471–3489. [CrossRef]
49. Kaniadakis, G.; Baldi, M.M.; Deisboeck, T.S. The κ-statistics approach to epidemiology. *Sci. Rep.* **2020**, *10*, 19949. [CrossRef]
50. Beck, C.; Cohen, E.G.D. Superstatistics. *Physical A* **2003**, *322*, 267–275. [CrossRef]
51. Ourabah, K.; Gougam, L.A.; Tribeche, M. Nonthermal and suprathermal distributions as a consequence of superstatistics. *Phys. Rev. E* **2015**, *91*, 012133. [CrossRef]
52. Chen, H.; Zhang, S.X.; Liu, S.Q. Jeans gravitational instability with κ-deformed Kaniadakis distribution. *Chin. Phys. Lett.* **2017**, *34*, 075101. [CrossRef]
53. Chen, H.; Zhang, S.X.; Liu, S.Q. The longitudinal plasmas modes of κ-deformed Kaniadakis distributed plasmas. *Phys. Plasmas* **2017**, *24*, 022125. [CrossRef]
54. He, K.R. Jeans analysis with κ-deformed Kaniadakis distribution in f (R) gravity. *Phys. Scr.* **2020**, *97*, 025601. [CrossRef]
55. Yang, W.H.; Xiong, Y.Z.; Chen, H.; Liu, S.Q. Jeans gravitational instability with κ-deformed Kaniadakis distribution in eddington-inspired born–infield gravity. *Chin. Phys. B* **2020**, *29*, 110401. [CrossRef]
56. Chen, Q.; Qin, H.; Liu, J. Photons, phonons, and plasmons with orbital angular momentum in plasmas. *Sci. Rep.* **2017**, *7*, 41731. [CrossRef]
57. Gougam, L.A.; Tribeche, M. Electron-acoustic waves in a plasma with a κ-deformed Kaniadakis electron distribution. *Phys. Plasmas* **2016**, *23*, 014501. [CrossRef]
58. Chen, X.C.; Li, X.Q. Comment on "Plasma oscillations and nonextensive statistics". *Phys. Rev. E* **2012**, *86*, 068401. [CrossRef]

entropy

MDPI

Article

Kaniadakis Entropy Leads to Particle–Hole Symmetric Distribution

Tamás S. Biró [1,2,3]

[1] Wigner Research Cenrer for Physics, H-1121 Budapest, Hungary; biro.tamas@wigner.hu; Tel.: +36-20-435-1283
[2] Hungarian Institute of Physics, University Babeş-Bolyai, RO-400084 Cluj, Romania
[3] Complexity Science Hub, A-1080 Vienna, Austria

Abstract: We discuss generalized exponentials, whose inverse functions are at the core of generalized entropy formulas, with respect to particle–hole (KMS) symmetry. The latter is fundamental in field theory; so, possible statistical generalizations of the Boltzmann formula-based thermal field theory have to take this property into account. We demonstrate that Kaniadakis' approach is KMS ready and discuss possible further generalizations.

Keywords: Kaniadakis entropy; kappa statistics; Tsallis distribution; KMS relation

Citation: Biró, T.S. Kaniadakis Entropy Leads to Particle–Hole Symmetric Distribution. *Entropy* **2022**, *24*, 1217. https://doi.org/10.3390/e24091217

Academic Editors: Antonio M. Scarfone, Dionissios T. Hristopulos and Sergio Luiz E. F. da Silva

Received: 2 August 2022
Accepted: 29 August 2022
Published: 30 August 2022

1. Introduction

Remembering when statistical physics passed the "classical" Boltzmann–Gibbs distribution view of exponential dependence on individual energies, one ought to formulate a few general statements. Certainly, a generalization [1,2], in addition to including the original classical formulas in some limits, can be infinite. In physics, however, nature gives us several clues as to which generalization is more useful, moving beyond a pure mathematical construction.

Generalizations of entropy formulas replace the logarithm with another function, and the change from an exponential function in the equilibrium or in other way stationary distributions to something else are the inverse operation to this. Informatics studies were pioneering in generalizing the entropy formula of Boltzmann in the 1950s and 1960s [3,4], while the thermodynamical consequences have been more vividly studied since the 1980s [5–7]. Here the power-law tailed distribution, originally considered an approximation to the exponential by Euler and in particle physics by Hagedorn as a "cut power-law", in the beginning did not have any physics rationale aside from its aimed application.

Another widespread nonexponential distribution, also extrapolating to power-law tails is given in the kappa statistics initiated by Kaniadakis. It is motivated by relativistic kinetics in plasmas, and its most renowned applications are also related to plasmas. While it can be mapped to an exponential of the rapidity, replacing the energy variable by a rapidity-like one as the argument, its high energy tail seems to show remarkable success in application to real world data. A general presentation of kappa statistics basics can be read in [8]. The relation to special relativity is discussed in [9], and to the Boltzmann equation in [10]. Fractional statistics in kappa statistics are dealt with in [11], nonlinear kinetics in [12], and a general review about the physical origins in [13].

In this paper we point out that based on a particular property of the mathematical formula appearing in kappa statistics, this form is able to reflect particle–hole symmetry, an important ingredient in field theory and particle physics. The underlying concept in field theory namely assumes a symmetry between particles and antiparticles, called CPT symmetry, changing charges, parity, and time direction to its opposites. The physical laws should not change in an antimatter world relative to the original one. The mathematical formulation of the time-dependent expectation values for elements of statistical ensembles in field theory is related to the use of a statistical operator. Whenever the exponential function of energy is generalized in the statistics, the corresponding statistical operator is also no more the Boltzmann–Gibbs exponential of the Hamiltonian operator. Still, the CPT reflection should not change the physical conclusions. Therefore, it is essential that the generalization of the exponential function shows similar reflection properties to the original Euler number-based function.

2. Kaniadakis' Generalized Exponential

There are several generailzed entropy formulas and corresponding canonical distributions [1–14]. At their core, they can be viewed as the generalization of the logarithm and exponential functions while keeping their inverse roles. However, the inverse relation between the exponential of x and $-x$ is, in general, lost.

The Boltzmann–Gibbs energy distribution at a fixed temperature utilizes the Euler exponential function, which has the property, $\exp(-x) = 1/\exp(x)$. Accordingly, its inverse, the logarithmic function, satisfying both $\ln(\exp(x)) = x$ and $\exp(\ln(x)) = x$, also satisfies

$$\ln\frac{1}{x} = -\ln x. \tag{1}$$

This is important in the use of the Boltzmannian entropy formula [15],

$$S/k_B = \sum_i p_i \ln\frac{1}{p_i} = -\langle \ln p_i \rangle \tag{2}$$

with the probability p_i of being in the i-th state, a real number between and including zero and one. The above formula is valid only if the probability set is normalized, i.e.,

$$\sum_i p_i = 1. \tag{3}$$

Otherwise, the leading order terms while applying the Stirling formula [16–19] to the permutation entropy [20–23] would not cancel. These basic features of this construction lead to an overall nonnegative entropy and to its concavity property [24–28].

When generalizing, such as in some axiomatic approaches, the properties have to be saved, while much less attention is paid to the $\ln(1/x) = -\ln x$ relation. In fact, some of the suggested extensions to the exponential and logarithm function satisfy such a relation, others do not. Let us review a few of them.

The Kaniadakis' κ-exponential [8],

$$e_\kappa(x) = \left(\kappa x + \sqrt{1+\kappa^2 x^2}\right)^{1/\kappa}, \tag{4}$$

satisfies the relation

$$e_\kappa(-x) = 1/e_\kappa(x). \tag{5}$$

On the other hand, the Tsallis q-exponential [5], designed to have a power-law tail relying on Euler's approximating formula for the exponential for $n = 1/(q-1)$,

$$e_q(x) = (1 + (q-1)x)^{\frac{1}{q-1}}, \tag{6}$$

behaves differently when reflecting the argument:

$$e_q(-x) = \frac{1}{e_{2-q}(x)} \neq \frac{1}{e_q(x)}. \tag{7}$$

Here $\lim\limits_{\kappa\to 0} e_\kappa(x) = e^x$ and $\lim\limits_{q\to 1} e_q(x) = e^x$ are the limits leading back to the traditional exponential.

It is easy to construct another class of functions based on a symmetric use of the Tsallis exponentials, which satisfies the product formula searched for in [29].

$$\tilde{e}_q(x) \;=\; \frac{e_q(x/2)}{e_q(-x/2)} \tag{8}$$

namely delivers

$$\tilde{e}_q(-x) \;=\; \frac{e_q(-x/2)}{e_q(x/2)} \;=\; \frac{1}{\tilde{e}_q(x)}. \tag{9}$$

This works only with the halved argument ratio definition.

The above sketched relation between the Kaniadakis' exponential and a symmetric ratio of Tsallis exponentials can be generalized. We construct a k-exponential class based on a general function, $f_k(x) = a_k(x) + kxb_k(x)$, with both $a_k(x)$ and $b_k(x)$ being even functions of x. Then,

$$e_k(x) \;=\; f_k(x)^{1/k} \;=\; (a_k(x) + kxb_k(x))^{1/k} \tag{10}$$

with its reflected pendant

$$e_k(-x) \;=\; (a_k(-x) - kxb_k(-x))^{1/k} \;=\; (a_k(x) - kxb_k(x))^{1/k} \tag{11}$$

satisfies

$$e_k(x) \cdot e_k(-x) \;=\; 1 \tag{12}$$

only if

$$a_k^2(x) \;=\; 1 + k^2x^2b_k^2(x). \tag{13}$$

Furthermore, having the traditional exponential in the $k \to 0$ limit, both a_0 and b_0 have to converge to unity. This leads to the following class of Kaniadakis type of deformed exponentials:

$$e_k(x) \;=\; \left(\sqrt{1 + k^2x^2b_k^2(x)} + kxb_k(x)\right)^{1/k}. \tag{14}$$

For a nontrivial $b_k(x)$ even function, we may consider an example:

$$b_k(x) \;=\; \frac{1}{\sqrt{1 - k^2x^2}}. \tag{15}$$

In this case, one obtains

$$e_k(x) \;=\; \left(\sqrt{\frac{1+kx}{1-kx}}\right)^{1/k} \;=\; \frac{e_q(x/2)}{e_q(-x/2)}, \tag{16}$$

with a power-law tail for an expression relating to the relativistic Doppler factor. On the other hand, this is equal to a symmetrized ratio at the half argument of Tsallis type deformed exponentials exactly with $q = 2k + 1$. In this interpretation, the $b_k(x)$ function is the Lorentz factor, with $kx = v/c = \tanh\eta$ being a velocity in units of the light speed. At the same time, $e_k(x) = e^{\eta/k}$. Hence, the rapidity variable η is additive due to the product of traditional exponential functions, and therefore, this delivers the mapping to the logarithm of the formal group: to the additive quantity belonging to the nonadditive rules generated by the deformed exponentials [22].

This additive variable, η, can also be constructed in the general case. Setting $kxb_k(x) = \sinh\eta$ and $kx = g(\eta)$ as a general function, one has

$$b_k(x) \;=\; \frac{\sinh\eta}{g(\eta)} \tag{17}$$

leading to $e_k(x) = e^{\eta/k}$. In our previous example, we had $g(\eta) = \tanh\eta$.

After reviewing examples and generalization paths, we turn to the question of why it is so important to have the property $e_k(x) \cdot e_k(-x) = 1$ in high energy physics and field theory in the next section. Further applications of Kaniadakis' exponential [30–32] and a general approach to group entropy [33] provide the reader with further information on generalizing the exponential function and its use in data processing and interpretation.

3. Particle–Hole Symmetry

The Kubo–Martin–Schwinger (KMS) relation is central in thermal field theory [34–36]. Physically, it reflects the reinterpretation of negative energy states of a quantum particle as the corresponding positive energy state of an antiparticle. A hole in the negative energy continuum is a positive energy propagating particle with opposite momentum and charges.

In this paper, we briefly review a somewhat generalized version of the KMS relation, in order to make it clear that its validity extends beyond thermal equilibrium. This presentation is based on Ref.[37].

Quantum packages of energy and charge do propagate according to field theory as solutions to the field equation Green functions, i.e., propagators. Such propagators have a few subtypes according to retarded and advanced options in their causality structure, reflected in pole positions on the complex energy plane. Since an interacting particle in a finite time can never have an energy which would exactly follow from the solution of the classical free field equation, quoted as the dispersion relation between the frequency and wave number vector, the off-mass-shell behavior is a mirror of its quantum nature. This deviation from the special relativistic energy–momentum formula for a free point particle is well comprised in the spectral function.

Spectral functions can be defined and investigated generally among two quantum field operators, say $\hat{A}$ and $\hat{B}$, in the presence of a statistical operator, $\hat{\rho}$, by a time-Fourier transform,

$$S_{AB}(\omega) = \int dt\, e^{-i\omega t} \mathrm{Tr}\big(\hat{\rho}\big[\hat{A}(t), \hat{B}(0)\big]\big). \tag{18}$$

The operators are taken in a time distant t from each other, utilizing the Heisenberg picture in field theory. The above definition tacitly assumes that the statistical operator, the statistical weight of states related to the Hamiltonian, is stationary. Whenever it contains a temperature parameter, such as $\beta = 1/T$, or further parameters, such as κ or q, the spectral function will be also parametrized by them.

In a stationary state including but not restricted to thermal equilibrium, the time reversal and energy reversal properties of the AB-generalized spectral function should be studied. Indeed, in the definition Equation (18) the time-shift invariance is also assumed, which is equivalent to the conservation of the total energy. Meanwhile, the operators $\hat{A}$ and $\hat{B}$ can be evaluated on the observed subsystem, whose spectral function we consider.

When the operators, correlated by the selected spectral function, are also time-shift invariant and Hermitean, then the following symmetry properties are ensured:

$$S^*_{AB}(\omega) = S_{B^\dagger A^\dagger}(\omega), \qquad S_{AB}(-\omega) = -S_{BA}(\omega). \tag{19}$$

Both properties utilize the time-shift invariance of the trace,

$$\mathrm{Tr}\big(\hat{\rho}\hat{A}(t)\hat{B}(0)\big) = \mathrm{Tr}\big(\hat{\rho}\hat{A}(0)\hat{B}(-t)\big). \tag{20}$$

The Wigner transform of the $[A, B] = AB - BA$ commutator's statistical expectation value in a given quantum state shows similar properties. The Wigner function definition

$$S_{AB}(x, p) = \int dq\, e^{\frac{i}{\hbar}p\cdot q} \left\langle \big[\hat{A}(x - q/2), \hat{B}(x + q/2)\big] \right\rangle, \tag{21}$$

extends the above concept from a simple time coordinate to the spacetime coordinates x and corresponding four-momenta p. The dot in the exponent denotes the Minkowski scalar product, $p \cdot q = Eq_0 - \vec{p} \cdot \vec{q}$. Now, in the 8-dimensional phase space, one has the properties:

$$S^*_{AB}(x,p) = S_{B^\dagger A^\dagger}(x,p), \qquad S_{AB}(x,-p) = -S_{BA}(x,p). \tag{22}$$

Analogous to the Wigner (spectral) function, a Keldysh function is defined, but it is based on the symmetric commutator (denoted by $\{A,B\} = AB + BA$):

$$iK_{AB}(x,p) = \frac{1}{2} \int dq \, e^{\frac{i}{\hbar} p \cdot q} \, \langle \{ \hat{A}(x - q/2), \hat{B}(x + q/2) \} \rangle. \tag{23}$$

As a consequence, the Keldysh function properties by inverting the energy and momentum in its argument are as follows:

$$iK^*_{AB}(x,p) = -iK_{B^\dagger A^\dagger}(x,p), \qquad iK_{AB}(x,-p) = iK_{BA}(x,p). \tag{24}$$

The behavior of the statistical expectation values of the number of particles, which is a particular case of using the creation and annihilation operators instead of A and B, follows some rules derived from the above. Considering bosons, for example, one uses Hermitean and scalar operators, $B = B^\dagger = A = A\dagger$, twice. We commonly denote them by Φ. In this case,

$$S^*_{\Phi\Phi}(x,p) = S_{\Phi\Phi}(x,p), \qquad S_{\Phi\Phi}(x,-p) = -S_{\Phi\Phi}(x,p). \tag{25}$$

So, the spectral Wigner function is real and antisymmetric for the change in the sign of the four-momentum. This quantity counts negative energy states as minus.

To translate this result to the particle number (occupation number) quantities, we utilize the general relation between the commutator and anti-commutator. In the special case of $A = a^\dagger$ and $B = a$ fulfilling elementary commutation relations, we have for the bosons $2iK \sim \{a^\dagger, a\} = 2\hat{n} + 1$ and $S \sim [a^\dagger, a] = 1$, while for the fermions, we have $2iK \sim 1$ and $S \sim 2\hat{n} - 1$. Based on this, we generalize the definition of occupation numbers by the relations

$$iK_{AB}(x,p) = \left(n_{AB}(x,p) \pm \frac{1}{2} \right)^{\pm 1} S_{AB}(x,p). \tag{26}$$

with the plus sign for bosons and the minus sign for fermions. One obtains the sought relation between the quantum field occupations of negative and positive energy states based on this as follows:

$$n_{AB}(x,-p) = \mp 1 - n_{BA}(x,p). \tag{27}$$

This is the particle–hole symmetry for bosons (upper sign) and fermions (lower sign). Complex conjugation leads to another relation,

$$n^*_{AB}(x,p) = n_{B^\dagger A^\dagger}(x,p). \tag{28}$$

This defined occupation number is real as long as $(AB)^\dagger = B^\dagger A^\dagger = AB$, i.e., the operator product AB is Hermitean. For the traditional quantum counting operator, $A = a^\dagger$, $B = a$, this is the case.

When the two operators coincide, $A = B$, then the particle–hole symmetry is expressed by containing the same quantity on the left and right hand side of the equation:

$$n_{AA}(x,-p) = \mp 1 - n_{AA}(x,p). \tag{29}$$

The antiparticle numbers are defined accordingly as

$$\overline{n}_{AB}(x,p) = \mp n_{BA}(x,-p), \tag{30}$$

in order to interpret the negative energy states.

Now comes the statistical part: we associate an exponential, eventually a generalized exponential function, to the ratio of particle and hole (antiparticle) states. The argument of the generalized exponential in a kinetic approach is usually the $\beta \cdot p = \beta u_\mu p^\mu$ Minkowski product for relativistic systems. The Jüttner distribution is generalized then by the ratio of our generalized occupation numbers:

$$\frac{n_{AB}(x,p)}{\overline{n}_{AB}(x,p)} = \frac{n_{AB}(x,p)}{1 \pm n_{AB}(x,p)} = e_k(-\beta \cdot p). \tag{31}$$

On the other hand, applying the same relation to a negative energy and opposite momentum state, the above formula by replacing p^μ with $-p^\mu$ reads as

$$\frac{n_{AB}(x,-p)}{\overline{n}_{AB}(x,-p)} = \frac{1 \pm n_{BA}(x,p)}{n_{BA}(x,p)} = e_k(+\beta \cdot p). \tag{32}$$

For the case $A = B$, self-correlation of an operator, this is only possible if

$$e_k(-\beta \cdot p)\, e_k(\beta \cdot p) = 1. \tag{33}$$

From Equation (31), it follows a given generalization of the Bose and Fermi distributions:

$$n_{AB}(x,p) = \frac{e_k(-\beta \cdot p)}{1 \mp e_k(-\beta \cdot p)}. \tag{34}$$

From its energy-momentum mirrored version, Equation (32), it follows another:

$$n_{BA}(x,p) = \frac{1}{e_k(\beta \cdot p) \mp 1}. \tag{35}$$

Again, these definitions coincide only if the deformed exponential, which is used to replace the original exponential function, fulfills the special product rule Equation (33).

This result underlines the fact that the particle–hole (in the vacuum particle–antiparticle) symmetry applies not only to the Boltzmannian statistics but is also a basic requirement for the generalized occupation number functions of energy, describing the statistics of elementary particles or other types of quantum excitations.

In conclusion, we selected a very particular property of Kaniadakis' generalized exponential function, namely its reciprocial property upon reflection of its argument, Equation (33), and emphasized its relation to the particle–hole symmetry, known in quantum field theory and reflected in the KMS relation. We also presented a generalization of this function class maintaining this special property and related it to another construction based on the Tsallis type generalization of the exponential function. By doing so, a slight generalization of the phase space occupation number density statistics revealed that more general correlation functions also satisfy a KMS-type relation, when taking into account the change in the order of non-identical operators.

Funding: This research was funded by NKFIH OTKA, Hungary, grant number K123815, and by the Romanian reserach project PN-III-P4-id-PCE-2020-0647, hosted at the University of Babeş-Bolyai, Cluj.

Institutional Review Board Statement: Not applicable.

Informed Consent Statement: Not applicable.

Data Availability Statement: Not applicable.

Acknowledgments: Discussions with Antal Jakovác on the KMS relation are acknowledged.

Conflicts of Interest: The author declares no conflict of interest.

Abbreviations

The following abbreviations are used in this manuscript:

MDPI	Multidisciplinary Digital Publishing Institute
KMS	Kubo–Martin–Schwinger

References

1. Tahir, M.H.; Cordiero, G.M.; Alizadeh, M.; Hausoor, M.; Zubair, M.; Hamedami, G.G. The odd generalized exponential family of distributions with applications. *J. Stat. Distrib. Appl.* **2015**, *2*, 1. [CrossRef]
2. Alizadeh, M.; Ghosh, I.; Yosouf, M.M.; Rusekhi, M.; Hamedami, G.G. The Generalized Odd Generalized Exponential Family of Distributions: Properties, Characteristics and Applications. *J. Data Sci.* **2017**, *16*, 443–465. [CrossRef]
3. Renyi, A. On measures of information and entropy. In *Proceedings of the 4th Berekeley Symposium on Mathematics, Statistics and Probability 1960*; Statistical Laboratory of the University of California: Berkeley, CA, USA, 1961; p. 547.
4. Zapirov, R.G. *New Measures and Methods in Information Theory*; Kazan State Technological University: Kazan, Russia, 2005. (In Russian)
5. Tsallis, C. Possible generalization of Boltzmann-Gibbs statistics. *J. Stat. Phys.* **1988**, *52*, 479–487. [CrossRef]
6. Tsallis, C. Nonadditive entropy: The concept and its use. *Eur. Phys. J.* **2009**, *40*, 257–266. [CrossRef]
7. Tsallis, C. *Introduction to Non-Extensive Statistical Mechanics: Approaching a Complex World*; Springer Science and Business Media LLC: New York, NY, USA, 2009.
8. Kaniadakis, G. Theoretical foundations and mathematical formalism of the power law tailed statistical distributions. *Entropy* **2013**, *15*, 3983–4010. [CrossRef]
9. Kaniadakis, G. Statistical mechanics in the context of special relativity. *Phys. Rev. E* **2002**, *66*, 056125. PhysRevE.66.056125. [CrossRef]
10. Biro, T.S.; Kaniadakis, G. Two generalizations of the Boltzmann equation. *Eur. Phys. J.-Condens. Matter Complex Syst.* **2006**, *50*, 3–6. [CrossRef]
11. Kaniadakis, G.; Lavagno, A.; Quarati, P. Kinetic approach to fractional exlcusion statistics. *Nucl. Phys. B* **1996**, *466*, 527–537. [CrossRef]
12. Kaniadakis, G. Non-linear kinetics underlying generalized statistics. *Phys. Stat. Mech. Its Appl.* **2001**, *296* 405–425. [CrossRef]
13. Kaniadakis, G. Physical origin of the power-law tailed statistical distribution. *Mod. Phys. Lett. B* **2012**, *26*, 1250061. [CrossRef]
14. Kaniadakis, G.; Scarfone, A.M. A new one-parameter deformation of the exponential function. *Phys. Stat. Mech. Its Appl.* **2002**, *305*, 69–75. [CrossRef]
15. Sharp, K.; Matschinsky, F. Translation of Ludwig Boltzmann's Paper "On the Relationship between the Second Fundamental Theorem of the Mechanical Theory of Heat and Probability Calculations Regarding the Conditions for Thermal Equilibrium" Sitzungberichte der Kaiserlichen Akademie der Wissenschaften. Mathematisch-Naturwissen Classe. Abt. II, LXXVI 1877, pp 373-435 (Wien. Ber. 1877, 76:373-435). Reprinted in Wiss. Abhandlungen, Vol. II, reprint 42, p. 164-223, Barth, Leipzig, 1909. *Entropy* **2015**, *17*, 1971–2009. [CrossRef]
16. Stirling, J. *Methodus Differentialis Sive Tractatus de Summation et Interpolation Serierum Infinitorum*; Typis Gul. Bowyer, impensis G. Strahan: London, UK, 1730.
17. Michel, R. On Stirling's Formula. *Am. Math. Mon.* **2002**, *109*, 388. [CrossRef]
18. Mortici, C. A substantial improvement of the Stirling formula. *App. Math. Lett.* **2011**, *24*, 1351–1354. [CrossRef]
19. Aissen, M.I. Some Remarks on Stirling's Formula. *Am. Math. Mon.* **2018**, *61*, 687–691. [CrossRef]
20. Jaynes, E.T. Gibbs vs Boltzmann entropies. *Am. J. Phys.* **1965**, *33*, 391–398. [CrossRef]
21. Riedl, M.; Müller, A.; Wessel, N. Practical considerations of permutation entropy: A tutorial review. *Eur. Phys. J. Spec. Top.* **2013**, *222*, 249–262. [CrossRef]
22. Biro, T.S. *Is There a Temperature? Conceptual Challenges at High Energy, Acceleration and Complexity*; Springer Science and Business Media LLC: New York, NY, USA, 2011.
23. Hanel, R.; Thurner, S.; Gell-Mann, M. How multiplicity determines entropy and the derivation of the maximum entropy principle for complex systems. *Proc. Natl. Acad. Sci. USA* **2004**, *111*, 6905-6910. [CrossRef]
24. Khinchin, A.J. On the book of B. V. Gnedenko and A. N. Kolmogorov, "Limit distributions for sums of independent random variables", awarded the Chebyshev Prize. *Uspekhi Mat. Nauk* **1953**, *7*, 239–241.
25. Khinchin, A.J. On the basic theorems of information theory. *Uspekhi Mat. Nauk* **1956**, *9*, 17.
26. Jaynes, E.T. Information theory and statistical mechanics I. *Phys. Rev.* **1957**, *106*, 620. [CrossRef]
27. Jaynes, E.T. Information theory and statistical mechanics II. *Phys. Rev.* **1957**, *108*, 171. [CrossRef]
28. Shore, J.E.; Johnson, R.W. Axiomatic derivation of the principle of maximum entropy and the principle of minimum cross-entropy. *IEEE Trans. Inf. Theory* **1980**, *26*, 26–37. [CrossRef]
29. Biro, T.S.; Shen, K.M.; Zhang, B.W. Non-Extensive Quantum Statistics with Particle-Hole Symmetry. *Phys. Stat. Mech. Its Appl.* **2015**, *428*, 410–415. [CrossRef]
30. Aliano, A.; Kaniadakis, G.; Miraldi, E. Bose-Einstein condensation in the framework of kappa-statistics. *Phys. Condens. Matter* **2003**, *325*, 35–40. [CrossRef]

31. Teweldeberhan, A.M.; Miller, H.G.; Tegans, R. κ-deformed statistics and the formation of quark-gluon plasma. *Int. J. Mod. Phys. E* **2003**, *12*, 669–673. [CrossRef]
32. Abul-Magd, A.Y. Nonextensive random-matrix theory based on Kaniadakis entropy. *Phys. Lett. A* **2007**, *361*, 450–454. [CrossRef]
33. Tempesta, P. Group entropies, conservation laws, and zeta functions. *Phys. Rev. E* **2011**, *84*, 021121. PhysRevE.84.021121. [CrossRef]
34. Kubo, R. Statistical Mechanical Theory of Irreversible Processes I. General Theory and Simple Applications to Magnetic and Conduction Problems. *J. Phys. Soc. Jpn.* **1957**, *12*, 570–586. [CrossRef]
35. Martin, P.C.; Schwinger, J. Theory of Many-Particle Sytems I. *Phys. Rev.* **1959**, *115*, 1342. [CrossRef]
36. Haag, R.; Winnink, H.; Hugenholtz, N.M. On the equilibrium states in quantum statistical meachanics. *Commun. Math. Phys.* **1967**, *5*, 215–236. [CrossRef]
37. Biro, T.S.; Jakovac, A. *Emergence of Temperature in Examples and Related Nuisances in Field Theory*; Springer: Cham, Switzerland, 2019.

entropy

MDPI

Article

A Bayesian Analysis of Plant DNA Length Distribution via κ-Statistics

Maxsuel M. F. de Lima [1], Dory H. A. L. Anselmo [1,2,*], Raimundo Silva [1,2], Glauber H. S. Nunes [3], Umberto L. Fulco [4], Manoel S. Vasconcelos [2] and Vamberto D. Mello [1]

1 Departamento de Física, Universidade do Estado do Rio Grande do Norte, Natal 59072-970, RN, Brazil
2 Departamento de Física, Universidade Federal do Rio Grande do Norte, Natal 59072-970, RN, Brazil
3 Departamento de Ciências Vegetais, Universidade Federal Rural do Semi-Árido, Mossoró 59625-900, RN, Brazil
4 Departamento de Biofísica e Farmacologia, Universidade Federal do Rio Grande do Norte, Natal 59072-970, RN, Brazil
* Correspondence: doryh@fisica.ufrn.br

Abstract: We report an analysis of the distribution of lengths of plant DNA (exons). Three species of *Cucurbitaceae* were investigated. In our study, we used two distinct κ distribution functions, namely, κ-Maxwellian and double-κ, to fit the length distributions. To determine which distribution has the best fitting, we made a Bayesian analysis of the models. Furthermore, we filtered the data, removing outliers, through a box plot analysis. Our findings show that the sum of κ-exponentials is the most appropriate to adjust the distribution curves and that the values of the κ parameter do not undergo considerable changes after filtering. Furthermore, for the analyzed species, there is a tendency for the κ parameter to lay within the interval $(0.27; 0.43)$.

Keywords: DNA; *cucurbitaceae*; non-additive statistics

Citation: de Lima, M.M.F.; Anselmo, D.H.A.L.; Silva, R.; Nunes, G.H.S.; Fulco, U.L.; Vasconcelos, M.S.; Mello, V.D. A Bayesian Analysis of Plant DNA Length Distribution via κ-Statistics. *Entropy* **2022**, *24*, 1225. https://doi.org/10.3390/e24091225

Academic Editors: Antonio M. Scarfone, Dionissios T. Hristopulos and Sergio Luiz E. F. da Silva

Received: 15 August 2022
Accepted: 31 August 2022
Published: 1 September 2022

Publisher's Note: MDPI stays neutral with regard to jurisdictional claims in published maps and institutional affiliations.

1. Introduction

There are 15 tribes in the family Cucurbitaceae [1]. The tribe Cucurbitae, which has an almost completely American distribution, consists of 11 genera, including the genus *Cucurbita*. The genus *Cucurbita* (Cucurbitaceae) has five major domesticated species: *Cucurbita moschata*, *Curcurbita pepo*, *Cucurbita maxima*, *Cucurbita argyrosperma*, and *Cucurbita ficifolia* [2,3].

The first three species cited are the most economically important as a popular food resource [4]. The fruits of the species are incredibly diverse, differing greatly in shape, surface topography, color, size, and color pattern [5]. Among them, *C. pepo* is the genus' most phenotypically variable species and has eight cultivar groups with edible fruits (groups) [6]. The second most diversified species in the genus is thought to be *C. moschata* [7].

All Cucurbita species have 20 pairs of chromosomes ($2n = 2x = 40$), making them all diploid. The theory that Cucurbiteae underwent one whole-genome duplication as a result of their high chromosome number has gained traction [8,9]. The tribe Cucurbiteae plant species, including the zucchini (*C. pepo*), pumpkin (*C. moschata* and *C. maxima*), and silver-seed gourd (*C. argyrosperma*), all suffered whole-genome duplication events, according to a number of studies [9–11].

There are few estimates of genome size in the genus *Cucurbita*. However, studies have shown relatively small genome sizes. The genome sizes of *C. maxima* and *C. moschata* were estimated to be 271.40 and 269.90 Mb, respectively, [9], while the genome size in *C. pepo* was estimated to be 263.0 Mb [10]. Concerning the number of genes, the estimated values for *C. maxima*, *C. moschata*, and *C. pepo* were 32.076; 32.205 [9]; and 27.868 genes [10], respectively.

On the other hand, numerous models based on statistical physics consistently attempt to represent statistical features, such as long-range and short-range correlations, in light of the large DNA sequence data. Some approaches used statistical tools in connection with random-walk simulations [12–14], wavelet transforms [15,16], $1D$ Ising models [17] (see e.g., [18] and references therein), and Tsallis' statistics together with Machine Learning [19]. Many live creatures' coding and non-coding sequence length distributions have been studied by some models in relation to long- and short-range correlations [20–23]. Non-additive entropy-based statistical physics methods have recently been actively advocated for use in complex system research [24,25]. In this case, the Kaniadakis entropy yields a power-law distribution rather than an exponential one and depends on a free parameter (the κ parameter) [26–28]. The κ-statistics arose as a useful statistical tool for many systems (see [29] and references therein). For problems associated with human DNA, see e.g., [30,31].

Additionally, the Bayesian inference has been effectively applied as a useful tool to investigate a number of issues in physics [32] and biophysics [33]. Which DNA models should be valid from the perspective of Bayesian inference is an intriguing subject. Additionally, the challenge in the context of this work would be to investigate an expansion of a model from Ref. [31], but this time in the context of other living structures, such as vegetables.

More recently in [34], statistical models of the Tsallis type provided the distribution of nucleotide chain lengths, successfully capturing the statistical correlations between the parts of the plant (for both coding and non-coding) DNA strands for two species of the Cucurbitaceae family. We expand the paradigm proposed in [31] in the context of vegetables in this article. We especially evaluate the distribution of nucleotide chain lengths measured in base pairs for *Cucurbita maxima*, *Cucurbita moschata*, and *Cucurbita pepo* utilizing κ-deformed statistics in light of the social and economic significance of cucurbits. The most practical model is then chosen using a Bayesian statistical analysis based on the κ-distributions. To the best of our knowledge, this is the first time the size distribution of plant DNA has been realized using a κ-statistical analysis.

2. Materials and Methods

We use the κ-statistics, developed by Kaniadakis [26–28], to analyze the correlations between the DNA length distributions of some species of the *Cucurbitaceae* family. There are some works in this direction using the Tsallis q-statistics [34–36]. The κ-entropy and power-law distribution functions naturally arise from the kinetic foundations of κ-statistics. Formally, the κ-framework is based on the κ-exponential and κ-logarithm functions (see Ref. [26]), defined as

$$\exp_\kappa(x) = \left[\sqrt{1+\kappa^2x^2}+\kappa x\right]^{\frac{1}{\kappa}} \tag{1}$$

$$\ln_\kappa(x) = \frac{x^\kappa - x^{-\kappa}}{2\kappa}. \tag{2}$$

The parameter κ is restricted to values belonging to the range $|\kappa| < 1$; for $\kappa = 0$, these expressions reduce to the usual exponential and logarithmic functions. From the optimization of entropy S_κ (see Ref. [37]),we can obtain the probability distributions ($P_{\kappa,1}(l)$) associated with the quantities of base pairs (bp) for each of the chromosomes of *Cucurbita maxima*, *Cucurbita moschata*, and *Cucurbita pepo*. Mathematically, the Kaniadakis entropy S_κ is given by

$$S_\kappa(l) = -\frac{1}{2\kappa}\int_\kappa\left[\frac{1}{1+\kappa}P_\kappa(l)^{(1+\kappa)} - \frac{1}{1-\kappa}P_\kappa(l)^{(1-\kappa)}\right]dl. \tag{3}$$

The optimization process is well described in Refs. [26,37–41] and gives us $P_{\kappa,1}(l)$

$$P_{\kappa,1}(l) = (1-\kappa^2)\beta\exp_\kappa[-\beta l]. \tag{4}$$

Rewriting (4) with the explicit form of $\exp_\kappa(-\beta l)$ given by (1), and using constraints as in Ref. [41], we get

$$P_{\kappa,1}(l) = \frac{(1-\kappa^2)}{L_\kappa}\left[\sqrt{1+\kappa^2\left(-\frac{l}{L_\kappa}\right)^2}+\kappa\left(-\frac{l}{L_\kappa}\right)\right]^{\frac{1}{\kappa}}. \quad (5)$$

Here, L_κ is an adjustable parameter that is related to the mean value of the length distribution, κ is the model's free parameter which measures the interaction between the nucleotides in the sample, and l is the chain of nucleotides' length, expressed in number of base pairs.

We employ the cumulative probability distribution because the probabilities for lengthy lengths l of the nucleotide chain are subject to significant fluctuations.

We employ the cumulative probability distribution because the probabilities for lengthy lengths l of the nucleotide chain are subject to significant fluctuations, (5) can be found by solving $\Phi(l) = p(l' < l) = \int_0^l p(l')dl'$, which provides

$$\Phi_{\kappa,1}(l) = 1 - \frac{1}{2}\left[G_\kappa^+(l) + G_\kappa^-(l)\right], \quad (6)$$

where

$$G_\kappa^{\pm}(l) = (1 \pm \kappa)\exp_\kappa^{1\mp\kappa}\left(-\frac{l}{L_\kappa}\right). \quad (7)$$

Here, $\Phi(l)$ denotes the probability of finding the sizes of the bases between 0 and l. In Ref. [34], it was proposed a comparison between the q-exponential and a sum of q-exponentials to explain the DNA length distribution of two species of cucurbits, *Cucumis melo* and *Cucumis sativus*. Based on this work, we propose an analysis of the same type but using the κ-*statistics*. We assume that the sum of Kaniadakis-type generalized probabilities (already normalized) is given by

$$P_{\kappa,2}(l) = (1-\kappa^2)\left[\frac{\gamma_1\gamma_2}{\gamma_1+\gamma_2}\right]\left[\exp_\kappa(-\gamma_1 l) + \exp_\kappa(-\gamma_2 l)\right], \quad (8)$$

where κ, γ_1, and γ_2 are adjustable parameters and l is the length of the nucleotides, respectively. By employing the identical steps as those leading to (6), the cumulative probability distribution is found to be

$$\Phi_{\kappa,2}(l) = 1 - \left[\frac{1}{\gamma_1}F_{1,\kappa}(l) + \frac{1}{\gamma_2}F_{2,\kappa}(l)\right], \quad (9)$$

where

$$F_{j,\kappa}(l) = \frac{\gamma_1\gamma_2}{\gamma_1+\gamma_2}\left[\frac{\exp_\kappa^{1-\kappa}(-\gamma_j l)}{1-\kappa} + \frac{\exp_\kappa^{1+\kappa}(-\gamma_j l)}{1+\kappa}\right], \quad j = 1,2. \quad (10)$$

Initial analyses indicate that, as occurred for the Tsallis' q-statistics [34], the κ-exponential sum model best fits the DNA length distributions of the species studied here. Therefore, we chose to make a comparison between the sum of κ-exponentials (9) and the κ-Maxwellian model (11) below, proposed in [31] to explain the length distribution of human DNA.

$$\Phi_{\kappa,3}(l) = 1 - \exp_\kappa\left(-\frac{l^2}{\sigma_\kappa^2}\right)\left[\sqrt{1+\kappa^2\frac{l^4}{\sigma_\kappa^4}} + \kappa^2\frac{l^2}{\sigma_\kappa^2}\right]. \quad (11)$$

The best model to describe the length distributions of the nucleotides for three species of the *Cucurbitaceae* family is obtained by comparing, via Bayesian analysis, the distributions $\Phi_{\kappa,2}(l)$ and $\Phi_{\kappa,3}(l)$, which are represented by Equations (9) and (11), respectively.

3. Results

We use the public database of the National Center for Biotechnology Information (NCBI) [42] and the Comparative Genomics (CoGe) [43]. They are databases that give users access to genetic and biological data. In our analysis, we considered only the coding bases (exons). We define a nucleotide sequence's length in terms of the l (bp) base pairs. All graphical and data modeling was written in R, a free statistical software [44].

By plotting the cumulative probability distribution function (CDF) and a box plot for chromosome 02 of one of the species studied here (Figure 1), we can see that some points are very far from the distribution and can be considered outliers. There are various techniques for defining, spotting, and dealing with outliers [45]. In this work, we decided to use the box plot approach. Outliers in this approach are points that are below the region $Q1 - 1.5 \times IQR$ and above $Q3 + 1.5 \times IQR$, where $Q1$, $Q2$, and $Q3$ are first, second, and third quartile, respectively, and IQR is the interquartile region defined as $IQR = Q3 - Q1$. To prevent these points from influencing the behavior of the proposed models, we decided to remove them. The cut was made around 1% of the cumulative distribution, designated by the hatched square in the lower right corner of Figure 1a. A similar approach has been proposed in [46] to analyze the length distribution of human DNA. Table A1 describes the statistical characteristics of some chromosomes of the three species of *Cucurbitaceae* after removing these outliers.

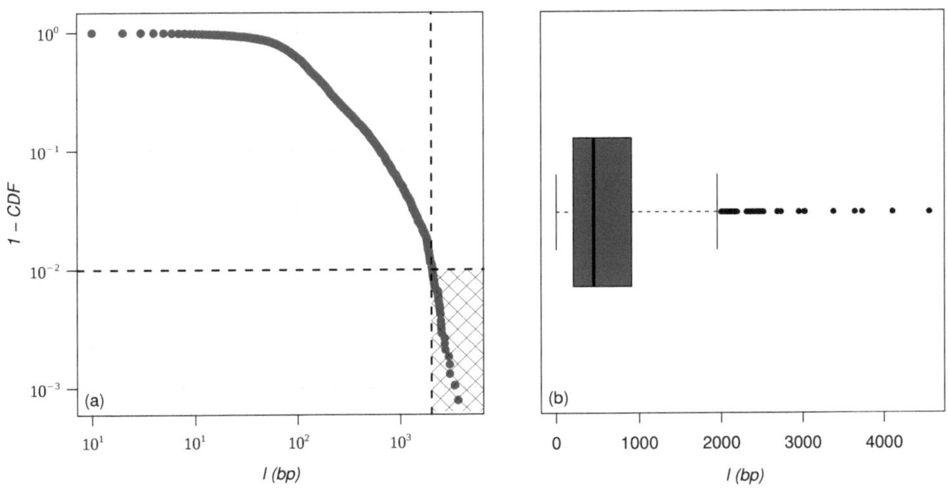

Figure 1. (**a**) Cumulative probability distribution function (CDF) and (**b**) box-plot for chromosome 02 of the species *Cucurbita maxima*. A similar analysis was performed for all chromosomes of the three species of *cucurbitaciae* studied in this paper.

We decided to analyze the impact this action had on the value of κ, taking into account the cumulative distribution functions (9) and (11). In Tables A2 and A3, we have the number of nucleotides (N) and the best fit values per κ. The subscripts 0 and f represent the values before and after the outliers are removed, and (RD) represents the relative difference between them. The values of RD are smaller than the errors associated with the values of κ in Tables A4–A6. This work deals with a statistical analysis of the distribution of DNA lengths in plants. Possible biological effects caused by removing nucleotides with large amounts of base pairs were not taken into account.

In Figures 2–4, we show the cumulative distributions, for exons, for some chromosomes of *Cucubita maxima*, *Cucurbita moschata*, and *Cucubita pepo*, with the other chromosomes behaving similarly. To get the best fit values for κ, the distribution functions (9) and (11) were fitted to the lengths (l). Tables A4–A6 show all numerical results for the parameters κ, γ_1 and γ_2 for distribution (9) in addition to κ and σ_κ for distribution (11).

Chromosome numbers are displayed in the first column (CHR), and the number of nucleotide chains is displayed in the second column (N) (exons). The correlations between the values of l are measured by the values of κ [26–28,39]. According to [36,47], the coding part of human DNA tends to present short-range correlations. The same behavior for plant DNA can be observed in [34]. This implies κ values close to zero. It is worth remembering that in the limit $\kappa \to 0$, we return to the well-known Boltzmann–Gibbs–Shannon statistics [26].

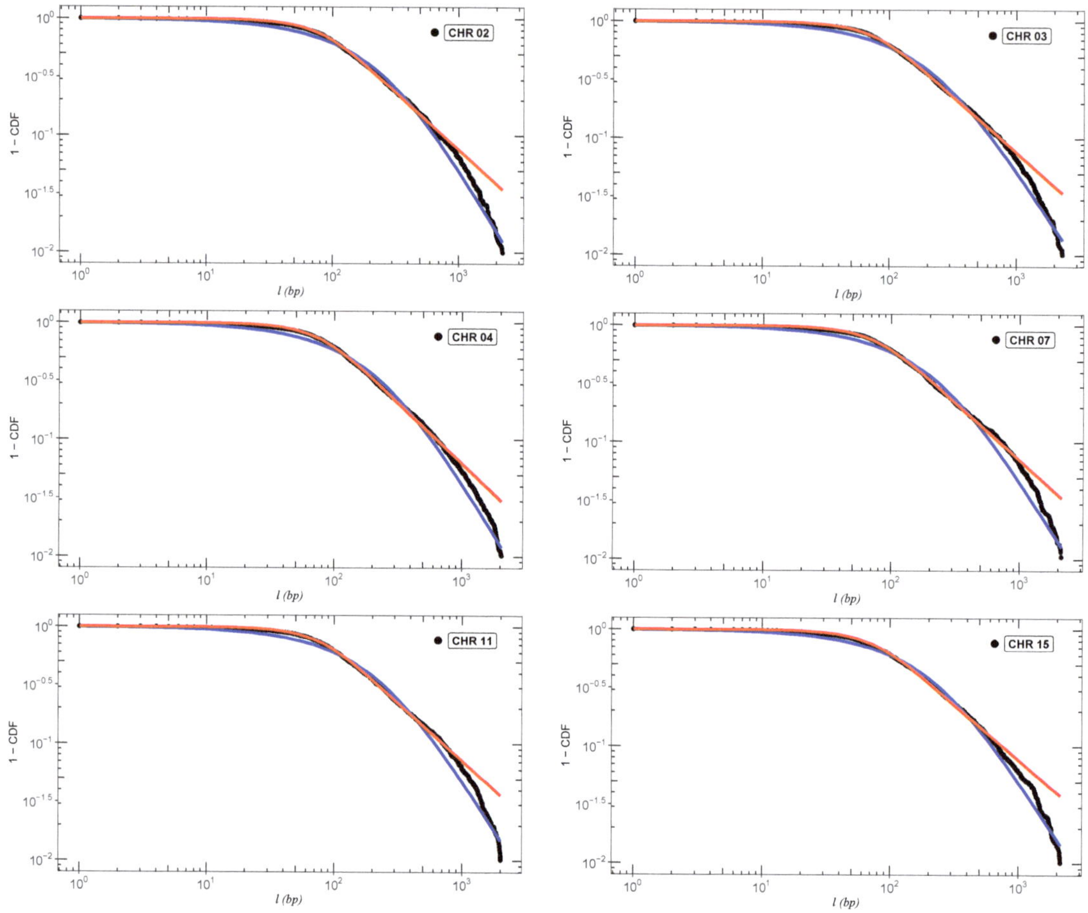

Figure 2. Best fit analysis for the exons of *Cucurbita maxima*. We can observe the adjustments for chromosomes (CHR) 02, 03, 04, 07, 11, and 15. The blue and red curves are, respectively, the distributions (9) and (11). The other chromosomes follow the same pattern.

The models that fit the length distribution $\Phi(l)$ the best are determined via Bayesian statistics. By taking into account the probability distribution of the hypotheses, conditioned on the evidence, Bayesian inference describes the relationship between the model and the data, and enables a rational and effective selection of one or more hypotheses [48]. The Bayes' theorem,

$$P(\Phi|D,M) = \frac{\mathcal{L}(D|\Phi,M) \cdot P(\Phi|M)}{\mathcal{E}(D|M)}, \tag{12}$$

offers us the likelihood that, given the data D, a posterior model Φ will be correct. For this, the probability of the prior model $P(\Phi|M)$ is multiplied by the likelihood function $\mathcal{L}(D|\Phi,M)$ and divided by the Bayesian evidence $\mathcal{E}(D|M)$. Here, we assume the pattern

$\chi^2 = (P(l^{obs}) - P(l^{the}))^2/\sigma^2_{obs}$ for the likelihood function, where $P(l^{obs})$, $P(l^{the})$ and σ_{obs} are the cumulative probabilities associated with the observed and the theoretical nucleotide lengths, and observed errors, respectively.

The input parameters used in the prior uniform distribution were obtained from the best fit found by the R-code. This approach, which defines the model parameters' potential range and significantly affects the Bayesian evidence, is a crucial phase in the study. This condition ensures that the parameters will fall inside the previously identified optimal adjustment range.

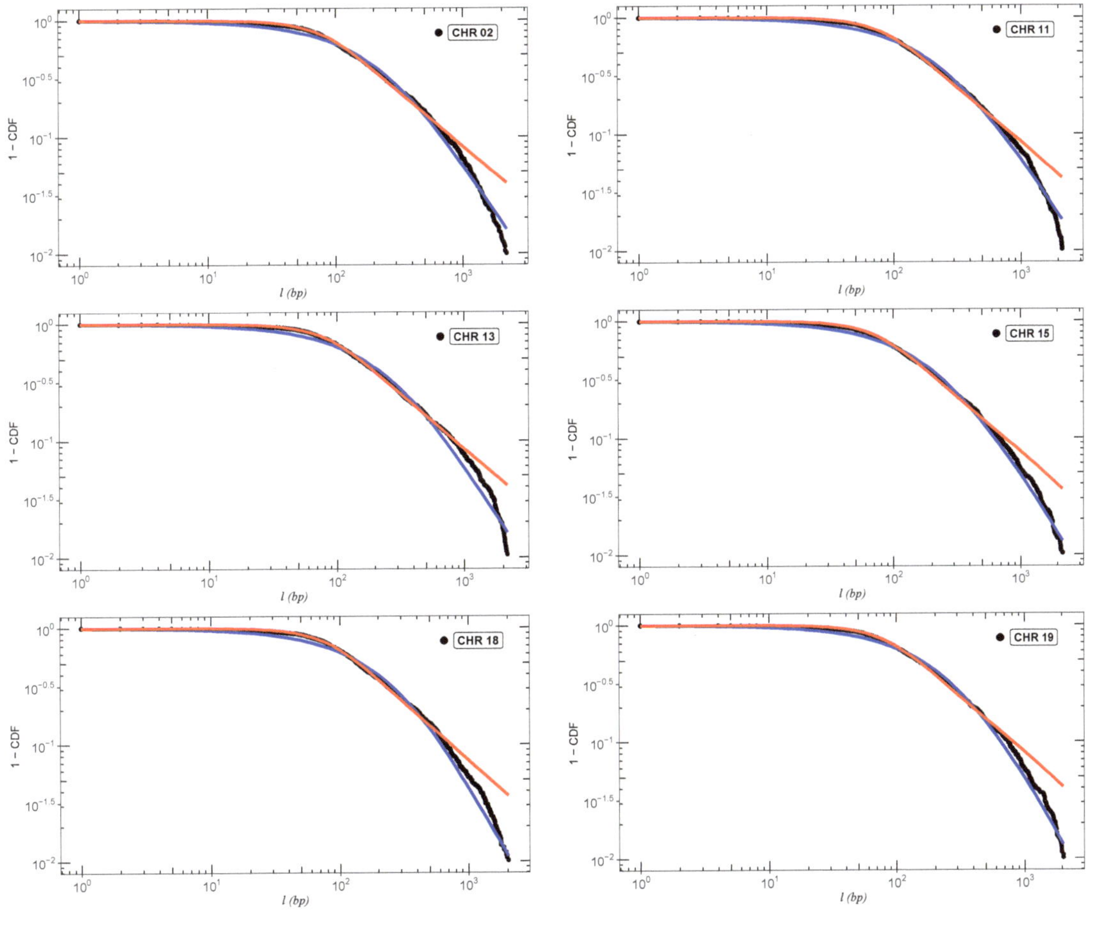

Figure 3. Best fit analysis for the exons of *Cucurbita Moschata*. We can observe the adjustments for chromosomes (CHR) 02, 11, 13, 15, 18, and 19. The blue and red curves are, respectively, the distributions (9) and (11). The other chromosomes follow the same pattern.

In Table A4, we have the parameter ranges for *Cucubita maxima*. Considering all chromosomes (CHR), $\kappa_M \sim U(0.64, 0.69)$, $\sigma_\kappa \sim U(91, 105)$, for cumulative distribution (11), and $\kappa_S \sim U(0.24, 0.39)$, $\gamma_1 \sim U(0.0041, 0.0087)$ and $\gamma_2 \sim U(0.0045, 0.0088)$ for cumulative distribution (9). The process is repeated for the species *Cucubita moschata* in Table A5 and *Cucubita pepo* in Table A6. The MULTINEST algorithm, a Bayesian inference tool that computes the evidence $\mathcal{E}(D|M)$ with an associated error estimate, is thus put into practice for each species and each model. It generates posterior samples from distributions that can contain multiple modes and pronounced degeneracy (curves) in high dimensions. More details can be seen in [49–53].

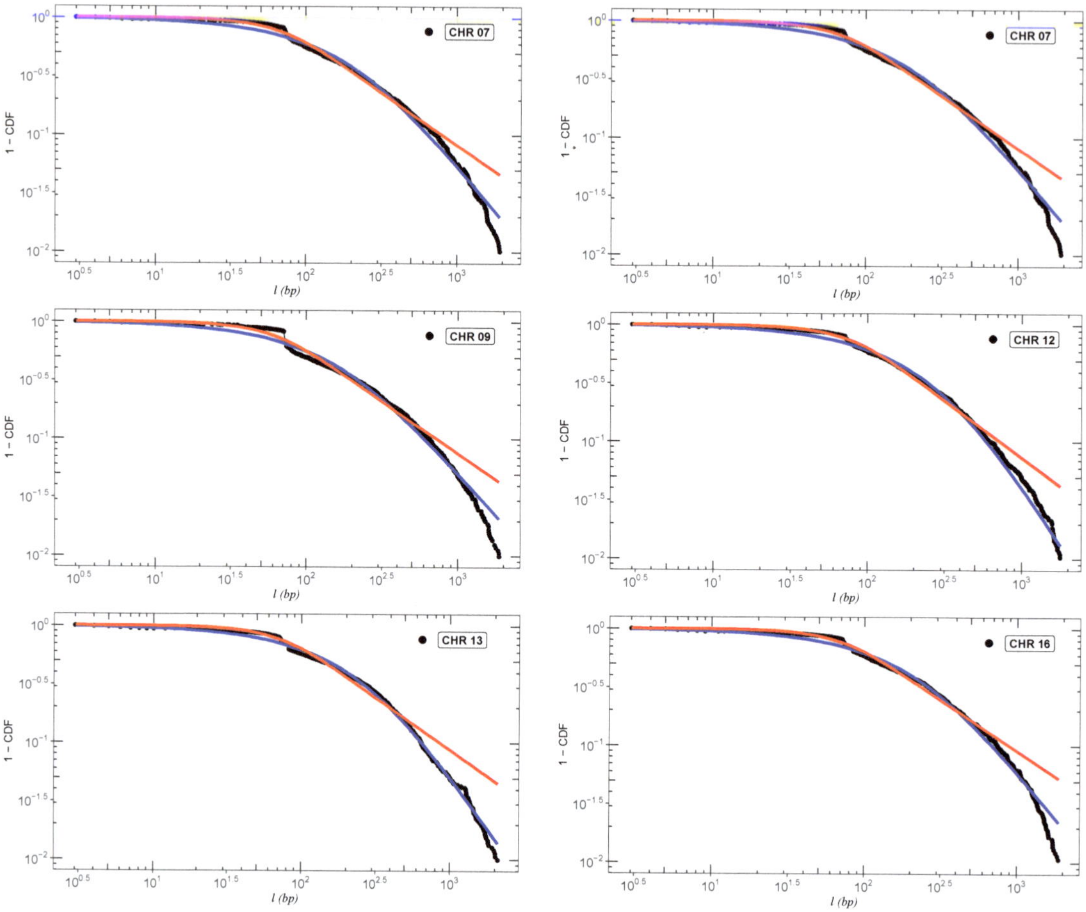

Figure 4. Best fit analysis for the exons of *Cucurbita Pepo*. We can observe the adjustments for chromosomes (CHR) 01, 07, 09, 12, 13, and 16. The blue and red curves are, respectively, the distributions (9) and (11). The other chromosomes follow the same pattern.

In order to compare the models, we make use of the Bayes factor, which is given by

$$B_{ij} = \frac{\mathcal{E}_i}{\mathcal{E}_j}. \tag{13}$$

Here, $\mathcal{E}_j$ is the evidence of the base model, which is used as a reference. In our case, this is the distribution (9), and $\mathcal{E}_i$ is the evidence of the model we want to compare, given by distribution (11). We employ the Bayes factor interpretation provided by Jeffrey's theory [35,54–56] to measure whether a model has favorable evidence in comparison to the base model. Table A7 contains the findings for each chromosome.

The Bayesian analysis is performed from each model's range of definite parameters. Therefore, the better we understand the behavior of the parameters, the more accurate our analysis will be, and we can guarantee that the evidence found will represent the curve with the best fit [48]. In Figures 5–7, we have scatter plots for the parameters of the models (9) (a) and (11) (b). For all chromosomes of all species analyzed here, we found strong correlations between the parameters γ_1 and γ_2 present in the distribution (9). This was expected, as this model appears as a variation of the model (6), as carried out in [34]. These two adjustable constants together (γ_1 and γ_2) have an inverse role to what L_κ has in the distribution (6), and when $\gamma_1 = \gamma_2$, we obtain the model (6) again. This implies that these parameters are related to the κ parameter in the same way, resulting in similar images for scattering but with different ranges. This behavior was repeated for all chromosomes.

The κ_S parameter (that is, the κ value that provides the best fit, when using the sum of κ-exponentials, Equation (9)) in Tables A4–A6, measures the correlation between lengths l, and belongs to the range $(0.27(4); 0.37(2))$ in the case of *Cucubita maxima*, $(0.28(3); 0.40(4))$ for *Cucubita moschata*, and $(0.32(3); 0.43(3))$ for *Cucubita pepo*. It can be seen in Figure 8 that the values of κ, for different species, seem to specify a universal behavior. Therefore, all of these findings lead us to the conclusion that for all the species under study, the model (9) (sum of κ-exponentials) is strongly preferred over the distribution model (11) (κ-Maxwellian).

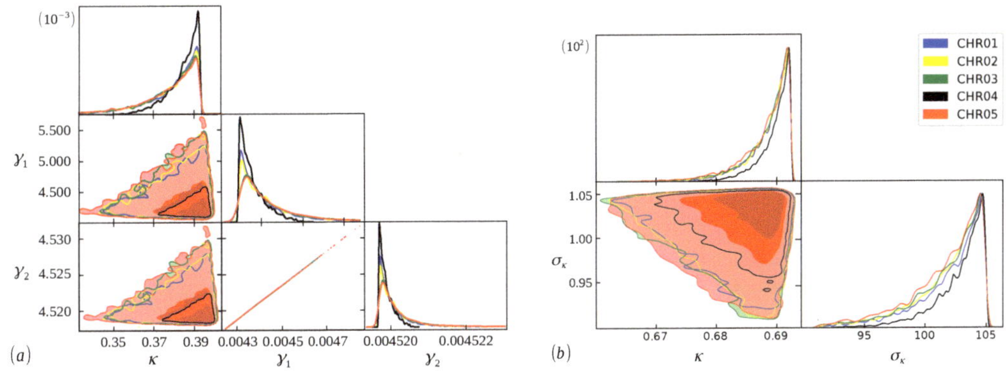

Figure 5. Bayesian analysis for the (9) (**a**) and (11) (**b**) distributions, using chromosomes 01, 02, 03, 04, and 05 of the coding part of *Cucurbita maxima* DNA. The rest of the sample follows a similar pattern.

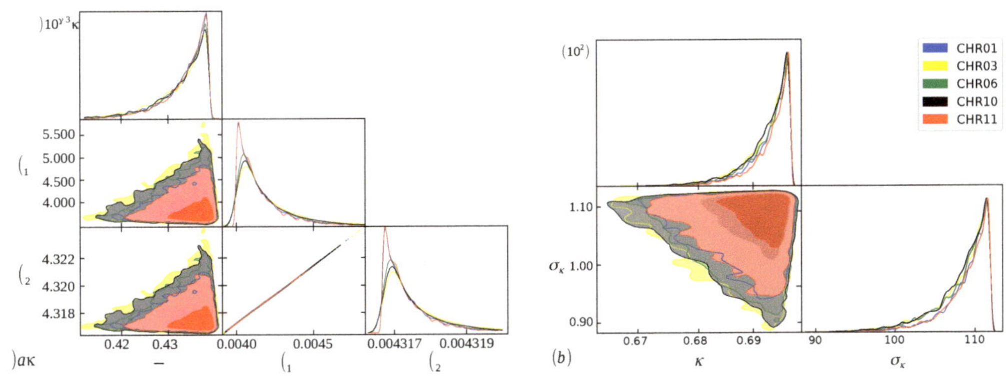

Figure 6. The same as Figure 5, but for chromosomes 01, 03, 06, 10, and 11 of the *Cucurbita moschata* species.

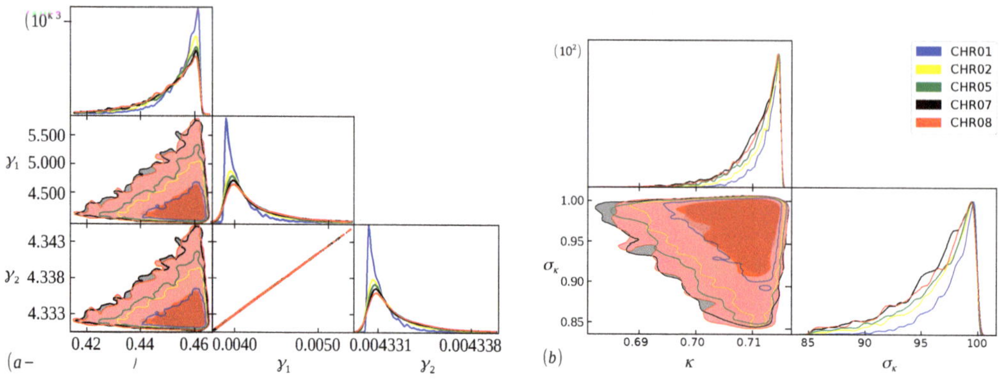

Figure 7. The same as Figure 5, but for chromosomes 01, 02, 05, 07 and 08 of the *Cucurbita pepo* species.

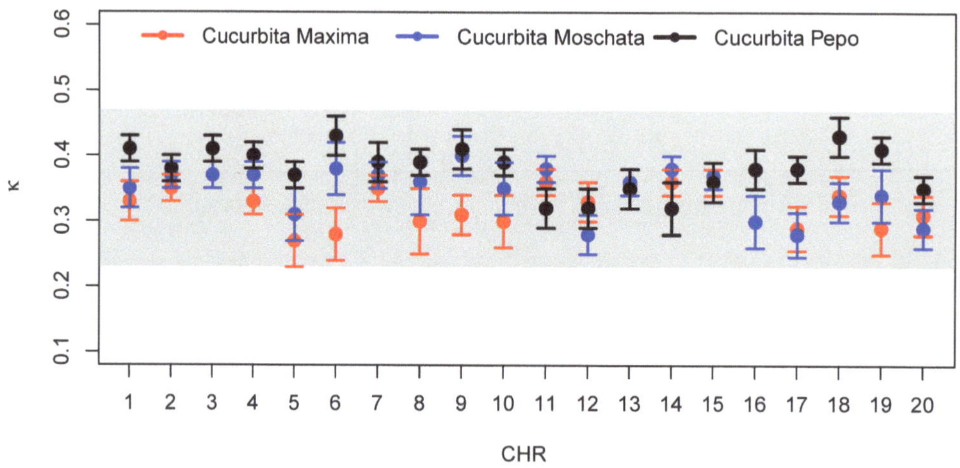

Figure 8. κ values, from the best fit model, Equation (9), for different species. In red, blue, and black, we have, respectively, *Cucurbita maxima*, *Cucurbita moschata*, and *Cucurbita pepo*.

4. Conclusions

A statistical model based on non-additive statistics was developed to describe the size distribution of nucleotide chains in the DNA of species belonging to the *Cucurbitaceae* family, namely *Cucurbita maxima*, *Cucurbita moschata*, and *Cucurbita pepo* [26–28,31]. Specifically, the proposed distribution, Equation (9), expands on a distribution studied in [41] through the sum of the κ-exponentials, which added the parameters γ_1 and γ_2 to capture the statistical correlations between the DNA strands. Another model investigated was the κ-Maxwellian distribution, Equation (11), proposed in [31] for human DNA. We tested the statistical feasibility of models, as well as methods based on Bayesian statistical analysis using the NCBI project database. The cumulative distribution function (9) best fitted the nucleotide base for all chromosomes, of the three species, with the parameter κ belonging to the range $(0.27(4); 0.37(2))$ for *Cucurbita maxima*, $(0.28(3); 0.40(4))$ for *Cucurbita moschata*, and $(0.32(3); 0.43(3))$ in the case of *Cucurbita pepo*. It can be seen in Figure 8 that the values of κ for different species of the coding parts (exons) of the DNA appear to be within a common and relatively narrow range.

Regarding the Bayesian analysis, we compared the κ-exponential-sum distribution with the κ-Maxwellian model. We demonstrated that the first has solid and favorable evidence compared to the κ-Maxwellian distribution. This was reasonably expected given that the distribution (9) has a free parameter for potential future adjustments. A general task should be to expand the model presented in this study to include additional species, determining whether they fall within the same range of κ for exons (0.35 ± 0.08) discovered for the species investigated here.

Author Contributions: M.M.F.d.L.: Methodology, Writing—Original draft preparation, Software, Visualization, Investigation. D.H.A.L.A.: Conceptualization, Supervision, Formal analysis, Methodology, Writing—Original draft preparation, Writing—Reviewing and editing. R.S.: Visualization, Investigation, Writing—Original draft preparation. G.H.S.N.: Visualization, Investigation, Writing—Original draft preparation. U.L.F.: Visualization, Investigation, Writing—Original draft preparation. M.S.V.: Visualization, Investigation, Writing—Original draft preparation. V.D.M.: Writing—Reviewing and editing. All authors have read and agreed to the published version of the manuscript.

Funding: This study was financed in part by CNPq (Conselho Nacional de Desenvolvimento Científico e tecnológico), and by the Coordenação de Aperfeiçoamento de Pessoal de Nível Superior—Brasil (CAPES). R.S. thanks CNPq (Grant No. 307620/2019-0) for financial support. D.H.A.L.A. thanks CNPq for financial support (CNPq grant No. 317464/2021-3). M.S.V. thanks CNPq (Grant No. 313207/2021-6).

Data Availability Statement: The DNA code data that support the findings of this study are available in NCBI [42].

Acknowledgments: The authors thank William J. da Silva for technical discussions on the Bayesian analysis.

Conflicts of Interest: The authors declare no conflict of interest. The funders had no role in the design of the study; in the collection, analyses, or interpretation of data; in the writing of the manuscript; or in the decision to publish the results.

Appendix A

Table A1. Statistical characteristics of the data after outliers are removed. The first column indicates the chromosome, the second the number of exons, the third and fourth the minimum and maximum lengths. Finally, the quartiles $Q1$, $Q2$, and $Q3$ are in the fifth, sixth, and seventh columns, respectively. In the later columns, we have indicated the same parameters for the other species of *cucurbitaceae*.

CHR	*Cucurbita maxima*						*Cucurbita moschata*						*Cucurbita pepo*					
	N	l_{min}	l_{max}	Q1	Q2	Q3	N	l_{min}	l_{max}	Q1	Q2	Q3	N	l_{min}	l_{max}	Q1	Q2	Q3
1	662	1	2034	166	350	738	667	1	1965	168	357	737	802	3	2049	208	451	833
2	666	1	2229	168	362	710	653	1	2169	165	363	741	663	3	1854	173	380	755
3	641	1	2316	162	352	744	610	1	2207	155	333	740	661	4	1804	172	381	721
4	790	1	2049	198	424	815	851	1	2070	214	458	871	572	4	2004	159	350	687
5	596	1	1967	151	321	650	605	1	1997	154	324	670	623	3	1953	162	352	685
6	654	1	1981	164	343	677	657	1	2147	165	369	747	503	4	1899	146	313	635
7	568	1	2127	143	309	692	608	1	2148	153	343	771	538	3	1917	145	315	650
8	534	2	1857	135	296	587	513	1	1932	130	284	609	578	4	1815	151	339	631
9	610	1	2361	153	320	647	599	1	2363	154	38	753	539	3	1866	152	330	647
10	532	1	2096	134	285	598	616	1	2346	157	342	701	529	4	1995	146	330	703
11	665	1	2001	167	353	753	714	1	2146	179	399	803	588	3	1773	158	337	618
12	557	1	2265	141	302	622	556	1	2133	141	293	610	570	3	1776	150	329	602
13	549	3	2037	141	307	675	605	1	2136	155	338	750	563	3	2092	153	329	606
14	723	1	2181	182	391	786	740	3	2382	188	410	832	534	3	1896	143	311	589
15	601	1	2121	152	325	675	630	1	2190	159	338	673	486	4	1896	137	321	603
16	597	1	2153	150	318	657	601	1	2188	155	333	660	512	3	1872	137	302	649
17	583	2	2052	148	312	614	641	2	1968	164	341	672	507	3	1782	140	333	695
18	574	1	2178	145	320	635	580	1	2016	146	317	624	445	4	2052	129	313	611
19	539	1	2124	137	301	648	555	1	2049	141	310	641	514	3	1884	151	347	697
20	564	1	2212	142	298	628	572	1	2427	145	300	683	501	4	1914	143	316	650

Table A2. Values of κ before and after removing outliers for model (9). (N) represents the number of nucleotides. κ are the best fit values. The subscripts 0 and f represent the values before and after the outliers are removed, (RD) represents the relative difference between them. This behavior is repeated for all chromosomes.

CHR	*Cucurbita maxima*					*Cucurbita moschata*					*Cucurbita pepo*				
	N_0	κ_0	N_f	κ_f	$\|RD\|$	N_0	κ_0	N_f	κ_f	$\|RD\|$	N_0	κ_0	N_f	κ_f	$\|RD\|$
1	687	0.3324	662	0.3336	0.0012	693	0.3455	667	0.3455	0.0000	834	0.4061	802	0.4073	0.0012
2	688	0.3532	666	0.3546	0.0014	675	0.3742	653	0.3749	0.0007	683	0.3812	663	0.3836	0.0024
3	661	0.3749	641	0.3747	0.0002	630	0.3641	610	0.3650	0.0009	682	0.4057	661	0.4076	0.0019
4	828	0.3576	790	0.3330	0.0246	893	0.3677	851	0.3680	0.0003	587	0.4004	572	0.4015	0.0011
5	615	0.2827	596	0.2835	0.0008	623	0.3127	605	0.3123	0.0004	643	0.3661	623	0.3673	0.0012

Table A3. The same as Table A2, but for model (11).

CHR	*Cucurbita maxima*					*Cucurbita moschata*					*Cucurbita pepo*				
	N_0	κ_0	N_f	κ_f	$\|RD\|$	N_0	κ_0	N_f	κ_f	$\|RD\|$	N_0	κ_0	N_f	κ_f	$\|RD\|$
1	687	0.6589	662	0.6590	0.0002	693	0.6648	667	0.6649	0.0001	834	0.6929	802	0.6930	0.0001
2	688	0.6704	666	0.6707	0.0003	675	0.6810	653	0.6811	0.0001	683	0.6919	663	0.6922	0.0003
3	661	0.6760	641	0.6762	0.0000	630	0.6733	610	0.6734	0.0001	682	0.6909	661	0.6911	0.0002
4	828	0.6588	790	0.6592	0.0096	893	0.6633	851	0.6633	0.0000	587	0.7067	572	0.7071	0.0004
5	615	0.6455	596	0.6459	0.0002	623	0.6555	605	0.6556	0.0001	643	0.6881	623	0.6885	0.0004

Table A4. The average of the best fit parameters for the *Cucurbita maxima* species. The sub-index S and M represent the κ-exponential sum function (9) and the κ-Maxwellian function (11), respectively. σ_κ, γ_1, and γ_2 are free parameters related to the length of the nucleotide chain. The numbers in parenthesis denote the calculated errors.

CHR	N	κ_M	σ_κ	κ_S	γ_1	γ_2
1	662	0.65(1)	97(2)	0.33(3)	0.0067(13)	0.0057(07)
2	666	0.67(1)	101(3)	0.35(2)	0.0054(09)	0.0062(14)
3	641	0.67(1)	95(3)	0.37(2)	0.0069(15)	0.0057(08)
4	790	0.65(1)	94(2)	0.33(2)	0.0067(13)	0.0056(08)
5	596	0.64(1)	102(3)	0.27(4)	0.0054(13)	0.0062(18)
6	654	0.64(1)	96(2)	0.28(4)	0.0056(08)	0.0069(15)
7	568	0.67(1)	92(3)	0.35(2)	0.0059(09)	0.0071(17)
8	534	0.65(1)	94(3)	0.30(5)	0.0060(18)	0.0066(23)
9	610	0.66(1)	103(2)	0.31(3)	0.0051(07)	0.0061(12)
10	532	0.66(1)	99(3)	0.30(4)	0.0053(08)	0.0066(16)
11	665	0.67(1)	96(3)	0.36(2)	0.0069(14)	0.0057(08)
12	557	0.66(1)	90(3)	0.33(3)	0.0059(09)	0.0072(16)
13	549	0.67(1)	95(3)	0.36(2)	0.0067(15)	0.0055(08)
14	723	0.66(1)	96(2)	0.36(2)	0.0068(13)	0.0057(07)
15	601	0.68(1)	91(2)	0.36(2)	0.0068(13)	0.0058(08)
16	597	0.65(1)	98(3)	0.30(4)	0.0067(16)	0.0055(08)
17	583	0.65(1)	102(3)	0.29(4)	0.0051(07)	0.0062(14)
18	574	0.67(1)	93(3)	0.34(3)	0.0070(17)	0.0057(09)
19	539	0.65(1)	98(3)	0.29(4)	0.0065(14)	0.0053(08)
20	564	0.66(1)	101(3)	0.31(3)	0.0052(08)	0.0063(13)

Table A5. The same as Table A4, but for the *Cucurbita moschata* species. The numbers in parenthesis denote the calculated errors.

CHR	N	κ_M	σ_κ	κ_S	γ_1	γ_2
1	667	0.66(1)	95(2)	0.35(3)	0.0069(14)	0.0057(08)
2	653	0.68(1)	97(2)	0.37(3)	0.0054(07)	0.0066(14)
3	610	0.67(1)	96(2)	0.37(3)	0.0068(14)	0.0056(08)
4	851	0.66(1)	96(2)	0.37(3)	0.0071(13)	0.0059(07)
5	605	0.65(1)	103(3)	0.31(2)	0.0052(08)	0.0063(14)
6	657	0.68(1)	97(2)	0.38(4)	0.0065(14)	0.0055(08)
7	608	0.67(1)	92(3)	0.37(3)	0.0074(19)	0.0059(09)
8	513	0.67(1)	88(3)	0.36(2)	0.0074(18)	0.0060(09)
9	599	0.69(1)	96(2)	0.40(4)	0.0054(07)	0.0064(12)
10	616	0.66(1)	98(3)	0.35(3)	0.0068(17)	0.0055(08)
11	714	0.68(1)	99(2)	0.38(2)	0.0054(06)	0.0064(11)
12	556	0.64(1)	99(3)	0.28(3)	0.0054(09)	0.0066(17)
13	605	0.67(1)	104(3)	0.36(3)	0.0051(07)	0.0062(14)
14	740	0.67(1)	95(2)	0.38(3)	0.0059(08)	0.0070(14)
15	630	0.67(1)	90(2)	0.37(4)	0.0070(13)	0.0059(07)
16	601	0.65(1)	107(3)	0.30(3)	0.0049(07)	0.0058(12)
17	641	0.64(1)	105(3)	0.28(3)	0.0061(13)	0.0051(08)
18	580	0.66(1)	99(3)	0.33(4)	0.0064(14)	0.0052(07)
19	555	0.67(1)	99(2)	0.34(3)	0.0051(07)	0.0062(12)
20	572	0.65(1)	110(2)	0.29(2)	0.0047(07)	0.0057(12)

Table A6. The same as Table A4, but for the *Cucurbita pepo* species. The numbers in parenthesis denote the calculated errors.

CHR	N	κ_M	σ_κ	κ_S	γ_1	γ_2
1	802	0.69(1)	88(2)	0.41(2)	0.0072(16)	0.0059(08)
2	663	0.69(1)	97(3)	0.38(2)	0.0053(07)	0.0062(12)
3	661	0.69(1)	84(3)	0.41(2)	0.0064(09)	0.0077(17)
4	572	0.70(1)	95(2)	0.40(2)	0.0062(14)	0.0051(07)
5	623	0.68(1)	96(3)	0.37(2)	0.0061(13)	0.0053(09)
6	503	0.70(1)	75(2)	0.43(3)	0.0070(13)	0.0091(31)
7	538	0.69(1)	87(3)	0.39(3)	0.0059(09)	0.0074(20)
8	578	0.70(1)	94(3)	0.39(2)	0.0061(12)	0.0052(08)
9	539	0.69(1)	77(2)	0.41(3)	0.0088(29)	0.0068(12)
10	529	0.70(1)	94(3)	0.39(2)	0.0063(14)	0.0051(08)
11	588	0.67(1)	109(3)	0.32(3)	0.0045(06)	0.0053(09)
12	570	0.67(1)	96(3)	0.32(3)	0.0063(16)	0.0053(10)
13	563	0.69(1)	92(3)	0.35(3)	0.0053(08)	0.0064(13)
14	534	0.67(1)	95(2)	0.32(4)	0.0053(09)	0.0067(17)
15	486	0.69(1)	88(2)	0.36(3)	0.0066(17)	0.0053(08)
16	512	0.69(1)	90(3)	0.38(3)	0.0055(08)	0.0067(16)
17	507	0.69(1)	94(3)	0.38(2)	0.0065(14)	0.0053(07)
18	445	0.71(1)	77(3)	0.43(3)	0.0074(27)	0.0065(12)
19	514	0.71(1)	93(3)	0.41(2)	0.0065(14)	0.0051(07)
20	501	0.69(1)	106(3)	0.35(2)	0.0045(06)	0.0053(09)

Table A7. Bayesian analysis for exons of each chromosome. The column $\ln(\mathcal{E})$ gives us the Baysian evidence for each of the models, Equation (11) for (i) and (9) for (j). The indices *max*, *mos*, and *pep* represent, respectively, the species *Cucurbita maxima*, *Cucurbita moschata*, and *Cucubita pepo*. The numbers in parenthesis indicate the calculated errors.

CHR	κ-Maxwellian			Sum κ-Exponentials			Bayes Factor		
	$\ln(\mathcal{E}_i^{max})$	$\ln(\mathcal{E}_i^{mos})$	$\ln(\mathcal{E}_i^{pep})$	$\ln(\mathcal{E}_j^{max})$	$\ln(\mathcal{E}_j^{mos})$	$\ln(\mathcal{E}_j^{pep})$	$\ln(B_{ij}^{max})$	$\ln(B_{ij}^{mos})$	$\ln(B_{ij}^{pep})$
1	−147.17(1)	−142.12(1)	−170.98(1)	−135.34(1)	−128.84(2)	−152.95(6)	−11.83(1)	−13.28(2)	−18.03(6)
2	−144.46(1)	−137.10(1)	−130.95(1)	−133.29(3)	−125.46(6)	−117.92(1)	−11.17(3)	−11.64(6)	−13.03(1)
3	−138.97(1)	−127.36(1)	−133.76(1)	−128.22(3)	−116.33(1)	−118.96(6)	−10.75(3)	−11.03(1)	−14.80(6)
4	−187.07(1)	−197.18(1)	−107.60(1)	−172.44(7)	−180.54(1)	−97.81(1)	−14.63(7)	−16.64(1)	−9.79 (1)
5	−128.49(1)	−126.69(1)	−121.06(1)	−117.14(3)	−115.10(4)	−108.52(2)	−11.35(3)	−11.59(4)	−12.54(2)
6	−146.70(1)	−137.90(1)	−94.60 (1)	−133.68(2)	−126.53(1)	−83.70(5)	−13.02(2)	−11.37(1)	−10.90(5)
7	−120.97(1)	−131.00(1)	−104.27(1)	−111.31(3)	−119.85(2)	−93.34(3)	−9.66 (3)	−11.15(2)	−10.93(3)
8	−112.37(1)	−103.10(1)	−105.71(1)	−102.26(2)	−93.35(2)	−94.89(2)	−10.11(2)	−9.75 (2)	−10.82(2)
9	−129.29(1)	−120.78(1)	−104.51(1)	−119.07(2)	−111.31(3)	−92.33(1)	−10.22(2)	−9.47 (3)	−12.18(1)
10	−110.92(1)	−132.10(1)	−100.74(1)	−101.87(1)	−120.77(2)	−91.92(3)	−9.05 (1)	−11.33(2)	−8.82 (3)
11	−146.58(1)	−152.99(1)	−108.48(1)	−135.32(1)	−140.62(1)	−97.11(3)	−11.26(1)	−12.37(1)	−11.37(3)
12	−116.96(1)	−114.18(1)	−110.76(1)	−106.53(3)	−102.74(2)	−98.23(6)	−10.43(3)	−11.44(2)	−12.53(6)
13	−114.54(1)	−125.34(1)	−105.07(1)	−105.62(2)	−115.35(3)	−93.44(3)	−8.92 (2)	−9.99 (3)	−11.63(3)
14	−163.94(1)	−163.55(1)	−103.71(1)	−151.50(4)	−150.13(3)	−92.43(1)	−12.44(4)	−13.42(3)	−11.28(1)
15	−126.73(1)	−129.25(1)	−90.10 (1)	−116.28(1)	−116.91(2)	−80.08(2)	−10.45(1)	−12.34(2)	−10.02(2)
16	−130.03(1)	−124.87(1)	−96.78 (1)	−119.19(3)	−114.18(1)	−87.58(1)	−10.84(3)	−10.69(1)	−9.20 (1)
17	−121.57(1)	−135.30(1)	−98.24 (1)	−111.25(2)	−122.31(5)	−89.39(1)	−10.32(2)	−12.99(5)	−8.85 (1)
18	−121.37(1)	−117.42(1)	−84.51 (1)	−111.18(2)	−106.33(1)	−76.01(2)	−10.19(2)	−11.09(1)	−8.50 (2)
19	−114.09(1)	−110.97(1)	−95.81 (1)	−104.59(1)	−101.15(1)	−87.73(1)	−9.50 (1)	−9.82 (1)	−8.08 (1)
20	−117.37(1)	−116.36(1)	−91.77 (1)	−108.02(1)	−106.56(2)	−83.96(1)	−9.35 (1)	−9.80 (2)	−7.81 (1)

References

1. Schaefer, H.; Renner, S.S. Phylogenetic relationships in the order Cucurbitales and a new classification of the gourd family (Cucurbitaceae). *Taxon* **2011**, *60*, 122–138. [CrossRef]
2. Eguiarte, L.; Hernández-Rosales, H.; Barrera-Redondo, J.; Castellanos-Morales, G.; Paredes-Torres, L.M.; Sánchez-de la Vega, G.; Ruiz-Mondragón, K.Y.; Vázquez-Lobo, A.; Montes-Hernández, S.; Aguirre-Planter, E.; et al. Domesticación, diversidad y recursos genéticos y genómicos de México: El caso de las calabazas. *TIP Rev. Espec. Cienc. Químico-Biológicas* **2018**, *21*, 85. [CrossRef]
3. Chomicki, G.; Schaefer, H.; Renner, S.S. Origin and domestication of Cucurbitaceae crops: insights from phylogenies, genomics and archaeology. *New Phytol.* **2020**, *226*, 1240–1255. [CrossRef] [PubMed]
4. Prohens, J.; Nuez, F.; Carena, M.J. *Handbook of Plant Breeding*; Springer: Berlin/Heidelberg, Germany, 2008; pp. 317–349. [CrossRef]
5. Paris, H.S. Genetic resources of pumpkins and squash, *Cucurbita* spp. In *Genetics and Genomics of Cucurbitaceae*; Springer: Berlin/Heidelberg, Germany, 2016; pp. 111–154. [CrossRef]
6. Paris, H.S., History of the Cultivar-Groups of *Cucurbita pepo*. In *Horticultural Reviews*; John Wiley & Sons, Ltd.: Hoboken, NJ, USA, 2000; Chapter 2, pp. 71–170. [CrossRef]
7. Lee, H.Y.; Jang, S.; Yu, C.R.; Kang, B.C.; Chin, J.H.; Song, K. Population structure and genetic diversity of *Cucurbita moschata* based on genome-wide high-quality SNPs. *Plants* **2020**, *10*, 56. [CrossRef] [PubMed]
8. Weiling, F. Genomanalytische Untersuchungen bei Kürbis (*Cucurbita* L.). *Der Züchter* **1959**, *29*, 161–179. [CrossRef]
9. Sun, H.; Wu, S.; Zhang, G.; Jiao, C.; Guo, S.; Ren, Y.; Zhang, J.; Zhang, H.; Gong, G.; Jia, Z.; et al. Karyotype stability and unbiased fractionation in the paleo-allotetraploid Cucurbita genomes. *Mol. Plant* **2017**, *10*, 1293–1306. [CrossRef] [PubMed]
10. Montero-Pau, J.; Blanca, J.; Bombarely, A.; Ziarsolo, P.; Esteras, C.; Martí-Gómez, C.; Ferriol, M.; Gómez, P.; Jamilena, M.; Mueller, L.; et al. De novo assembly of the zucchini genome reveals a whole-genome duplication associated with the origin of the Cucurbita genus. *Plant Biotechnol. J.* **2018**, *16*, 1161–1171. [CrossRef]
11. Barrera-Redondo, J.; Ibarra-Laclette, E.; Vázquez-Lobo, A.; Gutiérrez-Guerrero, Y.T.; de la Vega, G.S.; Piñero, D.; Montes-Hernández, S.; Lira-Saade, R.; Eguiarte, L.E. The genome of *Cucurbita argyrosperma* (silver-seed gourd) reveals faster rates of protein-coding gene and long noncoding RNA turnover and neofunctionalization within Cucurbita. *Mol. Plant* **2019**, *12*, 506–520. [CrossRef]
12. Peng, C.K.; Buldyrev, S.V.; Goldberger, A.L.; Havlin, S.; Sciortino, F.; Simons, M.; Stanley, H.E. Long-range correlations in nucleotide sequences. *Nature* **1992**, *356*, 168–170. [CrossRef]
13. Li, W.; Kaneko, K. Long-range correlation and partial $1/f\alpha$ spectrum in a noncoding DNA sequence. *Europhys. Lett.* **1992**, *17*, 655. [CrossRef]
14. Li, W. The study of correlation structures of DNA sequences: a critical review. *Comput. Chem.* **1997**, *21*, 257–271. [CrossRef]

15. Arneodo, A.; Bacry, E.; Graves, P.; Muzy, J.F. Characterizing long-range correlations in DNA sequences from wavelet analysis. *Phys. Rev. Lett.* **1995**, *74*, 3293. [CrossRef] [PubMed]
16. Audit, B.; Thermes, C.; Vaillant, C.; d'Aubenton Carafa, Y.; Muzy, J.F.; Arneodo, A. Long-Range Correlations in Genomic DNA: A Signature of the Nucleosomal Structure. *Phys. Rev. Lett* **2001**, *86*, 6. [CrossRef]
17. Colliva, A.; Pellegrini, R.; Testori, A.; Caselle, M. Ising-model description of long-range correlations in DNA sequences. *Phys. Rev. E* **2015**, *91*, 052703. [CrossRef]
18. Provata, A.; Almirantis, Y. Statistical dynamics of clustering in the genome structure. *J. Stat. Phys.* **2002**, *106*, 23–56. [CrossRef]
19. Karakatsanis, L.P.; Pavlos, E.G.; Tsoulouhas, G.; Stamokostas, G.L.; Mosbruger, T.; Duke, J.L.; Pavlos, G.P.; Monos, D.S. Spatial constrains and information content of sub-genomic regions of the human genome. *iScience* **2021**, *24*, 102048. [CrossRef]
20. Provata, A.; Almirantis, Y. Fractal Cantor patterns in the sequence structure of DNA. *Fractals* **2000**, *8*, 15–27. [CrossRef]
21. Katsaloulis, P.; Theoharis, T.; Provata, A. Statistical distributions of oligonucleotide combinations: Applications in human chromosomes 21 and 22. *Phys. A* **2002**, *316*, 380–396. [CrossRef]
22. Katsaloulis, P.; Theoharis, T.; Zheng, W.; Hao, B.; Bountis, A.; Almirantis, Y.; Provata, A. Long-range correlations of RNA polymerase II promoter sequences across organisms. *Phys. A* **2006**, *366*, 308–322. [CrossRef]
23. Provata, A.; Oikonomou, T. Power law exponents characterizing human DNA. *Phys. Rev. E* **2007**, *75*, 056102. [CrossRef]
24. Gell-Mann, M.; Tsallis, C. *Nonextensive Entropy: Interdisciplinary Applications*; Oxford University Press: Oxford, UK, 2004.
25. Kaniadakis, G. Maximum entropy principle and power-law tailed distributions. *Eur. Phys. J. B* **2009**, *70*, 3–13. [CrossRef]
26. Kaniadakis, G. Non-linear kinetics underlying generalized statistics. *Phys. A* **2001**, *296*, 405–425. [CrossRef]
27. Kaniadakis, G. Statistical mechanics in the context of special relativity. *Phys. Rev. E* **2002**, *66*, 056125. [CrossRef]
28. Kaniadakis, G. Statistical mechanics in the context of special relativity. II. *Phys. Rev. E* **2005**, *72*, 036108. [CrossRef]
29. Kaniadakis, G.; Baldi, M.M.; Deisboeck, T.S.; Hristopulos, G.G.D.T.; Scarfone, A.M.; Sparavigna, A.; Wada, T.; Lucia, U. The κ-statistics approach to epidemiology. *Sci. Rep.* **2020**, *10*, 19949. [CrossRef]
30. Souza, N.; Anselmo, D.; Silva, R.; Vasconcelos, M.; Mello, V. A κ-statistical analysis of the Y-chromosome. *EPL Europhys. Lett.* **2014**, *108*, 38004. [CrossRef]
31. Costa, M.; Silva, R.; Anselmo, D.; Silva, J. Analysis of human DNA through power-law statistics. *Phys. Rev. E* **2019**, *99*, 022112. [CrossRef]
32. Von Toussaint, U. Bayesian inference in physics. *Rev. Mod. Phys.* **2011**, *83*, 943. [CrossRef]
33. Hines, K.E. A primer on Bayesian inference for biophysical systems. *Biophys. J.* **2015**, *108*, 2103–2113. [CrossRef]
34. de Lima, M.M.F.; Silva, R.; Fulco, U.L.; Mello, V.D.; Anselmo, D.H.A.L. Bayesian analysis of plant DNA size distribution via non-additive statistics. *Eur. Phys. J. Plus* **2022**, *137*, 1–8. [CrossRef]
35. Silva, R.; Silva, J.; Anselmo, D.; Alcaniz, J.; da Silva, W.; Costa, M. An alternative description of power law correlations in DNA sequences. *Phys. A* **2020**, *545*, 123735. [CrossRef]
36. Oikonomou, T.; Provata, A.; Tirnakli, U. Nonextensive statistical approach to non-coding human DNA. *Phys. A* **2008**, *387*, 2653–2659. [CrossRef]
37. Oikonomou, T.; Kaloudis, K.; Bagci, G.B. The q-exponentials do not maximize the Rényi entropy. *Phys. A* **2021**, *578*, 126126. [CrossRef]
38. Clementi, F.; Gallegati, M.; Kaniadakis, G. κ-generalized statistics in personal income distribution. *Eur. Phys. J. B* **2007**, *57*, 187–193. [CrossRef]
39. Kaniadakis, G. Theoretical foundations and mathematical formalism of the power-law tailed statistical distributions. *Entropy* **2013**, *15*, 3983–4010. [CrossRef]
40. da Silva, S.L.E. κ-generalised Gutenberg–Richter law and the self-similarity of earthquakes. *Chaos Solitons Fractals* **2021**, *143*, 110622. [CrossRef]
41. Macedo-Filho, A.; Moreira, D.; Silva, R.; da Silva, L.R. Maximum entropy principle for Kaniadakis statistics and networks. *Phys. Lett. A* **2013**, *377*, 842–846. [CrossRef]
42. National Center for Biotechnology Information (NCBI). 2022. Available online: https://www.ncbi.nlm.nih.gov (accessed on 17 June 2021).
43. Comparative Genomics (CoGe). 2022. Available online: https://genomevolution.org (accessed on 8 June 2021).
44. R Core Team. *R: A Language and Environment for Statistical Computing*; R Foundation for Statistical Computing: Vienna, Austria, 2020.
45. Aguinis, H.; Gottfredson, R.K.; Joo, H. Best-practice recommendations for defining, identifying, and handling outliers. *Organ. Res. Methods* **2013**, *16*, 270–301. [CrossRef]
46. Correia, J.; Silva, R.; Anselmo, D.; da Silva, J. Bayesian inference of length distributions of human DNA. *Chaos Solitons Fractals* **2022**, *160*, 112244. [CrossRef]
47. Almirantis, Y.; Provata, A. Scaling properties of coding and non-coding DNA sequences. *J. Stat. Phys.* **1999**, *97*, 233–262. [CrossRef]
48. van de Schoot, R.; Depaoli, S.; King, R.; Kramer, B.; Märtens, K.; Tadesse, M.G.; Vannucci, M.; Gelman, A.; Veen, D.; Willemsen, J.; et al. Bayesian statistics and modelling. *Nat. Rev. Methods Prim.* **2021**, *1*, 1–26. [CrossRef]
49. Feroz, F.; Hobson, M.P. Multimodal nested sampling: an efficient and robust alternative to Markov Chain Monte Carlo methods for astronomical data analyses. *Mon. Not. R. Astron. Soc.* **2008**, *384*, 449–463. [CrossRef]

50. Feroz, F.; Hobson, M.; Bridges, M. MultiNest: an efficient and robust Bayesian inference tool for cosmology and particle physics. *Mon. Not. R. Astron. Soc.* **2009**, *398*, 1601–1614. [CrossRef]
51. Feroz, F.; Hobson, M.P.; Cameron, E.; Pettitt, A.N. Importance nested sampling and the MultiNest algorithm. *Open J. Astrophys.* **2013**. [CrossRef]
52. Skilling, J. Nested sampling. *AIP Conf. Proc.* **2004**, *735*, 395–405. [CrossRef]
53. Buchner, J.; Georgakakis, A.; Nandra, K.; Hsu, L.; Rangel, C.; Brightman, M.; Merloni, A.; Salvato, M.; Donley, J.; Kocevski, D. X-ray spectral modelling of the AGN obscuring region in the CDFS: Bayesian model selection and catalogue. *Astron. Astrophys.* **2014**, *564*, A125. [CrossRef]
54. Trotta, R. Bayes in the sky: Bayesian inference and model selection in cosmology. *Contemp. Phys.* **2008**, *49*, 71–104. [CrossRef]
55. Jeffreys, H. *The Theory of Probability*; OUP Oxford: Oxford, UK, 1998.
56. da Silva, W.; Silva, R. Cosmological perturbations in the Tsallis holographic dark energy scenarios. *Eur. Phys. J. Plus* **2021**, *136*, 1–19. [CrossRef]

entropy

Article

Kaniadakis Functions beyond Statistical Mechanics: Weakest-Link Scaling, Power-Law Tails, and Modified Lognormal Distribution

Dionissios T. Hristopulos [1,*,†] and Anastassia Baxevani [2,†]

[1] School of Electrical and Computer Engineering, Technical University of Crete, 73100 Chania, Greece
[2] Department of Mathematics and Statistics, University of Cyprus, Nicosia 1678, Cyprus
* Correspondence: dchristopoulos@tuc.gr or dchristopoulos@ece.tuc.gr; Tel.: +30-2821037688
† These authors contributed equally to this work.

Abstract: Probabilistic models with flexible tail behavior have important applications in engineering and earth science. We introduce a nonlinear normalizing transformation and its inverse based on the deformed lognormal and exponential functions proposed by Kaniadakis. The deformed exponential transform can be used to generate skewed data from normal variates. We apply this transform to a censored autoregressive model for the generation of precipitation time series. We also highlight the connection between the heavy-tailed κ-Weibull distribution and weakest-link scaling theory, which makes the κ-Weibull suitable for modeling the mechanical strength distribution of materials. Finally, we introduce the κ-lognormal probability distribution and calculate the generalized (power) mean of κ-lognormal variables. The κ-lognormal distribution is a suitable candidate for the permeability of random porous media. In summary, the κ-deformations allow for the modification of tails of classical distribution models (e.g., Weibull, lognormal), thus enabling new directions of research in the analysis of spatiotemporal data with skewed distributions.

Keywords: Kaniadakis exponential; modified lognormal distribution; earthquake recurrence times; Weibull distribution; power-law tail; precipitation; flow in random media; tensile strength

PACS: 02.50.Fz; 02.60.Ed; 89.60.-k; 92.60.Ry; 05.10.Ln

MSC: 60G15; 60G60; 62F40; 62H11; 62G05; 65C05

Citation: Hristopulos, D.T.; Baxevani, A. Kaniadakis Functions beyond Statistical Mechanics: Weakest-Link Scaling, Power-Law Tails, and Modified Lognormal Distribution. *Entropy* **2022**, *24*, 1362. https://doi.org/10.3390/e24101362

Academic Editor: Eun-jin Kim

Received: 1 August 2022
Accepted: 16 September 2022
Published: 26 September 2022

1. Introduction

Several physical processes exhibit asymmetric probability distributions which deviate from the Gaussian law (e.g., the exponential, gamma, Weibull, lognormal, Pareto, and generalized Pareto models) [1–8]. Skewed probability distributions describe various geophysical variables, including the amount and duration of precipitation over a certain time window [9–12], the waiting times (recurrence or interevent times) between consecutive earthquakes occurring over a given area [6–8], the fluid permeability of geological porous media [13–16], as well as the mechanical strength distribution of the earth's crust [17,18] and various technological brittle materials [19–22].

A feature of particular interest is the behavior of the tail(s) of a probability distribution, because the tails define the probabilities of extreme events. Distributions are characterized as sub-exponential (if their tail decays slower than the exponential) and super-exponential in the opposite case. The same function (e.g., the Weibull model) may exhibit transitions from sub-exponential to super-exponential by changing the value of a key parameter; in the Weibull case this is the modulus m: $m < 1$ leads to a sub-exponential and $m > 1$ to a super-exponential tail. Sub-exponential models are called heavy-tailed if the asymptotic behavior of the probability for large events decays algebraically, i.e., $P(X > x) \sim x^{-\alpha}$, where $\alpha > 0$.

In particular, probability distributions with power-law tails are ubiquitous in natural phenomena [23,24]. Such power laws can be generated by means of different physical mechanisms as described in ([23], Chap. 14). Multiple mechanisms including (but not limited to) phase transitions, self-organized criticality, optimization, multiplicative processes, and interdependence in complex systems [25–29] lead to power laws. This fact explains, to some extent, the omnipresence of power laws in physics, biology, earth science, cosmology, ecology, finance, and other disciplines. In addition, the detection of power-law distributions in data has significant impact on statistical analysis and forecasting [30].

Approximately twenty years ago, Kaniadakis introduced the κ-deformed exponential and κ-deformed logarithmic functions [31–34]. These functions provide a springboard for the construction of generalized, flexible probability distributions whose tails are controlled by the deformation parameter κ. Based on the deformed functions, extensions of the known generalized gamma, Weibull, generalized logistic, and exponential models can be constructed which exhibit power-law tails; explicit expressions for the probability functions of these models are given in [35]. The Kaniadakis functions have found applications in plasma physics [36,37], gravitational systems [38], income analysis [39–41], epidemiology [42], and other fields (see [35] for a more comprehensive list of applications). In particular, κ-statistical theory uses the κ-exponential function to generalize the Maxwell–Boltzmann distribution, leading to distributions with power-law tails. Distribution models with flexible tails are also needed in earth science, where datasets often exhibit tail behavior that is not adequately captured by classical distribution models [43–45].

The motivation for this work is the need for flexible statistical models that can adapt to the diversity of earth science data and also provide physical intuition. Our objective is to explore the possibilities created by the parameter κ which appears in the deformed exponential and logarithmic functions. This investigation leads to novel applications of the Kaniadakis functions in geostatistics, material fracture, precipitation, and fluid flow modeling.

The main contributions of this paper are as follows.

- We show (Section 3) that the κ-deformed exponential and logarithmic functions (henceforth, κ-exponential and κ-logarithm) can be used to define normalizing transforms for non-Gaussian data, which extend the well-known (in statistics) Box–Cox family of transformations [46].
- We formulate an autoregressive, intermittent precipitation model based on the κ-modified Box–Cox transform in Section 3.3. We show that the resulting precipitation time series has higher "peaks" than those obtained with the Box–Cox transform with the same parameter value.
- We review the κ-Weibull distribution focusing on its connection with weakest-link theory (Section 4). This demonstrates that the κ-Weibull is a physically motivated generalization of the classical Weibull distribution for the mechanical strength of brittle materials, unlike modified Weibull distributions which fail to satisfy the weakest-link principle.
- We show that for several physical quantities, including the thickness of magmatic sheet intrusions, the tensile strength of steel, earthquake waiting times, and precipitation amounts the κ-Weibull distribution provides a better fit than the Weibull according to model selection criteria.
- We introduce the κ-lognormal distribution, which provides a deformation of the lognormal with lighter tails than the latter in Section 5. The κ-lognormal can be used to model asymmetric data distributions which concentrate more probability mass around the median than the lognormal. We discuss the importance of the generalized mean (power mean) of the lognormal distribution for estimating the effective permeability of heterogeneous porous media, and we calculate the generalized mean of the κ-lognormal distribution.

2. Mathematical Preliminaries

Let $(\Omega, \mathcal{F}, P)$, where Ω is the sample space, $\mathcal{F}$ is a σ-field of subspaces of Ω, and P is a probability measure, define a probability space. A real-valued, scalar random variable $X(\omega)$ is defined by the mapping $X : \Omega \to \mathbb{R}$, where $\mathbb{R}$ is the set of real numbers [47]. Furthermore, a stochastic process $X(t;\omega)$ indexed by the time $t \in \mathbb{R}$ is defined by the mapping $X : \mathbb{R} \times \Omega \to \mathbb{R}$. In the following, the dependence on the state index ω is suppressed for convenience. In addition, the random variable X represents a load or the waiting time between consecutive "failure" events (e.g., earthquakes), or some other asymmetrically distributed variable.

The function $F(x) : P(X \leq x) : \mathbb{R} \to [0,1]$ defines the cumulative distribution function (CDF) of X, or the marginal CDF of a stationary stochastic process $\{X(t)\}$. The expectation of X, assuming it is mathematically well defined, is given by means of $\mathcal{E}[X] = \int \mathrm{d}F(x)\, x$.

Assuming that $F(x)$ is at least once differentiable, the probability density function (PDF) is given by the first derivative of the CDF, i.e., $f(x) = \mathrm{d}F(x)/\mathrm{d}x$.

The CDF is related to the so-called survival function, also known as the reliability function, which is given by $S(x) = 1 - F(x)$. Whereas $F(x)$ is a monotonically increasing function, $S(x)$ is monotonically decreasing. The term "survival function" comes from reliability engineering: if X represents the strength or critical loading of a given system, $F(x)$ is the probability that the system fails at loading level $X \leq x$; then, $S(x)$ is respectively the probability that the system remains intact (survives) at this loading level.

The quantile function, $Q(p)$, where $p \in [0,1]$, returns the value $x_p \in \mathbb{R}$ such that $F(x_p) = p$. Hence, $Q(\cdot)$ is the inverse of the CDF.

The hazard rate, also known as the hazard function, $h(x)$ represents the conditional probability that the system fails for $X \in (x, x+\delta x]$ where $\delta x \ll 1$, conditioned on the survival of the system for $X \leq x$. Let A denote the event that the system survives at level x and B denote the event of system failure in the interval $[x, x+\delta x]$. Then, by using the definition of conditional probabilities, $h(x) = P(B|A) = P(B \cap A)/P(A)$, the hazard rate is given by the following ratio:

$$h(x) = \lim_{\delta x \to 0} \frac{P(x < X \leq x + \delta x)}{S(x)\, \delta x} = \frac{f(x)\delta x}{S(x)\delta x} = \frac{f(x)}{S(x)}. \tag{1}$$

The asymptotic behavior of the hazard rate determines the probability of system failure with increasing load. The hazard rate for $x \to \infty$ is determined by the tail of the probability functions $f(x)$ and $S(x)$. For certain probability models, e.g., the exponential and the gamma, $h(x)$ tends to a constant as $x \to 0$; for the Weibull model with modulus $m > 1$, $h(x)$ tends to zero, whereas for models with power-law tails, the lognormal, and the Weibull model with $m < 1$, $h(x)$ diverges as $x \to \infty$. The hazard rate is an important factor in seismic risk assessment [48].

2.1. The κ-Exponential Function

The κ-generalized exponential is a one-parameter generalization of the exponential function, proposed by Kaniadakis [31,34]:

$$\exp_\kappa(x) = \left(\sqrt{1 + x^2\kappa^2} + x\kappa\right)^{1/\kappa}, \tag{2}$$

with $0 \leq \kappa < 1$ and $x \in \mathbb{R}$. The first few terms of the Taylor expansion of $\exp_\kappa(x)$, reported in [49], are given by

$$\exp_\kappa(x) = 1 + x + \frac{x^2}{2} + (1 - \kappa^2)\frac{x^3}{3!} + (1 - 4\kappa^2)\frac{x^4}{4!} + \dots. \tag{3}$$

The emerging pattern persists for higher orders, i.e., terms of $\mathcal{O}(x^3)$ consist of the ordinary exponential expansion and a κ-dependent correction. The κ-exponential is expressed as the following power series [49]:

$$\exp_\kappa(x) = \sum_{n=0}^{\infty} \xi_n(\kappa) \frac{x^n}{n!}. \tag{4}$$

The functions $\{\xi_n(\kappa)\}_{n=0}^{\infty}$ are polynomials of κ defined by the recurrence relations

$$\xi_0(\kappa) = \xi_1(\kappa) = 1, \quad n = 0, 1, \tag{5}$$

$$\xi_n(\kappa) = \prod_{j=1}^{n-1} [1 - (2j - n)\kappa] = \xi_{n-2}(\kappa) \left[1 - (n-2)^2 \kappa^2\right], \quad n > 1. \tag{6}$$

The polynomials $\xi_n(\kappa)$ for the first seven orders are given by

$$\xi_0(\kappa) = \xi_1(\kappa) = \xi_2(\kappa) = 1, \tag{7}$$
$$\xi_3(\kappa) = 1 - \kappa^2, \tag{8}$$
$$\xi_4(\kappa) = 1 - 4\kappa^2, \tag{9}$$
$$\xi_5(\kappa) = (1 - \kappa^2)(1 - 9\kappa^2), \tag{10}$$
$$\xi_6(\kappa) = (1 - 4\kappa^2)(1 - 16\kappa^2). \tag{11}$$

It follows from Equation (3) that when $x \to 0$ or $\kappa \to 0$, $\exp_\kappa(x)$ converges to the ordinary exponential, i.e.,

$$\exp_\kappa(x) \underset{x \to 0}{\sim} \exp(x), \tag{12}$$

$$\exp_\kappa(x) \underset{\kappa \to 0}{\sim} \exp(x). \tag{13}$$

Equation (2) shows that the asymptotic behavior of $\exp_\kappa(x)$ as $x \to \pm\infty$ follows a power law [34,49], i.e.,

$$\exp_\kappa(x) \underset{x \to \pm\infty}{\sim} \left| 2\kappa\, x \right|^{\pm 1/\kappa}. \tag{14}$$

Based on the above, for $x \to +\infty$ the modified exponential exhibits a heavy tail, i.e. $\exp_\kappa(-x) \sim (2\kappa x)^{-1/\kappa}$. Hence, $\exp_\kappa(x)$ can be used to model subexponential probability distributions which are suitable for heavy-tailed data.

The κ-exponential can also be introduced as the solution of a linear, first-order ordinary differential equation (ODE) with time-dependent rate [42]. Consider the ODE

$$\frac{\mathrm{d}f(x)}{\mathrm{d}x} = -r(x)\, f(x), \text{ with initial condition } f(0) = 1, \tag{15}$$

where $r(x)$ is the following x-dependent rate function

$$r(x) = \frac{\beta}{\sqrt{1 + \kappa^2 \beta^2 x^2}}. \tag{16}$$

The solution of the ODE (15) is given by the function $f(x) = \exp_\kappa(-\beta x)$. In case $\kappa = 0$, then $r(x) = \beta$, and the rate equation is solved by the standard exponential function $f(x) = \exp(-\beta x)$.

2.2. The κ-Logarithm Function

The inverse of the κ-exponential is the κ-logarithm, defined by the following function for $x > 0$:

$$\ln_\kappa(x) = \frac{x^\kappa - x^{-\kappa}}{2\kappa}. \tag{17}$$

The κ-logarithm satisfies the equation $\ln_\kappa \exp_\kappa(x) = x$. In addition, it respects the κ-symmetry property $\ln_\kappa(x) = \ln_{-\kappa}(x)$. A Taylor expansion of x^κ and $x^{-\kappa}$ around $\kappa = 0$ leads to

$$\lim_{\kappa \to 0} \ln_\kappa(x) = \ln x\,. \tag{18}$$

The first and second derivatives of the κ-logarithm are respectively given by

$$\frac{\mathrm{d} \ln_\kappa(x)}{\mathrm{d}x} = \frac{1}{2}\left(x^{\kappa-1} + x^{-\kappa-1}\right), \tag{19a}$$

$$\frac{\mathrm{d}^2 \ln_\kappa(x)}{\mathrm{d}x^2} = \frac{\kappa-1}{2}\,x^{\kappa-2} - \frac{\kappa+1}{2}\,x^{-\kappa-2}\,. \tag{19b}$$

Based on Equation (19a), the first derivative of the κ-logarithm is positive; therefore $\ln_\kappa(x)$ is a monotonically increasing function. Based on Equation (19b), the second derivative of $\ln_\kappa(x)$ is negative for $0 < \kappa < 1$; therefore, $\ln_\kappa(x)$ is a concave function for $\kappa \in (0,1)$.

3. Nonlinear Transformation of Data Based on the κ-Logarithm

Nonlinear, monotonic transformations are often applied to non-Gaussian data in order to restore normality [50–52]. This procedure, known as Gaussian anamorphosis, enables the use of data processing methods that are based on Gaussian assumptions. Various transforms are used in practice, including the Box–Cox [46] and Yeo–Johnson [53]. Such transforms can be generalized by means of the Kaniadakis functions. Below we focus on Box–Cox but the same arguments can be used for other transforms.

3.1. Box–Cox Transform and the Replica Trick

A widely used normalizing transformation in statistics is the so-called Box–Cox transform (BCT) [46]; the one-parameter version of BCT is given by the monotonic function

$$g_{BC}(x) = \frac{x^\lambda - 1}{\lambda}, \text{ where } \lambda \in \mathbb{R},\ x > 0. \tag{20}$$

The BCT is applied to skewed (non-Gaussian) data so that the transformed variable $y = g_{BC}(x)$ is better approximated by the Gaussian distribution. The BCT is applied to both time series and spatial data [52].

It is interesting to note that if $\lambda < 0$, $g_{BC}(x)$ takes positive (negative) values for $x < 1$ ($x > 1$), whereas if $\lambda > 0$, the $g_{BC}(x)$ takes positive (negative) values for $x > 1$ ($x < 1$). The BCT value for $\lambda = 0$ can be obtained by using either l'Hopital's rule or the Taylor expansion. The Taylor expansion of x^λ around $\lambda = 0$ shows that $x^\lambda = 1 + \lambda \ln x + \mathcal{O}(\lambda^2)$. Therefore,

$$\lim_{\lambda \to 0} \frac{x^\lambda - 1}{\lambda} = \ln x\,. \tag{21}$$

Equation (21) shows that the logarithmic transform is a special case of the BCT for $\lambda = 0$. The inverse BCT is given by $h_{BC}(y) \triangleq g_{BC}^{-1}(y) = (\lambda y + 1)^{1/\lambda}$.

In a different context, Edwards and Anderson [54] introduced the famous replica trick, which is also based on Equation (21), to study spin glasses. The replica trick is used to calculate the ensemble average (over the magnetic disorder) of the logarithm of the spin glass partition function, i.e., $\overline{\ln Z}$. By using the replica trick, $\overline{\ln Z}$ is calculated by first evaluating $\overline{g_{BC}(Z)}$, with $\lambda = n \in \mathbb{N}$ denoting the number of replicas (identical copies of the system), and then taking the limit $n \to 0$.

3.2. *The κ-Logarithmic Transform*

The κ-logarithm transform (KLT) is a nonlinear, monotonic transformation from $\mathbb{R}_+ \to R$. Therefore, it can be used like the BCT for the Gaussian anamorphosis of positive-valued data. The KLT takes the form

$$g_{KL}(x) = \ln_\kappa(x) = \frac{x^\kappa - x^{-\kappa}}{2\kappa}, \; \kappa \in \mathbb{R}, \; x > 0. \tag{22}$$

Equation (18) shows that the logarithmic transform is a special case of the KLT for $\kappa \to 0$, as it is a special case of the BCT. To understand how the transformation works, consider the special cases $\kappa = \lambda = 1$ shown in Figure 1. The choice $\lambda = 1$ for the BCT is a simple linear shift of x to $x - 1$, whereas setting $\kappa = 1$ in the KLT leads to the nonlinear transformation $\frac{x^2-1}{2x}$. The latter tends to the linear transformation $x/2$ for $x \gg 1$. Figure 2 shows the two transformations for different values of $\lambda = \kappa$. Notice that the KL transformation is symmetric with respect to $\kappa \mapsto -\kappa$. The inverse of the KLT is given by $h_{KL}(y) = \exp_\kappa(y)$.

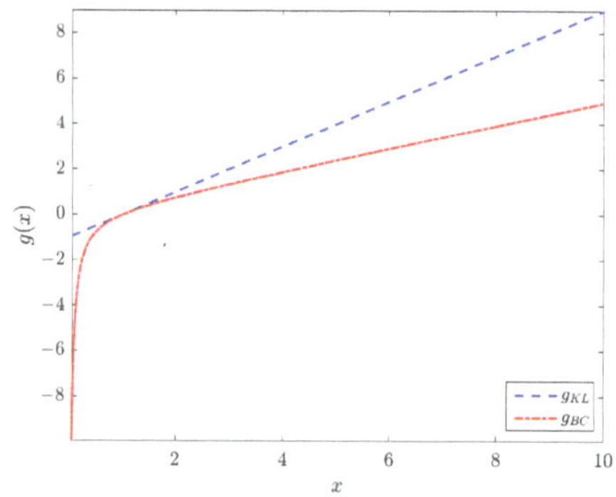

Figure 1. Plots of the Box–Cox and κ-logarithmic transform for $\lambda = \kappa = 1$ (λ is the Box–Cox parameter and κ is the deformation parameter of the Kaniadakis logarithm).

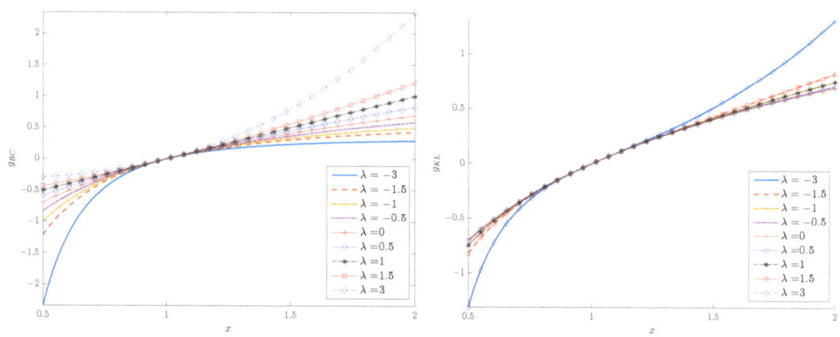

Figure 2. Plots of the Box–Cox (**left**) and κ-logarithmic (**right**) transform for different values of $\lambda = \kappa$ (λ is the Box–Cox parameter and κ is the deformation parameter of the Kaniadakis logarithm).

3.3. *Application to Precipitation Modeling*

Autoregressive (AR) models are used for different meteorological processes which exhibit memory. AR models exhibit short-term memory: the present depends on the past via the p most recent values of the process

$$y_t = m + \sum_{i=1}^{p} \phi_i \,(y_{t-i} - m) + \sigma_\epsilon \,\epsilon_t, \tag{23}$$

where $m = \mathcal{E}[y_t]$ is the expectation of y_t, the set $\{\phi_i\}_{i=1}^{p}$ comprises the real-valued auto-regressive coefficients, $\{\epsilon_t\}$ is the innovation process, and σ_ϵ is its standard deviation. The innovation is typically considered to be standard Gaussian white noise, i.e., $\epsilon_t \sim \mathcal{N}(0,1)$, leading to a normally distributed time series $\{y_t\}_{t=1}^{T}$.

Daily precipitation usually displays intermittency (i.e., intervals of zero precipitation) in addition to temporal variability of its intensity. These features can be modeled by using a censored autoregressive time series model [55]. Then, the amount of daily precipitation is given by $x_t = y_t\,\theta(y_t - y_c)$, where $\theta(\cdot)$ is the unit step function and y_c is a censoring threshold. The threshold is selected so that $F_y(y_c) = p_0$, where p_0 is the probability of observing a dry day. The censored AR model leads to a truncated normal distribution for x_t. Because this distribution does not adequately reflect the extreme values of precipitation, in practice the daily precipitation is given by means of the following censored and transformed autoregressive process,

$$x_t = h[\,y_t\,\theta(y_t - y_c)] - h(0), \tag{24}$$

where $h(\cdot) : \mathbb{R} \to \mathbb{R}$ is a monotonically increasing transform. The application of $h(\cdot)$ introduces skewness and increases the weight in the right tail of the distribution. The subtracted term $h(0)$ restores the zero precipitation values after application of the transform. The function $h(\cdot)$ could, for example, represent the inverse BC or KL transforms, i.e., $h(y) = (\lambda y + 1)^{1/\lambda}$ or $h(y) = \exp_\kappa(y)$, respectively.

Figure 3 presents six realizations of the censored AR(1) model of Equation (24) using both the inverse BCT and KLT as the nonlinear transformation $h(\cdot)$, with equal values of κ and λ in each frame. The AR(1) model of Equation (24) is applied with $\phi_1 = 0.5$ and $\sigma_\epsilon = 0.6$. The time series exhibit intermittent behavior due to censoring. The difference between BCT and KLT is negligible for small values of $\kappa = \lambda$. For $\kappa \approx 0$, both transforms yield a censored lognormal process, because $h_{KL}(y)$ and $h_{BC}(y)$ converge to the normal exponential for $\kappa = \lambda \to 0$. The KLT peaks become progressively higher compared to the BCT peaks as κ (and λ) increase. This behavior is due to the following inequality between the inverse transforms

$$h_{KL}(y) = \left(\sqrt{1+\kappa^2 y^2} + \kappa y\right)^{1/\kappa} > h_{BC}(y) = (1+\kappa y)^{1/\kappa},\ \kappa > 0.$$

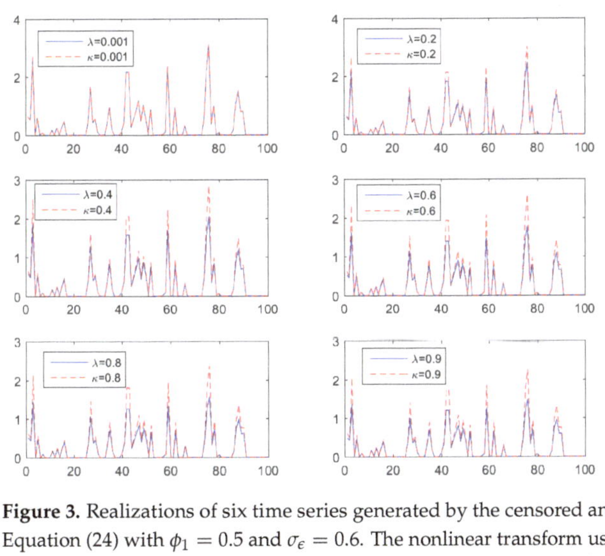

Figure 3. Realizations of six time series generated by the censored and transformed AR(1) model of Equation (24) with $\phi_1 = 0.5$ and $\sigma_\epsilon = 0.6$. The nonlinear transform uses BCT (blue, continuous lines) and KLT (red, broken lines) for $\kappa = \lambda \in \{0.001, 0.2, 0.4, 0.6, 0.8, 0.95\}$.

The increase in the relative height of KLT-based versus BCT-based peaks with increasing κ should not be confused with the fact that the peaks of x_t are highest for $\kappa = \lambda = 0$, i.e., when $h(\cdot)$ is the exponential function (herein $\kappa = 0$ implies the limit $\kappa \to 0$). This behavior is due to the inequality $\exp_\kappa(y) < \exp(y)$ for all $y \geq 0$ and for $0 < \kappa \leq 1$, e.g., [49].

4. The κ-Weibull Distribution and Its Applications

Complex systems involve collections of interacting units and often exhibit probability distributions with power-law tails [23]. A long (power-law) right tail of the PDF implies that the occurrence probability for extreme events decays slowly.

One mechanism that can lead to the emergence of long tails is due to limited size of the observation window as illustrated in Figure 4. Consider an observation domain (indicated by the square domain in the center) which is part of a larger interconnected system (denoted by the oval-shaped area). Let us assume that in the entire system, the failure events occur at times $t_1 < t_2 < t_3 < t_4$. Furthermore, assume that the interevent times $t_2 - t_1$, $t_3 - t_2$, and $t_4 - t_3$ follow a distribution which does not necessarily have a heavy tail. However, since the observed system involves only the square domain, the observed interevent times are $t_2 - t_1$ and $t_4 - t_2$; the latter results by adding the true interevent times $t_3 - t_2$ and $t_4 - t_3$, leading to a larger period of quiescence than if the entire system were taken into account. The repeated occurrence of events outside the observed domain can thus inflate waiting times and transfer probability weight from the low and middle range of the probability distribution to the right tail.

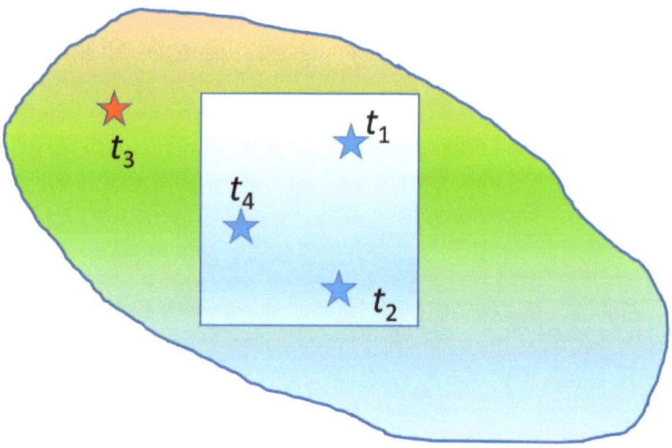

Figure 4. Schematic illustrating how long tails can emerge if the observation window (blue square) is a nested insider a larger, interacting system (see text for explanation). Blue stars indicate events inside the observation window, while the red star refers to an event outside the window.

In such cases, the classical Weibull distribution may not be adequate because it has a right tail which decays at best (i.e., for $m < 1$) as a stretched exponential. In contrast, the κ-Weibull distribution has a flexible power-law tail with exponent equal to $-m/\kappa$.

4.1. κ-Weibull Probability Functions

The Weibull distribution is most simply defined in terms of its survival function $S(x) = \exp\left[-(x/x_s)^m\right]$, for $x \geq 0$, where $x_s > 0$ is the scale parameter and $m > 0$ is the shape parameter or Weibull modulus. The Weibull distribution is used in many applications such as modeling the distributions of mechanical strength of materials [17,20,21,56], earthquake interevent times [1,2,6–8,57–61], and precipitation amounts [62–64].

The κ-Weibull distribution is a deformation of the Weibull model introduced in [39,40] to model the distribution of income in economy. The κ-Weibull has a power-law right tail which captures the observed Pareto law followed by income distributions. The κ-Weibull distribution was later applied to model the mechanical strength of materials and earthquake interevent times [43,44].

The κ-Weibull model admits explicit expressions for the main probability functions which are given by the following expressions [44]:

$$F_\kappa(x) = 1 - \exp_\kappa\left(-[x/x_s]^m\right), \tag{25a}$$

$$S_\kappa(x) = \exp_\kappa\left(-[x/x_s]^m\right), \tag{25b}$$

$$f_\kappa(x) = \frac{m}{x_s}\left(\frac{x}{x_s}\right)^{m-1} \frac{\exp_\kappa\left(-[x/x_s]^m\right)}{\sqrt{1+\kappa^2(x/x_s)^{2m}}}, \tag{25c}$$

$$h_\kappa(x) = \frac{m}{x_s}\frac{(x/x_s)^{m-1}}{\sqrt{1+\kappa^2(x/x_s)^{2m}}}, \tag{25d}$$

$$Q_\kappa(p) = \frac{1}{x_s}\left[\ln_\kappa\left(\frac{1}{1-p}\right)\right]^{1/m}. \tag{25e}$$

Note that due to the asymptotic behavior of the κ-exponential given by (14), the survival function $S_\kappa(x)$ of the κ-Weibull follows a power law with tail exponent $-m/\kappa$:

$$S_\kappa(x) \underset{x\to\infty}{\sim} 2\kappa\,(x/x_s)^{-m/\kappa}. \tag{26}$$

The power-law tail gives the κ-Weibull an advantage over the classical Weibull distribution for systems with algebraic decay of the right tail of the distribution. Figure 5 compares the tails of the survival functions for the Weibull and κ-Weibull models. Note that $S_\kappa(x)$ for $\kappa = 0.1$ is practically indistinguishable from the Weibull survival function.

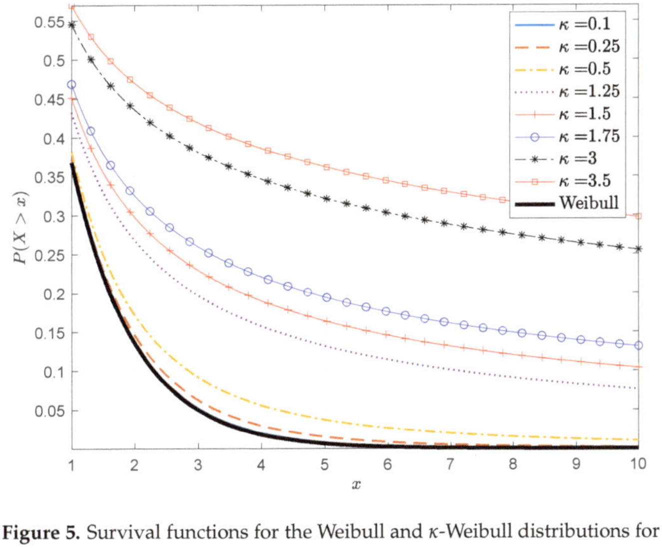

Figure 5. Survival functions for the Weibull and κ-Weibull distributions for different values of κ and $x_s = m = 1$.

4.2. Connection with Weakest-Link Scaling Theory

Weakest-link scaling theory underlies the classical Weibull distribution. Weakest-link scaling was proposed by Gumbel [65] and Weibull [66] in their works on the statistics of extreme values. This section provides (i) a brief review of weakest-link scaling in connection with the Weibull model and (ii) a demonstration that the κ-Weibull is also based on the same principle.

Weakest-link scaling treats a disordered system as a chain of critical clusters, also known as links or representative volume elements (RVEs). The term "weakest-link scaling" emphasizes the idea that the strength of an entire system is determined by the strength of its weakest link [67]. The concept of links implies the presence of critical subsystems. The failure of one such subsystem is presumably sufficient to cause failure of the entire system [21]. Thus, weakest-link scaling implies that the survival probability of the system is equal to the product of the link survival probabilities, i.e.,

$$S_{N_{\text{eff}}}(x) = \prod_{i=1}^{N_{\text{eff}}} S_1^{(i)}(x), \tag{27}$$

where N_{eff} is the number of links, $S_{N_{\text{eff}}}(x)$ is the system survival function, and $S_1^{(i)}(x)$ is the survival function of the i-th link, where $i = 1, \ldots, N_{\text{eff}}$. The above dependence of the survival function is characteristic of brittle fibers and ceramic materials [19,20,22] and justifies the use of the Weibull strength distribution. It has also been shown that if the strength of the earth's crust follows the Weibull distribution, then the latter is also justified for the distribution of recurrence times between earthquakes under the conditions specified in [18].

Assuming a uniform link survival function $S_1(x)$, the system's survival function, Equation (27), becomes

$$S_{N_{\text{eff}}}(x) = [S_1(x)]^{N_{\text{eff}}}. \tag{28}$$

The number of links can also be expressed as the ratio of the system's volume, V over the link volume, V_0, i.e., $N_{\text{eff}} = V/V_0$.

To obtain the classical Weibull distribution, the link's survival function is assumed to have the exponential form $S_1(x) = \exp[-(x/x_l)^m]$, where x_l is the link's scale parameter and $m > 0$ is the Weibull modulus or shape parameter [66]. This leads to the system survival function $S_{N_{\text{eff}}}(x) = \exp[-(x/x_s)^m]$, where $x_s = x_l/N_{\text{eff}}^{1/m}$ is the scale parameter for the entire system.

The κ-Weibull distribution can be obtained by simply replacing the exponential in the Weibull survival function with the deformed κ-exponential. However, this mathematically valid operation does not provide physical motivation for the κ-Weibull. The latter emerges in the framework of weakest-link scaling by using a modified link survival function. More precisely, let us assume that the link survival function depends on the parameter κ as shown in [43,44], i.e., that it satisfies

$$S_1(x) = \sqrt{1 + \left(\frac{x^m}{\tilde{x}_l^m}\right)^2} - \left(\frac{x}{\tilde{x}_l}\right)^m, \tag{29}$$

where $\tilde{x}_l^m = x_l^m/\kappa$. The parameter κ can be viewed as the inverse number of effective links, that is, $N_{\text{eff}} = 1/\kappa$. The link survival $S_1(x)$ thus depends on the number of links, which implies a degree of interactivity in the system. In addition, the asymptotic behavior of $S_1(x)$ for $x \to \infty$ is given by a power law, i.e., $S_1(x) \sim N_{\text{eff}}\,(2x/x_l)^{-m}$.

The link survival function for different system sizes N_{eff} is plotted in Figure 6 against the variable $z(x) = x^m/\tilde{x}_l^m$. The graphs exhibit the power-law asymptotic decline $S_1(x) \sim 1/z(x)$ as well as slower decrease of $S_1(x)$ for increasing N_{eff}.

Finally, the weakest-link scaling relation (28) in view of the link survival function (29) and the κ-exponential definition (2) leads to an $S_{N_{\text{eff}}}(x)$ which is given by the κ survival function (25b).

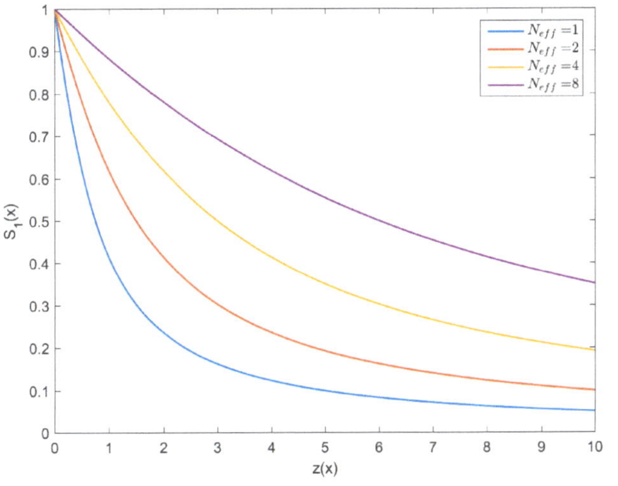

Figure 6. Link survival function for different effective system sizes. The horizontal axis denotes the variable $z(x) = x^m / \tilde{x}_l^m$. Larger values of $N_{\rm eff}$ correspond to slower decay of $S_1(x)$.

4.3. κ-Weibull Plot for Graphical Testing

If we define the function $\Phi_\kappa(x) = \ln \ln_\kappa(1/S_\kappa(x))$, it follows from Equation (25b) that $\Phi_\kappa(x) = m \ln(x/x_s)$. This relation suggests a graphical approach to test if a given dataset $\{x_n\}_{n=1}^N$ follows the κ-Weibull distribution: it suffices to test if the scatter plot of $\Phi_\kappa(x_n)$ versus $\ln x_n$ (for $n = 1, \ldots, N$) is concentrated around a straight line with slope equal to m. This property allows a quick visual test of the fit between the data and the κ-Weibull distribution which is analogous to the widely used Weibull plot [68].

The linear dependence of $\Phi_\kappa(x)$ on $\ln x$ is illustrated in the κ-Weibull plots of Figure 7. The graphs represent estimates of $\Phi_\kappa(x)$ derived from six samples of 500 random numbers; the latter are generated from the κ-Weibull distribution by using the inverse transform sampling method [43]. The samples are drawn from κ-Weibull distributions with $x_s = 10$, $m = 0.9$ and with $x_s = 10$, $m = 2$; the samples with the same m value differ with respect to κ which takes values in $\{0.1, 0.5, 0.9\}$. The function $\Phi_\kappa(x)$ is estimated by means of $\hat{\Phi}_{\hat{\kappa}}(x) = \ln \ln_{\hat{\kappa}}[1/\hat{S}(x)]$, where $\hat{S}(x)$ is the estimated survival function using the empirical staircase estimate of the CDF from the data, and $\hat{\kappa}$ is the maximum likelihood estimate of κ.

Although the use of graphical tools for estimating the tail exponent has an intuitive appeal, these tools can also be misleading. Thus, in the next section maximum likelihood estimation is used to determine the κ-Weibull parameters including the tail exponent. Statistical testing methods (e.g., Kolmororov–Smirnov test) can also be used to validate the hypothesis of a particular probability distribution model. Note that if the distribution parameters are not known a priori, but instead are estimated from the data (as is typically done in practice), Kolmororov–Smirnov testing must be implemented by using a Monte Carlo resampling approach as described in [27]. This testing approach was applied to probability models for earthquake recurrence times in [43].

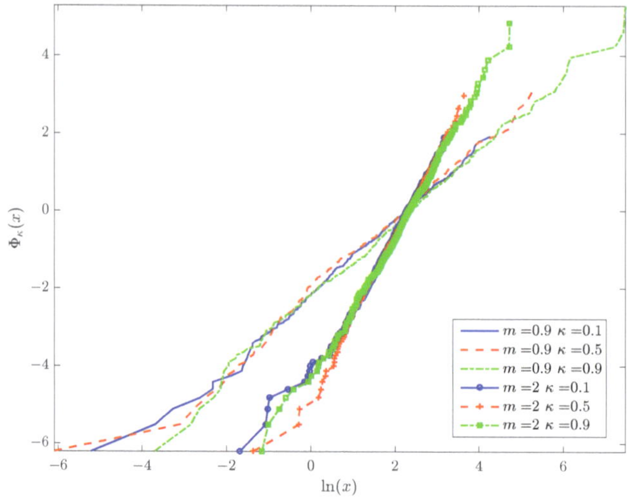

Figure 7. Plots of estimated $\Phi_\kappa(x)$ obtained from κ-Weibull synthetically generated samples with different values of the Weibull modulus m and the deformation parameter κ.

4.4. Application to Real Data

We investigate the κ-Weibull as an alternative to the Weibull distribution for different data. These include a dataset comprising measurements of tensile strength of carbon fibres [69], daily averaged wind speed in Cairo (Egypt) [70], thickness of magmatic sheet intrusions (dykes) for different tectonic settings [71], tensile strength of steel [72], and earthquake recurrence times [73,74].

The aforementioned datasets are fitted to the Weibull and κ-Weibull distributions by using maximum likelihood estimation. The Matlab code used to estimate the model parameters is publicly available [75]. The results of the fits are presented in Table 1. The entries include the maximum likelihood estimates of the model parameters as well as the optimal negative log-likelihood (NLL) values for each fit. The lower the NLL of a given distribution model is, the better its fit to a particular dataset. For all the cases listed in Table 1, the NLL is lower for the κ-Weibull than for the Weibull distribution. However, the κ-Weibull involves three parameters whereas the Weibull model involves only two parameters. To account for the different number of parameters, the selection of the optimal model can be performed by means of the Akaike information criterion (AIC) [76], i.e., $\mathrm{AIC} = 2k + 2\mathrm{NLL}$, where k is the number of free parameters for each model ($k = 2$ for the Weibull and $k = 3$ for the κ-Weibull). The best model has the lowest AIC value. Because $\mathrm{AIC}_{\kappa W} - \mathrm{AIC}_W = 2(1 + \mathrm{NLL}_{\kappa W} - \mathrm{NLL}_W)$, AIC favors the κ-Weibull only if $\mathrm{NLL}_{\kappa W} - \mathrm{NLL}_W < -1$. This condition is satisfied for all but the first two datasets.

A more stringent condition is provided by the Bayesian information criterion, $\mathrm{BIC} = k \log N + 2\mathrm{NLL}$, where N is the data size [77]. For $N > 8$, the BIC imposes a bigger penalty on model complexity than AIC. The difference in BIC values for the κ-Weibull and the Weibull models is given by $\mathrm{BIC}_{\kappa W} - \mathrm{BIC}_W = \log N + 2(\mathrm{NLL}_{\kappa W} - \mathrm{NLL}_W)$. Thus, under the BIC the κ-Weibull is optimal if $\mathrm{NLL}_{\kappa W} - \mathrm{NLL}_W < -(\log N)/2$. The BIC condition also favors the κ-Weibull distribution for the datasets 3–9. In order to allow easier comparison, the values of AIC and BIC (divided by the number of sample points) for both the Weibull and κ-Weibull distributions are listed for each dataset in Table 2.

Table 1. Results of maximum likelihood estimated fits to the Weibull and κ-Weibull distribution. 1. Tensile strength of carbon fibers. 2. Daily averaged wind speeds from 1 January 2009 to 4 October 2009 for Cairo, Egypt. 3–6. Thickness of magmatic sheet intrusions for different tectonic settings. 7. Tensile strength of low-alloy steels. 8. Recurrence times of aftershocks (A.R.T.) from 25 October 2018 until 31 May 2019, following the major M_w 6.9 Zakynthos earthquake (Greece). 9. Recurrence times of foreshocks (F.R.T.) preceding the Zakynthos earthquake (from 1 January 2014 until 25 October 2018). For more information regarding the data see the relevant sources. N, sample length; x_s, scale parameter; m, shape parameter; κ, Weibull deformation parameter. Values are rounded off to the second decimal digit. NLL, Negative log-likelihood.

		Weibull			κ-Weibull			
Data	N	x_s	m	NLL	x_s	m	κ	NLL
1. C fibers (GPa) [69]	100	2.94	2.79	141.53	2.90	2.98	0.285	141.23
2. Wind (mph) [70]	100	8.05	2.78	240.21	7.63	3.28	0.56	239.32
3. Dyrfjöll (m) [71]	487	0.90	1.26	378.05	0.84	1.46	0.42	368.40
4. Geitafell (m) [71]	546	0.57	1.02	233.88	0.52	1.17	0.43	225.62
5. Tenerife (m) [71]	550	1.83	1.02	875.18	1.65	1.18	0.45	867.38
6. La Palma (m) [71]	2093	0.43	1.14	206.51	0.37	1.53	0.66	83.98
7. Steel (MPa) [72]	915	548.58	1.98	6194.22	524.26	4.81	0.52	5753.13
8. A.R.T. (days) [74]	7822	0.027	0.94	−20207	0.024	1.19	0.49	−20374
9. F.R.T. (days) [74]	4731	0.28	0.68	−692.60	0.27	0.70	0.17	−698.37

Table 2. Measures of fit to the Weibull and κ-Weibull distributions for the datasets listed in Table 1. NLL, negative log-likelihood; AIC′, value of Akaike information criterion value per sample point, i.e., AIC′ = AIC/N = $2(k + \text{NLL})/N$; BIC′, value of Bayesian information criterion per sample point, i.e., BIC′ = BIC/N = $(k \log N + 2\,\text{NLL})/N$.

		Weibull			κ-Weibull		
Data	N	NLL	AIC′	BIC′	NLL	AIC′	BIC′
1. C fibers (GPa)	100	141.53	2.8706	2.9227	141.23	2.8846	2.9628
2. Wind (mph)	100	240.21	4.8842	4.8963	239.32	4.8464	4.9246
3. Dyrfjöll (m)	487	378.05	1.5608	1.5780	368.40	1.5253	1.5511
4. Geitafell (m)	546	233.88	0.8640	0.8798	225.62	0.8374	0.8611
5. Tenerife (m)	550	875.18	3.1897	3.2054	867.38	3.1650	3.1885
6. La Palma (m)	2093	206.51	0.1992	0.2046	83.98	0.0831	0.0912
7. Steel (MPa)	915	6194.22	13.5437	13.5542	5753.13	12.5817	12.5975
8. A.R.T. (days)	7822	−20207	−5.1662	−5.1644	−20374	−5.2086	−5.2060
9. F.R.T. (days)	4731	−692.60	−0.2919	−0.2892	−692.60	−0.2940	−0.2899

5. The κ-Lognormal Distribution

The lognormal distribution is often used to model long-tailed processes [23]. In this section we derive a generalization of the lognormal which is based on the κ-deformation of the exponential function. Let us assume that the random variable Y follows the normal distribution with marginal PDF given by

$$f_Y(y) = \frac{1}{\sqrt{2\pi}\sigma}\, \mathrm{e}^{-(y-m)^2/2\sigma^2}. \tag{30}$$

We then define the random variable $X = g_\kappa(Y)$ by means of the κ-exponential transformation

$$x = g_\kappa(y) \triangleq \exp_\kappa(y) = \left(\sqrt{1+\kappa^2 y^2} + \kappa y\right)^{1/\kappa}. \tag{31}$$

To determine the PDF of the random variable X we use the standard integral relation for the PDF under the nonlinear transformation $g_\kappa(y)$

$$f_X(x;\kappa) = \int_{-\infty}^{\infty} \delta(x - g_\kappa(y))\, f_Y(y)\, dy\,.$$

Because the nonlinear transform is monotonic and therefore invertible, the PDF of X is given by means of the following mapping [47,52]:

$$f_X(x;\kappa) = f_Y\left(y = g_\kappa^{-1}(x)\right) \left| \frac{dg_\kappa^{-1}(x)}{dx} \right|.$$

Taking into account the normal PDF of Equation (30), the inverse transform which is given by $g_\kappa^{-1}(x) = \ln_\kappa(x)$, and the first derivative of $\ln_\kappa(\cdot)$ which is given by Equation (19), it follows that the PDF of the κ-lognormal distribution is given by

$$f_X(x;\kappa) = \frac{1}{\sqrt{2\pi}\sigma\, x}\, e^{-(\ln_\kappa x - m)^2/2\sigma^2} \left(\frac{x^\kappa + x^{-\kappa}}{2} \right). \tag{32}$$

It is clear that $f_X(x;\kappa = 0)$ recovers the lognormal distribution because $\lim_{\kappa\to 0} \ln_\kappa(x) = \ln x$ and $\lim_{\kappa\to 0}\left(\frac{x^\kappa + x^{-\kappa}}{2}\right) = 1$. Moreover, the κ-lognormal PDF given by Equation (32) is symmetric with respect to κ.

The Box–Cox transform of the Gaussian PDF can be similarly obtained by means of the PDF transformation under a change of variables. The resulting PDF is given by

$$f_X(x;\lambda) = \frac{x^{\lambda-1}}{\sqrt{2\pi}\sigma}\, e^{-\left[(x^\lambda - 1)/\lambda - m\right]^2/2\sigma^2}\,. \tag{33}$$

The κ-lognormal PDF given by Equation (32) has heavier tails than the PDF in Equation (33) resulting from the Box–Cox transform of the Gaussian. This is confirmed by the parametric plots of the two parametric PDF families shown in Figure 8 for different values of $\lambda = \kappa$. Notice that for $\lambda = 0$ (left frame in Figure 8), and $\kappa = 0$ (right frame in Figure 8) the PDFs tend to the lognormal PDF.

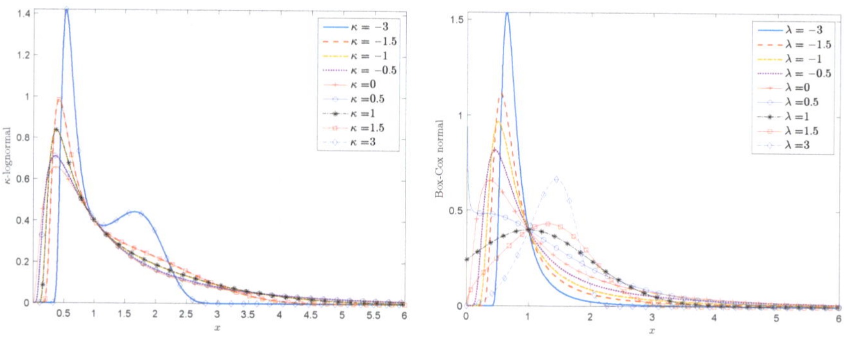

Figure 8. Probability density functions resulting from the κ-logarithmic (**left**) and Box–Cox (**right**) transformations of the standard normal distribution, given by Equations (32) and (33) respectively. The curves correspond to different values of $\lambda = \kappa$ (λ is the Box–Cox parameter and κ is the deformation parameter of the Kaniadakis logarithm).

In order to compare the tails of the κ-lognormal with the tails of the lognormal distribution, let us define the PDF ratio $R_f(x;\kappa)$:

$$R_f(x;\kappa) \triangleq \frac{f_X(x;\kappa)}{f_X(x;\kappa = 0)} = e^{-\left\{(\ln_\kappa x - m)^2 - (\ln x - m)^2\right\}/2\sigma^2} \left(\frac{x^\kappa + x^{-\kappa}}{2} \right). \tag{34}$$

First, we show that for $\kappa > 0$ and for $x > 1$ it holds that $\ln_\kappa(x) > \ln x$. Let us define the function $u(x) = \ln_\kappa(x) - \ln x$. (i) It holds that $u(1) = \ln_\kappa(1) - \ln 1 = 0$. It suffices to show that $u(x)$ is a monotonically increasing function for $x > 1$. (ii) The derivative of $u(x)$ is $\mathrm{d}u(x)/\mathrm{d}x = (x^{\kappa} + x^{-\kappa} - 2)/(2x)$, and it is positive if $x^{\kappa} + x^{-\kappa} > 2$. By multiplying both sides with x^{κ} ($x^{\kappa} > 0$) and setting $x^{\kappa} = y$, the inequality $x^{\kappa} + x^{-\kappa} > 2$ becomes equivalent to $(y-1)^2 > 0$, which is true for any $y \in \mathbb{R}$. Hence, it holds that $\lim_{x\to\infty} R_f(x;\kappa) = 0$, and thus the κ-lognormal has a lighter right tail than the lognormal distribution.

The proof of monotonicity of $u(x)$ also holds for $x < 1$: by replacing x with $x' = 1/x$ where $x' > 1$, it holds that $x^{\kappa} + x^{-\kappa} = (x')^{-\kappa} + (x')^{\kappa}$, and therefore $\mathrm{d}u(x)/\mathrm{d}x > 0$; that is $\ln_{\kappa}(x) > \ln x$ for $x < 1$. Therefore, $\lim_{x\to 0} R_f(x;\kappa) = 0$, and thus the left tail of the κ-lognormal is also lighter than the lognormal's tail. Hence, the κ-lognormal concentrates more density in the middle of the distribution than the lognormal. The relative transfer of density from the tails to the body of the distribution is controlled by κ and becomes more pronounced as κ increases. This is confirmed by the parametric plots of $R_f(x;\kappa)$ shown in Figure 9 (bottom panel) for different values of κ.

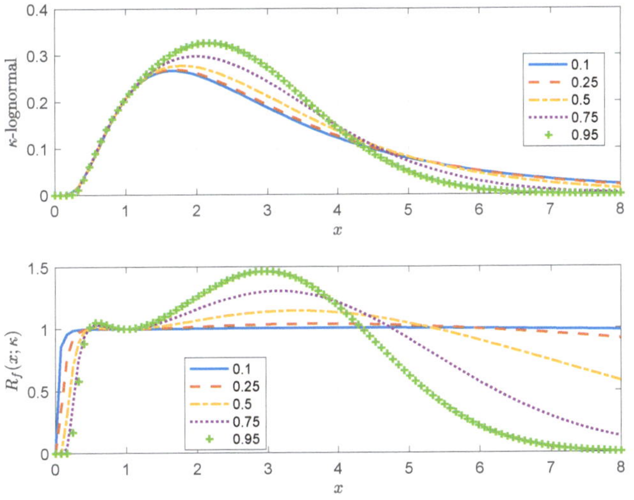

Figure 9. Parametric plots (versus x) of the κ-lognormal PDF defined in Equation (32) (**top**) and the ratio function $R_f(x;\kappa)$ defined in Equation (34); the latter compares the tails of the κ-lognormal relative to the lognormal distribution for different values of the deformation parameter κ (**bottom**).

5.1. Effective Permeability of Random Media

Single-phase, incompressible, steady-state flow in saturated random media is governed by the partial differential equation

$$\nabla \cdot K(\mathbf{s})\nabla P(\mathbf{s}) = 0, \quad \text{with suitable boundary conditions,} \tag{35}$$

where $\mathbf{s} \in \mathbb{R}^d$ is the position vector, $P(\mathbf{s})$ is the pressure, and $\nabla P(\mathbf{s})$ the pressure gradient, $\mathbf{a} \cdot \mathbf{b}$ denotes the inner product of the vectors $\mathbf{a}$ and $\mathbf{b}$, and $K(\mathbf{s})$ is the fluid permeability. The latter is assumed to be a scalar random field, i.e., a random function over the domain $D \subset \mathbb{R}^d$; (in the case of anisotropic media K becomes a tensor). Equation (35) is the continuity equation which expresses the conservation of mass.

In the case of slow viscous flow through random media, the fluid velocity is given by Darcy's law, i.e., $\mathbf{V}(\mathbf{s}) = -K(\mathbf{s})\nabla P(\mathbf{s})/\mu$ where μ is the fluid viscosity [78], and thus the continuity equation becomes $\nabla \cdot \mathbf{V}(\mathbf{s}) = 0$. The local variations of the velocity do not usually matter for the macroscopic flow behavior, i.e., for the average fluid velocity through a large domain. The macroscopic velocity is often determined in terms of an effective permeability; the latter connects the average pressure gradient to the average fluid velocity by means of $\mathcal{E}[\mathbf{V}(\mathbf{s})] = -K_{eff}\mathcal{E}[\nabla P(\mathbf{s})]$. The ensemble average here is evaluated over the joint probability distribution of $K(\mathbf{s})$, which represents the local variations (microstructure) of the medium. Similarly, effective measures can be defined for other properties (e.g., elasticity) of porous media by means of averages over the microstructural disorder [79–81].

The generalized mean $\langle K \rangle_\alpha$ with $\alpha = 1 - 2/d$, also known as the power-law mean, is used to estimate the effective flow permeability (for single-phase, saturated flow), K_{eff} of random porous media with lognormal disorder and short-range correlations [16]. For a given power-law exponent α, the generalized mean is defined by means of the following expectation (assuming that it is well-defined mathematically):

$$\langle K \rangle_\alpha \triangleq \{\mathcal{E}[K^\alpha]\}^{1/\alpha}, \quad -1 \leq \alpha \leq 1. \tag{36}$$

For $\alpha = -1$, Equation (36) yields the harmonic mean, whereas for $\alpha = 0$ the geometric mean, $K_G = \exp[\mathcal{E}(\ln K)]$, is obtained as the limit $\lim_{\alpha \to 0} \langle K \rangle_\alpha$.

Furthermore, assuming that $K = g(Y)$ where $Y \sim \mathcal{N}(m, \sigma^2)$ is a normally distributed random variable. In general, both Y and K can be random fields with spatial correlations. However, this does not affect the calculation of the generalized mean, which is a point (marginal) property. The generalized mean of $K = g(Y)$ is now given by

$$\langle K \rangle_\alpha \triangleq \{\mathcal{E}_Y[g^\alpha(Y)]\}^{1/\alpha}, \quad -1 \leq \alpha \leq 1. \tag{37}$$

Thus, if K follows the lognormal distribution, namely, $K = \exp(Y)$, it holds that

$$\langle K \rangle_\alpha = \{\mathcal{E}_Y[\exp(\alpha Y)]\}^{1/\alpha} = \mathrm{e}^{m + \frac{\alpha}{2}\sigma^2}, \quad -1 \leq \alpha \leq 1. \tag{38}$$

This equation, known as the Landau–Lifshitz–Matheron (LLM) ansatz, was first proposed for the dielectric permittivity of random dielectric mixtures [82]. In the case of electromagnetism, the continuity equation is embodied in Gauss's law; for zero free charge density the latter becomes $\nabla \cdot \mathbf{D}(\mathbf{s}) = 0$, where $\mathbf{D}(\mathbf{s}) = -\epsilon(\mathbf{s})\,\nabla\phi(\mathbf{s})$ is the dielectric displacement field, $\phi(\mathbf{s})$ is the applied electric potential, and $\epsilon(\mathbf{s})$ is the dielectric permittivity of the medium. Notation differences aside, the mathematical form of the electrostatic equations is identical to those of the fluid flow problem; in both cases the continuity principle results in an elliptic partial differential equation with suitable boundary conditions. This reason underlies LLM's applicability to both fluid permeability and dielectric permittivity; as Richard Feynman wrote ([83], Chap. 12.1): "there is a most remarkable coincidence: The equations for many different physical situations have exactly the same appearance".

For $d = 1$ (e.g., for pipe flows), LLM yields the harmonic mean $K_H = 1/\mathcal{E}[K^{-1}]$; in this case the flow is cut off if the permeability vanishes at a single point. For $d = 2$, $K_{eff} = \langle K \rangle_0$ yields the geometric mean, $K_G = \exp(\mathcal{E}[\ln K])$, which coincides with the exact solution in two dimensions [84]. Finally, in $d = 3$ the expression $\langle K \rangle_{1/3}$ follows from perturbative renormalization group analysis [16]. The LLM equation implies that the effect of disorder (as measured by the log-permeability variance σ^2) is reduced as the embedding dimension of the medium increases. The physical meaning of this dependence is that three-dimensional media include more permeable paths than one- or two-dimensional random media, thus enabling the bypass of flow bottlenecks.

5.2. Generalized Mean of the κ-Lognormal Distribution

This section focuses on the calculation of the generalized mean when K follows the κ-lognormal distribution. In this case, the respective ensemble average over the normal variable $Y \sim \mathcal{N}(m, \sigma^2)$, defined by Equation (37), is given by

$$\langle K \rangle_{\alpha;\kappa} = \left\{\mathcal{E}_Y\left[\{\exp_\kappa(Y)\}^\alpha\right]\right\}^{1/\alpha} = \left\{\mathcal{E}_Y\left[\exp_{\kappa/\alpha}(\alpha Y)\right]\right\}^{1/\alpha}, \quad -1 \leq \alpha \leq 1, \tag{39}$$

where in deriving the last equality above we used the identity

$$[\exp_\kappa(Y)]^\alpha = \exp_{\kappa/\alpha}(\alpha Y) = \exp_{\kappa'}(\alpha Y), \text{ where } \kappa' = \kappa/\alpha. \tag{40}$$

Note that for $\kappa = 0$ Equation (39) is equivalent to Equation (38) because the κ-exponential is reduced to the standard exponential.

If, on the other hand, $\kappa > 0$, it holds that $\kappa' \neq 0$. The expectation in Equation (39) can then be calculated by using the Taylor series expansion of $\exp_{\kappa/\alpha}(\alpha Y)$ around $\exp(\alpha Y)$; according to Equation (4), the expansion is given by

$$\exp_{\kappa'}(\alpha y) = \exp(\alpha y) + \sum_{n=0}^{\infty} [\xi_n(\kappa') - 1] \frac{(\alpha y)^n}{n!}, \text{ where } \kappa' = \kappa/\alpha \in \mathbb{R}, \tag{41}$$

and the polynomials $\xi_n(\cdot)$ are defined in Equation (5).

The power series in Equation (41) represents the correction of $\exp_{\kappa'}(\alpha y)$ with respect to $\exp(\alpha y)$. Based on Equation (7), it holds that $\exp_{\kappa'}(\alpha y) = \exp(\alpha y) + \mathcal{O}(\kappa'^2)$. Notice that $\xi_n(\kappa') - 1 = 0$ for $n = 0, 1, 2$, whereas $\xi_n(\kappa') - 1 < 0$ for $n = 3, 4$. Hence,

$$\exp_{\kappa'}(\alpha y) = \exp(\alpha y) + \sum_{n=3}^{\infty} [\xi_n(\kappa') - 1] \frac{(\alpha y)^n}{n!}. \tag{42}$$

Hence, the expectation of the above is given by means of the following expression:

$$\mathcal{E}_Y[\exp_{\kappa'}(\alpha Y)] = \mathcal{E}_Y[\exp(\alpha Y)] + \sum_{n=3}^{\infty} [\xi_n(\kappa') - 1] \frac{\alpha^n}{n!} \mathcal{E}_Y[Y^n], \quad \kappa' = \kappa/\alpha. \tag{43}$$

The expectation $\mathcal{E}_Y[Y^n]$ is given by means of the following expression by using the fluctuation $Y' = Y - m$ and Newton's binomial formula:

$$\mathcal{E}_Y[Y^n] = \mathcal{E}_Y\left[(m + Y')^n\right] = \sum_{j=0}^{n} \binom{n}{j} m^{n-j} \mathcal{E}_{Y'}\left[Y'^j\right].$$

The expectation over the fluctuations can be calculated by using the Wick–Isserlis theorem [52] $\mathcal{E}_{Y'}\left[Y^{2\ell}\right] = (2\ell)!\,\sigma^{2\ell}/(2^\ell \ell!)$, and $\mathcal{E}_Y\left[Y^{2\ell+1}\right] = 0$ for $\ell \in \mathbb{N}$.

Therefore the difference between the generalized mean of the κ-lognormal and the generalized mean of the lognormal (for the same value of α) is given by an infinite power series (correction factor), where $\delta_{j,2\ell}$ is the Kronecker delta, as follows:

$$\mathcal{E}_Y[\exp_{\kappa'}(\alpha Y)] = \mathcal{E}_Y[\exp(\alpha Y)] + \sum_{n=3}^{\infty} \alpha^n [\xi_n(\kappa') - 1] \sum_{j=0}^{n} m^{n-j} \frac{\sigma^j \delta_{j,2\ell}}{2^{j/2}(n-j)!(j/2)!}. \tag{44}$$

Finally, returning to Equation (39) and using Equation (44) the generalized mean of the κ-lognormal distribution is given by

$$\langle K \rangle_{\alpha;\kappa} = \langle K \rangle_\alpha \left[1 + \frac{\sum_{n=3}^{\infty} \sum_{j=0}^{n} \alpha^n [\xi_n(\kappa/\alpha) - 1] \frac{m^{n-j} \sigma^j \delta_{j,2\ell}}{2^{j/2}(n-j)!\,(j/2)!}}{\langle K \rangle_\alpha^\alpha} \right]^{1/\alpha}. \tag{45}$$

The convergence of the power series in Equation (45) should be further investigated mathematically. To gain some insight into Equation (45), consider the case $\alpha = 1$, which corresponds to the arithmetic (linear) mean. Then, the arithmetic mean of the κ-lognormal, i.e., $\overline{K}_\kappa \triangleq \langle K \rangle_{\alpha=1;\kappa}$ is given by

$$\overline{K}_\kappa = \overline{K} \left[1 + \frac{1}{\overline{K}} \sum_{n=3}^{\infty} \sum_{j=0}^{n} [\xi_n(\kappa) - 1]\, \delta_{j,2\ell} \frac{m^{n-j} \sigma^j}{2^{j/2}(n-j)!\,(j/2)!} \right]. \tag{46}$$

Hence, the arithmetic mean of the κ-lognormal is given by the standard arithmetic mean plus a correction factor which involves a double power series. Notice that when

$\kappa = 0$, according to Equation (5) it holds $\xi_n(\kappa = 0) = 1$ for all $n \in \mathbb{N}$; hence, Equation (46) recovers the arithmetic mean when $\kappa = 0$ because the double power series vanishes.

On a more practical note, a numerical calculation of the generalized mean shows that the infinite series in $\langle K \rangle_{\alpha;\kappa}$ converges for $-1 \leq \alpha \leq 1$ and $0 < \kappa < 1$. Figure 10 shows parametric plots of the generalized mean obtained by a numerical evaluation of Equation (39). The expectations (for different α and κ) are calculated by using an ensemble of 10^4 random variates drawn from the standard normal distribution, i.e., $Y \sim N(0,1)$.

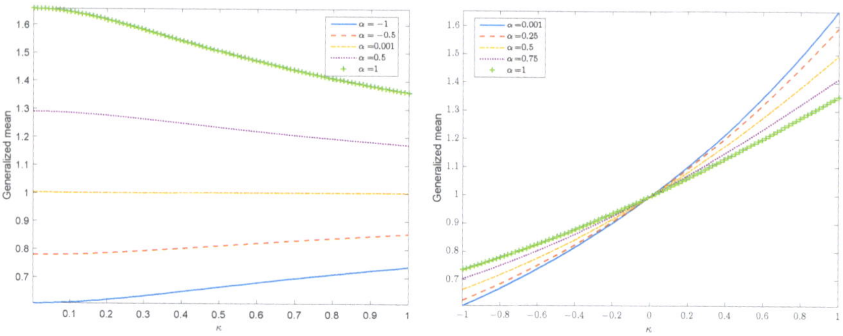

Figure 10. Parametric plots of the generalized mean versus κ for different values of the averaging exponent α (**left**) and the generalized mean versus α for different values of the deformation parameter κ (**right**).

The plots on the left of Figure 10 display the generalized mean as a function of κ for different values of the averaging exponent α. As evidenced in these plots, the difference between the harmonic mean ($\alpha = -1$) and the arithmetic mean ($\alpha = 1$) is reduced as κ increases. This behavior is due to the smaller tail weight of the κ-lognormal PDF for $\kappa \uparrow$. It is also observed that the geometric mean ($\alpha = 0$) is independent of κ (for numerical reasons we use $\alpha = 0.001$ instead of $\alpha = 0$). This is also understood in light of Equation (39). Finally, the arithmetic mean (more generally, the generalized mean for $\alpha > 0$) decays with increasing κ whereas the harmonic mean (more generally, the generalized mean for $\alpha < 0$) increases. This behavior is due to the fact that the arithmetic mean reflects the diminishing right tail of the κ-lognormal as κ increases. On the other hand, the harmonic mean is more strongly influenced by the lower part of the distribution; according to Figure 9 the κ-lognormal has higher density in the left tail (except for values very close to zero) for higher values of κ.

The plots on the right of Figure 10 display the generalized mean as a function of α for different values of κ. All the curves exhibit an increase of the generalized mean with increasing α. This reflects the progression from the harmonic to the arithmetic mean according to the well-known ordering $K_H \leq K_G \leq \overline{K}$. All the curves intersect at $\alpha = 0$, marking the independence of the generalized geometric mean on κ. Finally, the slope of the curves is reduced with increasing κ due to the respective shrinking of the tails of the κ-lognormal.

6. Discussion

Section 3 introduced and investigated the properties of a nonlinear normalizing transform which is based on the κ-logarithm. The proposal generalizes the Box–Cox transform, and it can be used for normalizing skewed data before geostatistical methods are applied. An application to precipitation time series modeling was presented. Note that the nonlinear κ-logarithm transform could also be used in spatial models of precipitation in the framework of the censored latent Gaussian field approach [10]. At fine spatiotemporal scales, the correlations of dry/wet spells as well as storm autocorrelation patterns can be better captured by means of two-state models that use copulas to simulate the dependence structure [85]. We believe that the κ-Weibull and κ-lognormal distributions discussed herein will be useful in the framework of two-state models as well, e.g., for modeling the intensity of wet spells.

Section 4 shows that the κ-Weibull model, which is a deformation of the classical Weibull with a power-law right tail, also respects the principle of weak-link scaling. Various extensions of the Weibull model have been proposed in the scientific literature, e.g., [86,87]. However, some of these models provide deformations of the Weibull expression that fail the weakest scaling equation (see Equation (28)) as pointed out by Zok [22]. Such modifications, although mathematically permissible, lack the physical justification of the classical Weibull model which is based on weakest-link scaling. Moreover, we established that the link statistical properties depend on the number of links in the system. This feature indicates a strongly interacting system; alternatively, it shows that the observed system is a part of a larger system.

The asymmetric κ-lognormal distribution introduced in Section 5 has lighter tails than the lognormal distribution. The deviation from the lognormal is controlled by the parameter κ. Smaller values of $\kappa \approx 0$ imply small deviations, whereas larger values of $\kappa \approx 1$ signify thinner tails than the lognormal. The κ-lognormal can be used as a model of fluid permeability for random porous media. We believe that the stochastic theory of single-phase, saturated fluid flow in κ-lognormal media can be derived, at least in the framework of perturbation analysis, and expressions for the effective permeability can be likewise obtained. An interesting question is how the parameter κ which controls tail behavior will impact flow properties.

We have reviewed the generalized mean for the lognormal distribution and its application in the estimation of the effective permeability of random media. We then studied the generalized mean for the κ-lognormal distribution. Note that our calculations do not prove that the effective permeability of random media with κ-lognormal disorder is given by the generalized mean. This intriguing hypothesis should be further explored in the framework of the stochastic theory of flow and transport [80].

7. Conclusions

We presented applications of the Kaniadakis κ-exponential and κ-logarithm functions in the modeling of mechanical strength and in earth science problems. In particular, we focused on κ deformations of classical distribution models such as the Weibull and the lognormal. The κ-Weibull distribution has a power-law tail which is useful for the modeling of mechanical strength, earthquake recurrence times, and properties of geological structures, among other applications. On the other hand, the κ-lognormal model has a tail lighter than the lognormal; this feature is of interest for skewed distributions which decline faster than the lognormal. The methodological applications of the Kaniadakis functions presented include the following:

- The modified Box–Cox transform given by Equation (22).
- Application of the modified Box–Cox transform to an autoregressive, intermittent model of precipitation as described in Section 3.3.
- Connection between the κ-Weibull probability model with the theory of weakest-link scaling as shown in Section 4.2.
- The study of the κ-lognormal distribution which is a generalization of the lognormal model with lighter tails. The PDF of this new model is given by Equation (32).
- The calculation of the power-mean (generalized mean) of the κ-lognormal as shown in Section 5.2.

Further study of probability models based on the deformed exponential and logarithmic functions will lead to significant advances in different fields of earth science. The most obvious applications at this time include (i) modeling the mechanical strength of technological materials and geologic media, earthquake recurrence times, wind speed, and precipitation amounts, (ii) nonlinear transforms used for Gaussian anamorphosis in geostatistical and ensemble Kalman filtering applications [88], and (iii) the permeability of random porous media.

Author Contributions: Conceptualization, D.T.H.; Formal analysis, D.T.H. and A.B.; Methodology, D.T.H. and A.B.; Software, D.T.H.; Validation, A.B.; Writing—original draft, D.T.H. and A.B.; Writing—review & editing, D.T.H. and A.B. All authors have read and agreed to the published version of the manuscript.

Funding: This research received no external funding.

Institutional Review Board Statement: Not applicable.

Informed Consent Statement: Not applicable.

Data Availability Statement: The datasets analyzed herein can be obtained from the respective publications cited in the paper. Matlab software for estimating the optimal κ-Weibull fits to the data is publicly available from [75].

Conflicts of Interest: The authors declare no conflict of interest.

Abbreviations

The following abbreviations are used in this manuscript:

AR	Autoregressive
AIC	Akaike Information Criterion
BCT	Box-Cox transform
BIC	Bayesian Information Criterion
CDF	Cumulative distribution function
KLT	κ-logarithm transform
LLM	Landau-Lifshitz-Matheron (ansatz)
NLL	Negative log-likelihood
PDF	Probability density function

References

1. Abaimov, S.G.; Turcotte, D.L.; Rundle, J.B. Recurrence-time and frequency-slip statistics of slip events on the creeping section of the San Andreas fault in central California. *Geophys. J. Int.* **2007**, *170*, 1289–1299. [CrossRef]
2. Abaimov, S.G.; Turcotte, D.; Shcherbakov, R.; Rundle, J.B.; Yakovlev, G.; Goltz, C.; Newman, W.I. Earthquakes: Recurrence and interoccurrence times. *Pure Appl. Geophys.* **2008**, *165*, 777–795. [CrossRef]
3. El Adlouni, S.; Bobée, B.; Ouarda, T.B. On the tails of extreme event distributions in hydrology. *J. Hydrol.* **2008**, *355*, 16–33. [CrossRef]
4. Akinsete, A.; Famoye, F.; Lee, C. The beta-Pareto distribution. *Statistics* **2008**, *42*, 547–563. [CrossRef]
5. Alava, M.J.; Phani, K.V.V.N.; Zapperi, S. Size effects in statistical fracture. *J. Phys. D Appl. Phys.* **2009**, *42*, 214012. [CrossRef]
6. Hagiwara, Y. Probability of earthquake occurrence as obtained from a Weibull distribution analysis of crustal strain. *Tectonophysics* **1974**, *23*, 313–318. [CrossRef]
7. Hasumi, T.; Akimoto, T.; Aizawa, Y. The Weibull-log Weibull transition of the interoccurrence time statistics in the two-dimensional Burridge-Knopoff earthquake model. *Phys. A Stat. Mech. Its Appl.* **2009**, *388*, 483–490. [CrossRef]
8. Hasumi, T.; Akimoto, T.; Aizawa, Y. The Weibull-log Weibull distribution for interoccurrence times of earthquakes. *Phys. A Stat. Mech. Its Appl.* **2009**, *388*, 491–498. [CrossRef]
9. Allard, D.; Bourotte, M. Disaggregating daily precipitations into hourly values with a transformed censored latent Gaussian process. *Stoch. Environ. Res. Risk Assess.* **2015**, *29*, 1436–3259. [CrossRef]
10. Baxevani, A.; Lennatsson, J. A spatiotemporal precipitation generator based on a censored latent Gaussian field. *Water Resour. Res.* **2015**, *51*, 4338–4358. [CrossRef]
11. Papalexiou, S.M.; Serinaldi, F. Random fields simplified: Preserving marginal distributions, correlations, and intermittency, with applications from rainfall to humidity. *Water Resour. Res.* **2020**, *56*, e2019WR026331. [CrossRef]
12. Papalexiou, S.M.; Serinaldi, F.; Porcu, E. Advancing space-time simulation of random fields: From storms to cyclones and beyond. *Water Resour. Res.* **2021**, *57*, e2020WR029466. [CrossRef]
13. Pickens, J.F.; Grisak, G.E. Scale-dependent dispersion in a stratified granular aquifer. *Water Resour. Res.* **1981**, *17*, 1191–1211. [CrossRef]
14. Sudicky, E.A. A natural gradient experiment on solute transport in a sand aquifer: Spatial variability of hydraulic conductivity and its role in the dispersion process. *Water Resour. Res.* **1986**, *22*, 2069–2082. [CrossRef]
15. Hess, K.M.; Wolf, S.H.; Celia, M.A. Large-scale natural gradient tracer test in sand and gravel, Cape Cod, Massachusetts: 3. Hydraulic conductivity variability and calculated macrodispersivities. *Water Resour. Res.* **1992**, *28*, 2011–2027. [CrossRef]

16. Hristopulos, D.T. Renormalization group methods in subsurface hydrology: Overview and applications in hydraulic conductivity upscaling. *Adv. Water Resour.* **2003**, *26*, 1279–1308. [CrossRef]
17. Amaral, P.M.J.; Cruz Fernandes, L.G.R. Weibull statistical analysis of granite bending strength. *Rock Mech. Rock Eng.* **2008**, *41*, 917–928. [CrossRef]
18. Hristopulos, D.T.; Mouslopoulou, V. Strength statistics and the distribution of earthquake interevent times. *Phys. A Stat. Mech. Its Appl.* **2013**, *392*, 485–496. [CrossRef]
19. Bazant, Z.P.; Pang, S.D. Activation energy based extreme value statistics and size effect in brittle and quasibrittle fracture. *J. Mech. Phys. Solids* **2007**, *55*, 91–131. [CrossRef]
20. Pang, S.D.; Bazant, Z.; Le, J.L. Statistics of strength of ceramics: Finite weakest-link model and necessity of zero threshold. *Int. J. Fract.* **2008**, *154*, 131–145. [CrossRef]
21. Bazant, Z.P.; Le, J.L.; Bazant, M.Z. Scaling of strength and lifetime probability distributions of quasibrittle structures based on atomistic fracture mechanics. *Proc. Natl. Acad. Sci. USA* **2009**, *1061*, 11484–11489. [CrossRef]
22. Zok, F.W. On weakest link theory and Weibull statistics. *J. Am. Ceram. Soc.* **2017**, *100*, 1265–1268. [CrossRef]
23. Sornette, D. *Critical Phenomena in Natural Sciences*, 2nd ed.; Springer: Berlin/Heidelberg, Germany, 2006.
24. Marković, D.; Gros, C. Power laws and self-organized criticality in theory and nature. *Phys. Rep.* **2014**, *536*, 41–74. [CrossRef]
25. Bak, P.; Christensen, K.; Danon, L.; Scanlon, T. Unified Scaling Law for Earthquakes. *Phys. Rev. Lett.* **2002**, *88*, 178501. [CrossRef]
26. Newman, M.E. Power laws, Pareto distributions and Zipf's law. *Contemp. Phys.* **2005**, *46*, 323–351. [CrossRef]
27. Clauset, A.; Shalizi, C.; Newman, M. Power-law distributions in empirical data. *SIAM Rev.* **2009**, *51*, 661–703. [CrossRef]
28. Kaniadakis, G. Maximum entropy principle and power-law tailed distributions. *Eur. Phys. J. B* **2009**, *70*, 3–13. [CrossRef]
29. Siegenfeld, A.F.; Bar-Yam, Y. An introduction to complex systems science and its applications. *Complexity* **2020**, *2020*, 6105872. [CrossRef]
30. Taleb, N.N. Statistical consequences of fat tails: Real world preasymptotics, epistemology, and applications. *arXiv* **2020**, arXiv:2001.10488.
31. Kaniadakis, G. H-theorem and generalized entropies within the framework of nonlinear kinetics. *Phys. Lett. A* **2001**, *288*, 283–291. [CrossRef]
32. Kaniadakis, G. Non-linear kinetics underlying generalized statistics. *Phys. A Stat. Mech. Its Appl.* **2001**, *296*, 405–425. [CrossRef]
33. Kaniadakis, G. Statistical mechanics in the context of special relativity. *Phys. Rev. E* **2002**, *66*, 056125. [CrossRef]
34. Kaniadakis, G. Statistical mechanics in the context of special relativity II. *Phys. Rev. E* **2005**, *72*, 036108. [CrossRef]
35. Kaniadakis, G. New power-law tailed distributions emerging in κ-statistics. *EPL (Europhys. Lett.)* **2021**, *133*, 10002. [CrossRef]
36. Leubner, M.P. A nonextensive entropy approach to kappa-distributions. *Astrophys. Space Sci.* **2002**, *282*, 573–579. [CrossRef]
37. Pierrard, V.; Lazar, M. Kappa distributions: Theory and applications in space plasmas. *Sol. Phys.* **2010**, *267*, 153–174. [CrossRef]
38. Klessen, R.; Burkert, A. The formation of stellar clusters: Gaussian cloud conditions. I. *Astrophys. J. Suppl. Ser.* **2000**, *128*, 287–319. [CrossRef]
39. Clementi, F.; Gallegati, M.; Kaniadakis, G. κ-generalized statistics in personal income distribution. *Eur. Phys. J. B* **2007**, *57*, 187–193. [CrossRef]
40. Clementi, F.; Di Matteo, T.; Gallegati, M.; Kaniadakis, G. The κ-generalized distribution: A new descriptive model for the size distribution of incomes. *Physica* **2008**, *387*, 3201–3208. [CrossRef]
41. Clementi, F.; Gallegati, M.; Kaniadakis, G. A κ-generalized statistical mechanics approach to income analysis. *J. Stat. Mech. Theory Exp.* **2009**, P02037.[CrossRef]
42. Kaniadakis, G.; Baldi, M.M.; Deisboeck, T.S.; Grisolia, G.; Hristopulos, D.T.; Scarfone, A.M.; Sparavigna, A.; Wada, T.; Lucia, U. The κ-statistics approach to epidemiology. *Sci. Rep.* **2020**, *10*, 19949. [CrossRef]
43. Hristopulos, T.D..; Petrakis, M.; Kaniadakis, G. Finite-size effects on return interval distributions for weakest-link-scaling systems. *Phys. Rev. E* **2014**, *89*, 052142. [CrossRef]
44. Hristopulos, D.T.; Petrakis, M.P.; Kaniadakis, G. Weakest-link scaling and extreme events in finite-sized systems. *Entropy* **2015**, *17*, 1103–1122. [CrossRef]
45. Nerantzaki, S.D.; Papalexiou, S.M. Tails of extremes: Advancing a graphical method and harnessing big data to assess precipitation extremes. *Adv. Water Resour.* **2019**, *134*, 103448. [CrossRef]
46. Box, G.E.P.; Cox, D.R. An analysis of transformations. *J. R. Stat. Soc. Ser. B (Methodol.)* **1964**, *26*, 211–243. [CrossRef]
47. Papoulis, A.; Pillai, S.U. *Probability Random Variables and Stochastic Processes*, 4th ed.; McGraw Hill: Boston, MA, USA, 2002.
48. Anagnos, T.; Kiremidjian, A.S. A review of earthquake occurrence models for seismic hazard analysis. *Probabilistic Eng. Mech.* **1988**, *3*, 3–11. [CrossRef]
49. Kaniadakis, G. Theoretical foundations and mathematical formalism of the power-law tailed statistical distributions. *Entropy* **2013**, *15*, 3983–4010. [CrossRef]
50. Wackernagel, H. *Multivariate Geostatistics*; Springer: Berlin/Heidelberg, Germany, 2003.
51. Chilès, J.P.; Delfiner, P. *Geostatistics: Modeling Spatial Uncertainty*, 2nd ed.; John Wiley & Sons: New York, NY, USA, 2012.
52. Hristopulos, D.T. *Random Fields for Spatial Data Modeling: A Primer for Scientists and Engineers*; Springer: Dordrecht, The Netherlands, 2020. [CrossRef]
53. Yeo, I.K.; Johnson, R.A. A new family of power transformations to improve normality or symmetry. *Biometrika* **2000**, *87*, 954–959. [CrossRef]

54. Edwards, S.F.; Anderson, P.W. Theory of spin glasses. *J. Phys. F Met. Phys.* **1975**, *5*, 965–974. [CrossRef]
55. Hannachi, A. Intermittency, autoregression and censoring: A first-order AR model for daily precipitation. *Meteorol. Appl.* **2014**, *21*, 384–397. [CrossRef]
56. Hristopulos, D.T.; Uesaka, T. Structural disorder effects on the tensile strength distribution of heterogeneous brittle materials with emphasis on fiber networks. *Phys. Rev. B* **2004**, *70*, 064108. [CrossRef]
57. Rikitake, T. Recurrence of great earthquakes at subduction zones. *Tectonophysics* **1976**, *35*, 335–362. [CrossRef]
58. Rikitake, T. Assessment of earthquake hazard in the Tokyo area, Japan. *Tectonophysics* **1991**, *199*, 121–131. [CrossRef]
59. Sieh, K.; Stuiver, M.; Brillinger, D. A more precise chronology of earthquakes produced by the San Andreas fault in Southern California. *J. Geophys. Res.* **1989**, *94*, 603–623. [CrossRef]
60. Yakovlev, G.; Turcotte, D.L.; Rundle, J.B.; Rundle, P.B. Simulation-based distributions of earthquake recurrence times on the San Andreas fault system. *Bull. Seismol. Soc. Am.* **2006**, *96*, 1995–2007. [CrossRef]
61. Holliday, J.R.; Rundle, J.B.; Turcotte, D.L.; Klein, W.; Tiampo, K.F.; Donnellan, A. Space-Time clustering and correlations of major earthquakes. *Phys. Rev. Lett.* **2006**, *97*, 238501. [CrossRef]
62. Wilks, D.S. Rainfall intensity, the Weibull distribution, and estimation of daily surface runoff. *J. Appl. Meteorol. Climatol.* **1989**, *28*, 52–58. [CrossRef]
63. Selker, J.S.; Haith, D.A. Development and testing of single-parameter precipitation distributions. *Water Resour. Res.* **1990**, *26*, 2733–2740. [CrossRef]
64. Papalexiou, S.M.; AghaKouchak, A.; Foufoula-Georgiou, E. A diagnostic framework for understanding climatology of tails of hourly precipitation extremes in the United States. *Water Resour. Res.* **2018**, *54*, 6725–6738. [CrossRef]
65. Gumbel, E.J. Les valeurs extrêmes des distributions statistiques. *Ann. De L'Institut Henri Poincaré* **1935**, *5*, 115–158.
66. Weibull, W. A statistical distribution function of wide applicability. *J. Appl. Mech.* **1951**, *18*, 293–297. [CrossRef]
67. Chakrabarti, B.K.; Benguigui, L.G. *Statistical Physics of Fracture and Breakdown in Disordered Systems*; Oxford University Press: New York, NY, USA, 1997.
68. Chambers, J.M.; Cleveland, W.S.; Kleiner, B.; Tukey, P.A. *Graphical Methods for Data Analysis*; CRC Press: Boca Raton, FL, USA, 2018.
69. Nichols, M.D.; Padgett, W. A bootstrap control chart for Weibull percentiles. *Qual. Reliab. Eng. Int.* **2006**, *22*, 141–151. [CrossRef]
70. Aslam, M. Testing average wind speed using sampling plan for Weibull distribution under indeterminacy. *Sci. Rep.* **2021**, *11*, 7532. [CrossRef]
71. Krumbholz, M.; Hieronymus, C.F.; Burchardt, S.; Troll, V.R.; Tanner, D.C.; Friese, N. Weibull-distributed dyke thickness reflects probabilistic character of host-rock strength. *Nat. Commun.* **2014**, *5*, 3272. [CrossRef]
72. MatNavi Mechanical Properties of Low Alloy Steels. Available online: https://www.kaggle.com/datasets/rohannemade/mechanical-properties-of-low-alloy-steels (accessed on 8 June 2020).
73. Mouslopoulou, V.; Bocchini, G.M.; Cesca, S.; Saltogianni, V.; Bedford, J.; Petersen, G.; Gianniou, M.; Oncken, O. Earthquake Swarms, Slow Slip and Fault Interactions at the Western-End of the Hellenic Subduction System Precede the Mw 6.9 Zakynthos Earthquake, Greece. *Geochem. Geophys. Geosyst.* **2020**, *21*, e2020GC009243. [CrossRef]
74. Mouslopoulou, V.; Bocchini, G.M.; Cesca, S.; Saltogianni, V.; Bedford, J.; Petersen, G.; Gianniou, M.; Oncken, O. Datasets for "Earthquake Swarms, Slow Slip and Fault Interactions at the Western-End of the Hellenic Subduction System Precede the Mw 6.9 Zakynthos Earthquake, Greece". *Zenodo* **2020**. [CrossRef]
75. Hristopulos, D. Matlab code for estimating the parameters of the kappa-Weibull distribution. *Zenodo* **2022**. [CrossRef]
76. Akaike, H. A new look at the statistical model identification. *IEEE Trans. Autom. Control* **1974**, *19*, 716–723. [CrossRef]
77. Schwarz, G. Estimating the Dimension of a Model. *Ann. Stat.* **1978**, *6*, 461–464. [CrossRef]
78. Scheidegger, A.E. *The Physics of Flow through Porous Media*, 3rd ed.; University of Toronto Press: Toronto, ON, Canada, 1974.
79. Torquato, S. Macroscopic behavior of random media from the microstructure. *Appl. Mech. Rev.* **1994**, *47*, S29–S37. [CrossRef]
80. Dagan, G. Stochastic modeling of flow and transport: The broad perspective. In *Subsurface Flow and Transport: A Stochastic Approach*; Dagan, G., Neuman, S.P., Eds.; Cambridge University Press: Cambridge, UK, 1997.
81. Torquato, S. *Random Heterogeneous Materials: Microstructure and Macroscopic Properties*; Springer: New York, NY, USA, 2002.
82. Landau, L.D.; Lifshitz, E.M.; Pitaevskii, L.P. *Electrodynamics of Continuous Media*; Course on Theoretical Physics; Pergamon Press: Oxford, UK, 1984; Volume 8.
83. Feynman, R.P.; Leighton, R.B.; Sands, M. *Lectures in Physics, Electromagnetism and Matter*; The New Millenium Edition; Basic Books; Perseus Books Group: Ney York, NY, USA, 2010; Volume 2.
84. Dykhne, A. Conductivity of a two-dimensional two-phase system. *Sov. Phys. JETP* **1971**, *32*, 63–65.
85. Papalexiou, S.M. Rainfall Generation Revisited: Introducing CoSMoS-2s and Advancing Copula-Based Intermittent Time Series Modeling. *Water Resour. Res.* **2022**, *58*, e2021WR031641. [CrossRef]
86. Lee, C.; Famoye, F.; Olumolade, O. Beta-Weibull distribution: Some properties and applications to censored data. *J. Mod. Appl. Stat. Methods* **2007**, *6*, 17. [CrossRef]
87. Alzaatreh, A.; Famoye, F.; Lee, C. Weibull-Pareto distribution and its applications. *Commun. Stat. -Theory Methods* **2013**, *42*, 1673–1691. [CrossRef]
88. Grooms, I. A comparison of nonlinear extensions to the ensemble Kalman filter. *Comput. Geosci.* **2022**, *26*, 633–650. [CrossRef]

entropy

MDPI

Article

Doppler Broadening of Neutron Cross-Sections Using Kaniadakis Entropy

Willian Vieira de Abreu *, João Márcio Maciel, Aquilino Senra Martinez, Alessandro da Cruz Gonçalves and Lucas Schmidt

Instituto Alberto Luiz Coimbra de Pós-Graduação e Pesquisa em Engenharia (COPPE/UFRJ), Programa de Engenharia Nuclear (PEN), Universidade Federal do Rio de Janeiro, Rio de Janeiro 21941-914, RJ, Brazil
* Correspondence: wabreu@coppe.ufrj.br; Tel.: +55-21992455819

Abstract: In the last seven years, Kaniadakis statistics, or κ-statistics, have been applied in reactor physics to obtain generalized nuclear data, which can encompass, for instance, situations that lie outside thermal equilibrium. In this sense, numerical and analytical solutions were developed for the Doppler broadening function using the κ-statistics. However, the accuracy and robustness of the developed solutions contemplating the κ distribution can only be appropriately verified if applied inside an official nuclear data processing code to calculate neutron cross-sections. Hence, the present work inserts an analytical solution for the deformed Doppler broadening cross-section inside the nuclear data processing code FRENDY, developed by the Japan Atomic Energy Agency. To do that, we applied a new computational method called the Faddeeva package, developed by MIT, to calculate error functions present in the analytical function. With this deformed solution inserted in the code, we were able to calculate, for the first time, deformed radiative capture cross-section data for four different nuclides. The usage of the Faddeeva package brought more accurate results when compared to other standard packages, reducing the percentage errors in the tail zone in relation to the numerical solution. The deformed cross-section data agreed with the expected behavior compared to the Maxwell–Boltzmann.

Keywords: Kaniadakis; κ-statistics; neutron cross-section; Doppler broadening function; Faddeeva function

Citation: de Abreu, W.V.; Maciel, J.M.; Martinez, A.S.; Gonçalves, A.d.C.; Schmidt, L. Doppler Broadening of Neutron Cross-Sections Using Kaniadakis Entropy. *Entropy* **2022**, *24*, 1437. https://doi.org/10.3390/e24101437

Academic Editors: Dionissios T. Hristopulos, Sergio Luiz E. F. da Silva and Antonio M. Scarfone

Received: 19 September 2022
Accepted: 4 October 2022
Published: 9 October 2022

1. Introduction

Over the last 20 years, the Kaniadakis entropy [1] and its power-law tailed statistical distributions have been applied in many different fields, such as finance [2], astrophysics [3–6], game theoretical equilibrium [7], gravitational physics [8,9], dusty plasma [10] and so many others.

In nuclear reactor physics, 2015 marked the first idealization of applying the κ-deformed statistics, intending to describe situations in non-thermal equilibrium inside a nuclear reactor, with the first article on this being published in 2017 [11]. The Doppler broadening function is utilized to represent the thermal nuclear movement. This function is commonly considered with a medium in thermal equilibrium with a temperature of T and using the Maxwell–Boltzmann distribution to describe the random velocities of the nuclei. However, to comprehend situations that lie outside the thermal equilibrium, Guedes et al. [11] proposed a very new expression for a deformed Doppler broadening function considering the Kaniadakis statistics:

$$\psi_\kappa(\xi,x) = \frac{\xi}{2\sqrt{\pi}}B(\kappa)\int_{-\infty}^{+\infty}\frac{1}{1+y^2}i\exp_\kappa\left[\frac{-\xi^2(x-y)^2}{4}\right]dy, \tag{1}$$

where,

$$x \equiv \frac{2}{\Gamma}(E - E_0); \tag{2}$$

$$\xi \equiv \frac{\Gamma}{\left(\frac{4E_0 K_B T}{A}\right)^{\frac{1}{2}}}; \tag{3}$$

κ is a deviation parameter that measures the deviation concerning the Maxwell–Boltzmann distribution [1,12], k_B is the Boltzmann constant, E_0 is the resonant energy, E is the energy of the incident neutron, A is the mass number, and Γ is the total width of the resonance as measured in the laboratory coordinates. Furthermore,

$$y \equiv \frac{2}{\Gamma}(E_{CM} - E_0); \tag{4}$$

$$B(\kappa) = (2|\kappa|)^{\frac{3}{2}}\left(1 + \frac{1}{2}3|\kappa|\right)\frac{\Gamma\left(\frac{1}{2|\kappa|} + \frac{3}{4}\right)}{\Gamma\left(\frac{1}{2|\kappa|} - \frac{3}{4}\right)}. \tag{5}$$

$$i\exp_\kappa(z) \equiv \left(\frac{\sqrt{1 + \kappa^2 z^2} - \kappa^2 z}{1 - \kappa^2}\right)\exp_\kappa(z); \tag{6}$$

$$z = \frac{-\xi^2 (x - y)^2}{4} \tag{7}$$

and $\exp_\kappa$ is the deformed exponential function, first introduced by Kaniadakis [1]:

$$\exp_\kappa(x) \equiv \left(\sqrt{1 + \kappa^2 x^2} + \kappa x\right)^{\frac{1}{\kappa}}. \tag{8}$$

However, the numerical calculation of Equation (1) can represent a considerable additional amount of computer processing time, especially when inserted in nuclear data processing codes. In order to surpass this issue, Abreu et al. [13] proposed an analytical solution based on obtaining a differential equation and its solution to represent the deformed Doppler broadening function using the Kaniadakis distribution [14]. This analytical solution proved to be up to five times faster than the numerical one [14]. Analytical solutions were also successfully applied in order to obtain faster methods for the Doppler broadening function considering the standard Maxwell-Boltzmann statistics [15] and Tsallis statistics [16].

The validation of the applicability of the Kaniadakis statistic can be performed in other areas through observational data, e.g., cosmic ray flux [12], stellar-residual-radial-velocities [6] and Stellar rotational velocities [4]. However, this kind of approach is not directly applicable to nuclear reactor physics, given the impossibility of observing and measuring the distribution of relative velocities between neutrons and nuclei in a nuclear reactor. Therefore, one of the possible ways to validate the use of κ-statistics is through numerical simulations, similar to in other scientific topics, e.g., relativistic plasmas under the effect of wave-particle interactions [3], non-extensive random matrix theories [17] and Jeans instability of self-gravitating systems [18].

Nevertheless, the accuracy and robustness of the developed solutions contemplating the κ distribution can only be appropriately verified if it is applied in a nuclear data processing code, e.g., FRENDY [19,20], NJOY [21], PREPRO [22] and NECP-Atlas [23]. These systems can process official evaluated nuclear data libraries such as ENDF [24], CENDL [25], JEFF [26] or JENDL [27]. Until the present work, the only results that have been presented for neutron cross-sections considering the Kaniadakis distribution were calculated without doing this [28,29].

Thus, this study's purpose is to calculate deformed neutron cross-sections of radioactive capture—in the resolved region—for the first time using the Kaniadakis distribution through an analytical solution inside a nuclear data processing code: the FRENDY.

Additionally, this work aims to apply an alternative numerical methodology to calculate the challenging error functions (with complex arguments) present in the analytical solution of the deformed Doppler broadening function.

2. Methodology

The Japan Atomic Energy Agency (JAEA) developed the nuclear data processing code FRENDY (From Evaluated Nuclear Data Library to any application) to treat the most recent nuclear data format, such as the evaluated nuclear library JENDL [27], also developed by JAEA. It was built using the object-oriented language C++ because of its modularity, maintainability, flexibility and portability [30].

Moreover, the FRENDY also intends to work in the future with the recent nuclear data format Generalized Nuclear Data Structure (GNDS), which the current processing codes cannot treat without a considerable amount of format revision [30].

To calculate neutron cross-sections in reactor physics, one can use different formalisms such as single-level Breit-Wigner (SLBW), Multi-level Breit-Wigner (MLBW), Adler–Adler and Reich-Moore [30]. To develop the integral formulation for the deformed Doppler broadening function, Equation (1), Guedes et al. [11] used the Single-level one, even though it is not the most recent method. According to the authors, that choice was made because it is easy to implement, it can use published resonance parameters, and it can be Doppler-broadened analytically. It also can be used in reactor physics calculations [11]. Furthermore, the SLBW is the only representation available, for instance, for the ENDF-6 format in the unresolved region [30,31].

2.1. Calculating Standard Neutron Cross-Sections with FRENDY

The FRENDY code, by default, uses the Kernel broadening method to more accurately calculate the Doppler-broadened cross-sections in the resolved resonance considering the standard Maxwell-Boltzmann distribution [30]. However, this method demands high computational effort, increasing computational time [30]. Considering this and the fact that the deformed Doppler broadening function using the Kaniadakis statistics adds an extra level of complexity and, consequently, higher computational times, the present work aims to use the single-level Breit-Wigner to calculate the deformed neutron cross-sections in the resolved region.

One of the advantages of the SLBW method is the possibility of using the $\psi - \chi$ method. Through this method, one can represent the standard radiative neutron cross-sections by [30]:

$$\sigma_\gamma(E,T) = \frac{4\pi}{k^2} \sum_j g_J \sum_r \Gamma_{\mathrm{nr}} \frac{\Gamma_{\gamma\mathrm{r}}}{{\Gamma_\mathrm{r}}^2} \psi(\xi, x), \tag{9}$$

σ_γ is the radiative capture cross-section, Γ_r the total width, $\Gamma_{\gamma\mathrm{r}}$ the radiative capture width, Γ_{nr} the neutron widths, k the neutron wave number, g_J a spin statistical factor and E an incident neutron energy.

To calculate the standard Doppler broadening function, $\psi(\xi, x)$, inside the FRENDY, Tada, Kunieda and Nagaya [30] adopted the four-pole Padé approximation to reduce the calculation time. By using this method, the expression for the standard Doppler broadening function is represented by:

$$\psi(s,x) \cong \frac{\xi\sqrt{\pi}}{2} Re[w(z)], \tag{10}$$

where,

$$\mathrm{s} = \frac{\xi}{2}(x-y). \tag{11}$$

$w(z)$ represents a scaled complex complementary error function, commonly known as the Fadeeva or Krump function [32]. It is defined by [33]:

$$w(z) = e^{-z^2}[1 - erf(-iz)] \tag{12}$$

$$erf(z) \equiv \frac{2}{\sqrt{\pi}} \int_0^z e^{-t^2} dt, \tag{13}$$

$$\mathrm{z} = \mathrm{u} + \mathrm{ih} \tag{14}$$

$$\mathrm{u} = (\xi/2) \cdot x \tag{15}$$

$$\mathrm{h} = \xi/2 \tag{16}$$

2.2. Calculating Deformed Neutron Cross-Sections with FRENDY

As mentioned earlier, the analytical solution of the Doppler broadening function using the Kaniadakis statistics, proposed by Abreu et al. [13] is obtained through a differential equation and its respective resolution [14], given by:

$$\psi_k(\xi, x) = \Lambda(x, \xi)\left[D(\xi, x) + \Omega_g(\xi, x)\right], \tag{17}$$

where,

$$\Lambda(\xi, x) = \exp\left(\frac{\xi^2 - \xi^2 x^2}{4}\right) \cdot \frac{\xi\sqrt{\pi}B(\kappa)}{4}; \tag{18}$$

$$D(\xi, x) \equiv [\Delta(\xi) \cdot cos(\Theta)]; \tag{19}$$

$$\Omega_g(\xi, x) \equiv \Pi(x, \xi) \cdot [i\Omega_1(\xi, x) + \Omega_2(\xi, x)]; \tag{20}$$

$$\Delta(\xi) = \frac{2 - 2erf\left(\frac{\xi}{2}\right)}{1 - \kappa^2}. \tag{21}$$

$$\Pi(\xi, x) = \frac{\sqrt{\xi^4 - 2\xi^2\kappa^2}}{-\xi^2 + 2\kappa^2} \cdot \exp\left(\frac{-\kappa^2}{2}\right); \tag{22}$$

$$\Omega_1(\xi, x) = sin(\Theta) \cdot \left[erf(P_1)\kappa^2 - erf(P_1) + erf(P_2)\kappa^2 - erf(P_2)\right]; \tag{23}$$

$$\Omega_2(\xi, x) = cos(\Theta) \cdot \left[2erf(P_3)\kappa^2 - 2erf(P_3) - erf(P_1)\kappa^2 + erf(P_1) + erf(P_2)\kappa^2 - erf(P_2)\right]; \tag{24}$$

$$P_1(\xi, x) = \frac{-i\xi^2 x + \sqrt{\xi^4 - 2\xi^2\kappa^2}}{2\xi}; \tag{25}$$

$$P_2(\xi, x) = \frac{-i\xi^2 x - \sqrt{\xi^4 - 2\xi^2\kappa^2}}{2\xi}; \tag{26}$$

$$P_3(\xi, x) = \frac{\sqrt{\xi^4 - 2\xi^2\kappa^2}}{2\xi}; \tag{27}$$

$$\Theta(\xi, x) = \frac{x}{2}\sqrt{\xi^4 - 2\xi^2\kappa^2}; \tag{28}$$

To calculate the deformed cross-sections using the Kaniadakis distribution, one needs to substitute the real part of the Faddeeva function for the analytical solution, Equation (17), in the definition of cross-sections in the code, represented by Equation (9), so that:

$$\sigma_\gamma(E, T) = \frac{4\pi}{k^2} \sum_j g_J \sum_r \Gamma_{\mathrm{nr}} \frac{\Gamma_{\gamma\mathrm{r}}}{\Gamma_{\mathrm{r}}^2} \psi_\kappa(\xi, x), \tag{29}$$

One of the main challenges in calculating the deformed analytical solution considering the Kaniadakis distribution is calculating the error functions with complex arguments represented in Equations (23) and (24).

The so-called Gaussian error function, $erf(x)$, is defined as follows [33]:

$$erf(x) = \frac{2}{\sqrt{\pi}} \int_0^x e^{-t^2} dt. \tag{30}$$

Though these error functions only mean relevance in the tails of the cross-section curves—far from the resonance peak—it is of great significance to implement suitable methodologies to elevate this region's precision. These regions—due to lower absolute values—usually present the most significant percentual errors.

The previous works that used the analytical solution implemented the default error functions present in the "special" module of the "scipy" library (Disponible in: https://docs.scipy.org/doc/scipy/reference/special.html accessed on 7 July 2022). However, one cannot find until the date of publishing this manuscript a similar module inside C++; i.e., there is not a unit that directly calculates error functions with complex arguments.

Therefore, the present paper implemented a new methodology to calculate these complex error functions to overcome this problem. The chosen method was the Faddeeva method, developed by Steven Johnson [34]. This methodology has the advantage of using different algorithms to calculate the erf function, Equation (26), according to the value of z. For sufficiently large values of $|z|$, the package uses a continued-fraction expansion for w(z), analogous to those described by Gautschi [35] and Pope and Wijers [36]. Meanwhile, for smaller values of $|z|$ or for z close to the real axis [34], Johnson used the algorithm 916, developed by Zaghloul and Ali [34]. According to Johnson, "algorithm 916 is competitive and faster for smaller values of $|z|$ and also has better relative accuracy in Re[z] for some regions near the real-z-axis" [34].

In fact, by using the Faddeeva method to calculate the complex error functions inside the deformed analytical Doppler broadening function, $\psi_\kappa(\xi, x)$, presented more accurate results in the tail region. In the next section, these results will be shown.

After conducting this modification, we used the FRENDY to calculate the deformed radiative capture neutron cross-section. Initially, we calculated the deformed cross-sections using the same adopted range of energy in the JENDL-4.0 library [27]. After that, two different resonance peaks were selected in order to compare with the results considering the Maxwell–Boltzmann distribution. Four nuclides in JENDL-4.0 are considered. The major calculation conditions are summarized as follows:

- Method: single-level Breit-Wigner;
- Nuclides: Pu238, Tc99, Gd155 and Gd157;
- Temperatures (K): 1500, 2000 and 2500;
- Maximum number of points (h_max): 10,000;
- Range of energy: 10^{-2} to 10^7 eV;
- Deformation in relation to the MB distribution: κ = 0.1.

3. Results and Discussion

Calculating the numerical deformed Doppler broadening function using the Kaniadakis entropy can be very computationally costly. In fact, in a recently published paper [14], the analytical solution provided by Equation (9) was approximately 4.6 times faster than the numerical one.

However, the analytical solution represented by Equation (9) presents higher values of percentual errors to the curve's tail, far from the resonance peak. These higher values are linked to the small values in these regions and the fact that the error functions, including those for P1 and P2, Equations (25) and (26), tend to present a more significant influence in these specific regions, as one can see in Figures 1 and 2.

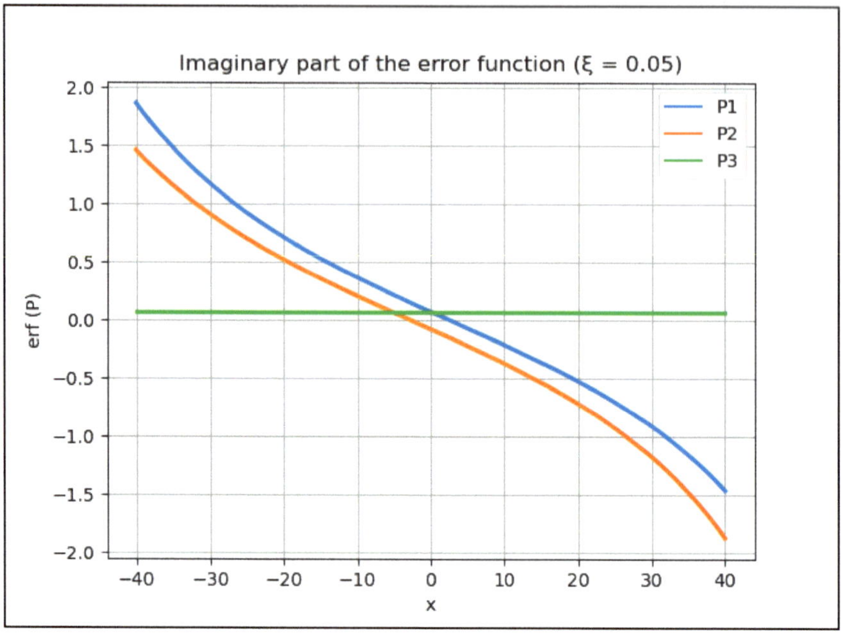

(a)

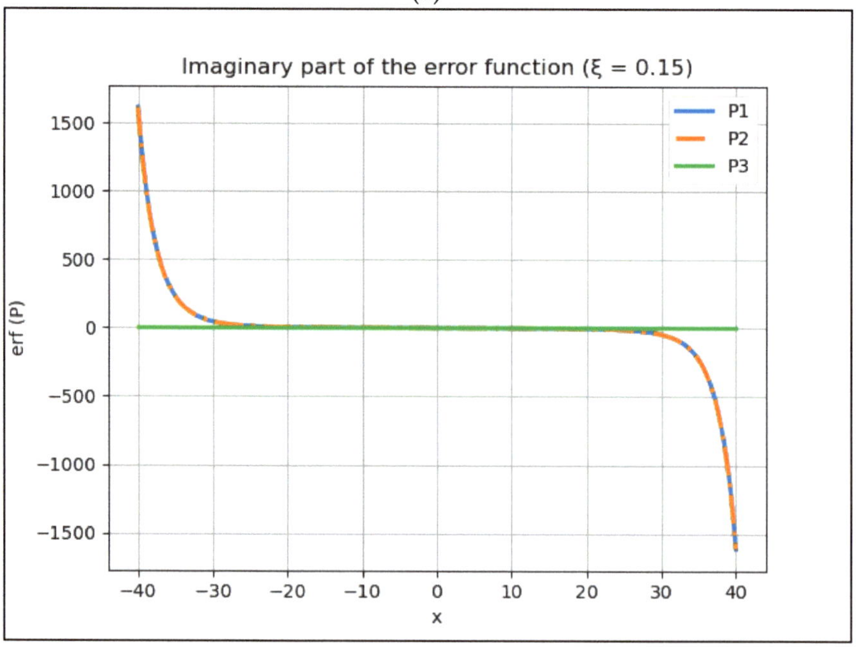

(b)

Figure 1. The imaginary part of the error function for P1, P2 and P3 considering two different values of ξ: (**a**) = 0.05 and (**b**) = 0.15. The real part of these error functions is close to zero ($\cong 10^{-18}$) and, consequently, presents little relevance.

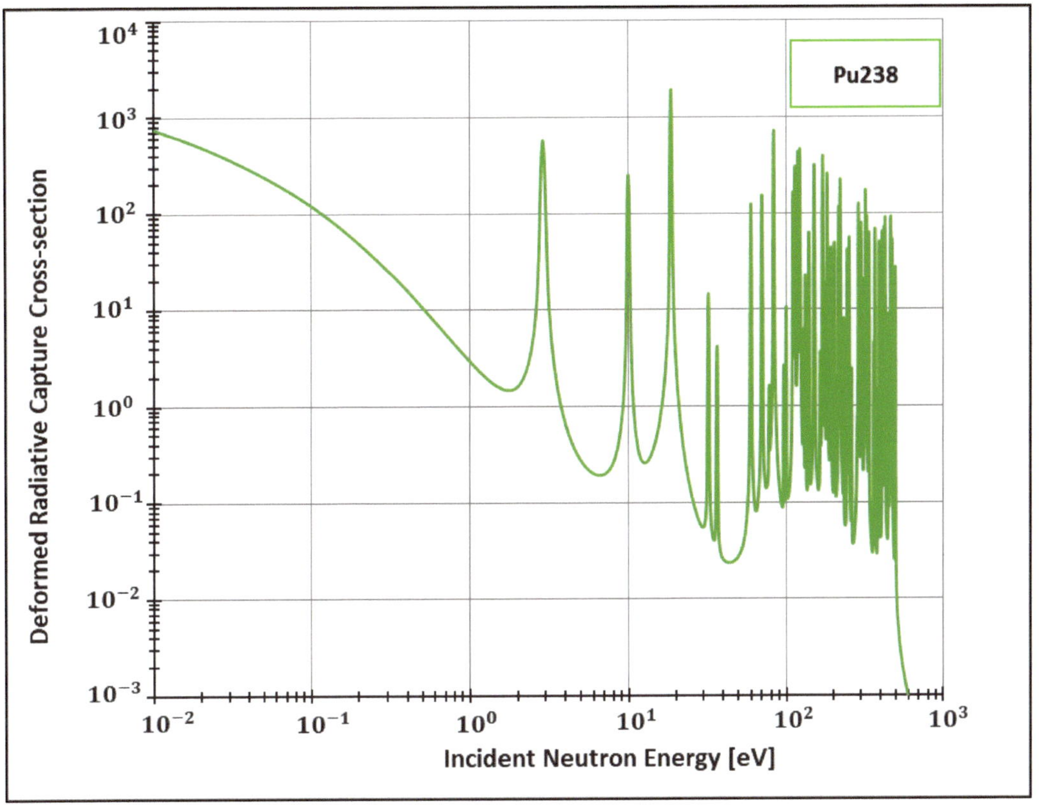

Figure 2. Deformed Radiative capture cross-section for Plutonium 238 considering 1500 K and $k = 0.1$.

Consequently, the application of a new, more robust method for the calculation of these functions could present an improvement to the deformed Doppler broadening functions. In fact, the Faddeeva package results showed better numbers, as seen in Tables 1–3, representing the percentual error of the analytical solution in relation to the numerical one.

Table 1. The percentual errors of the analytical solution in relation to the numerical one using the Numpy package (Python) to calculate the included error functions.

ξ	x = 0	x = 0.5	x = 1	x = 2	x = 4	x = 6	x = 8	x = 10	x = 20	x = 40
0.05	0.02	0.02	0.00	0.02	0.02	0.02	0.02	0.02	0.06	0.67
0.10	0.04	0.04	0.04	0.04	0.05	0.05	0.06	0.09	0.73	12.55
0.15	0.06	0.06	0.06	0.06	0.07	0.11	0.17	0.32	3.77	10.39
0.20	0.08	0.08	0.08	0.09	0.12	0.21	0.44	0.85	9.60	4.48
0.25	0.10	0.09	0.10	0.11	0.19	0.41	0.90	1.89	10.91	3.08
0.30	0.11	0.11	0.12	0.15	0.29	0.71	1.67	3.50	6.89	3.17
0.35	0.13	0.13	0.15	0.19	0.44	1.16	2.78	5.46	5.00	4.84
0.40	0.15	0.15	0.17	0.23	0.63	1.75	4.09	7.04	4.09	3.23
0.45	0.17	0.18	0.19	0.29	0.87	2.51	5.42	7.64	3.82	3.23
0.50	0.19	0.20	0.22	0.36	1.18	3.37	6.41	7.32	3.49	3.28

Table 2. The percentual errors of the analytical solution in relation to the numerical one using the Faddeeva package (C++) to calculate the included error functions.

ξ	x = 0	x = 0.5	x = 1	x = 2	x = 4	x = 6	x = 8	x = 10	x = 20	x = 40
0.05	0.02	0.02	0.02	0.02	0.02	0.02	0.02	0.02	0.06	0.67
0.10	0.04	0.04	0.04	0.04	0.04	0.05	0.06	0.08	0.68	10.61
0.15	0.06	0.06	0.06	0.06	0.07	0.09	0.15	0.26	3.40	8.00
0.20	0.08	0.08	0.08	0.08	0.10	0.17	0.34	0.71	8.17	3.08
0.25	0.10	0.10	0.10	0.10	0.15	0.31	0.73	1.59	8.89	2.55
0.30	0.11	0.11	0.12	0.13	0.22	0.55	1.37	2.96	5.73	2.41
0.35	0.13	0.13	0.14	0.16	0.32	0.89	2.28	4.55	3.72	2.34
0.40	0.15	0.15	0.16	0.19	0.45	1.37	3.36	5.82	2.97	2.30
0.45	0.17	0.17	0.18	0.23	0.63	1.97	4.42	6.28	2.68	2.28
0.50	0.19	0.19	0.20	0.27	0.85	2.63	5.18	5.92	2.54	2.26

Table 3. The percentual difference between Tables 1 and 2.

ξ	x = 0	x = 0.5	x = 1	x = 2	x = 4	x = 6	x = 8	x = 10	x = 20	x = 40
0.05	0.00	0.00	0.02	0.00	0.00	0.00	0.00	0.00	0.00	0.01
0.10	0.00	0.00	0.00	0.00	0.01	0.01	0.00	0.01	0.05	1.94
0.15	0.00	0.00	0.00	0.00	0.01	0.02	0.03	0.06	0.38	2.39
0.20	0.00	0.00	0.00	0.01	0.02	0.04	0.09	0.14	1.44	1.40
0.25	0.00	0.00	0.00	0.01	0.04	0.09	0.17	0.30	2.01	0.53
0.30	0.00	0.00	0.00	0.02	0.07	0.16	0.30	0.54	1.16	0.77
0.35	0.00	0.00	0.01	0.03	0.12	0.26	0.50	0.90	1.28	2.50
0.40	0.00	0.00	0.01	0.04	0.18	0.38	0.73	1.21	1.12	0.92
0.45	0.00	0.01	0.02	0.06	0.24	0.55	1.01	1.37	1.14	0.95
0.50	0.00	0.01	0.02	0.09	0.33	0.73	1.23	1.40	0.95	1.02

By analyzing Table 3, it is possible to note the lower values of percentual error when one uses the Faddeeva method to calculate the existing error functions in the deformed analytical solution of the Doppler broadening function using the Kaniadakis entropy. The maximum percentual reduction was 2.5%.

Deformed Cross-Sections with FRENDY

After implementing the deformed analytical solution for the Doppler broadening function inside FRENDY's test module, we were able to generate data for the deformed radiative cross-sections for different elements. Considering the adopted method for calculating these quantities (SLBW), FRENDY's default package offers the calculation of cross-sections for two important elements: Plutonium 238 and Technetium 99. The former element (Pu238) is of crucial importance, for instance, to space exploration [37] and Mars colonization [38,39]. In addition, 80% of the scans performed in nuclear medicine departments are made from the latter element [40]. Both can be produced in research nuclear reactors, such as the High Flux Isotope Reactor in the United States [41] and the Moly project of the recent Research Reactor Jules Horowitz (JHR), still under construction in France [40].

Additionally, the present work generated data for the isotopes 155 and 157 (Figures 2–5) of gadolinium, which is widely used for medical applications [42], radiation shielding [43], and also space exploration [44].

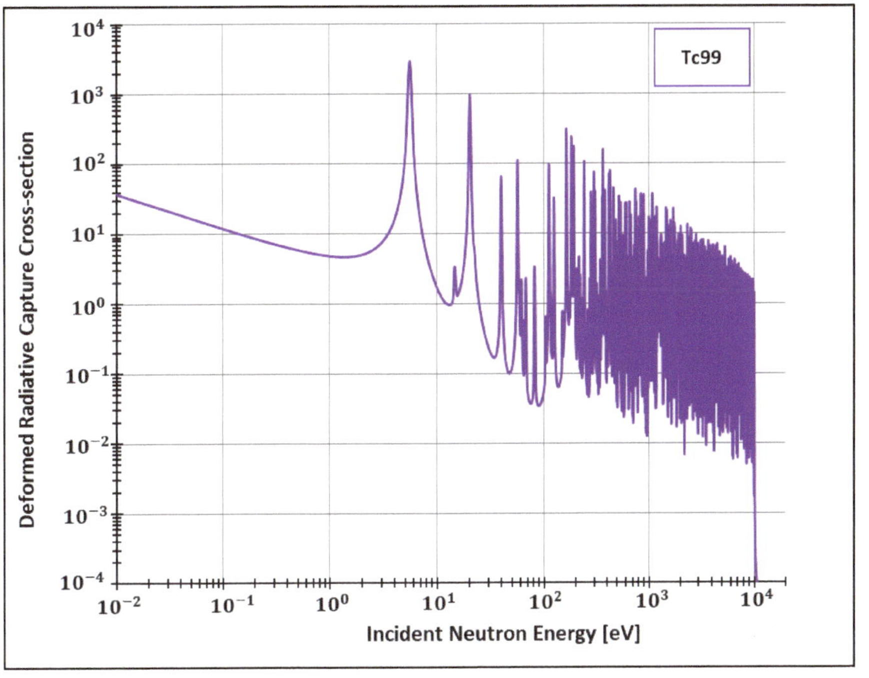

Figure 3. Deformed Radiative capture cross-section for Technetium 99 considering 1500 K and $k = 0.1$.

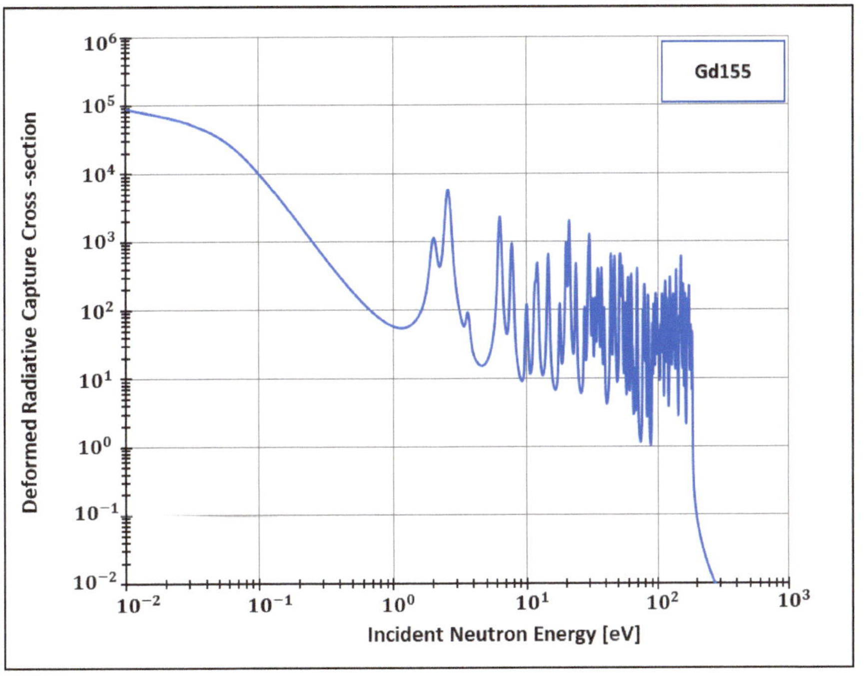

Figure 4. Deformed Radiative capture cross-section for Gadolinium 155 considering 1500 K and $k = 0.1$.

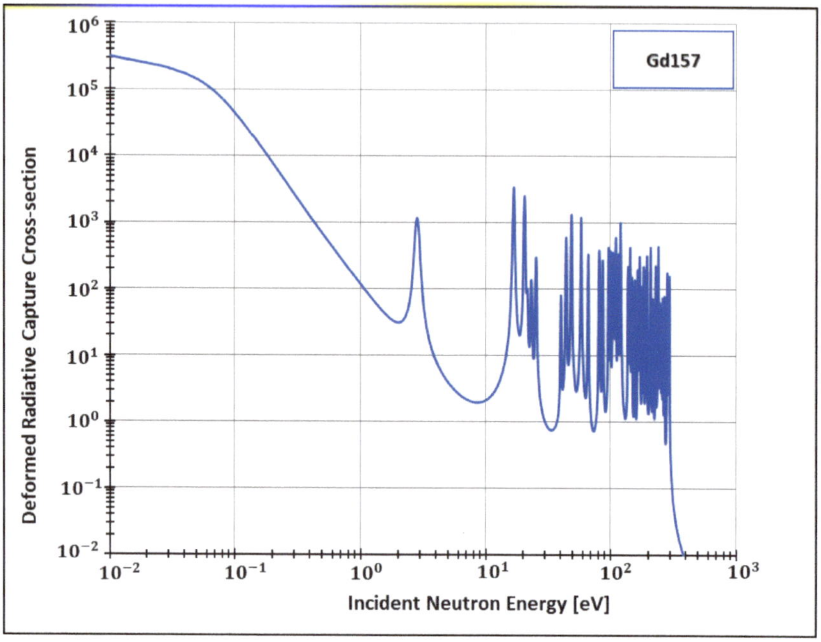

Figure 5. Deformed Radiative capture cross-section for Gadolinium 157 considering 1500 K and $k = 0.1$.

In order to see more closely and compare the standard Maxwell-Boltzmann behavior with the Kaniadakis, we selected two different resonance peaks—apart from each other—at three different temperatures (Figures 6–13) to confirm the expected curve attenuation illustrated in previous works [13,14,29,45]:

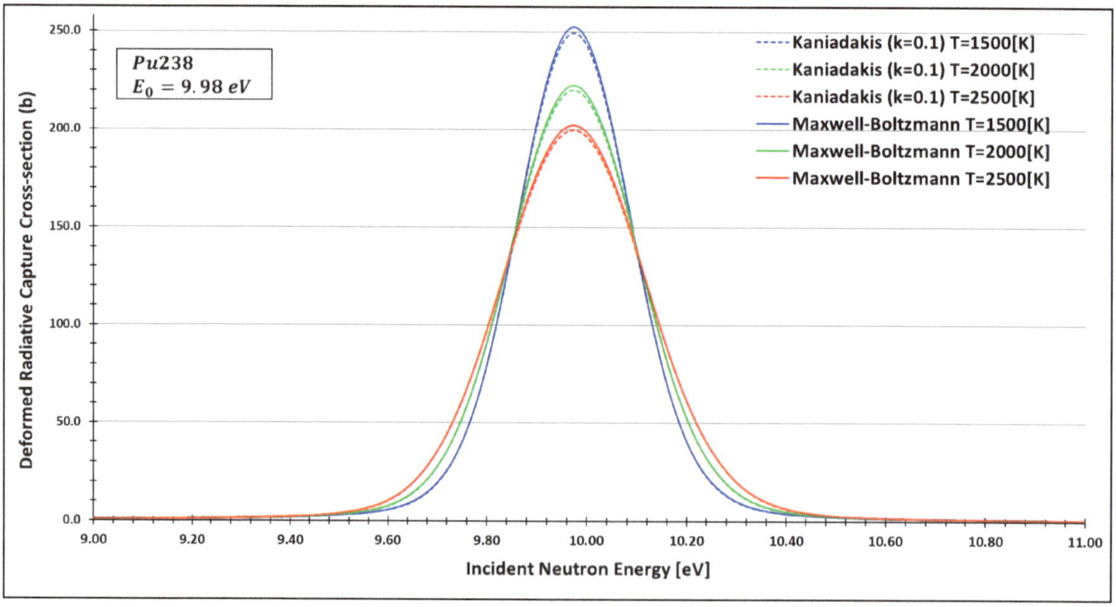

Figure 6. Deformed Radiative capture cross-section for Plutonium 238 considering $k = 0.1$ and the 9.98 eV peak.

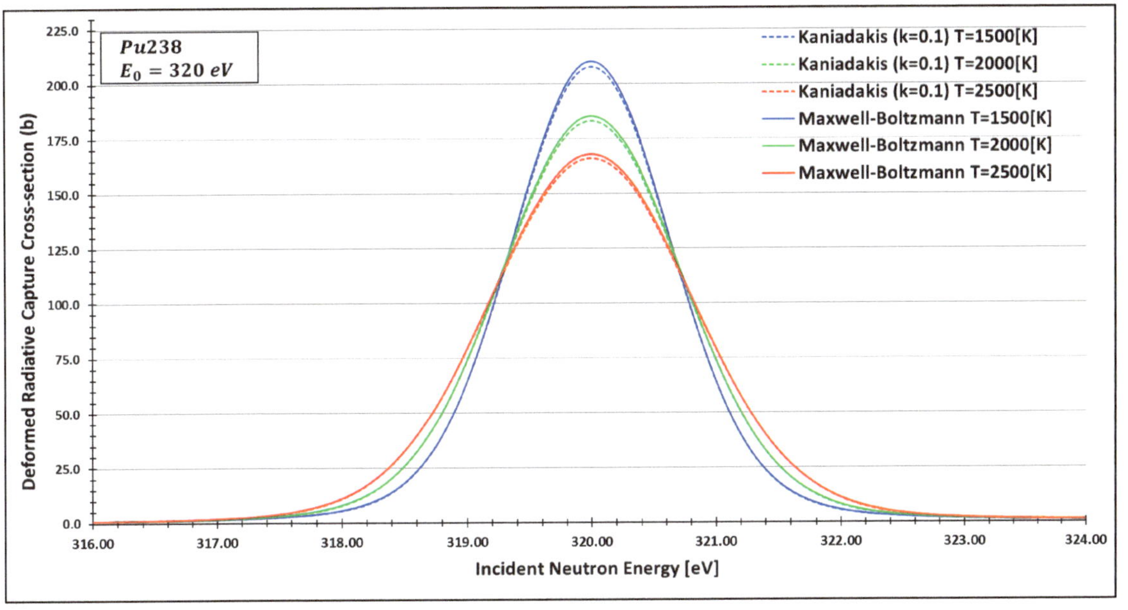

Figure 7. Deformed Radiative capture cross-section for Plutonium 238 considering $k = 0.1$ and the 320 eV peak.

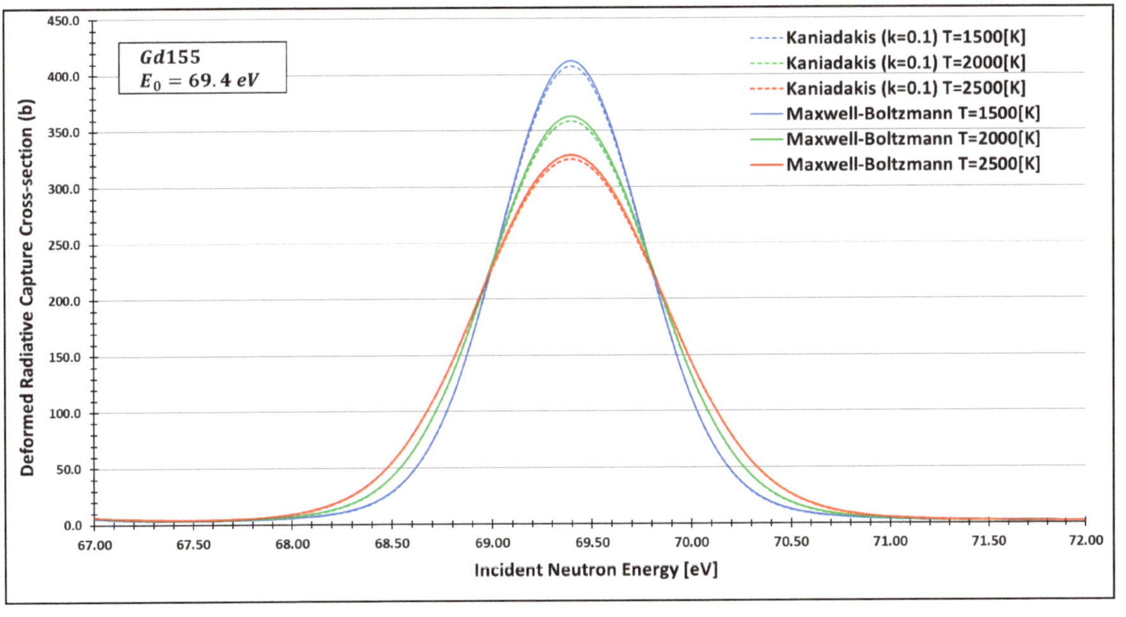

Figure 8. Deformed Radiative capture cross-section for Gadolinium 155 considering $k = 0.1$ and the 69.4 eV peak.

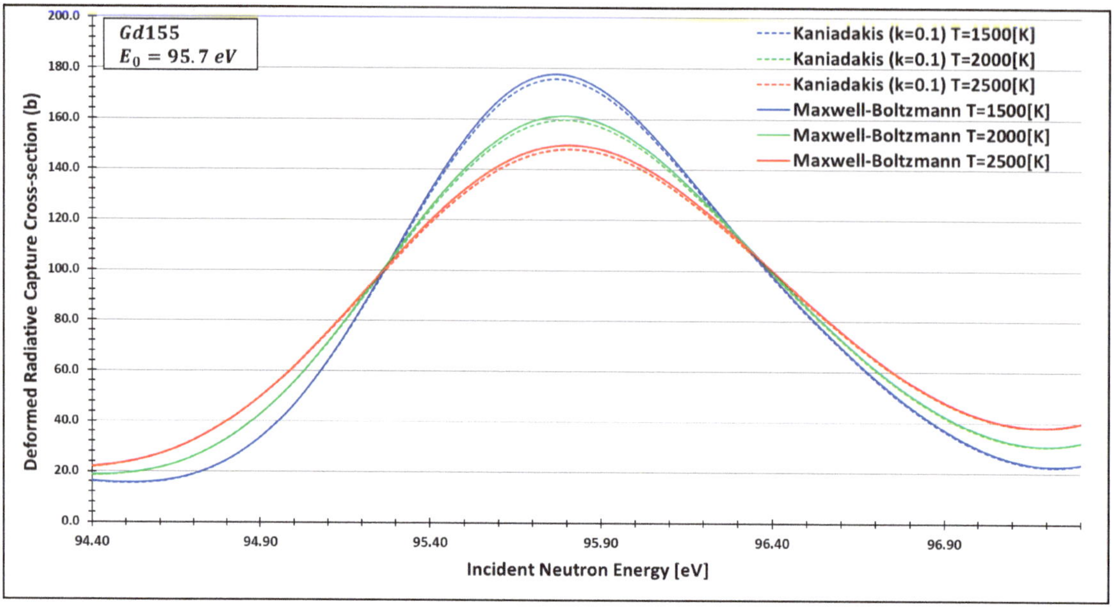

Figure 9. Deformed Radiative capture cross-section for Gadolinium 155 considering $k = 0.1$ and the 95.7 eV peak.

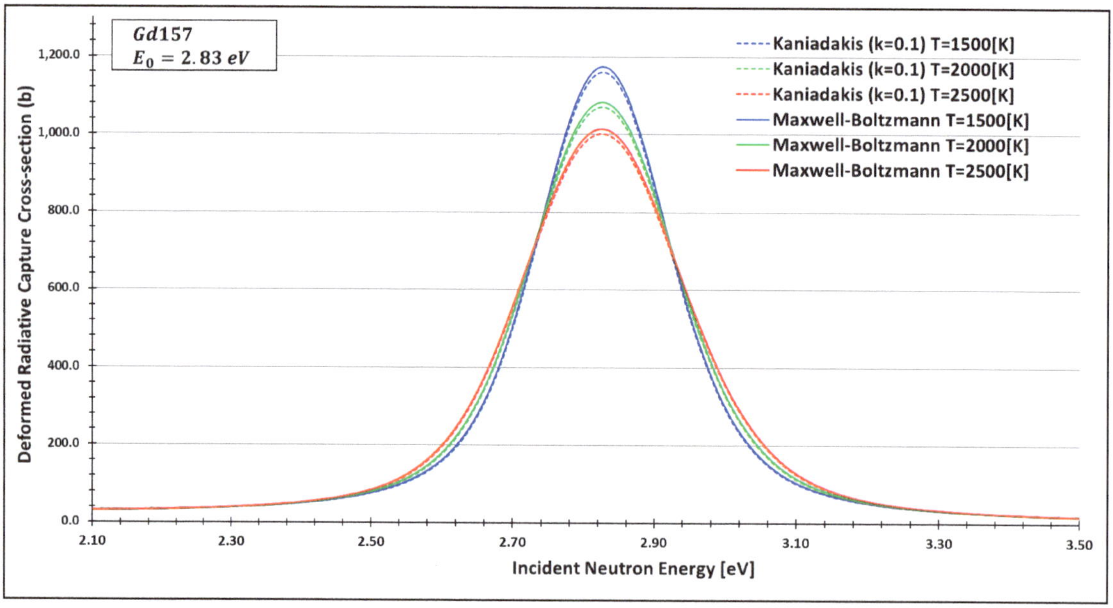

Figure 10. Deformed Radiative capture cross-section for Gadolinium 157 considering $k = 0.1$ and the 2.83 eV peak.

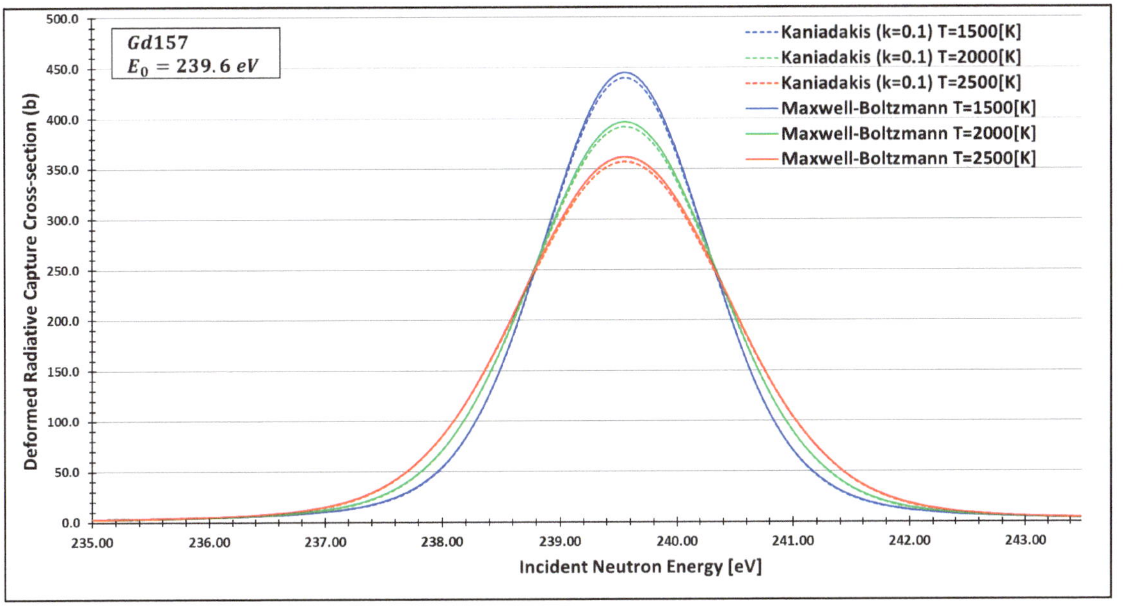

Figure 11. Deformed Radiative capture cross-section for Gadolinium 157 considering $k = 0.1$ and the 239.6 eV peak.

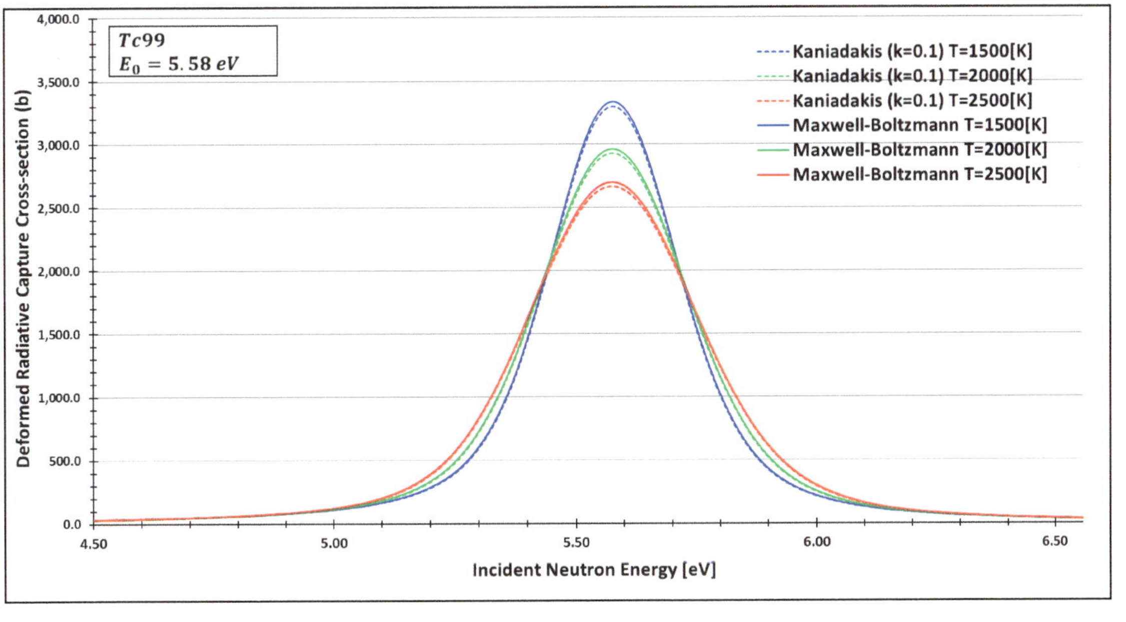

Figure 12. Deformed Radiative capture cross-section for Technetium 99 considering $k = 0.1$ and the 5.58 eV peak.

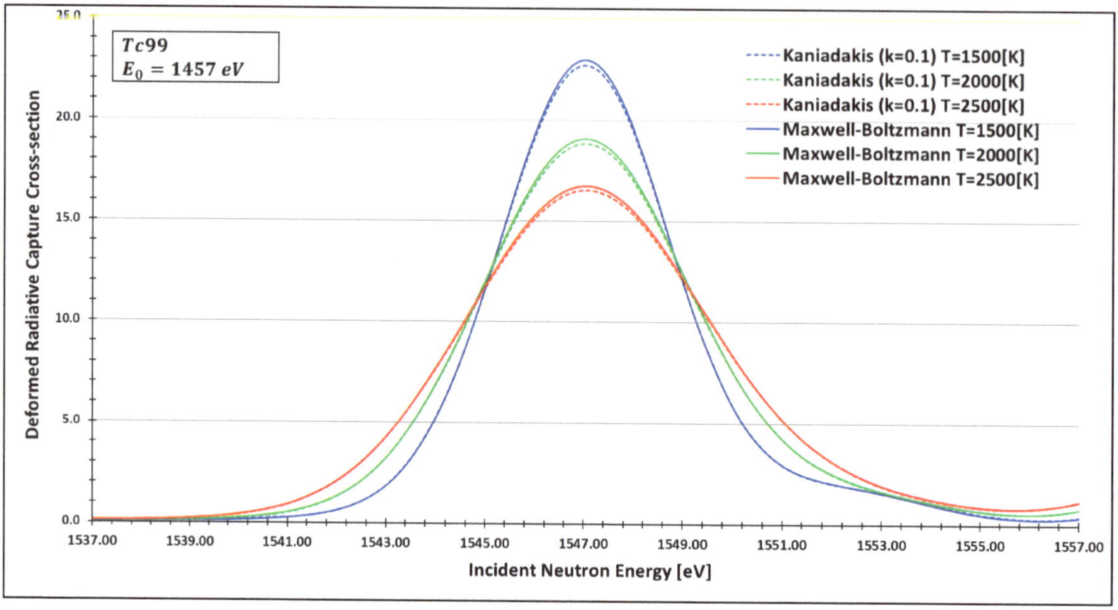

Figure 13. Deformed Radiative capture cross-section for Technetium 99 considering $k = 0.1$ and the 1457 eV peak.

As one can see, the deformed curves presented the expected behavior since there is an attenuation of the resonance curves, especially on the peaks, compared to the standard neutron radiative cross-section using the Maxwell–Boltzmann entropy. In fact, the relative error between the Maxwell–Boltzmann and Kaniadakis peaks is around 1%, which is the same order of magnitude (~1%) obtained in previous works for the calculations of deformed Doppler broadening functions using the Kaniadakis entropy.

4. Concluding Remarks

After 20 years of development of the Kaniadakis entropy and seven years of its application in nuclear reactor physics, this work presents for the first-time results for deformed neutron cross-sections considering the κ statistics using an official nuclear data processing code, FRENDY, and, consequently, official nuclear data (JENDL 4.0). This work was carried out by implementing the analytical solution for the deformed Doppler function using the Kaniadakis statistics, ψ_κ, inside the single-level Breit–Wigner module in the FRENDY. We used MIT's Faddeeva method to calculate the error functions inside the analytical solution. This implementation showed a percentual error reduction when compared to the numerical solution of ψ_κ.

With the implementation of ψ_κ inside FRENDY, it was possible to calculate deformed radiative capture cross-sections for four relevant nuclides: Pu238, Gd 155, Gd 157 and Tc99. Next, we selected two different resonance peaks of each nuclide to compare the data with the standard Maxwell-Boltzmann curves generated by the FRENDY code. The results agreed with previous calculations conducted out of a nuclear data processing code and without official nuclear data libraries.

Different from other areas, the evaluation of the viability of Kaniadakis entropy in the area of the nuclear reactor physics cannot be conducted observationally. Therefore, it is of great relevance to implement this methodology in nuclear data processing codes where it is possible to deal with accurate data. Thus, the present work can be interpreted as an essential step in validating the applicability of the Kaniadakis entropy in the nuclear fission area.

Author Contributions: Conceptualization, W.V.d.A., A.S.M., J.M.M. and A.d.C.G.; methodology, W.V.d.A., A.S.M., J.M.M. and A.d.C.G.; software, W.V.d.A., A.S.M., J.M.M. and L.S.; validation W.V.d.A. and J.M.M.; formal analysis, W.V.d.A., A.S.M., J.M.M. and A.d.C.G.; writing—original draft preparation, W.V.d.A.; writing—review and editing, W.V.d.A., A.S.M., J.M.M. and A.d.C.G.; supervision, A.S.M.; project administration, W.V.d.A. All authors have read and agreed to the published version of the manuscript.

Funding: This research was funded by Fundação Carlos Chagas Filho (FAPERJ/Pós-doutorado Nota 10), grant number E-26/204.444/2021, registration number 2021.00898.8 and by the Conselho Nacional de Desenvolvimento Científico e Tecnológico, grant number CNPq/304.580/2019-8.

Institutional Review Board Statement: Not applicable.

Acknowledgments: All authors thank Kenichi Tada (JAEA) for the support and contribution during the research. The first author thanks FAPERJ for his post-doctoral grant. The second author thanks FAPERJ and CNPq for the support.

Conflicts of Interest: The authors declare no conflict of interest.

References

1. Kaniadakis, G. Non-Linear Kinetics Underlying Generalized Statistics. *Phys. A Stat. Mech. Its Appl.* **2001**, *296*, 405–425. [CrossRef]
2. Trivellato, B. Deformed Exponentials and Applications to Finance. *Entropy* **2013**, *15*, 3471–3489. [CrossRef]
3. Lapenta, G.; Markidis, S.; Marocchino, A.; Kaniadakis, G. Relaxation of Relativistic Plasmas Under the Effect of Wave-Particle Interactions. *Astrophys. J.* **2007**, *666*, 949–954. [CrossRef]
4. Carvalho, J.C.; do Nascimento, J.D.; Silva, R.; Medeiros, J.R. Non-Gaussian statistics and stellar rotational velocities of main-sequence field stars. *Astrophys. J.* **2009**, *696*, L48–L51. [CrossRef]
5. Carvalho, J.C.; Silva, R.; do Nascimento, J.D., Jr.; Medeiros, J.R. Power Law Statistics and Stellar Rotational Velocities in the Pleiades. *EPL (Europhys. Lett.)* **2008**, *84*, 59001. [CrossRef]
6. Carvalho, J.C.; Silva, R.; do Nascimento, J.D., Jr.; Soares, B.B.; Medeiros, J.R. Observational Measurement of Open Stellar Clusters: A Test of Kaniadakis and Tsallis Statistics. *EPL (Europhys. Lett.)* **2010**, *91*, 69002. [CrossRef]
7. Topsøe, F. Entropy and Equilibrium via Games of Complexity. *Phys. A Stat. Mech. Its Appl.* **2004**, *340*, 11–31. [CrossRef]
8. Abreu, E.M.C.; Neto, J.A.; Barboza, E.M.; Nunes, R.C. Tsallis and Kaniadakis Statistics from the Viewpoint of Entropic Gravity Formalism. *Int. J. Mod. Phys. A* **2017**, *32*, 1750028. [CrossRef]
9. Abreu, E.M.C.; Neto, J.A.; Mendes, A.C.R.; Bonilla, A.; de Paula, R.M. Cosmological Considerations in Kaniadakis Statistics. *EPL (Europhys. Lett.)* **2018**, *124*, 30003. [CrossRef]
10. Lourek, I.; Tribeche, M. Dust Charging Current in Non Equilibrium Dusty Plasma in the Context of Kaniadakis Generalization. *Phys. A Stat. Mech. Its Appl.* **2019**, *517*, 522–529. [CrossRef]
11. Guedes, G.; Gonçalves, A.C.; Palma, D.A. The Doppler Broadening Function Using the Kaniadakis Distribution. *Ann. Nucl. Energy* **2017**, *110*, 453–458. [CrossRef]
12. Kaniadakis, G. Statistical Mechanics in the Context of Special Relativity. *Phys. Rev. E* **2002**, *66*, 56125. [CrossRef]
13. de Abreu, W.V.; Gonçalves, A.C.; Martinez, A.S. Analytical Solution for the Doppler Broadening Function Using the Kaniadakis Distribution. *Ann. Nucl. Energy* **2019**, *126*, 262–268. [CrossRef]
14. de Abreu, W.V.; Martinez, A.S.; do Carmo, E.D.; Gonçalves, A.C. A Novel Analytical Solution of the Deformed Doppler Broadening Function Using the Kaniadakis Distribution and the Comparison of Computational Efficiencies with the Numerical Solution. *Nucl. Eng. Technol.* **2022**, *54*, 1471–1481. [CrossRef]
15. Mamedov, B.A. Analytical Evaluation of Doppler Functions Arising from Resonance Effects in Nuclear Processes. *Nucl. Instrum. Methods Phys. Res.* **2009**, *608*, 336–338. [CrossRef]
16. Antunes, A.J.M.; Gonçalves, A.C.; Martinez, A.S. Analytical Solution for the Doppler Broadening Function Using the Tsallis Distribution. *Prog. Nucl. Energy* **2022**, *144*, 104071. [CrossRef]
17. Abul-Magd, A.Y. Nonextensive Random-Matrix Theory Based on Kaniadakis Entropy. *Phys. Lett. A* **2007**, *361*, 450–454. [CrossRef]
18. He, K.-R. Jeans Analysis with κ-Deformed Kaniadakis Distribution in f (R) Gravity. *Phys. Scr.* **2022**, *97*, 025601. [CrossRef]
19. Tada, K.; Nagaya, Y.; Kunieda, S.; Suyama, K.; Fukahori, T. Frendy: A New Nuclear Data Processing System Being Developed at JAEA. *EPJ Web Conf.* **2017**, *146*, 2028. [CrossRef]
20. Tada, K.; Nagaya, Y.; Kunieda, S.; Suyama, K.; Fukahori, T. Development and Verification of a New Nuclear Data Processing System FRENDY. *J. Nucl. Sci. Technol.* **2017**, *54*, 806–817. [CrossRef]
21. Macfarlane, R.; Muir, D.W.; Boicourt, R.M.; Kahler, I.A.C.; Conlin, J.L. *The NJOY Nuclear Data Processing System, Version 2016*; Los Alamos National Laboratory: Los Alamos, NM, USA, 2017.
22. Cullen, D.E. *PREPRO 2010-2010 ENDF-6 Pre-Processing Codes*; International Atomic Energy Agency: Vienna, Austria, 2010.
23. Zu, T.; Xu, J.; Tang, Y.; Bi, H.; Zhao, F.; Cao, L.; Wu, H. NECP-Atlas: A New Nuclear Data Processing Code. *Ann. Nucl. Energy* **2019**, *123*, 153–161. [CrossRef]

24. Chadwick, M.B.; Herman, M.; Obložinský, P.; Dunn, M.E.; Danon, Y.; Kahler, A.C.; Smith, D.L.; Pritychenko, B.; Arbanas, G.; Arcilla, R.; et al. ENDF/B-VII.1 Nuclear Data for Science and Technology: Cross Sections, Covariances, Fission Product Yields and Decay Data. *Nucl. Data Sheets* **2011**, *112*, 2887–2996. [CrossRef]
25. Zhuang, Y.; Liu, T.; Zhang, J.; Liu, P. CENDL-3–Chinese Evaluated Nuclear Data Library, Version 3. *J. Nucl. Sci. Technol.* **2002**, *39*, 37–39. [CrossRef]
26. Nuclear Energy Agency Joint Evaluated Fission and Fusion (JEFF) Nuclear Data Library. Available online: https://www.oecd-nea.org/dbdata/jeff/ (accessed on 5 July 2022).
27. Shibata, K.; Iwamoto, O.; Nakagawa, T.; Iwamoto, N.; Ichihara, A.; Kunieda, S.; Chiba, S.; Furutaka, K.; Otuka, N.; Ohsawa, T.; et al. JENDL-4.0: A New Library for Nuclear Science and Engineering. *J. Nucl. Sci. Technol.* **2011**, *48*, 1–30. [CrossRef]
28. de Abreu, W.V. *Solução Analítica Da Função de Alargamento Doppler Usando a Distribuição de Kaniadakis*; Universidade Federal do Rio de Janeiro: Rio de Janeiro, Brazil, 2020; p. 130. [CrossRef]
29. de Abreu, W.V.; Gonçalves, A.C.; Martinez, A.S. New Analytical Formulations for the Doppler Broadening Function and Interference Term Based on Kaniadakis Distributions. *Ann. Nucl. Energy* **2020**, *135*, 106960. [CrossRef]
30. Tada, K.; Kunieda, S.; Nagaya, Y. *Nuclear Data Processing Code FRENDY Version 1*; Japan Atomic Energy Agency: Tokai, Japan, 2018.
31. National Nuclear Data Center. *ENDF-6 Formats Manual*; Trkov, A., Herman, M., Brown, D.A., Eds.; Brookhaven National Laboratory: Upton, NY, USA, 2018.
32. Lehtinen, N.G. *Error Functions*. 2010. Available online: http://nlpc.stanford.edu/nleht/Science/reference/errorfun.pdf (accessed on 18 September 2022).
33. Arfken, G.B.; Weber, H.J.; Harris, F.H. *Mathematical Methods for Physicists*, 7th ed.; Academic Press: Cambridge, MA, USA; Elesvier: Amsterdam, The Netherlands, 2013; ISBN 978-0-12-384654-9.
34. Johnson, S.G. Faddeeva Package. Available online: http://ab-initio.mit.edu/wiki/index.php/Faddeeva_Package (accessed on 11 July 2022).
35. Gautschi, W. Efficient Computation of the Complex Error Function. *SIAM J. Numer. Anal.* **1970**, *7*, 187–198. [CrossRef]
36. Poppe, G.P.M.; Wijers, C.M.J. More Efficient Computation of the Complex Error Function. *ACM Trans. Math. Softw.* **1990**, *16*, 38–46. [CrossRef]
37. Gusev, V.V.; Pustovalov, A.A.; Rybkin, N.N.; Anatychuk, L.I.; Demchuk, B.N.; Ludchak, I.Y. Milliwatt-Power Radioisotope Thermoelectric Generator (RTG) Based on Plutonium-238. *J. Electron. Mater.* **2011**, *40*, 807–811. [CrossRef]
38. Werner, J.; Lively, K.; Kirkham, D. A Multi-Mission Radioisotope Thermoelectric Generator (MMRTG) for Mars 2020. In Proceedings of the 2017 IEEE Aerospace Conference, Big Sky, MT, USA, 7 March 2017; IEEE: Piscataway, NJ, USA; pp. 1–6.
39. Clarke, E.; Giglio, J.; Wahlquist, K.; Dees, C.; Gates, A.; Birch, J.; Davis, S.; Horkley, B.; Rich, L. Multi-Mission Thermoelectric Generator Fueling Testing and Integration Operations for Mars 2020. *Nucl. Technol.* 2022; *ahead of print*. [CrossRef]
40. Antony, M.; Coulon, J.-P.; Gay, S.; Bourrelly, F.; Tarabelli, D.; Drapeau, D.; Chapuis, C.; Derasse, F.; Aymard, N.; Mallet, R. *Moly Production in the Jules Horowitz Reactor: Capacity and Status of the Development*; IGORR: Sydney, Autralia, 2017.
41. Collins, E.D.; Morris, R.N.; McDuffee, J.L.; Mulligan, P.L.; Delashmitt, J.S.; Sherman, S.R.; Vedder, R.J.; Wham, R.M. Plutonium-238 Production Program Results, Implications, and Projections from Irradiation and Examination of Initial NpO2 Test Targets for Improved Production. *Nucl. Technol.* 2022; *ahead of print*. [CrossRef]
42. Toupin, S.; Pezel, T.; Bustin, A.; Cochet, H. Whole-Heart High-Resolution Late Gadolinium Enhancement: Techniques and Clinical Applications. *J. Magn. Reson. Imaging* **2022**, *55*, 967–987. [CrossRef] [PubMed]
43. Ozturk, S. Radiation Shielding Properties of Gadolinium-Doped Water. *Phys. Scr.* **2021**, *96*, 055402. [CrossRef]
44. Mani, V.; Prasad, N.S.; Kelkar, A. *Ultra High Molecular Weight Polyethylene (UHMWPE) Fiber Epoxy Composite Hybridized with Gadolinium and Boron Nanoparticles for Radiation Shielding*; Hughes, G.B., Ed.; SPIE: Bellingham, WA, USA, 2016; p. 99810D.
45. da Silva, M.V.; Martinez, A.S.; Gonçalves, A.C. Effective Medium Temperature for Calculating the Doppler Broadening Function Using Kaniadakis Distribution. *Ann. Nucl. Energy* **2021**, *161*, 108500. [CrossRef]

entropy

MDPI

Article

The κ-Deformed Calogero–Leyvraz Lagrangians and Applications to Integrable Dynamical Systems

Partha Guha

Department of Mathematics, Khalifa University of Science and Technology, Abu Dhabi P.O. Box 127788, United Arab Emirates; partha.guha@ku.ac.ae

Abstract: The Calogero–Leyvraz Lagrangian framework, associated with the dynamics of a charged particle moving in a plane under the combined influence of a magnetic field as well as a frictional force, proposed by Calogero and Leyvraz, has some special features. It is endowed with a Shannon "entropic" type kinetic energy term. In this paper, we carry out the constructions of the 2D Lotka–Volterra replicator equations and the $N = 2$ Relativistic Toda lattice systems using this class of Lagrangians. We take advantage of the special structure of the kinetic term and deform the kinetic energy term of the Calogero–Leyvraz Lagrangians using the κ- deformed logarithm as proposed by Kaniadakis and Tsallis. This method yields the new construction of the κ-deformed Lotka–Volterra replicator and relativistic Toda lattice equations.

Keywords: entropic kinetic energy; Lotka–Volterra; replicator equation; relativistic Toda lattice; Kaniadakis logarithm; Tsallis deformation; κ-deformed Lagrangian

MSC: 34C14; 34C20

Citation: Guha, P. The κ-Deformed Calogero–Leyvraz Lagrangians and Applications to Integrable Dynamical Systems. *Entropy* **2022**, *24*, 1673. https://doi.org/10.3390/e24111673

Academic Editors: Dionissios T. Hristopulos, Sergio Luiz E. F. da Silva and Antonio M. Scarfone

Received: 14 September 2022
Accepted: 26 October 2022
Published: 17 November 2022

1. Introduction

Recently, Calogero and Leyvraz [1,2] demonstrated a time-independent Hamiltonian description of the motion of a charged particle moving in a two dimensional space under the influence of a magnetic field perpendicular to the plane of motion and a frictional force proportional to the velocity. This motion may be viewed as a dynamics of cyclotron motion with friction; this model arises through the coupling of a particle to a large number of external degrees of freedom. The most interesting feature of this Lagrangian is the kinetic energy term—entropic type kinetic energy. This makes the Lagrangian a nonstandard one. We have explored the applications of this class of Lagrangians in our earlier papers [3,4].

It is worth noting that some physical systems cannot be described by the Boltzmann–Gibbs statistical mechanics, for example, systems such as long-range interactions, long-time memory and multifractal or hierarchical structures are some of them. To overcome at least some of these difficulties, Tsallis [5,6] proposed a generalized entropic form based on a κ-deformed logarithm. Later, an example of self-dual κ-deformed logarithmic functions is found in the work of Kaniadakis [7–11]. Over the last decade or so, scientists have observed that many physical and social phenomena often follow the so-called power law distributions (see for example, [12–15]). We demonstrated that many (generalized) power law distribution equations can be derived from Calogero–Leyvraz Lagrangian formalism using κ-deformation theory.

These are the two popular ways to deform logarithmic and exponential terms in physics. We propose a new Lagrangian where the logarithm term appearing in Calogero–Leyvraz is replaced by a deformed logarithmic function, and study the dynamics. The algebraic structures arising in this κ-deformed framework have been carefully grafted by Scarfone in [16]; in fact, the concept of generalized algebras has been employed constructively to study entropic forms in [16,17]. It is worth noting that the generalized entropies [18] play an important role in generalized distribution theory in complex systems [19] and they have been studied extensively from the information geometric point of view in [20,21]. The aim of this work is to carry out the formulation of the κ-deformed well-known dynamical systems, namely, the Lotka–Volterra replicator equation and $N = 2$ relativistic Toda lattice system.

Among ecological models, the Lotka–Volterra equation for predator–prey systems [22–24] has played a significant role in dynamical systems. It is known that the solutions to this conservative system in phase space are level curves of the energy function. In the mathematical investigations of ecological models, conservative dynamics are often very useful from the (geometrical) mechanics point of view. It is known that the replicator equation in the evolutionary game theory [25] is closely related to the Lotka–Volterra equation. The replicator equation is the first and most important game dynamics studied in connection with evolutionary game theory. It was originally developed for symmetric games with finitely many strategies. Evolutionary game theory [26] studies the behavior of large populations of agents who repeatedly engage in strategic interactions. Note that changes in behavior in these populations are driven either by natural selection via differences in birth and death rates, or by the application of myopic decision rules by individual agents.

The Hamiltonian of the two-dimensional motion of electrons in the presence of the periodic potential and the magnetic field perpendicular to the two-dimensional plane is described by:

$$H_{Hof} = e^{iq} + e^{-iq} + e^{ip} + e^{-ip} \quad [q, p] = i\hbar. \tag{1}$$

The spectrum of this system yields a butterfly like structure, known as Hofstadter's butterfly [27]. In a completely independent line of research, the string community investigated a system associated to Hamiltonian,

$$H = e^{q} + e^{-q} + e^{p} + e^{-p}, \tag{2}$$

when q and p are restricted to be purely imaginary; this equation reduces to Hofstadter's Hamiltonian (1). In general, q and p are complex coordinates, hence the equation determines a real two-dimensional Riemann surface, or equivalently a complex one-dimensional curve, whose shape is parameterized by the value of H [28,29]. This appears when mirror symmetry is applied to a non-compact Calabi–Yau geometry known as the local $\mathbb{P}_1 \times \mathbb{P}_1$ geometry. This curve is connected to the Seiberg–Witten curve, encoding the information on instantons in $N = 2$ supersymmetric pure $SU(2)$ gauge theory [30–32].

The system known nowadays as the relativistic Toda lattice (RTL) was invented by S.N.M. Ruijsenaars—the Hamiltonian of the periodic relativistic Toda lattice with just $N = 2$ particle, after removing the center-of-mass mode. We can illustrate this as follows [28,29]. The Ruijsenaars Hamiltonian [33] of $N = 2$ quantum Toda lattice is given by:

$$H_{RT} = e^{Rp_1} + e^{Rp_2} + R^2\left(e^{q_1 - q_2 + Rp_1} + e^{q_2 - q_1 + Rp_2}\right). \tag{3}$$

Let us consider center of mass frame $p_1 + p_2 = 0$ and define

$$p := Rp_1, \qquad q := q_1 - q_2 + Rp_1,$$

which yields an one parameter family of (2),

$$H_{RToda} = e^{p} + e^{-p} + R^2(e^{q} + e^{-q}). \tag{4}$$

Results of the paper The Calogero–Leyvraz Lagrangian has some interesting features; it is endowed with the Shannon entropic [34] kinetic energy term. The Legendre transform of this Lagrangian yields a Hamiltonian with the exponential momentum term. We have seen that the Calogero–Leyvraz Lagrangian/Hamiltonian allows us to formulate several features related to deformed dynamical systems, balanced loss–gain systems and generalized rate equations [3,4]. In particular, most of the power law distributions and rate equations can be manufactured from this class of Lagrangian. It has been explored that the Calogero–Leyvraz theory of cyclotron-friction motion is closely related to the "curl force" theory as proposed by Berry and Shukla [35,36], although the latter is a totally position-dependent nonconservative force with a nonvanishing curl, whereas the former is totally velocity dependent.

In this paper, we present a different formulation of the celebrated Lotka–Volterra equation [22,24] using the Calogero–Leyvraz Lagrangian. We also give a new derivation of the replicator equation using the Calogero–Leyvraz type Lagrangian. The final example is related to the $N = 2$ relativistic Toda lattice system. We formulate the latter equation using a different type of entropic Lagrangian; this entropic kinetic energy is described via a cross-entropy term, which yields a new formulation of the $N = 2$ relativistic Toda lattice system.

Since the kinetic energy term of the Calogero–Leyvraz Lagrangian involves a logarithmic term, we deform this logarithmic term using the Kaniadakis method [17] and obtain κ-deformed Lotka–Volterra, replicator and $N = 2$ relativistic Toda Lattice equations. In fact, the entire reason to shift the usual formalism to Calogero–Leyvraz formalism is to formulate κ-deformed integrable models. We also formulate these deformed equations using the Tsallis logarithm.

In the introductory section, we review a κ -deformed Liénard equation which satisfies the Chiellini integrability condition. This condition allows us to integrate the Liénard type equation using the Abel equation of the first kind. In general, the Liénard equation does not give a Lagrangian formulation, but with the imposition of the Chiellini condition, it yields a Lagrangian formulation.

This paper is *organized* as follows. In Section 2, we review the Calogero–Leyvraz Lagrangians and Hamiltonians and their applications. In particular, we also describe the κ-deformed Liénard type equation using Kaniadakis and Tsallis type deformation of the kinetic energy term and demonstrate that this Liénard equation admits the Chiellini integrability condition [37]. This integrability condition plays an important role in the formulation of Lagragian and solutions of the integrable class of the Liénard equation. A nonexhaustive list of applications includes, among others, those in [38–40]. After giving a gentle introduction to the Calogero–Leyvraz method, we apply this scheme for the construction of the Lotka–Volterra, replicator and $N = 2$ relativistic Toda lattice systems in Section 3. Our Section 4 is dedicated to the construction of κ-deformed equations. We give a formulation of the deformed Lotka–Volterra, replicator and $N = 2$ relativistic Toda lattice equations using the Kaniadakis and Tsallis methods.

2. Review of Calogero–Leyvraz's Lagrangian and Hamiltonian Formulation of the Dynamics of Cyclotron with Friction System

The Hamiltonian of the free particle moving against friction is given by:

$$H(p,z) = e^p + cz, \tag{5}$$

according to the Newtonian equation of motion $\ddot{z} = -\dot{z}$. The corresponding Lagrangian description of this system is given by:

$$L = \dot{z} \ln \dot{z} - cz. \tag{6}$$

A minor modification of the Hamiltonian $H(p,z) = e^p + \frac{\lambda p}{c} + cz$ yields the dynamics of a particle moving against friction in a constant force field λ.

Legendre transformation of Calogero–Leyvraz Lagrangian: Let us recall the Calogero–Leyvraz method first; the Lagrangian is given as $L = -\gamma q + v \ln v$, where $v = \dot{q}$. The equation of motion $\ddot{q} + \gamma\dot{q} = 0$ yields a constant of motion

$$C = v + \gamma q, \quad \text{where} \quad v = \dot{q}. \tag{7}$$

The corresponding momentum

$$p = \frac{\partial L}{\partial v} = (\ln v + 1) \;\Rightarrow\; v = e^{p-1}.$$

Substituting this in the Legendre transformation,

$$FL(L) = vp - L = v(1 + \ln\, v) - L = v + \gamma q; \tag{8}$$

thus we obtain the Calogero–Leyvraz Hamiltonian,

$$H_{CL} = e^{p} + \gamma q, \tag{9}$$

where we have scaled the momentum to ignore the constant term. Hence we establish the connection between the Calogero–Leyvraz Lagrangian and Hamiltonian via Legendre transformation. This construction can be extended to Lagrangian involving a time-dependent coefficient.

2.1. Calogero–Leyvraz Hamiltonian and Planar Systems

Calogero and Leyvraz straight-forwardly generalized this motion by complexification to describe motions taking place in a plane. Physically, this is connected to the motion against friction of a charged particle in the presence of a perpendicular constant magnetic field, or a constant electric field lying in that plane, or both these forces. If we set $c = \gamma + i\omega$ and go to the complex plane, the following pair of Poisson commuting Hamiltonians are obtained:

$$H_R = e^{p_x} \cos p_y + \gamma x - \omega y, \qquad H_I = -e^{p_x} \sin p_y + \omega x + \gamma y.$$

At first we consider a minor change; coefficients are considered to be time dependent. The two-dimensional Calogero–Leyvraz model is given by the following Hamiltonian:

$$H = e^{p_x} \cos p_y + \gamma(t)x - \omega(t)y, \tag{10}$$

where $\gamma(t), \omega(t)$ are parameters, (x, y) are coordinates and (p_x, p_y) are corresponding momenta. Here we note that the potential energy is a linear function of coordinates while the kinetic energy $\Psi = e^{p_x} \cos p_y$. The Hamiltonian (10) yields the following equations of motion:

$$\ddot{x} = -\gamma(t)\dot{x} + \omega(t)\dot{y}, \qquad \ddot{y} = -\gamma(t)\dot{y} - \omega(t)\dot{x}. \tag{11}$$

Calogero and Leyvraz reformulated (11) in a 3-dimensional context by introducing the 3-vector $r = (x, y, 0)$ in the xy-Cartesian plane and the unit vector $\hat{z} = (0, 0, 1)$ orthogonal to that plane; this yields:

$$\ddot{r} = -\gamma(t)\dot{r} + \omega(t)\dot{r} \times \hat{z}. \tag{12}$$

We obtain the sister (or mirror) equations of (11) if we consider a different K.E., viz., $\Phi(p_x, p_y) = e^{p_x} \sin p_y$, with the same potential energy $\gamma x - \omega y$, this is given by:

$$\ddot{x} = -\gamma(t)\dot{x} + \omega(t)\dot{y}, \qquad \ddot{y} = \gamma(t)\dot{y} - \omega(t)\dot{x}. \tag{13}$$

It is easy to check both (11) and (13).

The linear equation can be generalized to a nonlinear equation from the Calogero–Leyvraz Hamiltonian using generalized potential energy. Suppose we consider

$$H = e^{p_x} \cos p_y + \gamma(t)\phi(x, y) - \omega(t)\psi(x, y), \tag{14}$$

where ϕ and ψ are some functions of x and y. This yields

$$\ddot{x} = -\gamma(t)\phi_x(x, y)\dot{x} + \omega(t)\psi_y(x, y)\dot{y}, \quad \ddot{y} = -\gamma(t)\phi_x(x, y)\dot{y} - \omega(t)\psi_y(x, y)\dot{x}. \tag{15}$$

The complex kinetic energy is given by:

$$K_c = \Psi + i\Phi = e^{p_x}\cos p_y + ie^{p_x}\sin p_y = e^{p_x+ip_y} = e^P, \quad P = p_x + ip_y. \tag{16}$$

If we consider the (complex) potential energy $U_c = \Gamma(t)z$, then the equation of motion resulting from the Hamiltonian $H = e^P + \Gamma z$ is

$$\ddot{z} = \Gamma(t)\dot{z}. \tag{17}$$

Suppose $Re(P) = p$ and $Re(z) = a(t)x$, then the Hamiltonian becomes $H_R = e^p + q$ and this yields $\ddot{x} = -a(t)\dot{x}$. Note that a nonlinear potential energy with the same K.E. yields nonlinear equations, for example, $H_1 = e^p + a(t)\ln x$ and $H_2 = e^p + a(t)x^n$ yield the following equations of motion: *viz*

$$x\ddot{x} + a(t)\dot{x} = 0, \qquad \ddot{x} + na(t)x^{n-1}\dot{x} = 0. \tag{18}$$

respectively.

Equation (17) admits, for constant Γ, a Lagrangian:

$$L(z,\dot{z}) = -z + \Gamma^{-1}\dot{z}\log\dot{z}, \tag{19}$$

with (17) following from the associated Euler–Lagrange equation.

2.2. Illustration: Generalized Liénard Equation and the Calogero–Leyvraz Lagrangian

We have seen that the Calogero and Leyvraz construction yields interesting sets of dynamical equations. In this section, we formulate a nonlinear ODE belonging to the Liénard class of equations.

The Liénard type ordinary second order nonlinear differential equation is given by:

$$\ddot{q} + f(q)\dot{q} + g(q) = 0. \tag{20}$$

f and g are two continuously differentiable functions on $\mathbb{R}$. Since the Liénard equation itself is also an autonomous differential equation, the substitution, $y = \frac{dq}{dt}$ or $q = \int y(t)dt$, leads the Liénard equation to become a first order differential,

$$y\frac{dy}{dq} + f(q)y + g(q) = 0, \tag{21}$$

which belongs to the Abel equation of the second kind. This can also be expressed in terms of the Abel equation of the first kind, which we use later.

Let us define the following generalization of the Calogero–Leyvraz Lagrangian:

$$L = (\dot{q} + f(q))\ln(\dot{q} + f(q)) - f(q). \tag{22}$$

The Euler–Lagrange equation yields:

$$\ddot{q} + f'\dot{q} + f'\ln(\dot{q} + f(q)) = 0. \tag{23}$$

For small values of $(\dot{q} + f(q))$, the above equation reduces to the well-known nonstandard Lagrangian,

$$\ddot{q} + 2f'\dot{q} + ff'(q) = 0. \tag{24}$$

For different choices of f we get different types of equations. Let $f(q) = \lambda q^n$, then $n = 1$; this becomes a damped oscillator equation, $n = 2$; this maps to a second Riccati or modified Emden–Fowler equation. The corresponding Lagrangian is given by:

$$L_R = \ln(\dot{q} + f(q)). \tag{25}$$

2.3. *Deformations of Calogero–Leyvraz Lagrangians and κ-Deformed Oscillator Equations*

The most attractive feature of the Calogero–Leyvraz Lagrangian is the involvement of a logarithmic term. We grab this opportunity and deform the (entropic) kinetic energy term. We use primarily Kaniadakis and Tsallis logarithms. At first we deform (22) using the Tsallis logarithm.

Let us introduce the Tsallis logarithm. We assume $q > 0$ for practical purposes. The Tsallis q-logarithm and q-exponential functions are defined by:

$$\ln_q(x) = \frac{x^{1-q} - 1}{1-q}, \quad \exp_q(x) = (1 + (1-q)x)^{\frac{1}{1-q}}, \tag{26}$$

where $q \neq 1$ and $1 + (1-q)x \neq 0$. For $q \to 1$

$$(1 + \frac{x}{N})^N \approx e^x, \quad N(x^{\frac{1}{N}} - 1) \approx \ln x.$$

Proposition 1. *Let* $\ln_\kappa(\dot{q} + f(q))$ *be the Tsallis κ-deformed logarithm. The Euler–Lagrange equation of the Lagrangian*

$$L = (\dot{q} + f(q)) \ln_\kappa(\dot{q} + f(q)) - \mu f(q), \quad -1 < \kappa < 1, \tag{27}$$

yields

$$(\kappa+1)\Big((\ddot{q} + f'\dot{q}) - \frac{1}{\kappa} f'(q)(\dot{q} + f(q))\Big) - \frac{1}{\kappa}\Lambda f'(q)(\dot{q} + f(q))^{1-\kappa} = 0, \tag{28}$$

where $\Lambda = \mu \frac{\kappa}{\kappa+1} - 1$.

Proof. By direct computation we obtain:

$$\frac{d}{dt}\Big(\frac{\partial L}{\partial \dot{q}}\Big) = (\kappa+1)^2(\ddot{q} + f'\dot{q})(\dot{q} + f(q))^\kappa,$$

$$\begin{aligned}\frac{\partial L}{\partial q} &= f'(q)\big((1+\kappa)(\dot{q} + f(q))^{\kappa 1} + \ln_\kappa(\dot{q} + f(q)) + \mu f'(q) \\ &= \frac{\kappa+1}{\kappa} f'(q)\Big((1+\kappa)(\dot{q} + f(q))^\kappa - 1 + \mu\frac{\kappa}{\kappa+1}\Big).\end{aligned}$$

Here, we have tacitly used the formula of $\ln_\kappa x$. □

Corollary 1. *Suppose* $\mu = \frac{\kappa+1}{\kappa}$. *The Euler–Lagrange equation of the Lagrangian*

$$L = (\dot{q} + f(q)) \ln_\kappa(\dot{q} + f(q)) - \frac{\kappa+1}{\kappa} f(q), \quad -1 < \kappa < 1, \tag{29}$$

yields the

$$\ddot{q} + \frac{\kappa-1}{\kappa} f'(q)\dot{q} - \frac{1}{\kappa} f' f = 0. \tag{30}$$

Let us demonstrate this with a couple of examples. Suppose we take $\kappa = \frac{1}{2}$, then (28) reduces to

$$\frac{3}{2}(\ddot{q} - f'(q)\dot{q} - 2f' f(q)) - 2\Lambda f'(q)\sqrt{(\dot{q} + f(q))} = 0. \tag{31}$$

If we take $f(q) = \frac{a}{q}^2$, then Equation (30) yields the second-order Riccati (also known as the modified Emden) equation,

$$\ddot{q} + a\frac{\kappa-1}{\kappa} q\dot{q} - \frac{1}{\kappa} a^2 q^3 = 0. \tag{32}$$

2.4. Kaniadakis κ-Deformed Lagrangian, Liénard Equation and Chiellini Integrability Condition

We wish to repeat this calculation using the Kaniadikis κ-deformed logarithm. We obtain the following result. Let $\ln_\kappa q = \frac{1}{2\kappa}(q^\kappa - q^{-\kappa})$ be the Kaniadikis logarithm.

Proposition 2. *The Euler–Lagrange equation for the Lagrangian $L = (\dot{q} + f(q))\ln_\kappa(\dot{q} + f(q)) - \mu f(q)$, yields*

$$\begin{aligned}&(\ddot{q} + f'\dot{q} - \frac{\kappa}{\kappa+1}f'(q)) - \frac{\kappa-1}{\kappa+1}(\ddot{q} + f'\dot{q}\\&-\frac{\kappa}{\kappa-1}f'(q))(\dot{q} + f(q))^{-2\kappa} + \frac{2\kappa\mu}{\kappa+1}f'(q)(\dot{q}+f(q))^{-\kappa+1} = 0.\end{aligned}\tag{33}$$

Suppose we take $\kappa = \frac{1}{2}$ and $f(q) = \frac{3}{2}q^2$, then (33) reduces to

$$\ddot{q} + 3q\dot{q} - q + (\ddot{q} + 3q\dot{q} + 3q)(\dot{q} + \frac{3}{2}q^2)^{-1} + 2q\sqrt{\dot{q} + \frac{3}{2}q^2} = 0. \tag{34}$$

Consider the κ-deformed Lagrangian without the potential μ term.

Corollary 2. *The Euler–Lagrange equation corresponding to the Lagrangian L_κ yields a one parameter family of second-order equations,*

$$\begin{aligned}&(1+\kappa)(\ddot{q} + f'(q)\dot{q}) + (1-\kappa)(\ddot{q} + f'(q)\dot{q})(\dot{q} + f(q))^{-2\kappa}\\&-\frac{f'(q)}{\kappa}(\dot{q} + f(q))\big((1+\kappa) - (1-\kappa)(\dot{q}+f(q))^{-2\kappa}\big) = 0.\end{aligned}\tag{35}$$

This equation is a fractional damped system except for $\kappa = \pm\frac{n}{2}$, where $n \in \mathbb{Z}$.

The Liénard type ordinary second order nonlinear differential equation can be mapped to the first kind first order Abel differential equation,

$$\frac{dy}{dq} = f(q)y^3 + g(q)y^2. \tag{36}$$

This Abel equation allows us to find some exact general solutions of the Liénard type equations by using the integrability conditions of the Abel equation.

Lemma 1. *A first kind Abel type differential equation of the form (36) can be exactly integrated if the functions $q(x)$ and $p(x)$ satisfy the condition:*

$$\frac{d}{dq}\Big(\frac{g(q)}{f(q)}\Big) = \mu f(q), \quad \mu = \text{constant}, \quad \mu \neq 0. \tag{37}$$

Claim 1. *Equation (35) reduces to the Liénard equation for $\kappa = -1$,*

$$\ddot{q} + \frac{1}{2}f'(q)\dot{q} - \frac{1}{2}f'(q)f(q) = 0, \tag{38}$$

which satisfies the Chiellini condition.

One can also readily verify that for $\kappa = -1/2$, Equation (35) satisfies the generalized Liénard equation,

$$F(q,\dot{q})\ddot{q} + 2f'(q)\dot{q} - f'f(3f+2) = 0, \tag{39}$$

where $F(q,\dot{q}) = (1 + \dot{q} + 3f(q))$.

3. Entropic Lagrangian and Integrable Class of Systems

In this section, at first we give a new derivation of Lotka–Volterra and replicator equations using Calogero–Leyvraz Lagrangians endowed with the "Shannon entropic" [34] type kinetic energy terms. Then, using the cross entropy type kinetic energy term, we derive the $N = 2$ relativistic Toda lattice equation.

3.1. Calogero–Leyvraz Lagrangian and Lotka–Volterra Equation

We start with the derivation of the Lotka–Volterra equation. Consider the following logarithmic Lagrangian endowed with an entropic kinetic term:

$$L = (1 - \dot{q}) \ln (1 - \dot{q}) - aq - ae^{-q}. \tag{40}$$

The Euler–Lagrange equation yields:

$$\frac{\ddot{q}}{1 - \dot{q}} + 1 - e^{-q} = 0. \tag{41}$$

Let us write this equation as a system of first-order equations. Define:

$$1 - \dot{q} = z, \quad e^{-q} = y. \tag{42}$$

Then (41) equation can recasted as:

$$\dot{y} = (zy - y), \qquad \dot{z} = a(z - zy). \tag{43}$$

This is a standard form of the celebrated 2D Lotka–Volterra equation in non-dimensionalized form [23].

In the standard formalism of the Lotka–Volterra equation (43), the Hamiltonian is given by:

$$H = z - \ln z + ay - a \ln y. \tag{44}$$

The nonstandard Hamiltonian form,

$$\begin{pmatrix} \dot{y} \\ \dot{z} \end{pmatrix} = \begin{pmatrix} 0 & yz \\ -yz & 0 \end{pmatrix} \begin{pmatrix} \frac{\partial H}{\partial y} \\ \frac{\partial H}{\partial z}, \end{pmatrix}$$

yields Equation (43). The two Hamiltonians can be connected easily through exponential mapping.

3.2. Replicator Equation

In 1978, Taylor and Jonker [41] introduced a system of differential equations which were designated later on as the replicator equation. This equation plays an important role in evolutionary game theory. The replicator equation models the frequency evolution of certain strategic behaviors within a biological population. Hofbauer [25] unveiled an equivalence relation between the Lotka–Volterra equation and the replicator equation.

Consider the first population where individuals interact with each other according to a set of $n+1$ pure strategies $E_0, \cdots, E_n$ with relative frequencies $x_0, \cdots, x_n$, and the second population plays different $m+1$ pure strategies $F_0, \cdots, F_m$ with frequencies $y_0, \cdots, y_m$. After a contest E_i versus F_j, the payoff for the first player is a_{ij} whereas for the second player it is b_{ji}. Let $A = (a_{ij})$ be the matrix consisting of these a_{ij} and so also B, then for such games the evolutionary dynamics is given by:

$$\dot{x}_i = x_i\big((Ay)_i - x\dot{A}y\big), \quad i = 0, \cdots, n \tag{45}$$

$$\dot{y}_j = y_j\big((Bx)_j - y\dot{B}x\big), \quad j = 0, \cdots, m. \tag{46}$$

Adding or multiplying a (positive) constant to each column of A or B does not alter the dynamics.

In the case of $n = m = 1$, the above equations simplify to:

$$\dot{x} = x(1-x)\big(a-(a+b)y\big), \quad \dot{y} = y(1-y)\big(-c+(c+d)x\big). \tag{47}$$

We will use the Calogero–Leyvraz type Lagrangian to derive the planar replicator type equation.

Calogero–Leyvraz Lagrangian and Replicator Equation

We define the Lagrangian of the coupled system as:

$$L = \dot{q}_i \ln \dot{q}_i + (1-\dot{q}_i)\ln(1-\dot{q}_i) + \lambda \dot{q}_1 q_2 - \mu \dot{q}_2 q_1 + a q_1 - c q_2, \quad i = 1,2. \tag{48}$$

It is straight forward to see that the Lagrangian (48) yields:

$$\ddot{q}_1 = \dot{q}_1(1-\dot{q}_1)\big(a-(\lambda+\mu)\dot{q}_2\big), \quad \ddot{q}_2 = \dot{q}_2(1-\dot{q}_2)\big(-c+(\lambda+\mu)\dot{q}_1\big). \tag{49}$$

Let us define:

$$\dot{q}_1 = x, \quad \dot{q}_2 = y. \tag{50}$$

We obtain the replicator equation from (49):

$$\dot{x} = x(1-x)\big(a-(\lambda+\mu)y\big), \quad \dot{y} = y(1-y)\big(-c+(\lambda+\mu)x\big). \tag{51}$$

3.3. *Logarithmic Lagrangian Formulation of $N = 2$ Relativistic Toda Lattice Equation*

In this example, we consider a Lagrangian with the *cross entropic* type kinetic energy term,

$$L = \dot{q}\ln(\dot{q}+\sqrt{\dot{q}^2+1}) - \big((\dot{q}+\sqrt{\dot{q}^2+1}) - (\dot{q}+\sqrt{\dot{q}^2+1})^{-1}\big) - \cosh q. \tag{52}$$

Proposition 3. *The Euler–Lagrange equation, the Lagrangian (52), yields:*

$$\frac{\ddot{q}}{\sqrt{\dot{q}^2+1}} = -\sinh q. \tag{53}$$

Proof. After an elaborate calculation from the entropic K. E. term (L_1), it yields:

$$\frac{\partial L_1}{\partial \dot{q}} = \ln(\dot{q}+\sqrt{\dot{q}^2+1}) + \frac{\dot{q}}{\sqrt{\dot{q}^2+1}},$$

and the second K.E. term $\big((\dot{q}+\sqrt{\dot{q}^2+1}) - (\dot{q}+\sqrt{\dot{q}^2+1})^{-1}\big) = L_2$ yields:

$$\frac{\partial L_2}{\partial \dot{q}} = \frac{\dot{q}}{\sqrt{\dot{q}^2+1}}.$$

These two expressions lead to a magical cancellation of the term $\frac{\dot{q}}{\sqrt{\dot{q}^2+1}}$. Using the Euler–Lagrange equation we obtain the equation. □

The Ruijsenaars Hamiltonian is given by:

$$H = \sum_{n=1}^{N} \big(1 + q^{-1/2} R^2 e^{q_n - q_{n+1}}\big) e^{R p_n}, \tag{54}$$

where $q = e^{iR\hbar}$.

For $N = 2$ case this equation reduces to:

$$H_2 = e^{Rp_1} + e^{Rp_2} + R^2\left(e^{q_1 - q_2 + Rp_1} + e^{q_2 - q_1 + Rp_2}\right). \tag{55}$$

Consider the centre of mass frame $p_1 + p_2 = 0$. Let us define:

$$p := Rp_1, \quad q := q_1 - q_2 + Rp_1. \tag{56}$$

We express H_2 as:

$$\hat{H}_2 = e^p + e^{-p} + R^2\left(e^q + e^{-q}\right); \tag{57}$$

for practical purposes we scaled $R = 1$.

Let us express Hamiltonian (57) in terms of cosine hyperbolic function

$$H = \cosh p + \cosh q, \tag{58}$$

where we drop the factor 2. The Hamiltonian equation yields:

$$\dot{q} = \frac{\partial H}{\partial p} = \sinh p, \qquad \dot{p} = -\frac{\partial H}{\partial q} = -\sinh p, \tag{59}$$

which reduces to Equation (53).

3.3.1. Connection to Calabi–Yau Manifold

We must note that the energy function $E = e^p + e^{-p} + e^q + e^{-q}$ can be expressed as:

$$X + X^{-1} + Y + Y^{-1} = E. \tag{60}$$

This defines a genus one Riemann surface. The complex 3D space $V = X + X^{-1} + Y + Y^{-1} - E$ describes a Calabi–Yau manifold. This sets up a connection with the Calabi–Yau manifold. The Riemann surface has enough information to describe this Calabi–Yau manifold. The energy function $E(p,q)$ considered to be Hamiltonian appears in the string theory.

Mirror symmetry states that a CY manifold has its mirror dual. The Kähler structure of the original CY is mapped to the complex structure of the mirror CY, and vice versa. In our case, the mirror curve is given by:

$$e^p + \mu_1 e^{-p} + e^q + \mu_2 e^{-q} = 1, \tag{61}$$

where μ_1 and μ_2 are the complex moduli of the mirror CY.

The new equation is in the same form as the Lagrangian of the $N = 2$ relativistic Toda lattice equation. Our case is similar to the case of the quantized mirror curve for the local $\mathbb{P}_1 \times \mathbb{P}_1$, which is related to the quantum eigenvalue problem of the relativistic Toda lattice with just two particles.

3.3.2. Calogero–Leyvraz Type Lagrangian with Coupling Constant and Mirror Map

Consider the following map:

$$L = \dot{q}\ln(\dot{q} + \sqrt{\dot{q}^2 + 1}) - ((\dot{q} + \sqrt{\dot{q}^2 + 1}) - (\dot{q} + \sqrt{\dot{q}^2 + 1})^{-1}) - \gamma^{-1}\cosh q, \tag{62}$$

where γ^{-1} is a coefficient (or coupling) parameter. The corresponding Hamiltonian or energy function is given as:

$$E = (e^p + e^{-p}) + \gamma^{-1}(e^q + e^{-q}). \tag{63}$$

This yields:

$$\frac{1}{E} = \frac{1}{E^2}(e^p + e^{-p}) + \frac{1}{\gamma E}(e^q + e^{-q}).$$

Let us change the variable,

$$e^p \mapsto \frac{e^p}{E}, \quad e^q \mapsto \frac{\gamma^{-1}}{E}e^q. \tag{64}$$

4. The κ-Deformed 2D Lotka–Volterra, Replicator and Relativistic Toda Lattice Equations

By deforming the natural logarithm and exponential functions we present κ-deformed 2D Lotka–Volterra and relativistic Toda lattice equations in this section. We will use both Kaniadakis and Tsallis deformations to derive new sets of κ-deformed systems.

4.1. The κ Deformation of 2D Lotka–Volterra Equation

We describe two types of κ-deformed systems, semi-deformation and full deformation. In the first case we only deformed the kinetic part, whereas in the second case we consider both the kinetic energy (K.E.) and potential energy (P.E.) parts of the deformations.

Case 1: (deforming only K.E) Consider the following κ-deformed Lagrangian:

$$L_\kappa = (1-\dot{q})\ln_\kappa(1-\dot{q}) - q - e^{-q}. \tag{65}$$

Proposition 4. *The Euler–Lagrange equation of the deformed Lagrangian (65) reads:*

$$\frac{\ddot{q}}{1-\dot{q}} Exp_\kappa \ln(1-\dot{q}) + 1 - e^{-q} = 0. \tag{66}$$

Proof. It is easy to see that:

$$\frac{\partial L_\kappa}{\partial \dot{q}} = -\ln_\kappa(1-\dot{q}) - \frac{1}{2}((1-\dot{q})^\kappa + (1-\dot{q})^{-\kappa}),$$

$$= -\frac{1}{\kappa}\sinh \kappa \ln(1-\dot{q}) - \cosh \kappa \ln(1-\dot{q}).$$

Thus we obtain:

$$\frac{d}{dt}(\frac{\partial L_\kappa}{\partial \dot{q}}) = \frac{\ddot{q}}{1-\dot{q}}(\cosh \kappa \ln(1-\dot{q}) + \kappa \sinh \kappa \ln(1-\dot{q})) = \frac{\ddot{q}}{1-\dot{q}} Exp_\kappa \ln(1-\dot{q}),$$

where the generalized κ deformed exponential is given by

$$Exp_\kappa x = \cosh \kappa x + \kappa \sinh \kappa x. \tag{67}$$

The final result follows from the remaining part of the calculation. □

It is clear that when $\kappa \to 0$ we recover the ordinary Lotka–Volterra equation.

Case 2: (deforming both K.E. and P.E.) In this case we also deformed the exponential term e^{-q} in the potential. We take the following Lagrangian:

$$L^d_\kappa = (1-\dot{q})\ln_\kappa(1-\dot{q}) - q - exp_\kappa(-q). \tag{68}$$

Thus we obtain the following result from the straightforward computation.

Proposition 5. *The Euler–Lagrange equation of the deformed Lagrangian L^d_κ yields:*

$$\frac{\ddot{q}}{1-\dot{q}} Exp_\kappa ln(1-\dot{q}) + 1 - \frac{1}{\sqrt{1+\kappa^2 q^2}} exp_\kappa(-q) = 0. \tag{69}$$

4.1.1. Expressing κ-Deformed Equation

The inverse of the generalized κ deformed exponential $Exp_\kappa(x)$ is given by:

$$Exp_\kappa(x)^{-1} = Exp_{-\frac{1}{\kappa}}(x) = \cosh \kappa x - \frac{1}{\kappa}\sinh \kappa x. \tag{70}$$

We now express Equation (66) in a standard form. Let us define:

$$1 - \dot{q} = z, \quad y = e^{-q}. \tag{71}$$

Thus we obtain:

$$\dot{y} = -e^{-q}\dot{q} = -y(z-1), \quad \frac{-\dot{z}}{z} Exp_\kappa \ln z + 1 - y = 0.$$

This can be expressed as:

$$\dot{y} = y - zy, \qquad \dot{z} Exp_\kappa \ln z = z - yz. \tag{72}$$

We recover the original equation when $\kappa \to 0$.
Let $w = \ln z$ or $z = e^w$. Then the second equation becomes

$$\dot{w}w = Exp_{-\frac{1}{\kappa}}(w)(1-y). \tag{73}$$

A further change of variable $p = \frac{1}{2}w^2$ yields a modified set of deformed Lotka–Volterra equations:

$$\dot{q} = (e^{\sqrt{2p}} - 1), \quad \dot{p} = d(1 - e^{-q}), \quad \text{where} \quad d = Exp_{-\frac{1}{\kappa}}(\sqrt{2p}). \tag{74}$$

Hence we express the deformed Lotka–Volterra equation in a standard form using the generalized κ deformed exponential function.

4.1.2. Tsallis Logarithm and Deformed Lotka–Volterra System

In section we express the Tsallis logarithm and exponential in terms of κ, which are given as:

$$\ln_\kappa(q) = \frac{q^\kappa - 1}{\kappa}, \quad \exp_\kappa(q) = (1+\kappa q)^{\frac{1}{\kappa}}. \tag{75}$$

We now deform the Lotka–Volterra Lagrangian using the κ-deformed Tsallis logarithm and exponential. It is defined as:

$$L_\kappa^T = (1-\dot{q})\ln_\kappa(1-\dot{q}) - q - \exp_\kappa(-q), \tag{76}$$

with

$$\ln_\kappa(1-\dot{q}) = \frac{(1-\dot{q})^\kappa - 1}{\kappa}, \quad \exp_\kappa(-q) = (1-\kappa q)^{\frac{1}{\kappa}}, \tag{77}$$

where $1 - \kappa q > 0$.

Proposition 6. *The Euler–Lagrange equation associated with the Tsallis deformed Lagrangian L_κ^T yields:*

$$\ddot{q}(1+\kappa)(1-\dot{q})^{\kappa-1} + 1 - \frac{\exp_\kappa(-q)}{(1-\kappa q)} = 0. \tag{78}$$

Proof. Using the properties of the Tsallis logarithm and exponential functions we arrive at our desired result. □

One can readily check that, when $\kappa \to 0$, Equation (79) reduces to the usual Lotka–Volterra equation. If we assume only the deformation of the kinetic term using the Tsallis logarithm, then Equation (79) reduces to:

$$\ddot{q}(1+\kappa)^2(1-\dot{q})^{\kappa-1}+1-\exp(-q)=0. \tag{79}$$

4.2. The κ-Deformed Replicator Equation

We can deform the Lagrangian of the replicator Equation (48) using the Kaniadakis or Tsallis deformation of logarithm term. Using Kaniadakis deformation we obtain:

$$L=\dot{q}_i \ln_\kappa \dot{q}_i+(1-\dot{q}_i)\ln(1-\dot{q}_i)+\lambda\dot{q}_1 q_2-\mu\dot{q}_2 q_1+aq_1-cq_2, \quad i=1,2. \tag{80}$$

Proposition 7. *With the Euler–Lagrange equation related to the deformed Lagrangian (80), we obtain the Kaniadakis deformed coupled equation:*

$$\ddot{q_1}\big((1-\dot{q_1})\cosh\kappa\ln\dot{q_1}+\dot{q_1}\cosh\kappa\ln(1-\dot{q_1})\big)=\dot{q_1}(1-\dot{q_1})\big(a-(\lambda+\mu)\dot{q_2}\big), \tag{81}$$

$$\ddot{q_2}\big((1-\dot{q_2})\cosh\kappa\ln\dot{q}_2+\dot{q}_2\cosh\kappa\ln(1-\dot{q_2})\big)\big(-c+(\lambda+\mu)\dot{q}_1\big). \tag{82}$$

Corollary 3. *The Kaniadakis κ deformed replicator equation is given by:*

$$\dot{x}\big((1-x)\cosh\kappa\ln x+x\cosh\kappa\ln(1-x)\big)=x(1-x)\big(a-(\lambda+\mu)y\big), \tag{83}$$

$$\dot{y}\big((1-y)\cosh\kappa\ln y+y\cosh\kappa\ln(1-y)\big)=y(1-y)\big(-c+(\lambda+\mu)x\big), \tag{84}$$

where $x=\dot{q}_1$ *and* $y=\dot{q}_2$.

One can readily see when $\kappa \to 0$ the deformed replicator Equations (83) and (84) reduces to the original replicator equation.

We can repeat the same procedure using Tsallis deformation of the logarithm. The coupled equations are given by:

$$(1+\kappa)\ddot{q}_1\big(\dot{q}_1^{\kappa-1}+(1-\dot{q}_1)^{\kappa-1}\big)=a-(\lambda+\mu)\dot{q}_2, \quad (1+\kappa)\ddot{q}_2\big(\dot{q}_2^{\kappa-1}+(1-\dot{q}_2)^{\kappa-1}\big)=-c+(\lambda+\mu)\dot{q}_1, \tag{85}$$

which leads to the Tsallis deformed replicator equation,

$$(1+\kappa)\dot{x}\big(x^{\kappa-1}+(1-x)^{\kappa-1}\big)=a-(\lambda+\mu)y, \quad (1+\kappa)\dot{y}\big(y^{\kappa-1}+(1-y)^{\kappa-1}\big)=-c+(\lambda+\mu)x. \tag{86}$$

This again reduces to the original replicator equation for $\kappa \to 0$.

4.3. The κ-Deformed N = 2 Relativistic Toda Lattice System

In this section, at first we also modify the entropic kinetic energy term, keeping all other terms unchanged. The Kaniadakis κ-deformed Lagrangian for the $N=2$ relativistic Toda lattice system is defined as:

$$L_{1\kappa}=\dot{q}\ln_\kappa(\dot{q}+\sqrt{\dot{q}^2+1})-\frac{1}{2}(\dot{q}+\sqrt{\dot{q}^2+1})+\frac{1}{\dot{q}+\sqrt{\dot{q}^2+1}})-\gamma^{-1}\cosh q. \tag{87}$$

Proposition 8. *The Euler–Lagrange equation corresponding to the κ-deformed Lagrangian* $L_{1\kappa}$ *yields:*

$$\frac{\ddot{q}}{\sqrt{1+\dot{q}^2}}\cosh\kappa\ln(\dot{q}+\sqrt{1+\dot{q}^2})+\frac{d}{dt}\Big(\frac{\dot{q}}{\sqrt{1+\dot{q}^2}}\big(\cosh\kappa\ln(\dot{q}+\sqrt{1+\dot{q}^2})-1\big)\Big)+\gamma^{-1}\sinh q=0. \tag{88}$$

Proof. This proof follows from the direct computation. □

One can readily check that for $\kappa \to 0$ (89) reduces to the ordinary $N=2$ relativistic Toda lattice equation.

For the most general case, we deform the potential term $\cosh q$ too, which yields the following Lagrangian:

$$L =_{1\kappa}= \dot{q}\ln_\kappa(\dot{q}+\sqrt{\dot{q}^2+1}) - \frac{1}{2}(\dot{q}+\sqrt{\dot{q}^2+1}) + \frac{1}{\dot{q}+\sqrt{\dot{q}^2+1}}) - \gamma^{-1}\cosh_\kappa q.$$

We obtain κ-deformed $N = 2$ relativistic Toda lattice equation,

$$\frac{\ddot{q}}{\sqrt{1+\dot{q}^2}}\cosh\kappa\ln(\dot{q}+\sqrt{1+\dot{q}^2}) + \frac{d}{dt}\Big(\frac{\dot{q}}{\sqrt{1+\dot{q}^2}}(\cosh\kappa\ln(\dot{q}+\sqrt{1+\dot{q}^2})-1)\Big) + \frac{1}{\sqrt{1+\dot{q}^2}}\gamma^{-1}\sinh_\kappa q = 0. \quad (89)$$

This yields the most general Kaniadakis κ-deformation of the $N = 2$ relativistic Toda lattice equation which reduces to the original one when κ goes to zero.

Tsallis Deformed $N = 2$ Relativistic Toda Lattice Equation

In this section we present the deformation of the the $N = 2$ relativistic Toda lattice equation using Tsallis deformation. Let the entropic part of the kinetic term be given by:

$$L_{KE} = \dot{q}\ln_\kappa(\dot{q}+\sqrt{\dot{q}^2+1}) = \dot{q}\frac{(\dot{q}+\sqrt{\dot{q}^2+1})^\kappa - 1}{\kappa}.$$

We now compute the equation of motion using the Tsallis deformed kinetic energy.

Proposition 9. *The Euler–Lagrange equation of the Tsallis κ-deformed Lagrangian*

$$L_\kappa^{RT} = \dot{q}\ln_\kappa(\dot{q}+\sqrt{\dot{q}^2+1}) - \frac{1}{2}(\dot{q}+\sqrt{\dot{q}^2+1} + \frac{1}{\dot{q}+\sqrt{\dot{q}^2+1}}) - \gamma^{-1}\cosh q \quad (90)$$

yields

$$\frac{\ddot{q}}{\sqrt{\dot{q}^2+1}}(\dot{q}+\sqrt{\dot{q}^2+1})^\kappa + (\frac{\dot{q}}{\sqrt{\dot{q}^2+1}}(\dot{q}+\sqrt{\dot{q}^2+1})^\kappa - 1) + \gamma^{-1}\sinh q = 0. \quad (91)$$

5. Outlook

In this paper we considered a special class of Lagrangians proposed by Calogero and Leyvraz with an "exotic" kinetic energy term. This term has a close resemblance to the Shannon entropy function, $\dot{q}\ln\dot{q}$. Using this new type of Lagrangian, we derived the celebrated Lotka–Volterra and replicator equation. We then generalized the construction of Calogero and Leyvraz and considered a different type of kinetic energy term based on cross entropy. We then manufactured an $N = 2$ relativistic Toda lattice system. We also discussed the significance of this equation in modern physics. Different avatars of this equation appeared in string theory and theoretical high energy physics—purely imaginary position and momentum coordinates lead to the Hofstadter model.

The main goal is to express all these celebrated equations in terms of logarithmic kinetic energy using the deformation of the entropic kinetic energy term. We used the Kaniadakis κ-deformed logarithm and exponential functions to deform these Calogero–Leyvraz type Lagrangians to give a new formulation of κ -deformed Lotka–Volterra, replicator equation and $N = 2$ relativistic Toda lattice system. We also extended this deformation to the Tsallis class and derived Tsallis-deformed equations. All the original equations can be recovered from the deformed systems when $\kappa \to 0$.

In a nutshell, this paper elucidated the strength of the Calogero–Leyvraz formalism based on entropic kinetic terms. It would be interesting to derive more known and not so well known systems using this method. The predator–prey models are one of the best places to apply our scheme. We may try to apply this scheme to planar generalized Lotka–Volterra (GLV) equations; for example, consider two interacting populations with densities $x > O$

and $y > O$ with the simplest formal description of interaction with the linear dependence of the growth rates $\dot{x}/x$ and $\dot{y}/y$. This yields the following GLV equation:

$$\dot{x} = x(a + bx + cy), \quad \dot{y} = y(d + ex + fy).$$

We can generalize this construction and check whether we can manufacture this new equation using the Calogero–Leyvraz formalism. We then implement the κ - deformation of such equations and study their dynamics.

It would be worth investigating the Calabi–Yau manifold connected to the κ-deformed Hamiltonian of the $N = 2$ relativistic Toda lattice equation.

Funding: This research received funding from Khalifa University of Science and Technology under grant number FSU-2021-014.

Data Availability Statement: This article does not use data. No new data have been created or analysed in the present manuscript.

Acknowledgments: The author is immensely grateful to Francesco Calogero, Francois Leyvraz, Haret Rosu and Anindya Ghose Chaudhury for many valuable inputs in different stages of this work. He is also thankful to Giorgio Kaniadakis and Antonio Scarfone for their interest. Finally I would like to thank the anonymous reviewer for their suggestions and valuable comments.

Conflicts of Interest: The authors declare no conflict of interest.

References

1. Calogero, F.; Leyvrez, F. Time-independent Hamiltonians describing systems with friction: the "cyclotron with friction". *J. Nonlinear Math. Phys.* **2019**, *26*, 147–154. [CrossRef]
2. Leyvraz, F.; Calogero, F. A Hamiltonian yielding damped motion in an homogeneous magnetic field: Quantum treatment. *J. Nonlinear Math. Phys.* **2019**, *26*, 228–239. [CrossRef]
3. Guha, P. Balanced gain-loss dynamics of particle in cyclotron with friction, κ-deformed logarithmic Lagrangians and fractional damped systems. *Eur. Phys. J. Plus* **2022**, *137*, 64. [CrossRef]
4. Guha, P. The κ-deformed entropic Lagrangians, Hamiltonian dynamics and their applications. *Eur. Phys. J. Plus* **2022**, *137*, 932. [CrossRef]
5. Tsallis, C. Possible Generalization of Boltzmann–Gibbs Statistics. *J. Stat. Phys.* **1988**, *52*, 479–487. [CrossRef]
6. Schwämmle, V.; Tsallis, C. Two-parameter generalization of the logarithm andexponential functions and Boltzmann–Gibbs-Shannon entropy. *J. Math. Phys.* **2007**, *48*, 113301. [CrossRef]
7. Kaniadakis, G. Non-linear kinetics underlying generalized statistics. *Phys. A Stat. Mech. Appl.* **2001**, *296*, 405–425. [CrossRef]
8. Kaniadakis, G. Statistical mechanics in the context of special relativity. *Phys. Rev. E* **2002**, *66*, 056125. [CrossRef]
9. Kaniadakis, G. Theoretical foundations and Mathematical formalism of the power-law tailed statistical distributions. *Entropy* **2013**, *15*, 3983–4010. [CrossRef]
10. Kaniadakis, G. Relativistic kinetics and power-law tailed distributions. *Eur. Phys. Lett.* **2010**, 92, 35002. [CrossRef]
11. Kaniadakis, G.; Baldi, M.M.; Deisboeck, T.S.; Grisolia, G.; Hristopulos, D.T.; Scarfone, A.M.; Sparavigna, A.; Wada, T.; Lucia, U. The κ-statistics approach to epidemiology. *Sci. Rep.* **2020**, *10*, 19949. [CrossRef] [PubMed]
12. Kaniadakis, G. New power-law tailed distributions emerging in κ-statistics. *Eur. Phys. Lett.* **2021**, *133*, 10002. [CrossRef]
13. Newman, M.E.J. Power laws, Pareto distributions and Zipf's law. *Contemp. Phys.* **2005**, *46*, 323–351. [CrossRef]
14. Saichev, A.; Malevergne, Y.; Sornette, D. *Theory of Zipf's Law and Beyond*; Lecture Notes in Economics and Mathematical Systems; Springer: Berlin/Heidelberg, Germany, 2009; Volume 632.
15. Singh, S.K.; Maddala, G.S. A function for size distribution of Incomes. *Ecnometrica* **1976**, *44*, 963–970. [CrossRef]
16. Scarfone, A.M. Entropic Forms and Related Algebras. *Entropy* **2013**, *15*, 624–649. [CrossRef]
17. Kaniadakis, G.; Scarfone, A.M.; Sparavigna, A.; Wada, T. Composition law of κ-entropy for statistically independent systems. *Phys. Rev. E* **2017**, *95*, 052112. [CrossRef]
18. Ilic, V.; Korbel, J.; Gupta, S.; Scarfone, A. An overview of generalized entropic forms. *EPL* **2021**, *133*, 50005. [CrossRef]
19. Hanel, R.; Thurner, S. A comprehensive classification of complex statistical systems and an axiomatic derivation of their entropy and distribution functions. *Eur. Phys. Lett.* **2011**, *93*, 20006. [CrossRef]
20. Scarfone, A.M.; Matsuzoe, H.; Wada, T. Information Geometry of κ-Exponential Families: Dually-Flat, Hessian and Legendre Structures. *Entropy* **2018**, *20*, 436. [CrossRef]
21. Wada, T.; Scarfone, A.M. Information Geometry on the κ-Thermostatistics. *Entropy* **2015**, *17*, 1204–1217. [CrossRef]
22. Lotka, A.J. *Elements of Physical Biology*; Williams & Wilkins Co.: Baltimore, MD, USA, 1925.
23. Murray, J.D. *Mathematical Biology I: An introduction*, 3rd ed.; Springer: New York, NY, USA, 2002.

24. Volterra, V. Variazioni e fluttuazioni del numero d'individui in species animali conviventi. *Mem. R. Accad. Naz. Lincei. (Ser. VI)* **1926**, *2*, 31–113.
25. Hofbauer, J. Evolutionary dynamics for bimatrix games: A Hamiltonian system? *J. Math. Biol.* **1996**, *34*, 675–688. [CrossRef]
26. Hofbauer, J.; Sigmund, K. *The Theory of Evolution and Dynamical Systems*; Cambridge University Press: Cambridge, UK, 1988.
27. Hofstadter, D.R. Energy levels and wave functions of Bloch electrons in rational and irrational magneticfields. *Phys. Rev. B* **1976**, 142239–142249
28. Hatsuda, Y.; Mariño, M. Exact quantization conditions for the relativistic Toda lattice. *J. High Energy Phys.* **2016**, *5*, 133. [CrossRef]
29. Hatsuda, Y.; Katsura, H.; Tachikawa, Y. Hofstadter's butterfly in quantum geometry. *New J. Phys.* **2016**, *18*, 10, 103023. [CrossRef]
30. Seiberg, N.; Witten, E. Monopole condensation, and confinement in $N = 2$ supersymmetric Yang-Mills theory. *Nucl. Phys. B* **1994**, 42619–42652.
31. Nekrasov, N. Five-dimensional gauge theories and relativistic integrable systems. *Nucl. Phys. B* **1998**, *531*, 323–344. [CrossRef]
32. Mariño, M. Open string amplitudes and large order behavior in topological string theory. *J. High Energy Phys.* **2008**, *3*, 060. [CrossRef]
33. Ruijsenaars, S.N.M. Relativistic Toda systems. *Commun. Math. Phys.* **1990**, *133*, 217. [CrossRef]
34. Shannon, C.E. A mathematical theory of communication. *Bell Syst. Tech. J.* **1948**, *27*, 379–423. [CrossRef]
35. Berry, M.V.; Shukla, P. Classical dynamics with curl forces, and motion driven by time-dependent flux. *J. Phys. A* **2012**, *45*, 305201. [CrossRef]
36. Berry, M.V.; Shukla, P. Hamiltonian curl forces. *Proc. R. Soc. A* **2015**, *471*, 20150. [CrossRef]
37. Chiellini, A. Sull'integrazione dell'equazione differenziale $y' + Py^2 + Qy^3 = 0$. *Boll. Dell'unione Mat. Ital.* **1931**, *10*, 301-307.
38. Choudhury, A.G.; Guha, P. Chiellini integrability condition, planar isochronous systems and Hamiltonian structures of Liénard equation. *Discret. Contin. Dyn. Syst. B* **2017**, *22*, 2465–2478. [CrossRef]
39. Harko, T.; Lobo, F.S.N.; Mak, M.K. A class of exact solutions of the Liénard type ordinary non-linear differential equation. *J. Eng. Math.* **2014**, *89*, 193-205. [CrossRef]
40. Mancas, S.C.; Rosu, H.C. Integrable dissipative nonlinear second-order differential equations via factorizations and Abel equations. *Phys. Lett. A* **2013**, *377*, 1434. [CrossRef]
41. Taylor, P.D.; Jonker, L.B. Evolutionarily stable strategies and game dynamics. *Math. Biosci.* **1978**, *40*, 145–156. [CrossRef]

entropy

MDPI

Article

Gravity and Cosmology in Kaniadakis Statistics: Current Status and Future Challenges

Giuseppe Gaetano Luciano

Applied Physics Section of Environmental Science Department, Universitat de Lleida, Av. Jaume II, 69, 25001 Lleida, Spain

Abstract: Kaniadakis statistics is a widespread paradigm to describe complex systems in the relativistic realm. Recently, gravitational and cosmological scenarios based on Kaniadakis (κ-deformed) entropy have been considered, leading to generalized models that predict a richer phenomenology comparing to their standard Maxwell–Boltzmann counterparts. The purpose of the present effort is to explore recent advances and future challenges of Gravity and Cosmology in Kaniadakis statistics. More specifically, the first part of the work contains a review of κ-entropy implications on Holographic Dark Energy, Entropic Gravity, Black hole thermodynamics and Loop Quantum Gravity, among others. In the second part, we focus on the study of Big Bang Nucleosynthesis in Kaniadakis Cosmology. By demanding consistency between theoretical predictions of our model and observational measurements of freeze-out temperature fluctuations and primordial abundances of ^4He and D, we constrain the free κ-parameter, discussing to what extent the Kaniadakis framework can provide a successful description of the observed Universe.

Keywords: Kaniadakis entropy; relativistic theory; gravity; cosmology; big bang nucleosynthesis

Citation: Luciano, G.G. Gravity and Cosmology in Kaniadakis Statistics: Current Status and Future Challenges. *Entropy* **2022**, *24*, 1712. https://doi.org/10.3390/e24121712

Academic Editors: Dionissios T. Hristopulos, Sergio Luiz E. F. da Silva and Antonio M. Scarfone

Received: 3 November 2022
Accepted: 22 November 2022
Published: 24 November 2022

1. Introduction

In the last several decades, several approaches of statistical mechanics have been used in high energy physics to analyze cosmological models [1], particle interactions [2], Lorentz-violating extensions of the Standard Model [3], black holes and other gravitational systems [4–6]. Despite the different contexts, a common thread among all of these studies is the adoption of Boltzmann–Gibbs–Shannon (BGS) entropy, which conducts the celebrated Maxwell–Boltzmann exponential distribution according to the Jaynes maximum entropy principle. However, it is well-known that Boltzmann–Gibbs formalism exhibits severe restrictions when applied to many complex systems, such as out-of-equilibrium, long-interacting and thermally fluctuating systems [7]. This motivates the introduction of a more general setting that contains the Maxwell–Boltzmann distribution measure as a special case.

Among the most popular generalizations, evidence from relativistic particle systems [8,9] has suggested a non-exponential distribution function with power tails that originated from *Kaniadakis entropy* [10–12]

$$S_\kappa = -\sum_i n_i \ln_\kappa n_i \tag{1}$$

where the κ-deformed logarithm is defined by (here, and henceforth, we use natural units $k_B = \hbar = G = c = 1$:)

$$\ln_\kappa x \equiv \frac{x^\kappa - x^{-\kappa}}{2\kappa} \qquad -1 < \kappa < 1 \tag{2}$$

and the generalized Boltzmann factor for the i-th level of the system takes the form

$$n_i = \alpha \exp_\kappa[-\beta(E_i - \mu)]. \tag{3}$$

Here, the κ-deformed exponential is given by

$$\exp_\kappa(x) \equiv \left(\sqrt{1+\kappa^2 x^2} + \kappa x\right)^{1/\kappa} \tag{4}$$

while

$$\alpha = [(1-\kappa)/(1+\kappa)]^{1/2\kappa} \qquad 1/\beta = \sqrt{1-\kappa^2}\, T\,. \tag{5}$$

As shown in [11], the statistical model developed from the generalized functions (2) and (4) emerges naturally and unequivocally within the framework of special relativity. Notice that the κ parameter is not fixed by the theory and should be constrained via theoretical and/or observational analyses. It is straightforward to check that the classical $\kappa \to 0$ limit reproduces the ordinary (Maxwell–Boltzmann) statistical mechanics, thus making Equation (1) a self-consistent relativistic generalization of BGS entropy formula.

By using the definition of microcanonical ensemble, it has been argued that, for the case of black holes, the κ-entropy (1) can be rewritten in the form [13,14]

$$S_\kappa = \frac{1}{\kappa}\sinh(\kappa\, S_{BH}) \tag{6}$$

where $S_{BH} = A/(4)$ is the standard Bekenstein–Hawking entropy, which is still recovered for $\kappa \to 0$. Since $S_\kappa = S_{-\kappa}$, in what follows, we can restrict to the $\kappa \geq 0$ case without a loss of generality.

While being expressly formulated for black holes, Equation (6) can also be used within the cosmological framework in the lines of gravity-thermodynamic conjecture. For instance, in [15], Drepanou et al. have shown that Holographic Dark Energy based on Kaniadakis entropy (6) leads to interesting cosmological behavior, retracing the standard thermal history of the Universe in good agreement with observations [16,17]. Similarly, in [18,19] (and references therein), Kaniadakis statistics has been used in gravity scenarios to address the Jeans instability and simulate dark matter-like effects, respectively. Applications of κ-entropy can also be found in plasma physics, astrophysics, information theory, fluid dynamics and other fields (see [20] for a recent review). All of this makes Kaniadakis statistics a very flexible framework that can potentially adapt to the diversity of relativistic physical contexts where Maxwell–Boltzmann distribution fails, thus motivating a careful analysis of the subject.

In passing, we mention that, besides Kaniadakis formulation, there exist many other generalized entropies which are relevant and commonly used in physics, for instance Tsallis, Abe, Landsberg–Vedral, Sharma–Mittal, Rény and Barrow entropies, among others (see [21] for a detailed review), all containing the classical BGS entropy as a special case. While exhibiting some mathematical similarities with κ-deformation, these entropy measures are better suited to describe statistical properties of long-interacting, dissipative or large-scale fluctuating systems, but do not work properly in the relativistic realm, which is the framework of this review. Clearly, one could in principle extend the present study to the above family of generalized entropies to see how temperature fluctuations, non-extensive (Tsallis-like) or quantum gravity (Barrow-like) corrections affect ensuing scenarios in comparison with Kaniadakis conjecture. Some of these research lines have already been explored in literature [22–26]. However, a systematic investigation of these aspects goes beyond the scope of the present review and will be considered elsewhere.

Starting from the above premises, in the present manuscript, we focus on the study of Kaniadakis statistics applied to Gravity and Cosmology. The structure of the work can be basically divided into two parts: the first one (Section 2) contains a review of recent advances in the literature. Special care is devoted to examine:

- implications on open stellar clusters: such systems are physically related group of stars held together by the mutual gravitational attraction. Since their constituents typically have similar age and chemical composition, they provide very important laboratories where stellar properties compared to isolated field stars can be studied. In [22], Carvalho et al. have shown that the characteristic relaxation mechanism associated with radial orbital instability cannot be understood in the classical Maxwell–Boltzmann framework, emphasizing the need of non-Gaussian (Kaniadakis-like) statistics to fit the distribution of stellar residual radial velocity in some baseline stellar open clusters.

- Jeans instability and gravitational collapse: the dynamical stability of a self-gravitating system can be described by the Jeans criterion, which states that, if the wavelength of a density fluctuation inside the system is greater than a certain threshold given by the Jeans wavelength, then the density will grow in time exponentially, and the system becomes gravitationally unstable. In [27], this criterion has been revisited in the context of Kaniadakis statistics, obtaining a κ-deformed critical wavelength larger than the standard expression. Similar studies have also been developed in Eddington-inspired Born–Infeld [28] and $f(R)$ [29] gravity, and the dark-baryonic matter model [30], among others.
- Holographic Cosmology: Holographic Dark Energy (HDE) is a theoretical framework that arises from the attempt of applying the holographic principle to the dark energy problem [31]. A crucial ingredient in the construction of this model is the relationship between the entropy of the Universe (conceived as a thermodynamic system) and its geometrical properties, such as its radius. The standard HDE scenario is built upon Bekenstein–Hawking entropy, which arises as the black hole application of the BGS one. However, in [13–15,32], a generalized HDE based on Kaniadakis entropy has been investigated along with its implications on the cosmic evolution and thermal properties of the Universe (see also [23,24,33–35] for further applications). Remarkably, it has been shown that Kaniadakis dark energy exhibits peculiar features that do not have any correspondence in the traditional HDE, potentially providing a way to alleviate the Hubble tension.
- Entropic gravity: Verlinde's conjecture of entropic gravity [36] presents gravitational force as an emergent (rather than fundamental) force caused by changes in the information associated with the positions of material bodies. Starting from this idea, an effective gravitational constant can be derived and used to introduce Kaniadakis statistics, the ensuing method being a simpler alternative to the usual procedure employed in non-Gaussian statistics. In [25], such a formalism has been applied to infer Kaniadakis-induced corrections to the Jeans criterion for self-gravitating systems, as well as to establish a connection with deviations of Newton's law arising in a submillimeter range (for the sake of transparency, it must be said that the issue of whether Newton's law exhibits deviations from inverse square behavior in submillimeter regime is quite controversial. For instance, in [37], it has been found that there are still no deviations in separation down to $\mathcal{O}(10^2)$ µm.) [38].
- Black hole thermodynamics: inspired by a dual Rény entropy [39,40], in [41], Abreu et al. have suggested and applied a dual Kaniadakis entropy to black hole thermodynamics. In this way, a generalized equipartition theorem has been derived, leading to a κ-modified black hole temperature and heat capacity. In addition, it has been argued that black holes in Kaniadakis statistics could exhibit a thermally stable phase, thus opening new glimpses into the study of black hole thermodynamics at both theoretical and phenomenological levels [41].
- Loop Quantum Gravity (LQG): this is a well-known non-perturbative and background independent theory of gravity which aims to merge Quantum Mechanics and General Relativity [42]. One of the characteristic parameters of LQG is the so-called Immirzi parameter [43], which is an arbitrary number that measures the size of the quantum of area in Planck units. By using Kaniadakis statistics, Abreu et al. have derived a non-trivial relation between the Immirzi parameter, the κ deformation parameter and the area of a punctured surface [44], which is a topological two-sphere with defects carrying spin quantum numbers endowed by the edges of the spin network that represents the bulk quantum geometry. The question arises as to whether Kaniadakis statistics might play any role in the context of quantum gravity.

On the other hand, the second part of the manuscript (Section 3) provides the original contribution of this work. Here, we study consequences of Kaniadakis Holographic Dark Energy (KHDE) on Big Bang Nucleosynthesis (BBN). Specifically, we constrain the deformation κ-parameter by requiring consistency between theoretical predictions of our model and observational data of primordial abundances and freeze-out temperature fluctuations, which only allow for very tiny deviations from General Relativity. Finally, in Section 4, we summarize results and discuss some possible future challenges of Kaniadakis statistics aimed at both broadening the current research lines and opening novel prospects in this field.

2. Gravity and Cosmology in Kaniadakis Statistical Theory: Recent Advances

In this section, we discuss some recent findings pertaining to Gravity and Cosmology in Kaniadakis statistics. Our aim is to highlight the advantages of Kaniadakis model in describing some phenomena which are not well framed (or not even understood) in Maxwell–Boltzmann theory.

2.1. Open Stellar Clusters

Open stellar clusters are a type of star cluster made of up to a few thousand stars that have roughly the same age and composition, being formed from the same giant molecular cloud. The knowledge of the different properties of these clusters, such as the distribution of dispersion velocity and phase density, is needed to establish the statistical laws and the relaxation mechanisms that rule their evolution.

There are essentially three main mechanisms: collisional relaxation, which is characterized by a Maxwell-like distribution; the Lynden–Bell relaxation, leading to a Fermi distribution; and a relaxation associated with radial orbit instability that attains a non-monotonic distribution. While the first two mechanisms are well described by the standard statistical mechanics, the last one is not well-understood yet.

The above issue has been examined in [22] in the background of Kaniadakis statisticsSpecifically, Carvalho et al. have investigated the effects of non-Gaussianity on the distribution of stellar residual radial velocity in some open clusters' samples. The generalized κ-distribution function they find for the radial velocity v_r has the form

$$\begin{aligned} \phi_\kappa(v_r) &= A_\kappa\left[\sqrt{1+\kappa^2\left(\frac{v_r^2}{\sigma_\kappa^2}\right)^2}-\kappa\frac{v_r^2}{\sigma_\kappa^2}\right]^{1/\kappa} \\ &= A_\kappa\,\exp_\kappa\left(-\frac{v_r^2}{\sigma^2}\right), \end{aligned} \tag{7}$$

where A_κ is a constant, and σ_κ denotes the characteristic distribution width. In the second step, we have used the definition (4). For $\kappa \to 0$, this expression reduces to the standard Gaussian distribution

$$\phi(v_r) = A\,\exp\left(-\frac{v_r^2}{\sigma^2}\right), \tag{8}$$

as expected.

By using the Kolmogorov–Smirnov statistical test, the best $\phi_\kappa(v_r)$ has been obtained for each observed cumulative distribution of the residual radial velocities. As a result, it has been shown that the generalized Kaniadakis distribution fits data much better than the standard Gaussian does, provided that one allows the κ parameter to be varying with the stellar-cluster ages. Below, we will show that a similar running behavior is supported by completely independent arguments in the framework of Kaniadakis Cosmology.

2.2. Jeans Instability and Gravitational Collapse

Jeans criterion provides a condition to establish whether a self-gravitating system is stable under the effects of its internal gas pressure. The so-called Jeans length that represents the watershed between stable and unstable systems is given by [45]

$$\lambda_J = \sqrt{\frac{\pi T}{\mu m_H \rho_0}} \tag{9}$$

where T is the temperature, μ the mean molecular weight, m_H the hydrogen atomic mass and ρ_0 the equilibrium mass density of the system, respectively.

According to Jeans instability, if the wavelength λ of a density perturbation is higher than λ_J, then the density grows exponentially, giving rise to an unstable system. Otherwise, stability is kept. The same criterion can also be expressed in terms of a critical mass for self-gravitating systems (see Section 2.4).

The condition (9) follows from the canonical equipartition theorem in Maxwell–Boltzmann statistics. Nevertheless, motivated by relativistic considerations, in [27], Abreu et al. have shown that the critical density λ_J gets non-trivially modified in the context of Kaniadakis statistics. In particular, by using the κ-generalized equipartition theorem

$$E_\kappa = \frac{1}{2}N\frac{\left(1+\frac{\kappa}{2}\right)}{\left(1+\frac{3}{2}\kappa\right)2\kappa}\frac{\Gamma\left(\frac{1}{2\kappa}-\frac{3}{4}\right)\Gamma\left(\frac{1}{2\kappa}+\frac{1}{4}\right)}{\Gamma\left(\frac{1}{2\kappa}+\frac{3}{4}\right)\Gamma\left(\frac{1}{2\kappa}-\frac{1}{4}\right)}T \tag{10}$$

and Verlinde formalism of entropic gravity [36], the following expression for the κ-deformed critical wavelength has been exhibited

$$\lambda_c^\kappa = \sqrt{\frac{\left(1+\frac{\kappa}{2}\right)}{\left(1+\frac{3}{2}\kappa\right)2\kappa}\frac{\Gamma\left(\frac{1}{2\kappa}-\frac{3}{4}\right)\Gamma\left(\frac{1}{2\kappa}+\frac{1}{4}\right)}{\Gamma\left(\frac{1}{2\kappa}+\frac{3}{4}\right)\Gamma\left(\frac{1}{2\kappa}-\frac{1}{4}\right)}}\,\lambda_J \tag{11}$$

where Γ is the Gamma function. From this equation, we infer that Jeans instability is modified in such a way that:

- for $\kappa = 0 \Longrightarrow \lambda_c^\kappa = \lambda_J$, i.e., the classical criterion is restored;
- for $0 < \kappa < 2/3 \Longrightarrow \lambda_c^\kappa > \lambda_J$;
- for $\kappa \to 2/3^- \Longrightarrow \lambda_c^\kappa \to \infty$, which means that the self-gravitating systems are always stable (notice that for $\kappa \geq 2/3$ the modified equipartition law based on Kaniadakis statistics diverges, thus making the derivation of the generalized Jeans criterion meaningless.).

Apart from the limit case $\kappa = 0$, we then see that the modified critical wavelength is always larger than the corresponding Maxwell–Boltzmann value. In other terms, Kaniadakis statistics predicts self-gravitating systems to be more stable compared to the classical scenario.

Furthermore, one can show that the κ-deformed entropy also affects the physical temperature of gravitating systems and the velocity of the propagation of sound inside them. The resulting expressions are [27]

$$T_\kappa = \frac{\left(1+\frac{\kappa}{2}\right)}{\left(1+\frac{3}{2}\kappa\right)2\kappa}\frac{\Gamma\left(\frac{1}{2\kappa}-\frac{3}{4}\right)\Gamma\left(\frac{1}{2\kappa}+\frac{1}{4}\right)}{\Gamma\left(\frac{1}{2\kappa}+\frac{3}{4}\right)\Gamma\left(\frac{1}{2\kappa}-\frac{1}{4}\right)}T \qquad v_\kappa^s = \sqrt{\frac{T_\kappa}{m}}\,. \tag{12}$$

These results have been tested considering 16 galaxy clusters. It has been found that Boltzmann–Gibbs statistics are consistent with data, although non-Gaussian effects cannot be completely ruled out, constraining $0 \leq \kappa \leq 0.034$ at the 2σ confidence level.

It is worth mentioning that Jeans criterion in Kaniadakis statistics has also been addressed in other different contexts. For instance, in [28], Jeans instability has been revisited in the framework of Eddington-inspired Born–Infeld gravity, showing that the κ-deformed distribution may have non-negligible effects on the Jeans modes of the collisionless Eddington-inspired Born–Infeld gravitational systems. In a similar fashion, the influence of the κ-generalized Jeans criterion has been examined in [29] in $f(R)$ gravity for both high and low frequency density perturbations. As a result, it has been proven that the range of the unstable modes and the growth rates decrease with increasing values of κ. A further step forward has been taken in [30], where implications of Kaniadakis statistics have been explored on gravitational systems composed also by dark and baryonic matter. The analysis of the κ-modified dispersion equation for such systems has pointed out that Jeans instability is suppressed comparing to the standard case, implying that Kaniadakis corrections oppose the gravitational collapse. Clearly, all of the above outcomes disclose a new class of phenomena and/or mechanisms, which potentially allow us to highlight signatures of Kaniadakis statistics in gravitational systems.

2.3. Holographic Dark Energy

Holographic Dark Energy is a dynamical model of dark energy built on the usage of the holographic principle and Bekenstein–Hawking area law. Although cosmological applications of HDE have been extensively considered in the past literature [46–48], its shortcomings in reproducing the thermal history of the Universe have motivated some tentative changes over the years [14,15,49–53]. Among these generalizations, promising results have been provided by HDE based on Kaniadakis entropy (Kaniadakis Holographic Dark Energy, KHDE). Several models of KHDE have been proposed: here, we refer to the approach of [14], which correctly reduces to the usual HDE in the $\kappa \to 0$ limit and does not involve large deviations from standard entropy to describe the Universe evolution, as it should be according to Equation (2). For other possible extensions, see also [13,32,33].

Starting from the entropy (6) and using the gravity-thermodynamic conjecture, in [14], Lymperis et al. have derived modified Friedmann equations ruling the evolution of a homogeneous and isotropic Friedmann–Robertson–Walker (FRW) geometry filled with matter and dark energy fluids and bounded by the apparent horizon (see Section 3). In turn, these equations allow for computing characteristic cosmic parameters, such as the Equation of State (EoS) parameter, the deceleration parameter, the squared speed of sound and the Hubble parameter, to be compared with the theoretical predictions of the Λ-CDM model of Cosmology. A more detailed experimental analysis of KHDE has been carried out in [15–17], showing that KHDE predictions do agree with observational data and might contribute to alleviate the H_0 tension too. In particular, concerning the dark energy EoS parameter, it has been found to exhibit a phenomenology richer than standard HDE, being quintessence-like, phantom-like, or experiencing the phantom-divide crossing in the past or in the future. In addition, observations from Supernovae type Ia and Baryon Acoustic Oscillations data enable constraining the Kaniadakis parameter around the vanishing value, consistently with the expectation of small deviations from Gaussian-like statistics in nature.

2.4. Entropic Gravity

In the entropic gravity formalism by Verlinde [36], gravitational force is conceived as an entropic force caused by changes in the information associated with the positions of material bodies. This conjecture combined with the generalized Kaniadakis' equipartition law gives an effective gravitational constant in the form [25]

$$G_\kappa = \frac{(1+\frac{3}{2}\kappa)2\kappa}{(1+\frac{\kappa}{2})} \frac{\Gamma\left(\frac{1}{2\kappa}+\frac{3}{4}\right)\Gamma\left(\frac{1}{2\kappa}-\frac{1}{4}\right)}{\Gamma\left(\frac{1}{2\kappa}-\frac{3}{4}\right)\Gamma\left(\frac{1}{2\kappa}+\frac{1}{4}\right)} G \tag{13}$$

where here we have restored the gravitational constant G for the sake of clarity.

In the same spirit as Section 2.2, Equation (13) can be used to describe Kaniadakis statistics effects on Jeans mass criterion in self-gravitating systems. Specifically, Abreu et al. have derived an expression for the modified Jeans critical mass as [25]

$$M_J^\kappa = \left[\frac{\left(1+\frac{\kappa}{2}\right)}{\left(1+\frac{3}{2}\kappa\right)2\kappa} \frac{\Gamma\left(\frac{1}{2\kappa}-\frac{3}{4}\right)\Gamma\left(\frac{1}{2\kappa}+\frac{1}{4}\right)}{\Gamma\left(\frac{1}{2\kappa}+\frac{3}{4}\right)\Gamma\left(\frac{1}{2\kappa}-\frac{1}{4}\right)} \right]^{3/2} M_J, \tag{14}$$

where

$$M_J = \left(\frac{5T}{m}\right)^{3/2} \left(\frac{3}{4\pi\rho}\right)^{1/2} \tag{15}$$

is the usual Jeans mass in Maxwell–Boltzmann framework, while all other quantities are defined as in Section 2.2. As before, one can distinguish three possible regimes given by: (*i*) $\kappa = 0$, which implies instability for $M > M_J^\kappa = M_J$; (*ii*) $0 < \kappa < 2/3$, which entails $M > M_J^\kappa > M_J$; and (*iii*) $\kappa \to 2/3^-$, in which $M_J^\kappa \to \infty$, giving rise to an always stable gravitational system. This confirms the previous result that Kaniadakis entropy opposes gravitational collapse.

Another important physical quantity considered in [25] is the free fall time

$$t_{FF} = \sqrt{\frac{3}{2\pi\rho}}, \tag{16}$$

which is defined as the time necessary to the system to finally collapse. In the case of Kaniadakis statistics, this turns out to be modified as [25]

$$t_{FF}^\kappa = \left[\frac{\left(1+\frac{\kappa}{2}\right)}{\left(1+\frac{3}{2}\kappa\right)2\kappa} \frac{\Gamma\left(\frac{1}{2\kappa}-\frac{3}{4}\right)\Gamma\left(\frac{1}{2\kappa}+\frac{1}{4}\right)}{\Gamma\left(\frac{1}{2\kappa}+\frac{3}{4}\right)\Gamma\left(\frac{1}{2\kappa}-\frac{1}{4}\right)} \right]^{1/2} t_{FF}. \tag{17}$$

which indicates that $t_{FF}^\kappa > t_{FF}$, except for the $\kappa = 0$ case, where equality is recovered. Thus, the self-gravitating system in Kaniadakis scenario takes more time to collapse comparing to the Gaussian framework.

As a further application, in [25], a connection has been studied between the modifications of Newton's law induced by the generalized gravitational constant (13) and possible deviations arising in a submillimeter range, which are parameterized by [38]

$$G(r) = G\left[1+\alpha\left(1+\frac{r}{\lambda}\right)e^{-r/\lambda}\right], \tag{18}$$

where α is a dimensionless parameter and λ gives the energy (or length) scale at which departures of Newton's law from the standard inverse square behavior should occur.

In so doing, Abreu et al. have obtained a relationship between α and the Kaniadakis parameter in the form

$$\alpha = \frac{\left(1+\frac{3}{2}\kappa\right)2\kappa}{\left(1+\frac{\kappa}{2}\right)} \frac{\Gamma\left(\frac{1}{2\kappa}+\frac{3}{4}\right)\Gamma\left(\frac{1}{2\kappa}-\frac{1}{4}\right)}{\Gamma\left(\frac{1}{2\kappa}-\frac{3}{4}\right)\Gamma\left(\frac{1}{2\kappa}+\frac{1}{4}\right)} - 1. \tag{19}$$

This result opens up the tantalizing possibility of probing Kaniadakis-like deviations from Gaussianity via tests of the gravitational inverse square law in experimentally accessible regions.

2.5. Black Hole Thermodynamics

Inspired by a novel type of Rény entropy proposed in [40], Abreu et al. have introduced a dual Kaniadakis entropy in the form [41]

$$\begin{aligned} S_\kappa^* &= \frac{1}{\kappa}\log\left(\kappa\, S_{BH} + \sqrt{1+\kappa^2\, S_{BH}^2}\right) \\ &= \log[\exp_\kappa(S_{BH})], \end{aligned} \tag{20}$$

where S_{BH} is the Bekenstein–Hawking entropy defined below Equation (6).

Based on the above equation, one can derive a modified Hawking temperature and heat capacity of black holes as [41]

$$T = \frac{\sqrt{1+16\kappa^2\pi^2M^4}}{8\pi M} \tag{21}$$

$$C_{BH} = -\frac{8\pi M^2\left(1+16\kappa^2\pi^2M^4\right)^{1/2}}{1-16\kappa^2\pi^2\,M^4} \tag{22}$$

where M denotes the black hole mass. Notice that both of the above expressions reproduce the semiclassical Hawking results

$$T = 1/(8\pi M) \qquad C_{BH} = -8\pi M^2 \tag{23}$$

in the $\kappa \to 0$ limit, as it should be.

Remarkably, from Equation (22), we see that C_{BH} takes negative (positive) values for $M < [2(\kappa\pi)^{1/2}]^{-1}$ ($M > [2(\kappa\pi)^{1/2}]^{-1}$), leading to a thermally unstable (stable) black hole. Such a result implies that it is possible for a phase transition to occur in the dual Kaniadakis statistics framework—a result which has no correspondence in the Maxwell–Boltzmann scenario. This points out the potential relevance of dual Kaniadakis entropy in the analysis of black hole thermodynamics.

2.6. Loop Quantum Gravity

Implications of Kaniadakis statistics have also been analyzed in Loop Quantum Gravity, which arises from the effort to grasp what quantum spacetime is at the fundamental level. More specifically, this formalism is characterized by quantum operators for areas and volumes that exhibit discrete spectra.

One of the peculiar parameters of LQG is the so-called Immirzi parameter, which is a free dimensionless quantity that provides the size of a quantum of area in Planck units. A way to compute this parameter is by counting the number of microstates of a given system in LQG. For black holes, this is typically accomplished by use of the Bekenstein–Hawking entropy area law, which roots its origin in the BGS entropy. As a result, one obtains [43]:

$$\gamma = \frac{\log 2}{\pi\sqrt{3}}. \tag{24}$$

This expression can be straightforwardly generalized to the background of Kaniadakis statistics by using the κ-deformed entropy in the microcanonical ensemble. Calculations have been carried out in [44] for a generic system of surface area A, leading to

$$\begin{aligned} \gamma_\kappa &= \gamma\frac{\frac{\kappa A}{4}}{\log\left[\frac{\kappa A}{4}+\sqrt{1+\frac{\kappa^2A^2}{16}}\right]} \\ &= \gamma\frac{\frac{A}{4}}{\log\left[\exp_\kappa\left(\frac{A}{4}\right)\right]} \end{aligned} \tag{25}$$

which correctly reduces to γ for $\kappa \to 0$. Since the extra factor appearing in the above relation is greater than unity, we have $\gamma_\kappa > \gamma$, resulting in a larger size of the quantum of area in Planck units.

The outlined κ-dependence of the Immirzi parameter reveals a non-trivial interplay between LQG and Kaniadakis statistics. Hence, much effort is needed to better understand the potential rôle of Kaniadakis entropy within the framework of quantum gravity.

3. Big Bang Nucleosynthesis in Kaniadakis Statistics

In physical Cosmology, Big Bang Nucleosynthesis (BBN) refers to the sequence of nuclear processes which synthesized primordial light elements, such as Hydrogen H, Deuterium D, Helium isotopes ^3He and ^4He and Lithium isotope ^7Li [54]. Clearly, since BBN drives the whole evolution of Universe's chemical composition, BBN parameters must be very tightly constrained to be consistent with current observations. Therefore, this phenomenon provides an unparalleled arena to test cosmological models and constrain related parameters with great accuracy.

An interesting issue to address is how relativistic degrees of freedom of the early Universe affect BBN when described in the more proper framework of Kaniadakis statistics. To solve the problem, let us first derive modified Friedmann equations in Kaniadakis Cosmology. For this purpose, we mainly refer to [14,15], though we feature κ-induced corrections in a slightly different way.

Consider a homogeneous and isotropic FRW flat geometry of metric

$$ds^2 = -dt^2 + a^2(t)\left(dr^2 + r^2 d\Omega^2\right), \tag{26}$$

where $a(t)$ is the scale factor and t the cosmic time. In addition, we assume that the Universe is filled up with with a matter perfect fluid of equilibrium mass density ρ_0 and pressure $p_0 = w\rho_0$, where $-1 \le w \le 1/3$ is the EoS parameter.

Invoking the gravity-thermodynamic conjecture, we can think of our Universe as a thermodynamic system bounded by an apparent horizon $r_a = 1/H = a/\dot{a}$ and endowed with a temperature and entropy obeying the same rules as for black holes (in our notation, the dot indicates a derivative with respect to the cosmic time) [55]. In this scenario, by using the energy–momentum tensor of matter content, the continuity equation and BH entropy area law, we are led to the canonical Friedmann equations

$$-4\pi\left(\rho_0 + p_0\right) = \dot{H} \tag{27}$$

$$\frac{8\pi}{3}\rho_0 = H^2 \tag{28}$$

where cosmological constant effects have been neglected.

The question now arises as how Equations (27) and (28) get modified when using the κ-deformed entropy (6) instead of BH entropy. Following the same recipe as above, one arrives to the Kaniadakis–Friedmann equations [14]

$$-4\pi(\rho + p) = \cosh\left(\kappa\frac{\pi}{H^2}\right)\dot{H} \tag{29}$$

$$\frac{8\pi}{3}\rho = \cosh\left(\kappa\frac{\pi}{H^2}\right)H^2 - \kappa\pi\,\mathrm{shi}\left(\kappa\frac{\pi}{H^2}\right), \tag{30}$$

where ρ and p now denote the total energy density and pressure including Kaniadakis corrections, and

$$\mathrm{shi}(x) \equiv \int_0^x \frac{\sinh(x')}{x'}\,dx'. \tag{31}$$

Notice that the classical scenario is easily recovered for $\kappa \to 0$.

To make the κ-dependence of the l.h.s. in Equations (29) and (30) explicit, let us recast the total energy density and pressure as

$$\rho = \rho_0 + \delta\rho_\kappa \tag{32}$$

$$p = p_0 + \delta p_\kappa \tag{33}$$

where we have separated out Kaniadakis-induced corrections $\delta\rho_\kappa$ and δp_κ, respectively.

We then expand $\cosh(x)$ and $\mathrm{shi}(x)$ in Equations (29) and (30) for small κ, which is indeed the case, since departures from Maxwell–Boltzmann statistics are expected to be small. We obtain the leading order

$$-4\pi(\rho_0 + p_0 + \delta\rho_\kappa + \delta p_\kappa) \simeq \dot{H}\left(1 + \frac{\pi^2}{2}\frac{\kappa^2}{H^4}\right), \tag{34}$$

$$\frac{8\pi}{3}(\rho_0 + \delta\rho_\kappa) \simeq H^2 - \frac{\pi^2}{2}\frac{\kappa^2}{H^2} \tag{35}$$

which gives [56]

$$\delta\rho_\kappa \simeq -\frac{9}{128}\frac{\kappa^2}{\rho_0}, \tag{36}$$

$$\delta p_\kappa \simeq \frac{21}{128}\frac{\kappa^2}{\rho_0}. \tag{37}$$

For our next purposes, it is now convenient to rewrite the modified Friedmann Equation (35) in the equivalent form

$$H(\rho) \equiv H(\rho_0)\, Z_\kappa(\rho) \tag{38}$$

where $H(\rho_0)$ is the unmodified Hubble parameter obeying Equation (28) and

$$Z_\kappa(\rho) \equiv 1 + \frac{9}{256}\frac{\kappa^2}{\rho^2}. \tag{39}$$

This can be further manipulated by resorting to Equations (32) and (36) and expressing the equilibrium energy density ρ_0 as a function of the temperature according to

$$\rho_0(T) = \frac{\pi^2 g(T)}{30}\, T^4 \tag{40}$$

where $g(T)$ denotes the effective number of degrees of freedom of the Universe at temperature T. Since in the following we shall focus on the radiation dominated epoch, we can roughly set $g(T) \simeq 10$. In so doing, we obtain

$$H(\rho) \to H(T) = \frac{2\pi}{3}\sqrt{\frac{\pi g(T)}{5}}\, T^2\, Z_\kappa(T) \tag{41}$$

where

$$Z_\kappa(T) \approx 1 + \frac{2025}{64\,\pi^4}\frac{\kappa^2}{g^2(T)\, T^8}. \tag{42}$$

Before going further, we point out that, in the ordinary Cosmology based on GR and BH entropy, the Z function takes unit value, as it can be seen by considering the $\kappa \to 0$ limit. In general, departures of $Z - 1$ from zero could emerge from either extended formulations of gravity, including alternative geometric frameworks and/or different entropic scenarios, or by introducing extra particle degrees of freedom in the standard theory. Since in the present work we are interested in effects induced by Kaniadakis statistics, we focus on the first setting, neglecting corrections brought about by exotic particles. A similar analysis has been recently proposed in [57,58] in the context of Tsallis statistics and generalizations of the Heisenberg principle induced by gravity, respectively.

3.1. Freeze-Out Mechanism

According to the BBN model, the current abundances of the first very light atomic nuclei were already nearly defined few minutes after the initial Big Bang, when the energy and number density were still dominated by relativistic degrees of freedom—leptons and photons [59]. Owing to their continuous and repeated collisions, such particles were in thermal equilibrium. In turn, protons and neutrons were kept in equilibrium through the reactions

$$(a)\ \ \nu_e + n \longleftrightarrow p + e^- \tag{43a}$$

$$(b)\ \ e^+ + n \longleftrightarrow p + \bar{\nu}_e \tag{43b}$$

$$(c)\ \ n \longleftrightarrow p + e^- + \bar{\nu}_e\,. \tag{43c}$$

In this scenario, neutron abundance can be estimated by computing the rate of conversion $\lambda_{pn}(T)$ of protons into neutrons and its inverse

$$\lambda_{np}(T) = e^{-\mathcal{Q}/T}\lambda_{pn}(T) \tag{44}$$

where $\mathcal{Q} = m_n - m_p \simeq 1.29\,\text{MeV}$, with $m_{n(p)}$ being the neutron (proton) mass. The rate $\lambda_{np}(T)$ is given by the sum of the three rates for the processes *(a)*, *(b)* and *(c)*, i.e.,

$$\lambda_{np}(T) = \lambda_a(T) + \lambda_b(T) + \lambda_c(T)\,. \tag{45}$$

The reactions (43) went on until the decreasing temperature and density content of the Universe caused them to become too slow, at about the *freeze-out* temperature $T_{0f} \simeq 0.6\,\text{MeV}$. In compliance with [59], we require that T during the freeze-out period was low enough compared to the typical energy scale for the processes (43). In addition, we assume to neglect the electron mass m_e with respect to the electron and neutrino energies. Under these hypotheses, one obtains [59]

$$\lambda_a(T) \simeq qT^5 + \mathcal{O}\left(\frac{\mathcal{Q}}{T}\right) = \lambda_b(T) \tag{46}$$

where $q \simeq 10^{-10}\,\text{GeV}^{-4}$. On the other hand, $\lambda_c(T)$ is roughly three orders of magnitude lower than $\lambda_a(T)$ and can in principle be neglected.

Let us now observe that the ${}^4He$ mass fraction of the total baryonic mass is [60]

$$Y_p \equiv \gamma\,\frac{2x(t_f)}{1 + x(t_f)} \tag{47}$$

where

$$\gamma = e^{-(t_n - t_f)/\tau} \simeq 1. \tag{48}$$

Here, $t_f \simeq 1\,\text{s}$ and $t_n \simeq 20\,\text{s}$ are the freeze-out and nucleosynthesis times, respectively, while $\tau \simeq 877\,\text{s}$ is the neutron mean lifetime. Moreover, we have denoted the neutron-to-proton equilibrium ratio by $x(t_f) = e^{-\mathcal{Q}/T(t_f)}$.

Fluctuations of Y_p are related to variations of the freeze-out temperature δT_F by (see [57] and references therein)

$$\delta Y_p = Y_p\left[\left(1 - \frac{Y_p}{2\gamma}\right)\log\left(\frac{2\gamma}{Y_p} - 1\right) - \frac{2t_f}{\tau}\right]\frac{\delta T_f}{T_f}\,. \tag{49}$$

Observational measurements from ^4He emission lines in extragalactic HII regions enable estimating [61]

$$Y_p = 0.2449 \qquad |\delta Y_p| \lesssim 10^{-4}. \tag{50}$$

By plugging these values into Equation (49) and solving for δT_f, we are led to the following variation of the freeze-out temperature

$$\left|\frac{\delta T_f}{T_f}\right| \lesssim 10^{-4}. \tag{51}$$

We now have all the ingredients to constrain the Kaniadakis parameter. Indeed, following [57], we can evaluate the freeze-out temperature in the Kaniadakis framework by imposing that the interaction rate (45) is of the same order as (or small than) the Hubble rate (41), and setting $\delta T_f = T_{f,\kappa} - T_{0f} = T_{f,\kappa} - 0.6\,\mathrm{MeV}$, where we have denoted by $T_{f,\kappa}$ the Kaniadakis-corrected freeze-out temperature. The resulting equation has the form

$$y\,T_{f,\kappa}^{11} = \kappa^2\,x + z\,T_{f,\kappa}^{8} \tag{52}$$

where

$$y \equiv 384\sqrt{5}\pi^4\,g^2\,q \tag{53}$$

$$x \equiv 4050\pi^{3/2}\sqrt{g} \tag{54}$$

$$z \equiv 128\pi^{11/2}\,g^{5/2}. \tag{55}$$

Equation (52) cannot be solved analytically. However, we can infer an upper bound on the Kaniadakis parameter by resorting to numerical evaluation. In order for $T_{f,\kappa}$ to satisfy Equation (51), we must have

$$|\kappa| \lesssim 10^{-92} \tag{56}$$

which shows that the Kaniadakis parameter must be tightly constrained around zero to be consistent with experimental measurements of freeze-out temperature. More comments on the obtained result can be found at the end of the next subsection.

3.2. Primordial Abundances of ^4He and Deuterium D

Based on the previous considerations, let us now investigate implications of Kaniadakis statistics on the primordial abundances of Helium isotope ^4He and Deuterium D. To this aim, we recall that the sequence of nuclear processes leading to the generation of these elements is

$$n + p \rightarrow D + \gamma \tag{57}$$

$$D + D \rightarrow {}^3He + n \tag{58}$$

$$D + D \rightarrow T + p. \tag{59}$$

In the final stage, Deuterium and Tritium T or Deuterium and Helium isotope ^3He combine to give

$$D + T \rightarrow {}^4He + n \tag{60}$$

$$D + {}^3He \rightarrow {}^4He + p. \tag{61}$$

From [62], we know that the primordial ^4He abundance is constrained by the numerical best fit to the value

$$Y_p = 0.2485 \pm 0.0006 + 0.0016[(\eta_{10} - 6) + 100(Z - 1)], \tag{62}$$

where the baryon density number is given by $\eta_{10} \equiv 10^{10}\eta_B \simeq 6$, with η_B being the baryon-to-photon ratio [62]. Of course, here we have to consider Z equal to Equation (42) to study Kaniadakis entropy effects on ^4He abundance. For $Z = 1$ (standard GR value), we get back $Y_p = 0.2485 \pm 0.0006$, according to the predictions of the traditional BBN model.

Now, as shown in [57], the requirement of consistency between Equation (62) and observational measurements of ^4He abundance gives

$$\delta Z \equiv Z - 1 \lesssim \mathcal{O}(10^{-2}). \tag{63}$$

By taking $Z = Z_\kappa$, the above equation allows us to constrain the κ deformation parameter to

$$|\kappa| \lesssim 10^{-88} \tag{64}$$

where we have considered $T \simeq 10\,\text{MeV}$.

We can repeat the same considerations as above for the case of D abundance. The best numerical fit from [63] gives in this case

$$y_{D_p} = 2.6(1 \pm 0.06)\left(\frac{6}{\eta_{10} - 6(Z-1)}\right)^{1.6} \tag{65}$$

which still leads to the standard BBN prediction $y_{D_p} = 2.6 \pm 0.16$ for $\eta_{10} = 6$ and $Z = 1$. Observational measurements of D abundance combined with Equation (65) set again $\delta Z \lesssim \mathcal{O}(10^{-2})$ [57,64], thus leading to the same bound as in Equation (64).

It is worth noting that the constraint (64) is less tight than both the bound in Equation (56) and than results obtained in [16] via cosmic chronometers/Supernovae type Ia ($|\kappa| \simeq 10^{-124}$) and Baryon Acoustic Oscillations ($|\kappa| \simeq 10^{-125}$) measurements. Although not contemplated in the original formulation by Kaniadakis, such an apparent incompatibility could be understood by allowing Kaniadakis parameter to be running in time. This scenario would not be surprising in Kaniadakis Cosmology: in fact, it is legitimate to expect that the Universe degrees of freedom encoded by holographic entropy may evolve from an initial description obeying relativistic (Kaniadakis-like) laws to a classical (Boltzmann-like) picture at present time, just as it happens for matter–energy degrees of freedom. In this framework, departures from the standard BGS entropy would be quantified by a time-dependent, or equivalently, temperature-dependent parameter $\kappa \equiv \kappa(t) \equiv \kappa(T)$, such that κ substantially differs from zero at high T (early stages of the Universe), while it recovers the classical $\kappa \to 0$ behavior as the Universe cools down. This would justify why the bound (56), which has been derived for the phase of the Universe corresponding to $T \simeq 0.6\,\text{MeV}$, is more stringent than that in Equation (64), associated with $T = 10\,\text{MeV}$.

We emphasize that the possibility of a running κ has already been discussed in [56]. Here, we have found further confirmation of this hypothesis. In addition, we point out that a similar proposal has been recently put forward in the context of Tsallis generalized statistics in [65–69], among others.

4. Discussion and Conclusions

In this work, we have discussed recent advances of Gravity and Cosmology in Kaniadakis statistical theory. Special care has been devoted to review effects of generalized κ entropy on open stellar clusters, Jeans instability and gravitational collapse, Holographic Dark Energy, Entropic Gravity formalism, Black hole thermodynamics and Loop Quantum Gravity. For each of these frameworks, we have shown that Kaniadakis statistics manifests through non-trivial modifications of characteristic theoretical predictions, such as the distribution of residual radial velocity of open clusters (Equation (7)) [22], the critical Jeans wavelength/mass (Equations (11) and (14)) [27], the gravitational constant (Equation (13)) and free fall time (Equation (17)) [25], the black hole temperature and heat capacity (Equation (21)) [41] and the Immirzi parameter (Equation (25)) [43]. The ensuing κ-dependent expressions unveil potential mechanisms to test Kaniadakis-induced deviations from Boltzmann statistics in Gravity and Cosmology scenarios.

On the other hand, we have focused on the study of Holographic Dark Energy in Kaniadakis Cosmology. Although this model is well-established in literature at both theoretical and observational levels [13–17], here we have followed an alternative procedure to treat the κ-modified Friedmann equations ruling the evolution of the Universe in Kaniadakis Cosmology (see Equations (34) and (35)). These equations have been used to analyze BBN and, in particular, the freeze-out mechanism and the generation of primordial elements. By demanding consistency between theoretical predictions of our model and observational constraints on freeze-out temperature fluctuations and abundances of ^4He and D, we have constrained departures from BGS entropy, showing that the κ parameter must be tightly bounded around the vanishing value to be consistent with phenomenology. Remarkably, it has been found that different stages of the Universe evolution correspond to different upper bounds on κ (see Equations (56) and (64)). This result opens up the possibility that a realistic description of the history of the Universe in Kaniadakis Cosmology is allowed, provided that one considers a running κ. Clearly, in order to substantiate this paradigm, the above analysis should be carried out by assuming an ab initio time- (or temperature-) dependent κ. This requires further investigation and will be presented elsewhere.

Other aspects are to be explored. Here, we present a list of some possible future challenges:

- as a first extension of the above analysis, it would be interesting to search for signatures of inflationary perturbations propagated during the hypothetical Kaniadakis cosmic epoch in present/upcoming experiments on primordial gravitational waves, such as VIRGO, LIGO or LISA. This work is already in progress.
- It has been recently argued that Holographic Dark Energy construction might alleviate the H_0 tension [70], the reason being that it could lead to the phantom regime for dark energy. Since KHDE has been shown to exhibit this feature [15], it is worth going more deeply into the problem to understand whether KHDE may provide a good candidate toward a solution to the H_0 tension.
- In [58], BBN has been studied by using the Generalized Uncertainty Principle (GUP), which emerges from the phenomenological attempt to embed gravity corrections in quantum mechanics so as to predict a minimal length at Planck scale (see [71–73] and references therein). Specifically, it has been shown that GUP enters Friedmann equations through a deformation of the entropy area law, which in turn modifies the density/temperature dependence of Hubble constant. Primordial abundances evaluated in this way exhibit a non-trivial dependence on the GUP deformation parameter. Given the formal analogies between such a result and those obtained in the present context, the question naturally arises as to whether any kind of connection between Kaniadakis statistics and GUP can be established at a more fundamental level. We expect that this study could also pave the way toward formulating a relativistic model of GUP.
- As argued in Section 2.5, the analysis of black holes thermodynamics from a dual Kaniadakis entropy reveals a possible thermally stable phase in Kaniadakis statistics, a fact that cannot be noticed when working in Boltzmann theory. This shows that κ-deformed entropy not only generalizes standard results, but also predicts features that do not have any correspondence in the ordinary black hole thermodynamics. Without a doubt, a thorough examination of this framework is needed to find out all peculiarities of black holes in Kaniadakis statistics.
- In the recent study of [30], effects of Kaniadakis statistics on the Jeans instability have been analyzed for gravitational systems composed by dark and baryonic matter. As a result, it has been found that instability is suppressed in comparison with the background of Maxwell distribution and, thus, opposes the gravitational collapse of such systems. An appealing extension of this work is to consider Kaniadakis implications on Jeans instability of partially ionized dusty plasma and discuss their

relevance in the formation of planetesimals and collapse of interstellar clouds in star forming regions.

- Based on the quantum tunnelling concept and Boltzmann statistics, one can derive the critical Gamow temperature T_G at which the star-burning process occurs. The problem has been recently addressed in the context of Kaniadakis statistics [26], showing that Gamow temperature decreases with respect to the standard value. Therefore, stars whose burning temperature differs from T_G might be signals of deviations from Gaussian statistics in stellar sciences. This provides a challenging framework where to test the Kaniadakis theory experimentally.
- In [34], Abreu et al. have derived a κ-modified version of the Tully–Fisher relation, which connects the rotation velocity of galaxies to their mass. In contrast to the classical formula, this new relation contains a dependence on the distance of the star to the center of the galaxy. By virtue of this result, it would then be interesting to study whether Kaniadakis statistics can shed any new light on the dark-matter problem.
- Kaniadakis entropy has also been applied to the context of quantum information. In particular, in [74], it has been found to exhibit suitable properties to be a candidate for a generalized quantum information theory. Along this direction, a demanding perspective is to explore the possible relevance of Kaniadakis information theory in solving some puzzles arising in quantum gravity scenarios, such as the black hole information paradox and the firewall paradox (see [75] for a recent review on the topic).

The investigation of these and other issues will be performed in separate projects.

Funding: This research received no external funding.

Data Availability Statement: Not applicable.

Acknowledgments: The author is grateful to Giorgio Kaniadakis for useful comments on the original manuscript. He also acknowledges the Spanish "Ministerio de Universidades" for the awarded Maria Zambrano fellowship and funding received from the European Union–Next, GenerationEU. He is grateful for participation in the COST Association Action CA18108 "Quantum Gravity Phenomenology in the Multimessenger Approach" and LISA Cosmology Working group. He would finally like to express his thanks for the kind hospitality of Hotel Goya, where most of this work has been conceived and written.

Conflicts of Interest: The author declares no conflict of interest.

Abbreviations

The following abbreviations are used in this manuscript:

KHDE	Kaniadakis Holographic Dark Energy
BGS	Boltzmann–Gibbs–Shannon
BBN	Big Bang Nucleosynthesis
GR	General Relativity
BH	Bekenstein–Hawking

References

1. Bernstein, J. *Kinetic Theory in the Expanding Universe*; Cambridge University Press: Cambridge, UK, 1988.
2. Giovannini, A.; Lupia, S.; Ugoccioni, R. Thermodynamics of clan production. *Phys. Rev. D* **2002**, *65*, 094028. [CrossRef]
3. Colladay, D.; McDonald, P. Statistical mechanics and Lorentz violation. *Phys. Rev. D* **2004**, *70*, 125007. [CrossRef]
4. Lee, H.K. Thermodynamic constraint on the primordial black hole formation in the radiation dominated epoch. *Phys. Rev. D* **2022**, *66*, 063001. [CrossRef]
5. Husain, V. Probing entropy bounds with scalar field spacetimes. *Phys. Rev. D* **2004**, *69*, 084002. [CrossRef]
6. Bhaduri, R.K.; Tran, M.N.; Das, S. Microcanonical entropy of a black hole. *Phys. Rev. D* **2004**, *69*, 104018. [CrossRef]
7. Yalcin, G.C.; Beck, G. Generalized statistical mechanics of cosmic rays. *Sci. Rep.* **2018**, *8*, 1764. [CrossRef] [PubMed]
8. Vasyliunas, V.M. A survey of low-energy electrons in the evening sector of the magnetosphere with OGO 1 and OGO 3. *J. Geophys. Res.* **1968**, *73*, 2839–2884. [CrossRef]

9. Hasegawa, A.; Kunioki, A.M.; Duong-va, M. Plasma Distribution Function in a Superthermal Radiation Field. *Phys. Rev. Lett.* **1985**, *54*, 2608. [CrossRef]
10. Kaniadakis, G. Non Linear Kinetics underlying Generalized Statistics. *Phys. A* **2001**, *296*, 405–425. [CrossRef]
11. Kaniadakis, G. Statistical mechanics in the context of special relativity. *Phys. Rev. E* **2002**, *66*, 056125. [CrossRef]
12. Kaniadakis, G. Statistical mechanics in the context of special relativity. II. *Phys. Rev. E* **2005**, *72*, 036108. [CrossRef] [PubMed]
13. Moradpour, H.; Ziaie, A.H.; Kord Zangeneh, M. Generalized entropies and corresponding holographic dark energy models. *Eur. Phys. J. C* **2020**, *80*, 732. [CrossRef]
14. Lymperis, A.; Basilakos, S.; Saridakis, E.N. Modified cosmology through Kaniadakis horizon entropy. *Eur. Phys. J. C* **2021**, *81*, 1037. [CrossRef]
15. Drepanou, N.; Lymperis, A.; Saridakis, E.N.; Yesmakhanova, K. Kaniadakis holographic dark energy and cosmology. *Eur. Phys. J. C* **2022**, *82*, 449. [CrossRef]
16. Hernández-Almada, A.; Leon, G.; Magaña, J.; García-Aspeitia, M.A.; Motta, V.; Saridakis, E.N.; Yesmakhanova, K. Kaniadakis-holographic dark energy: observational constraints and global dynamics. *Mon. Not. R. Astron. Soc.* **2022**, *511*, 4147–4158. [CrossRef]
17. Hernández-Almada, A.; Leon, G.; Magaña, J.; García-Aspeitia, M.A.; Motta, V.; Saridakis, E.N.; Yesmakhanova, K.; Millano, A.D. Observational constraints and dynamical analysis of Kaniadakis horizon-entropy cosmology. *Mon. Not. R. Astron. Soc.* **2022**, *512*, 5122–5134. [CrossRef]
18. Chen, H.; Zhang, S.; Liu, S. Jeans Gravitational Instability with kappa-Deformed Kaniadakis Distribution. *Chin. Phys. Lett.* **2017**, *34*, 075101. [CrossRef]
19. Sadeghnezhad, N. Gravity and Cosmology in Kaniadakis Statistics. *arXiv* **2021**, arXiv:2111.13623.
20. Kaniadakis, G. New power-law tailed distributions emerging in κ-statistics. *EPL (Europhys. Lett.)* **2021**, *133*, 10002. [CrossRef]
21. Beck, C. Generalised information and entropy measures in physics. *Contemp. Phys.* **2009**, *50*, 495. [CrossRef]
22. Carvalho, J.C.; Silva, R.; do Nascimiento, J.D., Jr.; Soares, B.B.; De Medeiros, J.R. Observational measurement of open stellar clusters: A test of Kaniadakis and Tsallis statistics. *EPL (Europhys. Lett.)* **2010**, *91*, 69002. [CrossRef]
23. Abreu, E.M.C.; Neto, J.A.; Mendes, A.C.R.; Bonilla, A. Tsallis and Kaniadakis statistics from a point of view of the holographic equipartition law. *EPL (Europhys. Lett.)* **2018**, *121*, 45002. [CrossRef]
24. Jawad, A.; Sultan, A.M. Cosmic Consequences of Kaniadakis and Generalized Tsallis Holographic Dark Energy Models in the Fractal Universe. *Adv. High Energy Phys.* **2021**, *2021*, 5519028. [CrossRef]
25. Abreu, E.M.C.; Ananias Neto, J.; Barboza, E.M.; Nunes, R.C. Tsallis and Kaniadakis statistics from the viewpoint of entropic gravity formalism. *Int. J. Mod. Phys. A* **2017**, *32*, 1750028. [CrossRef]
26. Moradpour, H.; Javaherian, M.; Namvar, E.; Ziaie, A.H. Gamow Temperature in Tsallis and Kaniadakis Statistics. *Entropy* **2022**, *24*, 797. [CrossRef] [PubMed]
27. Abreu, E.M.C.; Ananias Neto, J.; Barboza, E.M.; Nunes, R.C. Jeans instability criterion from the viewpoint of Kaniadakis' statistics. *EPL (Europhys. Lett.)* **2016**, *114*, 55001. [CrossRef]
28. Jang, W.; Xiong, Y.; Chen, H.; Liu, S. Jeans gravitational instability with κ-deformed Kaniadakis distribution in Eddington-inspired Born–Infeld gravity. *Chin. Phys. B* **2020**, *29*, 11041.
29. He, K.R. Jeans analysis with κ-deformed Kaniadakis distribution in f (R) gravity. *Phys. Scr.* **2022**, *97*, 025601. [CrossRef]
30. Jang, W.; Xiong, Y.; Chen, H.; Liu, S. Jeans instability of dark-baryonic matter model in the context of Kaniadakis' statistic distribution. *J. Taibah Univ. Sci.* **2022**, *16*, 337–343.
31. Li, M. A Model of holographic dark energy. *Phys. Lett. B* **2004**, *603*, 1. [CrossRef]
32. Sharma, U.K.; Dubey, V.C.; Ziaie, A.H.; Moradpour, H. Kaniadakis holographic dark energy in non-flat universe. *Int. J. Mod. Phys. D* **2022**, *31*, 2250013. [CrossRef]
33. Rani, S.; Jawad, A.; Sultan, A.M.; Shad, M. Cosmographic and thermodynamic analysis of Kaniadakis holographic dark energy. *Int. J. Mod. Phys. D* **2022**, *31*, 2250078. [CrossRef]
34. Abreu, E.M.C.; Neto, J.A.; Mendes, A.C.; Bonilla, A.; De Paula, R.M. Cosmological considerations in Kaniadakis statistics. *EPL (Europhys. Lett.)* **2018**, *124*, 30003. [CrossRef]
35. Ghaffari, S. Kaniadakis holographic dark energy in Brans–Dicke cosmology. *Mod. Phys. Lett. A* **2022**, *37*, 2250152. [CrossRef]
36. Verlinde, E.P. On the Origin of Gravity and the Laws of Newton. *JHEP J. High Energy Phys.* **2011**, *4*, 029. [CrossRef]
37. Tan, W.H.; Yang, S.Q.; Shao, C.G.; Li, J.; Du, A.B.; Zhan, B.F.; Wang, Q.-L.; Luo, P.-S.; Tu, L.-C.; Luo, J. New Test of the Gravitational Inverse-Square Law at the Submillimeter Range with Dual Modulation and Compensation. *Phys. Rev. Lett.* **2016**, *116*, 131101. [CrossRef] [PubMed]
38. Adelberger, E.G.; Heckel, B.R.; Nelson, A.E. Tests of the Gravitational Inverse-Square Law. *Ann. Rev. Nucl. Part Sci.* **2003**, *53*, 77. [CrossRef]
39. Dong, X. The Gravity Dual of Renyi Entropy. *Nat. Commun.* **2016**, *7*, 12472. [CrossRef]
40. Czinner, V.G.; Iguchi, H. Rényi Entropy and the Thermodynamic Stability of Black Holes. *Phys. Lett. B* **2016**, *752*, 306. [CrossRef]
41. Abreu, E.M.C.; Neto, J.A. Black holes thermodynamics from a dual Kaniadakis entropy. *EPL (Europhys. Lett.)* **2021**, *133*, 49001. [CrossRef]
42. Rovelli, C. *Quantum Gravity*; Cambridge University Press: Cambridge, UK, 2004.
43. Immirzi, G. Quantum gravity and Regge calculus. *Nucl. Phys. B Proc. Suppl.* **1997**, *57*, 65. [CrossRef]

44. Abreu, E.M.C.; Ananias Neto, J.; Mendes, A.C.R.; de Paula, R.M. Loop Quantum Gravity Immirzi parameter and the Kaniadakis statistics. *Chaos Solitons Fractals* **2019**, *118*, 307. [CrossRef]
45. Lang, K.R. *Astrophysical Formulae*; Springer: Berlin/Heidelberg, Germany, 1974.
46. Enqvist, K.; Hannestad, S.; Sloth, M.S. Searching for a holographic connection between dark energy and the low-l CMB multipoles. *JCAP J. Cosmol. Astropart. Phys.* **2005**, *2*, 004. [CrossRef]
47. Setare, M.R. Holographic tachyon model of dark energy. *Phys. Lett. B* **2007**, **653**, 116–121. [CrossRef]
48. Srivastava, S.; Sharma, U.K. Barrow holographic dark energy with Hubble horizon as IR cutoff. *Int. J. Geom. Meth. Mod. Phys.* **2021**, *18*, 2150014. [CrossRef]
49. Tavayef, M.; Sheykhi, A.; Bamba, K.; Moradpour, H. Tsallis Holographic Dark Energy. *Phys. Lett. B* **2018**, *781*, 195. [CrossRef]
50. Saridakis, E.N. Barrow Holographic Dark Energy. *Phys. Rev. D* **2020**, *102*, 123525. [CrossRef]
51. Luciano, G.G.; Gine, J. Baryogenesis in non-extensive Tsallis Cosmology. *Phys. Lett. B* **2022**, *833*, 137352. [CrossRef]
52. Luciano, G.G. Cosmic evolution and thermal stability of Barrow holographic dark energy in nonflat Friedmann-Robertson-Walker Universe. *Phys. Rev. D* **2022**, *106*, 083530. [CrossRef]
53. Luciano, G.G.; Giné, J. Generalized interacting Barrow Holographic Dark Energy: Cosmological predictions and thermodynamic considerations. *arXiv* **2022**, arXiv:2210.09755.
54. Cyburt, R.H.; Fields, B.D.; Olive, K.A.; Yeh, T.H. Big bang nucleosynthesis: Present status. *Rev. Mod. Phys.* **2016**, *88*, 015004. [CrossRef]
55. Jacobson, T. Thermodynamics of Spacetime: The Einstein Equation of State. *Phys. Rev. Lett.* **1995**, *75*, 1260. [CrossRef] [PubMed]
56. Luciano, G.G. Modified Friedmann equations from Kaniadakis entropy and cosmological implications on baryogenesis and ^7Li-abundance. *Eur. Phys. J. C* **2022**, *82*, 314. [CrossRef]
57. Ghoshal, A.; Lambiase, G. Constraints on Tsallis Cosmology from Big Bang Nucleosynthesis and Dark Matter Freeze-out. *arXiv* **2021**, arXiv:2104.11296.
58. Luciano, G.G. Primordial big bang nucleosynthesis and generalized uncertainty principle. *Eur. Phys. J. C* **2021**, *81*, 1086. [CrossRef]
59. Bernstein, J.; Brown, L.S.; Feinberg, G. Cosmological helium production simplified. *Rev. Mod. Phys.* **1989**, *61*, 25. [CrossRef]
60. Kolb, E.W.; Turner, M.S. The Early Universe. *Front. Phys.* **1990**, *69*, 1.
61. Aver, E.; Olive, K.A.; Skillman, E.D. The effects of He I λ10830 on helium abundance determinations. *JCAP* **2015**, **07**, 011. [CrossRef]
62. Kneller, J.P.; Steigman, G. BBN for pedestrians. *New J. Phys.* **2004**, *6*, 117. [CrossRef]
63. Steigman, G. Neutrinos In addition, Big Bang Nucleosynthesis. *Adv. High Energy Phys.* **2012**, *2012*, 268321. [CrossRef]
64. Bhattacharjee, S.; Sahoo, P.K. Big bang nucleosynthesis and entropy evolution in $f(R,T)$ gravity. *Eur. Phys. J. Plus* **2020**, *135*, 350. [CrossRef]
65. Nojiri, S.; Odintsov, S.D.; Saridakis, E.N. Modified cosmology from extended entropy with varying exponent. *Eur. Phys. J. C* **2019**, *79*, 242. [CrossRef]
66. Luciano, G.G. Tsallis statistics and generalized uncertainty principle. *Eur. Phys. J. C* **2021**, *81*, 672. [CrossRef]
67. Luciano, G.G.; Blasone, M. q-generalized Tsallis thermostatistics in Unruh effect for mixed fields. *Phys. Rev. D* **2021**, *104*, 045004. [CrossRef]
68. Luciano, G.G.; Blasone, M. Nonextensive Tsallis statistics in Unruh effect for Dirac neutrinos. *Eur. Phys. J. C* **2021**, *81*, 995. [CrossRef]
69. Jizba, P.; Lambiase, G.; Luciano, G.G.; Petruzziello, L. Decoherence limit of quantum systems obeying generalized uncertainty principle: New paradigm for Tsallis thermostatistics. *Phys. Rev. D* **2022**, *105*, L121501. [CrossRef]
70. Abdalla, E.; Abellán, G.F.; Aboubrahim, A.; Agnello, A.; Akarsu, Ö.; Akrami, Y.; Alestas, G.; Aloni, D.; Amendola, L.; Anchordoqui, L.A.; et al. Cosmology intertwined: A review of the particle physics, astrophysics, and cosmology associated with the cosmological tensions and anomalies. *J. High Energy Astrophys.* **2022**, *34*, 49. [CrossRef]
71. Kempf, A.; Mangano, G.; Mann, R.B. Hilbert space representation of the minimal length uncertainty relation. *Phys. Rev. D* **1995**, *52*, 1108. [CrossRef]
72. Scardigli, F. The deformation parameter of the generalized uncertainty principle. *J. Phys. Conf. Ser.* **2019**, *1275*, 012004. [CrossRef]
73. Luciano, G.G.; Petruzziello, L. Generalized uncertainty principle and its implications on geometric phases in quantum mechanics. *Eur. Phys. J. Plus* **2021**, *136*, 179. [CrossRef]
74. Ourabah, K.; Hamici-Bendimerad, A.H.; Tribeche, M. Quantum entanglement and Kaniadakis entropy. *Phys. Scr.* **2015**, *90*, 045101. [CrossRef]
75. Almheiri, A.; Hartman, T.; Maldacena, J.; Shaghoulian, E.; Tajdini, A. The entropy of Hawking radiation. *Rev. Mod. Phys.* **2021**, *93*, 035002. [CrossRef]

entropy

MDPI

Article

On the Kaniadakis Distributions Applied in Statistical Physics and Natural Sciences

Tatsuaki Wada [1,*] and Antonio Maria Scarfone [2]

[1] Region of Electrical and Electronic Systems Engineering, Ibaraki University, Nakanarusawa-cho, Hitachi-shi 316-8511, Japan
[2] Istituto dei Sistemi Complessi, Consiglio Nazionale delle Ricerche (ISC-CNR), c/o Politecnico di Torino, Corso Duca degli Abruzzi, 24, 10129 Torino, Italy
* Correspondence: tatsuaki.wada.to@vc.ibaraki.ac.jp

Abstract: Constitutive relations are fundamental and essential to characterize physical systems. By utilizing the κ-deformed functions, some constitutive relations are generalized. We here show some applications of the Kaniadakis distributions, based on the inverse hyperbolic sine function, to some topics belonging to the realm of statistical physics and natural science.

Keywords: κ-deformed functions; constitutive relations; Gompertz rule; Lotka–Volterra equations; contact density dynamics

Citation: Wada, T.; Scarfone, A.M. On the Kaniadakis Distributions Applied in Statistical Physics and Natural Sciences. *Entropy* **2023**, *25*, 292. https://doi.org/10.3390/e25020292

Academic Editor: Christian Beck

Received: 27 December 2022
Revised: 31 January 2023
Accepted: 2 February 2023
Published: 4 February 2023

1. Introduction

The κ-exponential function [1–3] is defined by:

$$\exp_\kappa(x) := \left(\kappa x + \sqrt{1+\kappa^2 x^2}\right)^{\frac{1}{\kappa}} = \exp\left[\frac{1}{\kappa}\,\mathrm{arsinh}(\kappa x)\right], \tag{1}$$

for a real deformation parameter κ. The inverse function, i.e., the κ-deformed logarithmic function, is defined by:

$$\ln_\kappa x := \frac{x^\kappa - x^{-\kappa}}{2\kappa} = \frac{1}{\kappa}\sinh[\kappa \ln x]. \tag{2}$$

Both κ-deformed functions are important ingredients of the generalized statistical physics based on κ-entropy [1–3]. This influences a wide range of scientific fields, and, based on the κ-deformed functions (Appendix A), several basic fields developed over two decades. Kaniadakis [4] provided the theoretical foundations and mathematical formalism generated by the κ-deformed functions, and some references, including many fields of applications. Recently, the usefulness of the κ-statistics was demonstrated for the analysis [5] of epidemics and pandemics.

Constitutive relations are fundamental and essential to characterize physical systems. They are combined with the other equations of the physical laws in order to solve physical problems. There are well-known examples of linear constitutive relations, such as the following: Hooke's law $F = k_s x$, for the tensile, or compressive, force F of a spring with a spring constant k_s against the change in its length x; Ohm's law $V = RI$ for the voltage V of an electrical conductor with resistance R under an electric current I, and so on. However, as a real spring deviates from Hooke's law, we know that any linear constitutive relation describes an idealized situation, and it is merely a linearized- and/or approximated-relation to describe some real physical properties. Hence, in general, non-linearity plays a crucial role to describe more realistic physical systems.

The κ-exponential function (1) can be regarded as a useful tool (or device) to make such non-linear constitutive relations for a better description of real physical systems. For example, consider the following κ-deformation of Hooke's law:

$$F_\kappa := k_s \ln\left[\exp_\kappa(x)\right] = \frac{k_s}{\kappa} \ln\left(\kappa x + \sqrt{1+\kappa^2 x^2}\right), \tag{3}$$

which reduces to the original Hooke's law $F = k_s x$ in the limit of $\kappa \to 0$. For any linear constitutive relation, we can apply this type of the κ-deformation. For example, Ohm's law can be cast into the following form: $V = RI = R\ln[\exp(I)]$. By changing the exponential function with the κ-exponential function, we obtain the κ-deformed version of Ohm's law: $V_\kappa = R\ln\left[\exp_\kappa(I)\right]$. In this research, we focused on this type of the κ-deformation of a physical quantity (say A), i.e.,

$$A \quad \Rightarrow \quad \ln\left[\exp_\kappa(A)\right] = \frac{1}{\kappa}\,\mathrm{arsinh}(\kappa A). \tag{4}$$

Throughout this paper, we call this κ-deformation *the* arsinh-*type deformation* of a physical quantity A.

Another type of the κ-deformation can be:

$$A \quad \Rightarrow \quad \ln_\kappa[\exp(A)] = \frac{1}{\kappa}\sinh(\kappa A), \tag{5}$$

which is called here *the* sinh-*type deformation*. In Reference [6], the thermodynamic stability of the κ-generalization S_κ^{B} of Boltzmann entropy S^{B} was studied. The κ-generalization S_κ^{B} was rewritten in the form:

$$S_\kappa^{\mathrm{B}} := k_{\mathrm{B}} \ln_\kappa W = k_{\mathrm{B}} \ln_\kappa[\exp(\ln W)] = k_{\mathrm{B}} \ln_\kappa[\exp(S^{\mathrm{B}})], \tag{6}$$

which could be regarded as the sinh-type deformation of Boltzmann entropy S^{B}. Recently, in cosmology, Lymperis et al. [7] modified Bekenstein–Hawking entropy S^{BH} as follows:

$$S_\kappa^{\mathrm{BH}} = \frac{1}{\kappa}\sinh\left(\kappa S^{\mathrm{BH}}\right), \tag{7}$$

which was obviously the sinh-type deformation of S^{BH}.

In this paper we considered the arsinh-type deformations against some constitutive relations in the field of statistical physics and natural sciences. In our previous work [8] we studied a thermal particle under a velocity-dependent potential which could be regarded as a deformation of Rayleigh's dissipation function [9] and showed that the probability distribution function (pdf) for the stationary-state of this thermal particle was a κ-deformed Gaussian pdf. It was considered the canonical pdf $\rho(v)$, in the velocity space, of a thermal particle with unit mass ($m = 1$) in the κ-deformed confining potential $U_{\kappa\beta}(v)$:

$$U_{\kappa\beta}(v) := \frac{1}{\kappa\beta}\,\mathrm{arsinh}\left(\kappa\beta \frac{v^2}{2}\right), \tag{8}$$

where $\beta := 1/k_{\mathrm{B}}T$ is a coldness (or inverse temperature). This κ-deformed potential $U_{\kappa\beta}(v)$ was rewritten, in the momentum–space, as:

$$U_{\kappa\beta}(p) = \frac{1}{\kappa\beta}\,\mathrm{arsinh}\left(\kappa\beta \frac{p^2}{2}\right) = \frac{1}{\beta} \ln\left[\exp_\kappa\left(\beta\frac{p^2}{2}\right)\right], \tag{9}$$

which was the arsinh type deformation of the quantity $\beta p^2/2$ (the ratio of the kinetic energy to the mean thermal energy $k_{\mathrm{B}}T = 1/\beta$). In other words, we considered the following

κ-deformation $Q_\kappa(U)$ of the Boltzmann factor $\exp(-\beta U)$ for an equilibrium state with the energy U:

$$Q_\kappa(U) := \exp_\kappa(-\beta U) = \exp\left[\frac{1}{\kappa}\operatorname{arsinh}(-\kappa\beta U)\right]. \tag{10}$$

One may wonder why the inverse hyperbolic sine function (arsinh) plays a role. In many different fields of sciences, there is no doubt that the exponential and logarithmic functions are important and fundamental. Since the inverse hyperbolic sine function and logarithmic function are mutually related as:

$$\operatorname{arsinh} x = \ln\left[x + \sqrt{1+x^2}\right], \quad \ln x = \operatorname{arsinh}\left[\frac{1}{2}\left(x - \frac{1}{x}\right)\right], \tag{11}$$

for a positive real x, we think both functions are important. By using the second relation, for any real parameter $\kappa \neq 0$, we have:

$$\ln x = \frac{1}{\kappa}\ln x^\kappa = \frac{1}{\kappa}\operatorname{arsinh}\left[\frac{1}{2}\left(x^\kappa - x^{-\kappa}\right)\right] = \frac{1}{\kappa}\operatorname{arsinh}[\kappa \ln_\kappa x]. \tag{12}$$

Note that this relation corresponds to the arsinh-type deformation of $\ln_\kappa x$ and is equivalent to definition (2) of the κ-deformed logarithmic function that can be regarded as the sinh-type of κ-deformation of $\ln x$. Kaniadakis already discussed this issue in section II of Reference [2] from the viewpoint of deformed algebra.

On the other hand, Pistone [10] was the first one to study the κ-exponential model in the field of information geometry [11], and later, through our research activities [8,12,13], we realized that there exist some relations among statistical physics, thermodynamics, mathematical biology, and information geometry. Harper [14,15] pointed out that the replicator equation (RE) [16] in mathematical biology or in an evolutional game theory [17] is related with information geometry and a general form of the Lotka–Volterra (gLV) equation as briefly explained in Appendix B. The gLV equations [14,15,18,19]:

$$\frac{dy_i}{dt} = y_i\, f_i(\boldsymbol{y}), \tag{13}$$

are used to model the competition dynamics of the populations $y_1, y_2, \ldots, y_n$ of n biological species. The Gompertz function [20] is a type of mathematical model for time evolution. Historically, he studied human mortality and proposed his law of human mortality in which he assumed that a person's resistance to death decreases as his or her years increase. His law is now called *Gompertz rule* (or law) and we would like to point out the relation of his function and his rule to some important quantities concerning statistical physics.

The rest of the paper is organized as follows. In Section 2, we briefly explain Gompertz function, and the gLV equations, which are important in mathematical biology (or evolutional game theory). Their relations to thermal physics are pointed out. Section 3 considers the thermal density operator, which is characterized by the so-called Bloch equation [21,22] for thermal states, and we show that the Bloch equation can be regarded as a Gompertz rule after the parameter transformation β to $t = -\ln\beta$. In Section 4, we discuss the arsinh-type deformation from the viewpoint of the κ-addition. In Section 5, we study the numerical simulations of the thermostat algorithm for the Hamiltonian with the κ-deformed kinetic energy, which can be regarded as the arsinh type of the κ-deformation of the ratio $\beta p^2/2$ as shown in (10). The final section is devoted to our conclusions.

2. Gompertz Functions and Gompertz Rule

Here we would like to point out that there exist relations between evolutional game dynamics and thermal physics. In evolutional game theory [17], evolutional game dynamics is described by a RE. The gLV equations are related to REs, as shown in Appendix B. On the other hand, Gompertz function is a mathematical model describing an evolutional curve.

Gompertz function (or Gompertz curve) [20] is a type of mathematical model for a time series. Gompertz function $f_G(t)$ is a sigmoid function and is given by:

$$f_G(t) := K \exp\left[C \exp(-t)\right], \tag{14}$$

where C and K are positive constants. A distinctive feature of Gompertz function is its double exponential t-dependency. His function is nowadays used in many different areas to model time evolution of populations where growth is slowest at the start and end of a period. For example, Reference [23] applied Gompertz model to describe the growth dynamics of the COVID-19 pandemic. Gompertz [20] studied human mortality by working out a series of mortality tables, and this suggested to him his law of human mortality, in which he assumed that a person's resistance to death decreases as age increases. The rule of his model is called *Gompertz rule* which states that:

$$\frac{d}{dt} f_G(t) = -f_G(t) \ln \frac{f_G(t)}{K}. \tag{15}$$

The solution of the Gompertz rule is the Gompertz function (14), if we set $K = \lim_{t\to\infty} f_G(t)$ and $C = \ln(f_G(0)/K)$.

If we choose $f_i(\boldsymbol{y}(t)) = -\ln y_i(t)$ and assume $\lim_{t\to\infty} y_i(t) = 1$, the gLV Equation (13) becomes:

$$\frac{dy_i(t)}{dt} = -y_i(t)\ \ln y_i(t), \tag{16}$$

which can be regarded as the Gompertz rule (15) with $K = 1$ for each $y_i(t)$. Consequently, its solution $y_i(t)$ is the Gompertz function:

$$y_i(t) = \exp\left[\ln y_i(0) \exp(-t)\right]. \tag{17}$$

Now, by changing the parameter t to $\beta = \exp(-t)$, we have $d\beta = -\beta dt$ so that the limit $t \to 0$ corresponds to $\beta \to 1$, and each constant E_i is introduced as:

$$-E_i = \lim_{t\to 0} \ln y_i(t) = \lim_{\beta\to 1} \ln y_i(\beta), \tag{18}$$

where $y_i(\beta)$ is the shorthand notation of $y_i(t(\beta))$ with $t(\beta) = -\ln\beta$. Then, the solution $y_i(\beta)$ in (17) can be expressed as a quantity very familiar to statistical physics:

$$y_i(\beta) = \exp(-\beta E_i), \tag{19}$$

that is the Boltzmann factor. The corresponding Gompertz rule (15) for $y_i(\beta)$ is equivalent to:

$$\frac{d}{d\beta} y_i(\beta) = -E_i\, y_i(\beta). \tag{20}$$

Having described the relation between the Gompertz rule and the Boltzmann factor $\exp(-\beta E_i)$ in statistical physics, in the next section we discuss a κ-deformation of the Bloch equation for thermal states.

3. Bloch Equation for Thermal States

For a given Hamiltonian $\hat{H}$ and the corresponding eigenvalues E_i and eigenstate $|\psi_i\rangle$, which are related in:

$$\hat{H}|\psi_i\rangle = E_i|\psi_i\rangle, \tag{21}$$

and assuming the completeness relation $\sum_i |\psi_i\rangle\langle\psi_i| = \hat{1}$, the density operator $\hat{\rho}(\beta)$ for a canonical ensemble is constructed as:

$$\hat{\rho}(\beta) := \sum_i \exp(-\beta E_i)|\psi_i\rangle\langle\psi_i| = \exp(-\beta\hat{H}). \tag{22}$$

In order to determine the canonical density matrix, we have to solve the eigenvalue Equation (21) and to sum over all the states. This needs heavy calculations in general. Note that $\hat{\rho}(\beta)$ is un-normalized and its trace is $\mathrm{Tr}\,\hat{\rho}(\beta) = Z(\beta)$, which is the partition function.

The Bloch equation [21,22] for thermal states is known as:

$$-\frac{\partial}{\partial\beta}\hat{\rho}(\beta) = \hat{H}\,\hat{\rho}(\beta), \tag{23}$$

which can be regarded as the diffusion equation in imaginary time β, and it has a similar form as Schrödinger equation and diffusion equation. Bloch Equation (23) offers an alternative route to determine the density operator $\hat{\rho}(\beta)$. The initial ($\beta = 0$) condition is provided if we know the eigenstates in the high-temperature limit.

Now, by multiplying β to both sides of (23), we have:

$$-\beta\frac{\partial}{\partial\beta}\hat{\rho}(\beta) = \beta\hat{H}\,\hat{\rho}(\beta) = -\ln[\hat{\rho}(\beta)]\,\hat{\rho}(\beta). \tag{24}$$

Changing the parameter β to $t = -\ln\beta$, it follows:

$$\frac{d}{dt}\hat{\rho}(t) = -\beta\frac{d}{d\beta}\hat{\rho}(\beta) = -\ln[\hat{\rho}(t)]\,\hat{\rho}(t). \tag{25}$$

This is the same form of the Gompertz rule (15). In this way, the Bloch equation can be considered as a sort of Gompertz rule.

Next, let us consider the κ-deformed density operator:

$$\hat{\rho}_\kappa(\beta) := \sum_i \exp_\kappa(-\beta E_i)|\psi_i\rangle\langle\psi_i| = \exp_\kappa(-\beta\hat{H}). \tag{26}$$

This leads to the following κ-deformation of the Bloch equation:

$$-\frac{\partial}{\partial\beta}\hat{\rho}_\kappa(\beta) = \sum_i E_i \frac{\exp_\kappa(-\beta E_i)}{u_\kappa\big[(\exp_\kappa(-\beta E_i)\big]}|\psi_i\rangle\langle\psi_i| = \frac{\hat{H}}{u_\kappa\big[\exp_\kappa(-\beta\hat{H})\big]}\hat{\rho}_\kappa(\beta). \tag{27}$$

Again, by changing the parameter β to $t = -\ln\beta$ and using the relation (A3), we have:

$$\frac{d}{dt}\hat{\rho}_\kappa(t) = -\frac{\ln_\kappa[\hat{\rho}_\kappa(t)]}{u_\kappa[\hat{\rho}_\kappa(t)]}\,\hat{\rho}_\kappa(t), \tag{28}$$

which can be regarded as a κ-deformation of the Gompertz rule.

Differentiating (27), again with respect to β, we obtain the following nonlinear differential equation:

$$(1+\kappa^2\beta^2\hat{H}^2)\frac{\partial^2\hat{\rho}_\kappa(\beta)}{\partial\beta^2} + \kappa^2\beta\hat{H}^2\frac{\partial\hat{\rho}_\kappa(\beta)}{\partial\beta} - \hat{H}^2\,\hat{\rho}_\kappa(\beta) = 0. \tag{29}$$

This differential equation reminds us of the research work [24] on the quantum free particle on the two-dimensional hyperbolic plane. The relevant two-dimensional Schrödinger equation was separable in the κ-dependent coordinate system (z_x, y) with $z_x := x/\sqrt{1+\kappa^2 y^2}$.

The Schrödinger equation $\hat{H}_1\Psi = e_1\Psi$ for the first partial Hamiltonian $\hat{H}_1$ leads to the following differential equation with the variable z_x alone:

$$(1+\kappa^2 z_x^2)\frac{d^2\Psi(z_x)}{dz_x^2} + \kappa^2 z_x \frac{d\Psi(z_x)}{dz_x} + \mu\Psi(z_x) = 0, \quad \mu := \frac{2m}{\hbar^2}e_1. \tag{30}$$

In the limit of $\kappa \to 0$, this differential equation reduces to the standard time-independent Schrödinger equation: $d^2\Psi(x)/dx^2 + \mu\Psi(x) = 0$. Cariñena et al. [24] obtained the solution of the differential Equation (30) as the κ-deformed plane wave (in our notations):

$$\Psi(z_x) = \exp\left[\pm i\,\frac{\mu}{\kappa}\,\mathrm{arsinh}(\kappa\, z_x)\right], \tag{31}$$

which is regarded as an arsinh-type deformation.

4. The κ-Addition and the Law of Large Number

Next, we considered the κ-addition from the viewpoint of the law of large numbers (LLN), which plays a central role in probability, statistics, and statistical physics [25]. The κ-addition [4] is defined by:

$$x \overset{\kappa}{\oplus} y := x\sqrt{1+\kappa^2 y^2} + y\sqrt{1+\kappa^2 x^2}. \tag{32}$$

This deformation of the additive rule comes from the addition rule of the inverse hyperbolic sine function as follows. For $a, b \in \mathbb{R}$, the addition rule is written as:

$$\mathrm{arsinh}(a) + \mathrm{arsinh}(b) = \mathrm{arsinh}\left(a\sqrt{1+b^2} + b\sqrt{1+a^2}\right). \tag{33}$$

By setting $a = \kappa x$ and $b = \kappa y$, we obtain:

$$\begin{aligned} \mathrm{arsinh}(\kappa x) + \mathrm{arsinh}(\kappa y) &= \mathrm{arsinh}\left(\kappa x\sqrt{1+\kappa^2 y^2} + \kappa y\sqrt{1+\kappa^2 x^2}\right) \\ &= \mathrm{arsinh}\left[\kappa(x \overset{\kappa}{\oplus} y)\right]. \end{aligned} \tag{34}$$

This relation is equivalent to the definition (32). The additive relation (34) is readily generalized to:

$$\sum_{i=1}^{n} \mathrm{arsinh}(\kappa x_i) = \mathrm{arsinh}\left[\kappa(x_1 \overset{\kappa}{\oplus} x_2 \overset{\kappa}{\oplus} \cdots \overset{\kappa}{\oplus} x_n)\right]. \tag{35}$$

By applying this relation to the Boltzmann factor $\exp\left[-\beta\sum_{i=1}^{n} K_{\kappa\beta}(p_i)\right]$ with respect to the κ-deformed kinetic energy [8] with $m = 1$:

$$\sum_{i=1}^{n} K_{\kappa\beta}(p_i) := \sum_{i=1}^{n} \frac{1}{\kappa\beta}\,\mathrm{arsinh}\left(\kappa\beta\frac{p_i^2}{2}\right), \tag{36}$$

we have:

$$
\begin{aligned}
\exp\left[-\beta\sum_{i=1}^{n}K_{\kappa\beta}(p_i)\right] &= \exp\left[-\frac{1}{\kappa}\,\mathrm{arsinh}\left\{\kappa\left(\beta\frac{p_1^2}{2}\overset{\kappa}{\oplus}\beta\frac{p_2^2}{2}\overset{\kappa}{\oplus}\ldots\overset{\kappa}{\oplus}\beta\frac{p_n^2}{2}\right)\right\}\right] \\
&= \exp_\kappa\left[\left(-\beta\frac{p_1^2}{2}\right)\overset{\kappa}{\oplus}\left(-\beta\frac{p_2^2}{2}\right)\overset{\kappa}{\oplus}\ldots\overset{\kappa}{\oplus}\left(-\beta\frac{p_n^2}{2}\right)\right] \\
&= \exp_\kappa\left[-\beta\frac{p_1^2}{2}\right]\exp_\kappa\left[-\beta\frac{p_2^2}{2}\right]\ldots\exp_\kappa\left[-\beta\frac{p_n^2}{2}\right] = \prod_{i=1}^{n}\exp_\kappa\left[-\beta\frac{p_i^2}{2}\right].
\end{aligned} \tag{37}
$$

Note that the κ-exponential of the κ-summation of each term $-\beta\frac{p_i^2}{2}$ in the second line is expressed as a factorized form in the last line.

It is well known that LLN plays a fundamental role in statistical physics [25]. Łapiński [26] showed that the standard LLN yielded the most probable state of the system, which equaled the point of maximum of the entropy and this point could be either Maxwell–Boltzmann statistics or Bose–Einstein statistics, or Zipf–Mandelbort law. McKeague [27] studied the central limit theorems under the special theory of relativity based on the κ-additivity. Scarfone [28] studied the κ-deformation of Fourier transform and discussed the limiting distribution of the κ-sum of statistically independent variables. The κ-additivity extension of the strong LLN was shown in [27] and it stated that if X_i were iid with finite mean, then:

$$
\frac{X_1}{n}\overset{\kappa}{\oplus}\frac{X_2}{n}\overset{\kappa}{\oplus}\ldots\overset{\kappa}{\oplus}\frac{X_n}{n} \quad\to\quad \frac{1}{\kappa}\,\mathrm{arsinh}[\kappa\langle X\rangle]_{a.s.}, \tag{38}
$$

where a.s. stands for almost surely, i.e., the above sequence of the random variables X_i converges almost surely, and $\langle X\rangle$ is the standard average of the random variable X. Of course, in the limit of $\kappa\to 0$, the relation (38) reduced to the standard strong LLN. Note that the converged value in (38) was the arsinh-type deformation of the average $\langle X\rangle$. In this way, the κ-additivity extension of the strong LLN supports the arsinh-type deformation of the average of a stochastic variable X.

5. Contact Density Dynamics

Nosé-Hoover (NH) thermostat [29,30] is a famous deterministic algorithm for constant-temperature molecular dynamics simulations. Based on the idea of NH thermostat, several improved versions were proposed. Among them, contact density dynamics (CDD) [31] is an algorithm based on contact Hamiltonian systems and generates any prescribed target distribution in physical phase space. The dynamical equations of CDD are the following:

$$
\begin{cases}
\dfrac{dq^i}{dt} = \dfrac{\partial h(q,p,S)}{\partial p_i}, & (39a)\\[2ex]
\dfrac{dp_i}{dt} = -\dfrac{\partial h(q,p,S)}{\partial q^i} + \dfrac{\partial h(p,q,S)}{\partial S}p_i, & (39b)\\[2ex]
\dfrac{dS}{dt} = -p_i\dfrac{\partial h(q,p,S)}{\partial p_i} + h(q,p,S), & (39c)
\end{cases}
$$

where S is the thermostatting variable, q_i and p_i are the i-th component ($i = 1,2,\cdots,n$) of n-dimensional vectors, respectively. Here $h(q,p,S)$ denotes the contact Hamiltonian which is formed as:

$$
h(q,p,S) = (\rho_t(q,p)f(S))^{-\frac{1}{n+1}}, \tag{40}
$$

with a target distribution $\rho_t(q,p)$ on $2n$-dimensional Γ-space and a normalized distribution $f(S)$ for the thermostatting variable S. As in the case of Reference [29,30], we also chose $f(S)$ as the logistic distribution with scale 1 and mean $c = 0.0$:

$$f(S) = \frac{\exp(S-c)}{(1+\exp(S-c))^2}. \tag{41}$$

Utilizing this CDD algorithm, the κ-deformed exponential distributions were simulated. The target distribution $\rho_t(q,p)$ was the one-dimensional ($n = 1$) κ-deformed Gaussian function:

$$\rho_t(q,p) = \frac{1}{Z_\kappa(\beta)}\exp[-\beta H_\kappa(q,p)] = \frac{1}{Z_\kappa(\beta)}\exp\left[-\frac{1}{\kappa}\operatorname{arsinh}\left(\kappa\beta\frac{p^2}{2}\right)\right]\exp\left[-\beta\frac{q^2}{2}\right], \tag{42}$$

where the associated Hamiltonian was:

$$H_\kappa(q,p) = \frac{1}{\kappa\beta}\operatorname{arsinh}\left(\kappa\beta\frac{p^2}{2}\right) + \frac{q^2}{2}, \tag{43}$$

and the normalization factor $Z_\kappa(\beta)$ [4] was:

$$Z_\kappa(\beta) = \frac{\pi}{\beta}\frac{\sqrt{\frac{2}{\kappa}}\Gamma\left(\frac{1}{2\kappa}-\frac{1}{4}\right)}{\left(\frac{\kappa}{2}+1\right)\Gamma\left(\frac{1}{4}+\frac{1}{2\kappa}\right)}. \tag{44}$$

In general, the kinetic energy can be defined by:

$$K(p) := \int_0^p v(p)dp, \tag{45}$$

where $v(p)$ denotes the constitutive relation between the velocity v and the canonical momentum p. In the standard case of $v(p) = p/m$ with $m = 1$, we have $K(p) = p^2/2$. In the case of the Hamiltonian (43), from (39a) we have:

$$v_\kappa(p) := \frac{dq}{dt} = \frac{\partial H_\kappa(q,p)}{\partial p} = \frac{p}{u_\kappa\left[\exp_\kappa\left(-\beta\frac{p^2}{2}\right)\right]} = \frac{p}{\sqrt{1+\kappa^2\left(\beta\frac{p^2}{2}\right)^2}}. \tag{46}$$

It is worthwhile to note that the $v_\kappa(p)$ had a β (or temperature) dependency when $\kappa \neq 0$. Then the corresponding kinetic energy $K_\kappa(p)$ was the first term $\frac{1}{\kappa\beta}\operatorname{arsinh}\left(\kappa\beta\frac{p^2}{2}\right)$ in (43), which could be regarded as a κ-deformation of the standard kinetic energy $p^2/2$.

We performed a number of CDD simulations for the target state (42) with different parameters and initial conditions. As an example, Figure 1 shows the phase space orbit and the histogram of the frequencies of the momentum p for a typical result of the CDD simulation of the target state (42) with $\beta = 0.2$, $\kappa = 0.4$. The initial conditions used are also denoted in the figure captions.

The CDD simulated result obeys ergodicity, as can be seen from the well distributed points in the phase space in Figure 1a. Note that the momentum distribution in the histogram of Figure 1b was well fitted with the κ–Gaussian distribution, which was cased by the arsinh-type deformation of the kinetic energy $p^2/2$.

Note also that for the κ-deformed Hamiltonian (42), we have [8]:

$$\left\langle p\frac{\partial}{\partial p}H_\kappa(q,p)\right\rangle = \frac{1}{\beta} \tag{47}$$

which reminds us of *a generalization of equipartition theorem* [32]: $\left\langle p\frac{\partial}{\partial p}\mathcal{H}\right\rangle = k_{\rm B}T$, where $\mathcal{H}$ is the Hamiltonian of a system in thermal equilibrium with the temperature T.

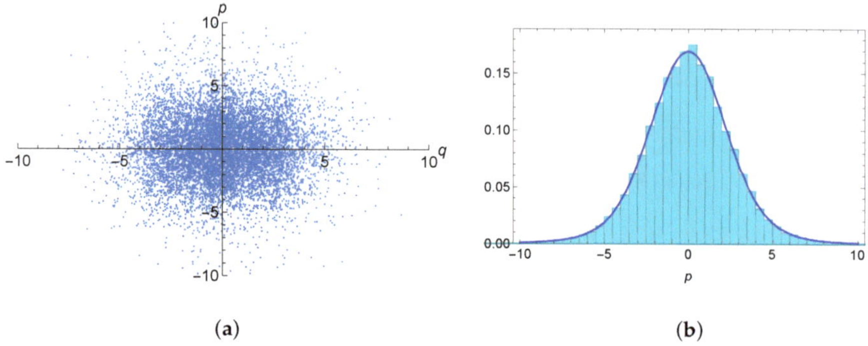

Figure 1. The simulated results of the CDD simulations of the target distribution (42) with $\kappa = 0.4$ and $\beta = 0.2$. (**a**) the phase (q-p) space orbit of the κ-deformed distribution. The 1.5×10^4 points of a simulated orbit with the initial condition ($q_0 = 0.1, p_0 = 0.1$, and $S_0 = 0.9$ are shown. (**b**) the histogram of the frequencies for p and the corresponding momentum κ-distribution (blue solid curve).

6. Conclusions

We considered the κ-deformations of some quantities concerning statistical physics and pointed out some unexpected relations among different fields, such as statistical mechanics, mathematical biology and evolutional game theory. We especially focused on the arsinh-type deformation of the ratio $\beta p^2/2$ of kinetic energy to the average thermal energy $k_{\rm B}T = 1/\beta$. With the help of the thermostat (CDD) algorithm we performed the relevant numerical simulations for the Hamiltonian with the arsinh-type deformation of kinetic energy term and showed the resultant momentum distribution was the κ–Gaussian distribution.

Finally, we would like to point out a relation which might be suggested for future research. Let us consider the κ-deformed energy density of state $\Omega_\kappa(U)$:

$$\Omega_\kappa(U) := \exp_\kappa\left(\frac{U}{k_{\rm B}T_c}\right) = \exp\left[\frac{1}{\kappa}\,\mathrm{arsinh}\left(\kappa\frac{U}{k_{\rm B}T_c}\right)\right], \tag{48}$$

which is the κ-deformation of the energy density of state $\exp(U/k_{\rm B}T_c)$ for the thermal reservoir with a constant-temperature T_c (Boltzmann reservoir [33]). In other words, $\ln\Omega_\kappa(U)$ is regarded as the arsinh-type deformation of the ratio $U/(k_{\rm B}T_c)$. The Boltzmann temperature $T(U)$ for this κ-deformed thermal reservoir is given by:

$$\frac{1}{k_{\rm B}T(U)} := \frac{d\ln\Omega(U)}{dU} = \frac{\frac{1}{k_{\rm B}T_c}}{\sqrt{1+\kappa^2\left(\frac{U}{k_{\rm B}T_c}\right)^2}}. \tag{49}$$

Rearranging this relation leads to:

$$k_{\rm B}T(U) = \sqrt{(\kappa U)^2 + (k_{\rm B}T_c)^2}, \tag{50}$$

which reminds us of the relativistic energy–momentum relation: $E(p) = \sqrt{(cp)^2 + (mc^2)^2}$.

Author Contributions: Conceptualization, T.W. and A.M.S.; methodology, T.W.; software, T.W.; validation, T.W. and A.M.S.; formal analysis, T.W.; investigation, T.W. and A.M.S.; resources, T.W.; data curation, T.W.; writing—original draft preparation, T.W.; writing—review and editing, T.W. and A.M.S.; visualization, T.W.; supervision, T.W.; project administration, T.W.; funding acquisition, T.W. All authors have read and agreed to the published version of the manuscript.

Funding: The first named author (T.W.) was partially supported by the Japan Society for the Promotion of Science (JSPS) Grants-in-Aid for Scientific Research (KAKENHI) Grant Number 22K03431.

Informed Consent Statement: Not applicable.

Data Availability Statement: Not applicable.

Acknowledgments: The authors thank anonymous referees for their valuable comments.

Conflicts of Interest: The authors declare no conflict of interest.

Appendix A. Basics of the κ-Deformed Functions

Here we briefly review some κ-deformed functions and the associated useful relations [2,3]. Because all κ-deformed functions are symmetric under the sign change of the deformation parameter κ, i.e., changing κ to $-\kappa$, throughout this paper we assume $\kappa > 0$. In the $\kappa \to 0$ limit, the κ-exponential function (1) and the κ-logarithmic function (2) reduce to the standard exponential function $\exp(x)$ and logarithmic function $\ln(x)$, respectively

$$\lim_{\kappa \to 0} \exp_\kappa(x) = \exp(x), \quad \lim_{\kappa \to 0} \ln_\kappa x = \ln x. \tag{A1}$$

We next introduce another κ-deformed function:

$$u_\kappa(x) \equiv \frac{x^\kappa + x^{-\kappa}}{2} = \cosh\left[\kappa \ln(x)\right], \tag{A2}$$

which is the conjugate (or co-function) of $\ln_\kappa x$, as similar as that $\cos(x)$ is the co-function of $\sin(x)$. In the $\kappa \to 0$ limit, this κ-deformed function reduces to the unit constant function $u_0(x) = 1$. By using $u_\kappa(x)$, the derivative of the κ-exponential is expressed as

$$\frac{d}{dx} \exp_\kappa(x) = \frac{\exp_\kappa(x)}{u_\kappa\left[\exp_\kappa(x)\right]} = \frac{\exp_\kappa(x)}{\sqrt{1 + \kappa^2 x^2}}, \tag{A3}$$

and the derivative of κ-logarithm is expressed as

$$\frac{d}{dx} \ln_\kappa(x) = \frac{u_\kappa(x)}{x}, \tag{A4}$$

respectively.

When $\kappa \neq 0$, the inverse function of $u_\kappa(x)$ exists, and given by

$$u_\kappa^{-1}(x) = \exp\left[\frac{1}{\kappa} \operatorname{arcosh}(x)\right], \tag{A5}$$

which is the co-function of $\exp_\kappa(x)$.

The κ-entropy S_κ [2,3] is a κ-generalization of the Gibbs-Shannon entropy $S^{\mathrm{GS}} = -k_{\mathrm{B}} \sum_i p_i \ln p_i$ by replacing the standard logarithm with the κ-logarithm, i.e.,

$$S_\kappa = -k_{\mathrm{B}} \sum_i p_i \ln_\kappa p_i. \tag{A6}$$

Appendix B. Replicator Equations and the General Form of Lotka-Volterra Equations

We here summarize some known important facts in mathematical biology and evolutional game theory according to Ref. [14,15,19]. Consider a discrete probability distribution described by a set of n positive variables $\boldsymbol{x} = (x_1, x_2, \ldots, x_n)$ with the normalization $\sum_i^n x_i = 1$, where each x_i denotes the proportion of the i-th type in the total population. The RE for this distribution is given by

$$\frac{d}{dt} x_i = x_i\Big(f_i(\boldsymbol{x}) - \bar{f}(\boldsymbol{x})\Big), \tag{A7}$$

where $f(x) = (f_1(x), \ldots, f_n(x))$ is a fitness landscape and $\bar{f}(x) = \sum_{i=1}^{n} x_i f_i(x)$ is the mean fitness. Replicator dynamics can be described as a time evolutional curve on the simplex $\Delta^n := \{x \in \mathbb{R}_+^n \mid x_i \geq 0, \sum_i x_i = 1\}$ with the matrix component $g^{ij}(x)$ of *Shahshahani metric* [16] g as

$$g^{ij}(x) = \frac{\delta_{ij}}{x_i}, \tag{A8}$$

The inverse matrix is $g_{ij}(x) = x_i \delta_{ij}$. Note that the n-simplex Δ^n is $(n-1)$-dimensional and the Shahshahani metric diverges on the boundary of the simplex. So this metric is valid only on the interior S^n of Δ^n.

There is a natural mapping: $(p_1, p_2, \ldots, p_n) \to (x_1, x_2, \ldots, x_n)$. Fisher metric is induced by the Shahshahani metric under this mapping.

$$(g^{\mathrm{F}})^{ij}(x) = \mathbb{E}\left[\frac{\partial \ln x}{\partial x_i}\frac{\partial \ln x}{\partial x_j}\right] = \sum_{k=1}^{n} x_k \frac{\delta_{ik}}{x_i}\frac{\delta_{ik}}{x_i} = \frac{\delta_{ij}}{x_i}. \tag{A9}$$

It is known that the Shahshahani manifolds yields an interpretation of the RE. Theorem 1 in [14]: if the differential equation $dx_i/dt = f_i(x)$ is a Euclidean gradient with $f_i = \partial V/\partial x_i$, the RE (A7) is a gradient with respect to Shahshahani metric. A brief explanation is as follows. The gradient with respect to Shahshahani metric is

$$(\nabla_g V)_i = \sum_j g_{ij}\frac{\partial V}{\partial x_j} = \sum_j x_i \delta_{ij} f_j = x_i f_i, \tag{A10}$$

which is the first term in the left hand side of the RE (A7). The variable x_i in the RE has to satisfy the normalization constraint ($\sum_i x_i = 1$), i.e., the dynamics of each x_i is restricted on the simplex Δ^n. Recall that Shahshahani metric is valid only on the interior S^n of Δ^n. Indeed, the normalization constraint is satisfied during an time evolution as follows

$$\frac{d}{dt}\sum_i x_i = \sum_i \frac{dx_i}{dt} = \sum_i x_i(f_i - \bar{f}) = \sum_i x_i f_i - \bar{f} = 0. \tag{A11}$$

The state $\hat{x}$ is said to be *evolutionarily stable state* if for all $x \neq \hat{x}$ in some neighborhood of $\hat{x}$,

$$x \cdot f(x) < \hat{x} \cdot f(x). \tag{A12}$$

Let the potential $V(x) = \mathrm{D}(\hat{x}\|x) = \sum_i \hat{x}_i \ln \hat{x}_i - \sum_i \hat{x}_i \ln x_i$, then we have

$$\frac{d}{dt}V(x) = -\sum_i \hat{x}_i \frac{1}{x_i}\frac{dx_i}{dt} = -\sum_i \hat{x}_i(f_i - \bar{f}) = -\sum_i \hat{x}_i f_i + \bar{f} = -(\hat{x}\cdot f - x \cdot f) < 0. \tag{A13}$$

Hence the Kullback-Leibler divergence $\mathrm{D}(\hat{x}\|x)$ is a local Lyapunov function for the RE.

Next, if $x_i = \exp(v_i(x) - \psi)$ with $dv_i(x)/dt = f_i(x)$ and $\psi(x)$ a normalization constant. From the normalization $\sum_i x_i = 1$, we have

$$0 = \sum_i \frac{d}{dt}x_i = \sum_i \left(\frac{d}{dt}v_i(x) - \frac{d}{dt}\psi(x)\right)x_i = \sum_i x_i f_i(x) - \frac{d}{dt}\psi(x) = \bar{f}(x) - \frac{d}{dt}\psi(x). \tag{A14}$$

As a result we see that $d\psi(x)/dt = \bar{f}(x)$, and x_i satisfies

$$\frac{d}{dt}x_i = x_i\left(\frac{d}{dt}v_i(x) - \frac{d}{dt}\psi(x)\right) = x_i(f_i(x) - \bar{f}(x)). \tag{A15}$$

Consequently, the exponential families $x_i = \exp(v_i(x) - \psi)$ are solutions of the RE.

If there is no constraint the corresponding dynamics is described by the gLV Equation (13). The gLV equations and REs are related as follows. Let each y_i satisfies the gLV Equation (13). Changing the variable y_i to x_i as

$$x_i = \frac{y_i}{\sum_{j=1}^{n} y_j}, \tag{A16}$$

which lead to the new normalized variables $\{x_i\}$, i.e., $\sum_j x_j = 1$. Then, we see that

$$\frac{dx_i}{dt} = \frac{\frac{dy_i}{dt}}{\sum_j y_j} - y_i \frac{\sum_k \frac{dy_k}{dt}}{\left(\sum_j y_j\right)^2} = \frac{y_i f_i}{\sum_j y_j} - \frac{y_i}{\left(\sum_j y_j\right)} \frac{\sum_k y_k f_k}{\left(\sum_j y_j\right)} = x_i(f_i - \bar{f}). \tag{A17}$$

Thus, the transformed variable x_i in (A16) satisfies the RE.

References

1. Kaniadakis, G.; Scarfone, A.M. A new one-parameter deformation of the exponential function. *Physica A* **2002**, *305*, 69–75. [CrossRef]
2. Kaniadakis, G. Statistical mechanics in the context of special relativity. *Phys. Rev. E* **2002**, *66*, 56125. [CrossRef] [PubMed]
3. Kaniadakis, G. Statistical mechanics in the context of special relativity II. *Phys. Rev. E* **2005**, *72*, 036108. [CrossRef] [PubMed]
4. Kaniadakis, G. Theoretical foundations and mathematical formalism of the power-law tailed statistical distributions. *Entropy* **2013**, *15* 3983–4010. [CrossRef]
5. Kaniadakis, G.; Baldi, M.M.; Deisboeck, T.S.; Grisolia, G.; Hristopulos, D.T.; Scarfone, A.M.; Sparavigna, A.; Wada, T.; Lucia, U. The κ-statistics approach to epidemiology. *Sci. Rep.* **2020**, *10*, 19949. [CrossRef] [PubMed]
6. Wada, T. Thermodynamic stabilities of the generalized Boltzmann entropies. *Physica A* **2004**, *340*, 126–130. [CrossRef]
7. Lymperis, A.; Basilakos, S.; Saridakis, E.N. Modified cosmology through Kaniadakis horizon entropy. *Eur. Phys. J. C* **2021**, *81*, 1037. [CrossRef]
8. Wada, T.; Scarfone A.M.; Matsuzoe H. On the canonical distributions of a thermal particle in a generalized velocity-dependent potential. *Physica A* **2020**, *541*, 123273. [CrossRef]
9. Strutt (Lord Rayleigh), J.W. Some general theorems relating to vibrations. *Proc. Lond. Math. Soc.* **1871**, *s1–s4*, 357–368. [CrossRef]
10. Pistone, G. κ-exponential models from the geometrical viewpoint. *Eur. Phys. J. B* **2009**, *70*, 29–37. [CrossRef]
11. Amari, S.-I. *Information Geometry and Its Applications*; Springer: Tokyo, Japan, 2016; Volume 194,
12. Wada, T.; Scarfone A.M. Information geometry on the κ-thermostatistics. *Entropy* **2015**, *17*, 1204–1217. [CrossRef]
13. Wada, T.; Scarfone A.M.; Matsuzoe H. An eikonal equation approach to thermodynamics and the gradient flows in information geometry. *Physica A* **2021**, *570*, 125820. [CrossRef]
14. Harper, M. Information geometry and evolutionary game theory. *arXiv* **2009**, arXiv:0911.1383.
15. Harper, M. Escort evolutionary game theory. *Physica D* **2011**, *240*, 1411–1415. [CrossRef]
16. Sigmund, K. Gradients for replicator systems. In *Dynamical Systems and Environmental Models: Proceedings of an International Workshop, Eisenach (GDR), Germany, 17–21 March 1986*; Bothe, H.G., Ebeling, W., Kurzhanski, A.B., Peschel, M., Eds.; De Gruyter: Berlin, Germany, 1987; pp. 186–195.
17. Hofbauer J.; Sigmund, K. *Evolutionary Games and Population Dynamics*; Cambridge University Press: Cambridge, UK, 1998.
18. Hernández-Bermejo, B.; Fairén, V. Lotka-Volterra representation of general nonlinear systems. *Math. Biosci.* **1997**, *140*, 1–32. [CrossRef]
19. Baez, J.C. The fundamental theorem of natural selection. *Entropy* **2021**, *23*, 1436. [CrossRef]
20. Gompertz, B. On the Nature of the Function Expressive of the Law of Human Mortality, and on a New Mode of Determining the Value of Life Contingencies. *Philos. Trans. R. Soc. Lond.* **1825**, *115*, 513–583. [CrossRef]
21. Bloch, F. Zur Theorie des Austauschproblems und der Remanenzerscheinung der Ferromagnetika. *Zeitschrift für Physik* **1932**, *74*, 295. [CrossRef]
22. Kirkwood, J.G. Quantum Statistics of Almost Classical Assemblies. *Phys. Rev.* **1933**, *44*, 31. [CrossRef]
23. Pelinovsky, E.; Kokoulina, M.; Epifanova, A.; Kurkin, A.; Kurkina, O.; Tang, M.; Macau, E.; Kirillin, M. Gompertz model in COVID-19 spreading simulation. *Chaos Solitons Fractals* **2022**, *154*, 111699. [CrossRef]
24. Cariñena, J.F.; Rañada, M.F.; and Santander, M. The quantum free particle on spherical and hyperbolic spaces: A curvature dependent approach. *J. Math. Phys.* **2011**, *52*, 072104. [CrossRef]
25. Lewis, J.T.; Pfister C-E.; Sullivan W.G. Entropy, concentration of probability and conditional limit theorems. *Markov Process. Relat. Fields* **1995**, *1*, 319–386.
26. Łapiński, T.M. Law of large numbers unifying Maxwell-Boltzmann, Bose-Einstein and Zipf-Mandelbrot distributions, and related fluctuations. *Physica A* **2021**, *572*, 125909. [CrossRef]
27. McKeague, I.W. Central limit theorems under special relativity. *Stat. Probab. Lett.* **2015**, *99*, 149–155. [CrossRef]
28. Scarfone, A.M.; Matsuzoe, H. κ-deformed Fourier transform. *Physica A* **2017**, *480*, 63. [CrossRef]

29. Nosé, S. A unified formulation of the constant temperature molecular-dynamics methods. *J. Chem. Phys.* **1984**, *81*, 511–519. [CrossRef]
30. Hoover, W.G. Canonical dynamics: Equilibrium phase-space distributions. *Phys. Rev. A* **1985**, *31*, 1695–1697. [CrossRef]
31. Bravetti, A.; Tapias, D. Thermostat algorithm for generating target state. *Phys. Rev. E* **2016**, *93*, 022139. [CrossRef]
32. Tolman, R.C. A General Theory of Energy Partition with Applications to Quantum Theory. *Phys. Rev.* **1918**, *11*, 261–275. [CrossRef]
33. Leff, H.S. The Boltzmann reservoir: A model constant-temperature environment. *Am. J. Phys.* **2000**, *68*, 521. [CrossRef]

entropy

MDPI

Review

The Scientific Contribution of the Kaniadakis Entropy to Nuclear Reactor Physics: A Brief Review

Aquilino Senra Martinez and Willian Vieira de Abreu *

Instituto Alberto Luiz Coimbra de Pós-Graduação e Pesquisa em Engenharia (COPPE/UFRJ), Programa de Engenharia Nuclear (PEN), Universidade Federal do Rio de Janeiro, Rio de Janeiro 21941-914, Brazil
* Correspondence: wabreu@coppe.ufrj.br; Tel.: +55-21992455819

Abstract: In nuclear reactors, tracking the loss and production of neutrons is crucial for the safe operation of such devices. In this regard, the microscopic cross section with the Doppler broadening function is a way to represent the thermal agitation movement in a reactor core. This function usually considers the Maxwell–Boltzmann statistics for the velocity distribution. However, this distribution cannot be applied on every occasion, i.e., in conditions outside the thermal equilibrium. In order to overcome this potential limitation, Kaniadakis entropy has been used over the last seven years to generate generalised nuclear data. This short review article summarises what has been conducted so far and what has to be conducted yet.

Keywords: Kaniadakis entropy; Doppler broadening function; nuclear reactor physics

Citation: Martinez, A.S.; de Abreu, W.V. The Scientific Contribution of the Kaniadakis Entropy to Nuclear Reactor Physics: A Brief Review. *Entropy* **2023**, *25*, 478. https://doi.org/10.3390/e25030478

Academic Editors: Dionissios T. Hristopulos, Sergio Luiz E. F. da Silva and Antonio M. Scarfone

Received: 15 December 2022
Revised: 12 January 2023
Accepted: 1 March 2023
Published: 9 March 2023

1. Introduction

In 2001, Giorgio Kaniadakis presented [1] a new one-parameter deformation for the exponential function, which describes power-law asymptotic behaviour. He did this in order to obtain a novel distribution that generalises the Maxwell–Boltzmann (MB) one. This new deformed exponential is provided by:

$$\exp_\kappa(x) \equiv \left(\sqrt{1+\kappa^2 x^2}+\kappa x\right)^{1/\kappa}, \tag{1}$$

where the κ parameter represents the level of deformation concerning the standard exponential. Hence, when $\kappa \to 0$, the deformed exponential reduces to $\exp(x)$.

The $\exp_\kappa(x)$, function obeys the following condition:

$$\exp_\kappa(x)\exp_\kappa(-x) = 1. \tag{2}$$

By considering the deformed exponential function, one can obtain a new statistical distribution, starting from a particle system in the velocity space and postulating a generalised density of entropy, given by [1,2]:

$$\sigma_\kappa(f) = -\int df \ln_{\{\kappa\}}(\alpha f). \tag{3}$$

With that, after some calculations, one can arrive at the following distribution function [1–3]:

$$f_\kappa(V,T) = A(\kappa)\exp_\kappa\left(-\frac{MV^2}{2k_B T}\right), \tag{4}$$

where in nuclear reactor physics terms, M is the nucleus mass, V is the velocity of the target nucleus, k_B the Boltzmann constant, T is the temperature of the medium, and $A(\kappa)$ is defined as [2,3]:

$$A(\kappa) = \left(\frac{|\kappa|M}{\pi k_B T}\right)^{n/2}\left(1 + \frac{3|\kappa|}{2}\right)\frac{\Gamma\left(\frac{1}{2|\kappa|} + \frac{3}{4}\right)}{\Gamma\left(\frac{1}{2|\kappa|} - \frac{3}{4}\right)} \tag{5}$$

Two significant advantages of using deformed statistics, such as those of Kaniadakis or Tsallis [4], are the capacity to describe systems with long-term time correlations [5] and the capacity to describe physical phenomena outside thermal equilibrium [3].

Given this considerable versatility, one can find several scientific papers that have applied the Kaniadakis-type deformed statistics. To illustrate this usefulness, one can cite works in astrophysics [6], cosmology [7], DNA analysis [8], quark–gluon plasma [9], game theory [10], error theory [11], information theory [12], random matrices [13], fractal systems [14], dusty-type plasmas [15], gravitational physics [16,17], epidemiology [18], Jeans instability of self-gravitating systems [19], and nuclear reactor physics [3,20–23].

In the present work, we will review the seven years of application of Kaniadakis statistics in the nuclear reactor physics area, presenting the theories developed until now and reflecting on their future prospects.

2. Microscopic Cross Sections and the Doppler Broadening Function

In nuclear reactors, the power generation is guided by neutrons being absorbed by unstable nuclides (uranium, for instance), which, in turn, split into fast-moving lighter nuclides. This fission process generates energy plus additional neutrons, which, again, split other nuclides, starting the nuclear chain reaction.

To understand the neutron–nucleus reaction, it is necessary to present the concept of microscopic cross sections, which are characterized by the probability that a certain reaction will occur and is represented by σ.

Specifically, from a nuclear reactor physics point of view, one can think in cross sections in the diversity of nuclear reactions, which occur inside a nuclear reactor. These reactions are usually divided in two types. The first one is scattering, σ_s, which has the subdivisions of inelastic scattering, σ_{in}, and elastic scattering, σ_e. The second division is the absorption types, σ_a, such as radiative capture, σ_γ, (n,α) reactions, $\sigma_{(n,\alpha)}$, and, of course, fission, σ_f. Considering all the possible reactions, one can define a conception of a total microscopic cross section, σ_t, which is, consequently, the sum of all of types and summarizes the probability that any type of neutron–nuclear reaction will occur:

$$\sigma_t = \sigma_s + \sigma_a = \sigma_e + \sigma_{in} + \sigma_f + \sigma_\gamma + \sigma_{(n,\alpha)} + \dots \tag{6}$$

The nuclear cross sections will vary strongly according to the incident neutron energy and weakly from the incident beam angle. For this reason, the beam angle is usually ignored in nuclear reactor applications.

Of all these types of nuclear cross sections, the radiative capture, σ_γ, is significantly relevant for nuclear reactor analysis since this is an important factor for removing neutrons from the chain reactions. This process occurs when the incident neutron is absorbed by the nucleus, forming a compound one. After this absorption, this new compound nucleus will decay by the emission of high-energy gamma radiation.

The process of compound nucleus formation occurs only at those energies at which the centre of mass (CM) incident neutron energy plus the neutron binding energy matches the exact energy level of the compound nucleus. Due to this, by adding the Heisenberg's uncertainty principle considerations, one can obtain a resonance behaviour for the functional dependence of the capture cross sections on the neutron kinetic energy.

In nuclear reactors, neutrons are generated in a highly energetic state in thermal nuclear reactors, far from the ideal energy range to fission U235 nuclei. Consequently, it is necessary to decrease the neutron energies. This process—called neutron slowing down—is conducted by multiple collisions (scattering) in moderators, such as water, graphite, or heavy water. Nevertheless, these scattering processes can lead to absorption, generating undesirable neutron losses.

These absorptions occur in an energy range called the resonance region and are represented by the neutron cross sections. These cross sections, in turn, depend on the relative speed between the neutron and the nuclei. Considering that the nuclei are in thermal motion, the relative velocity can be different (greater or less) than the neutron speed. That fact causes a Doppler effect in the existent neutron cross sections in the resonance region [24].

A way to represent this Doppler effect is by utilising the nuclear cross-section through the single-level Breit–Wigner (SLBW) resonance cross section formula. Although there are more modern methodologies, the SLBW method is easier to implement. Besides, it can present analytical functions for reactor physics calculations and, consequently, produce faster processing times [3,25].

One of the cross-section expression components is the Doppler broadening function, which considers a medium in thermal equilibrium at a temperature T. Additionally, the different velocities of the target nuclei are described by the Maxwell–Boltzmann (MB) distribution [26]. The averaged capture resonance cross section formula using the Breit–Wigner methodology represents the thermal nuclei movement and is expressed by [24]:

$$\overline{\sigma}_{\gamma} = \sigma_0 \frac{\Gamma_{\gamma}}{\Gamma} \left(\frac{E_0}{E} \right)^{1/2} \psi(x, \xi). \tag{7}$$

where:

$\Gamma_{\gamma}/\Gamma \equiv$ the probability that, once formed, the compound nucleus decays to the ground state of the original nucleus by gamma emission;

$$\xi \equiv \frac{\Gamma}{\left(\frac{4E_0 k_B T}{A} \right)^{\frac{1}{2}}}; \tag{8}$$

$$x \equiv \frac{2}{\Gamma}(E - E_0); \tag{9}$$

A is the mass number, k_B is the Boltzmann constant, σ_0 is the value of the total cross section, Γ_{γ} is radiative line width, Γ is total width of the resonance as measured in the laboratory coordinates, E_0 is the resonant energy, E is the energy of the incident neutron, and $\psi(x, \xi)$ is the so-called Doppler broadening function.

The Doppler broadening function, after the Bethe and Placzek [27] approximation, can be represented by:

$$\psi(\xi, x) \equiv \frac{\xi}{2\sqrt{\pi}} \int_{-\infty}^{+\infty} \frac{dy}{1 + y^2} exp\left[\frac{-\xi^2 (x - y)^2}{4} \right] \tag{10}$$

where:

$$y \equiv \frac{2}{\Gamma}(E_{CM} - E_0) \tag{11}$$

and E_{CM} is the centre-of-mass energy.

Its formulation usually takes into consideration the Maxwell–Boltzmann distribution. By doing a formal analysis, it is possible to note that $\psi(\xi, x)$ is affected by the temperature, i.e., the higher the temperature, the broader the resonance curve, with a peak attenuation, resulting in a higher probability of loss of neutrons by absorption. This effect is represented in Figure 1:

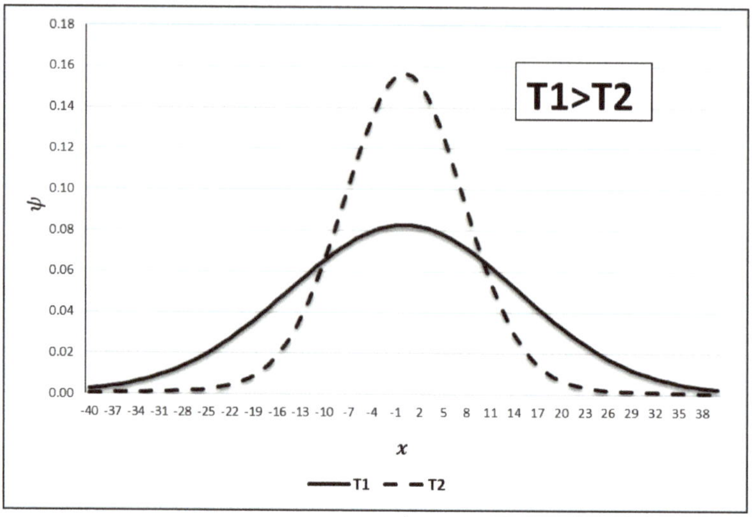

Figure 1. The temperature rise causes the Doppler broadening effect in a resonance curve [28]. Source: de Abreu (2020, p. 14).

As shown in Figure 1, there is an enlargement in the energy range—illustrated by the x variable—caused by the temperature rise (T1 to T2), which affects the curve's width. This leads to the probability increase in absorption.

The Doppler broadening function is also known as the first Voigt function. It represents an even function, i.e., it has symmetry with respect to the energy variable, x.

The Doppler broadening phenomenon is of crucial importance for the control of thermal nuclear reactors. This relevance can be observed if one analyzes the expressions for the average cross sections of resonance capture, Equation (7), and scattering, given by:

$$\overline{\sigma}_S = \sigma_0 \frac{\Gamma_n}{\Gamma} \psi(\xi, x) + \sigma_0 \frac{\mathrm{R}}{\lambda_0} \chi(\xi, x) + 4\pi R^2 \tag{12}$$

where:

$\Gamma_\mathrm{n}/\Gamma \equiv$ the probability that, once formed, the compound nucleus decays to the ground state of the original nucleus by neutron emission;

$\chi(\xi, x) \equiv$ interference term;

$4\pi R^2 \equiv$ potential scattering term.

Another way of realizing the importance of the Doppler broadening phenomenon is through the Doppler temperature coefficient. When the fuel temperature of a thermal reactor is increased, there is a growth in the energy range of the nuclear resonances. Hence, there is a decrease in the probability of resonance escape (p), i.e., a reduction in the fraction of fission neutrons moderated to the thermal range without suffering resonant absorption. As the resonance escape probability decreases, the multiplication factor, $\left(k_{eff}\right)$, and, consequently, the reactivity, (ρ), will also be reduced. This relationship is represented in the fuel temperature coefficient—also known as the Doppler temperature coefficient [29]:

$$\alpha_{T_f} = \frac{\partial \rho}{\partial T_f} = \frac{1}{p} \frac{\partial p}{\partial T_f}, \tag{13}$$

where T_f is the fuel temperature.

2.1. Literature Review of Doppler Broadening Functions Approximations Using the Maxwell–Boltzmann Distribution

Over the years, several methods have been developed to obtain approximations of the Doppler broadening function using standard Maxwell–Boltzmann statistics. Therefore, it is relevant to make a brief review of the literature regarding the main methods used, as these methods should also be natural candidates in obtaining generalized solutions.

To conduct this review, we focused on works that considered the approaches proposed by Bethe and Placzek in the year 1936 and which were crucial for the development of the deformed solution for $\psi_\kappa(\xi, x)$.

2.1.1. Beynon and Grant Methods (1963)

In 1963, T. D. Beynon and I. S. Grant published, in the journal *Nuclear Science and Engineering*, an article that proposed two different methods for calculating the Doppler broadening function and for the interference term ($\psi(\xi, x)$ and $\chi(\xi, x)$), calculated with the aid of a "digital machine".

The first of these proposed methods use Chebychev polynomial expansions, while the second uses the Gauss–Hermite method. According to the authors, this second method was developed as an alternative to the first due to the excessive memory space spent [30]. Considering that this justification no longer represents a concern for most researchers due to the advancement of technology, the present work will only present the first method, which is more widespread in the literature [3,22,23,25,28,31].

As previously mentioned, the first method proposed by Beynon and Grant, in their 1963 work, carried out the expansion of the exponential part of the integrands of the Doppler broadening function in Chebyshev polynomials. Once this was conducted, the terms were integrated one by one. Before making the expansion, the authors presented the following definitions:

$$a = \frac{\xi}{2} \tag{14}$$

$$b = \xi \cdot x \tag{15}$$

Thus, the proposed series are represented as follows:

$$\psi(a,b) = a\left\{\sqrt{\pi}\cdot cos(ab)\cdot[1 - E_2(a)]\cdot e^{a^2} + J(a,b)\right\}\cdot e^{-\frac{1}{4}b^2} \tag{16}$$

where:

$$\begin{aligned} J(a,b) = \; & \tfrac{1}{a}\left\{\tfrac{1}{2!}(ab)^2 - \tfrac{1}{4!}(ab)^4 + \tfrac{1}{6!}(ab)^6\cdots\right\} + \tfrac{1}{2a^3}\left\{\tfrac{1}{4!}(ab)^4 - \tfrac{1}{6!}(ab)^6\cdots\right\} \\ & + \cdots + \tfrac{1}{\sqrt{\pi}a^{2n+1}}\Gamma\left(\tfrac{2n+1}{2}\right)\cdot\left\{\tfrac{1}{[2(n+1)]!}(ab)^{2n+1}\cdots\right\} + \cdots \end{aligned} \tag{17}$$

and

$$E_2(a) = \frac{2}{\sqrt{\pi}}\int_0^a e^{-y^2}dy. \tag{18}$$

The $\Gamma(z)$ term represents the gamma function, which is one of the most common special functions in discussions of physical problems. One of the possible ways to express this function is [32]:

$$\Gamma(z) \equiv \int_0^\infty e^{-t}e^{z-1}dt, \;\; \mathbb{R}e(z) > 0. \tag{19}$$

Still, in their article, the authors of the work argued that the proposed approximation works well for values where $b < 6$. For values where $b \geq 6$, the authors suggested using the asymptotic expansions method [33].

The following method consists of the Taylor series expansion of the term $1/(1+y^2)$ of the Doppler broadening function around $y = x$, given by [34]:

$$\frac{1}{(1+y^2)} = \frac{1}{(1+x^2)} - \frac{2x}{(1+x^2)^2}(y-x) + \frac{-1+3x^2}{(1+x^2)^2}(y-x)^2 \cdots. \tag{20}$$

By substituting the above-mentioned equation in the approximation proposed by Bethe and Placzek and then performing the integration of all terms, the expansion is obtained:

$$\Psi(\xi, x) = \frac{1}{(1+x^2)}\left\{1 + \frac{2(-1+3x^2)}{\xi^2(1+x^2)^2} + \frac{12(1-10x^2+5x^4)}{\xi^4(1+x^2)^4} \cdots\right\}. \tag{21}$$

For large values of x, Equation (21) tends to:

$$\Psi(\xi, x) \approx \frac{1}{(1+x^2)}. \tag{22}$$

Through Equation (21), it is possible to obtain results for the Doppler broadening function considering temperatures close to zero Kelvin [34].

The consideration of alternating methods for different ranges of b ($b < 6$ and $b \geq 6$) is a strategy that was applied in later works [22,31].

The tables of values for the Doppler broadening function and for the interference term generated by the work of Beynon and Grant were essential for the reactor physics field, being present even in two of the main books in this field [24,29].

2.1.2. Campos and Martinez Method (1987)

In 1987, Campos and Martinez published, in the journal *Annals of Nuclear Energy*, the article "The dependence of practical width on temperature". This work, among other topics, presented a new procedure for calculating the Doppler broadening function. To do that, the authors used the definitions of Beynon and Grant for the first Voigt function, $\psi(\xi, x)$, and the interference term, $\chi(\xi, x)$, also known as second Voigt function.

For this, the authors differentiated both the integral definition of the Doppler broadening function and the interference term in relation to x, arriving at the following differential equation [31]:

$$\frac{4}{\xi^2}\frac{\partial^2\psi(\xi, x)}{\partial x^2} + 4x\frac{\partial\psi(\xi, x)}{\partial x} + \frac{\xi^2}{4}\left[2 + \xi^2 x^2 + \xi^2\right]\psi(\xi, x) = \xi^2 \tag{23}$$

subject to the following initial conditions:

$$\psi(\xi, x)|_{x=0} = \psi_0 = \frac{\xi\sqrt{\pi}}{2}e^{\frac{\xi^2}{4}}\left[1 - \text{erf}\left(\frac{\xi}{2}\right)\right] \tag{24}$$

$$\left.\frac{\partial\psi(\xi, x)}{\partial x}\right|_{x=0} = 0. \tag{25}$$

To solve the proposed differential equation, Campos and Martinez (1987) used the power series expansion method, presenting the following result:

$$\psi(\xi, x) = \sum_{n=0}^{\infty} c_n x^{2n}, \tag{26}$$

with the coefficients generated from the recurrence formulas:

$$c_0 = \psi_0; \tag{27}$$

$$c_1 = \frac{\xi^2}{8}\left[\xi^2 - \left(\xi^2 + 2\right)\psi_0\right]; \tag{28}$$

$$c_{n+1} = -\frac{\xi^2}{4}\left[\frac{(4n + \xi^2 + 2)c_n + \xi^2 c_{n-1}}{(n+2)(n+1)}\right] \tag{29}$$

Similarly to Beynon and Grant, Campos and Martinez also used the alternation of methods for different ranges of b ($b < 6$ and $b \geq 6$), using the same Taylor expansion method, demonstrated by Equation (21) for values of $b \geq 6$.

Additionally, in this 1987 work, an expression for the temperature-dependent practical width was developed, which, in turn, was used later in developing the Campos–Martinez model for resonance integrals for isolated ones [31,35].

2.1.3. The Palma, Martinez, and Silva Method (2006)

In 2006, Palma, Martinez, and Silva published, in the *Journal of Nuclear Science and Technology*, the article "The derivation of the Doppler broadening function using Frobenius method" [36].

In this work, the authors solved Equation (23) using a different methodology from the one proposed by Campos and Martinez: utilizing the Frobenius method for the resolution of the homogeneous part of the equation and the method of variation of parameters to obtain the particular solution.

Taking this into account, the solution for the obtained Doppler broadening function was:

$$\psi(\xi, x) = \frac{\xi\sqrt{\pi}}{2} e^{-\frac{\xi^2}{4}(x^2-1)} cos\left(\frac{\xi^2 x}{2}\right)\left[1 + Re\phi(\xi, x) + tan\left(\frac{\xi^2 x}{2}\right) Im\phi(\xi, x)\right] \tag{30}$$

where

$$\phi(\xi, x) = erf\left(\frac{i\xi x - \xi}{2}\right) \tag{31}$$

To measure the method's accuracy, the authors compared the results of Equation (30) with results obtained from Padé's four-pole method [37,38], which is widely used in calculating the function of Doppler broadening and for calculating resonance integrals. According to Palma et al., approximations of this type are better than using the Taylor series to describe functions with poles.

This comparison concluded that the proposed solution presented relative percentage errors lower than the Padé method [36].

3. Kaniadakis Entropy in the Context of Nuclear Reactor Physics

In the context of the current thermal nuclear reactors scenario, it is appropriate to apply the Maxwell–Boltzmann statistics to treat the nuclei velocities distribution. However, considering the perspective of a new generation of nuclear reactors with different characteristics, the MB approach will not be enough, especially if one considers describing situations outside the thermal equilibrium [3].

Guedes et al. [3] proposed using the Kaniadakis statistics to obtain an expression for the Doppler broadening function, which could be capable of contemplating situations other than the regular nuclear reactors, for instance, in the context of thermal non-equilibrium.

To do this, the authors proposed a new integral expression, $\psi_\kappa(\xi, x)$, starting from the Doppler broadening function without any approximations, $\Psi(\xi, x)$, given by [3]:

$$\Psi(\xi, x) \equiv \pi\sqrt{\frac{2k_B T}{M}}\xi \int_{-\frac{2}{\Gamma}E_O}^{+\infty} \frac{dy}{1+y^2} \int_{v(x)-v_r(y)}^{v(x)+v_r(y)} dV V f(V, T), \tag{32}$$

where v is the neutron velocity, v_r the relative velocity between the neutron and the nucleus, and $f(V,T)$ is the chosen nuclei velocities distribution. By substituting the Kaniadakis statistics, $f_\kappa(V,T)$, into the equation above, one has [3]:

$$\Psi(\xi,x) \equiv \pi\sqrt{\frac{2k_BT}{M}}A(\kappa)\xi \int_{-\frac{2}{\Gamma}E_O}^{+\infty} \frac{dy}{1+y^2}\int_{v(x)-v_r(y)}^{v(x)+v_r(y)} dVV\exp_\kappa\left(-\frac{MV^2}{2k_BT}\right). \quad (33)$$

After some algebraic manipulation to perform the integration of the V variable, Guedes et al. [3] presented a new function, called $iexp_\kappa$, since:

$$\int \exp_\kappa(x)dx = i\exp_\kappa(x) + C, \quad (34)$$

where

$$i\exp_\kappa(x) \equiv \left(\frac{\sqrt{1+\kappa^2x^2}-\kappa^2x}{1-\kappa^2}\right)\exp_\kappa(x); \quad (35)$$

Noting that:

$$\lim_{\kappa\to 0} i\exp_\kappa(x) = e^x, \quad (36)$$

Equation (33) becomes:

$$\Psi(\xi,x) \equiv \frac{\xi}{2\sqrt{\pi}}B(\kappa)\int_{-\frac{2}{\Gamma}E_O}^{+\infty}\frac{dy}{1+y^2}\left[i\exp_\kappa\left(-\frac{1}{2k_BT}M(v(x)-v_r(y))^2\right) - i\exp_\kappa\left(-\frac{1}{2k_BT}M(v(x)+v_r(y))^2\right)\right]. \quad (37)$$

where:

$$B(\kappa) = (2|\kappa|)^{3/2}\left(1+\frac{3|\kappa|}{2}\right)\frac{\Gamma\left(\frac{1}{2|\kappa|}+\frac{3}{4}\right)}{\Gamma\left(\frac{1}{2|\kappa|}-\frac{3}{4}\right)}. \quad (38)$$

Finally, by using the Bethe–Placzek [27] approximations, Guedes et al. provided the integral solution for the Doppler broadening function using the Kaniadakis distribution:

$$\Psi_\kappa(\xi,x) \approx \psi_\kappa(\xi,x) = \frac{\xi}{2\sqrt{\pi}}B(\kappa)\int_{-\infty}^{+\infty}\frac{dy}{1+y^2}i\exp_\kappa\left[\frac{-\xi^2(x-y)^2}{4}\right]. \quad (39)$$

Noting again that, when κ tends towards zero, one can obtain the integral solution for the Doppler broadening function using the Maxwell–Boltzmann distribution.

The primary behaviour with the usage of the Kaniadakis statistical distribution is the peak attenuation for the resonance curves as the value of κ rises, as one can see in the Figures 2 and 3 below:

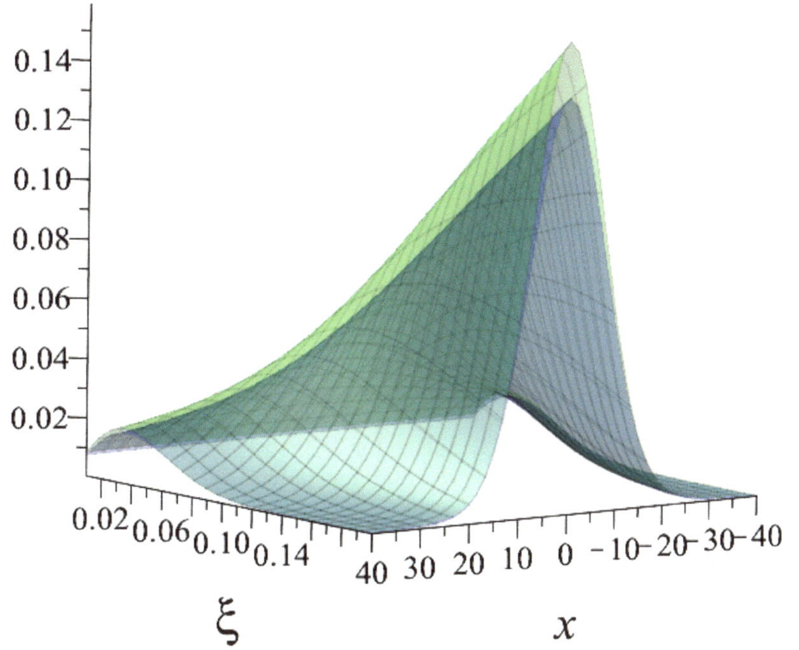

Figure 2. The comparison between the $\psi_{\kappa=0.3}(\xi, x)$ (blue) and $\psi(\xi, x)$ (green) curves [28]. It is possible to see the peak's attenuation.

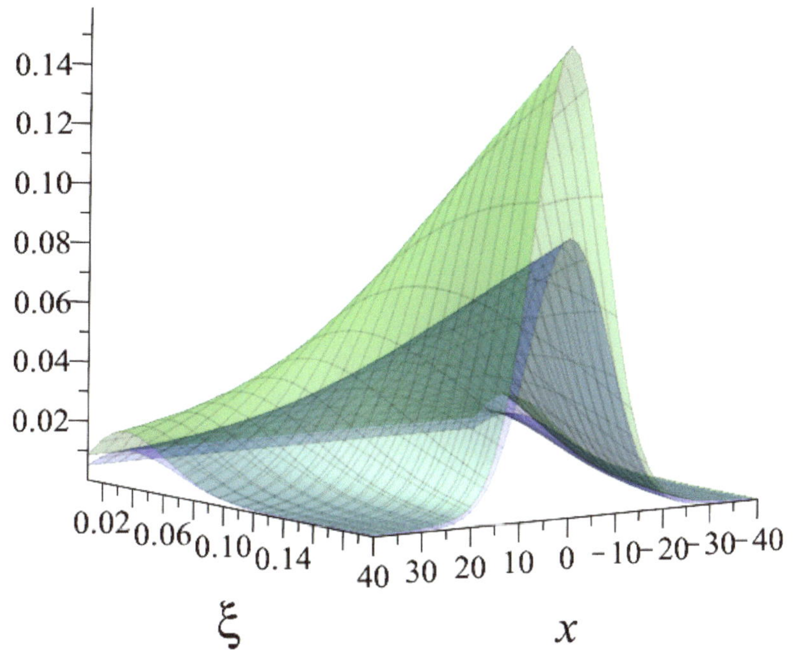

Figure 3. The comparison between the $\psi_{\kappa=0.5}(\xi, x)$ (blue) and $\psi(\xi, x)$ (green) curves. It is possible to see the peak's attenuation.

An animation of this behaviour for different values of κ is available at the following link: "https://youtu.be/wAeBAETbU1c (Accessed on 1 March 2023)".

4. Analytical Solutions for the Doppler Broadening Function Using the Kaniadakis Distribution

Guedes et al. [3] presented the results of the integral formulation for the Doppler broadening function using the Kaniadakis distribution, Equation (39), by using a numerical calculation, which, in turn, can present very high processing times when inserted into complex systems of calculations, such as nuclear reactor simulations.

In the already existing computer codes for nuclear core design, the Doppler broadening process requires a relatively long computing time to calculate Equation (10). Therefore, more time will be needed to calculate the deformed function, Equation (39). One should keep in mind that the Doppler broadening function for neutron cross sections should be calculated for each energy mesh point, isotope, and temperature.

In order to surpass this potential problem, de Abreu et al. proposed an analytical solution for the integral expression. To do that, the authors started by obtaining a differential equation for $\psi_\kappa(\xi,x)$ from the integral expression, using the same methodology by Campos and Martinez for obtaining an analytical expression for ψ considering the Maxwell–Boltzmann distribution.

$$\frac{\partial^2\psi_\kappa(\xi,x)}{\partial x^2} - \frac{\xi^2 x}{2}\left[\frac{\left(\kappa^2-1\right)^2+1}{\left(\kappa^2-1\right)}\right]\frac{\partial\psi_\kappa(\xi,x)}{\partial x} + \frac{\xi^2}{4}\left[-2\left(\kappa^2-1\right)+\xi^2x^2+\xi^2\right]\psi_\kappa(\xi,x) = -\frac{\xi^4}{4}\left(\kappa^2-1\right)B(\kappa). \tag{40}$$

One relevant thing to mention is that when κ tends to zero, the differential equation proposed by Abreu et al. [21,22] becomes the equation proposed by Campos and Martinez, proving the validity of the solution.

Initially, to solve the deformed differential equation using the Kaniadakis distribution, the authors used the same method by Palma, Martinez, and Gonçalves, i.e., the Frobenius method, for the homogeneous part and the parameter variation method for the particular solution. However, to eliminate some of the approximations and assumptions made, a new paper was published later using the dependent variable method to solve the homogeneous part of the equation.

The solution of the deformed differential equation for ψ_κ is divided in two based on the above-mentioned works of Beynon and Grant [30] and Campos and Martinez [31]: one for $|x{\cdot}\xi| < 6$ and other for $|x{\cdot}\xi| \geq 6$. The first one can be represented by:

$$\psi_\kappa(\xi,x) = \Lambda(x,\xi)\left[D(\xi,x)+\Omega_g(\xi,x)\right], \tag{41}$$

where

$$D(\xi,x) \equiv [\Delta(\xi){\cdot}cos(\Theta)], \tag{42}$$

$$\Omega_g(\xi,x) \equiv \Pi(x,\xi){\cdot}[i\Omega_1(\xi,x)+\Omega_2(\xi,x)]; \tag{43}$$

$$\Lambda(\xi,x) = \exp\left(\frac{\xi^2-\xi^2x^2}{4}\right){\cdot}\frac{\xi\sqrt{\pi}B(\kappa)}{4}; \tag{44}$$

$$\Pi(\xi,x) = \frac{\sqrt{\xi^4-2\xi^2\kappa^2}}{-\xi^2+2\kappa^2}{\cdot}\exp\left(\frac{-\kappa^2}{2}\right); \tag{45}$$

$$\Delta(\xi) = \frac{2-2erf\left(\frac{\xi}{2}\right)}{1-\kappa^2}. \tag{46}$$

$$\Omega_1(\xi,x) = sin(\Theta){\cdot}\left[erf(P_1)\kappa^2 - erf(P_1) + erf(P_2)\kappa^2 - erf(P_2)\right]; \tag{47}$$

$$\Omega_2(\xi,x) = cos(\Theta)\cdot\left[2erf(P_3)\kappa^2 - 2erf(P_3) - erf(P_1)\kappa^2 + erf(P_1) + erf(P_2)\kappa^2 - erf(P_2)\right]; \quad (48)$$

$$P_1(\xi,x) = \frac{-i\xi^2 x + \sqrt{\xi^4 - 2\xi^2\kappa^2}}{2\xi}; \quad (49)$$

$$P_2(\xi,x) = \frac{-i\xi^2 x - \sqrt{\xi^4 - 2\xi^2\kappa^2}}{2\xi}; \quad (50)$$

$$P_3(\xi,x) = \frac{\sqrt{\xi^4 - 2\xi^2\kappa^2}}{2\xi}; \quad (51)$$

$$\Theta(\xi,x) = \frac{x}{2}\sqrt{\xi^4 - 2\xi^2\kappa^2}; \quad (52)$$

The solution for the $|x\cdot\xi| \geq 6$ domain is obtained by the conduction of asymptotic expansions in the Taylor series. Thus, the expression for $\psi_\kappa(\xi,x)$ in this case is:

$$\psi_\kappa[\![|x\cdot\xi| \geq 6]\!] \cong B(\kappa)\cdot\left[\frac{1}{(1-\kappa^2)(1+x^2)} + \frac{-3\kappa^2\xi^2 - \kappa^2\xi^2x^4 + 4 - 12x^2}{2\xi^2(\kappa^2-1)(1+x^2)^3} + \ldots\right] \quad (53)$$

In a more recent paper [25], the authors also calculated the computational processing times. They obtained an average value of 4.6 for the ratio (numerical/analytical), confirming the predicted faster processing times for the analytical solution.

In parallel with the development of analytical solutions, da Silva et al. [39] proposed a method to establish a relation of equivalence among the standard Doppler broadening function, $\psi(\xi,x)$, and the deformed one, $\psi_\kappa(\xi,x)$.

The effective medium temperature model consists of determining the temperature, T_{eff}, which will have the Doppler broadening function from the perspective of the Maxwell–Boltzmann distribution reproducing the same value for the function using Kaniadakis distribution in the actual temperature (T) of the medium, that is:

$$\psi_\kappa(x,\xi) \cong \psi_\kappa\left(x,\tilde{\xi}\right) \quad (54)$$

where:

$$\tilde{\xi} \equiv \frac{\Gamma}{\left(4E_0\, k_B T_{eff}/A\right)^{1/2}} \quad (55)$$

and

$$\xi \equiv \frac{\Gamma}{(4E_0\, k_B T/A)^{1/2}} \quad (56)$$

Applying the polynomial regression method optimised by the genetic algorithm technique, the values of $\tilde{\xi}$ and, consequently, of the effective medium temperature, T_{eff}, were obtained.

Thus, the deformed Doppler broadening function, $\psi_\kappa(x,\xi)$, according to the Kaniadakis distribution, can be obtained directly via the original Doppler broadening function by replacing the medium temperature, T, with the effective temperature of the medium, T_{eff}.

According to the authors, the model proved to be a much faster and simpler method for calculating deformed Doppler broadening functions concerning the numerical approach [39].

In this 'twenty years of Kaniadakis entropy' issue, a paper was published [23], applying the analytical solution for ψ_k to calculate deformed nuclear cross sections in the nuclear data processing code FRENDY, developed by the Japan Atomic Energy Agency.

One of the contributions of this work was to study the relevance of the so-called Gaussian error function, erf, in the proposed analytical formulation for ψ_k. The erf function is defined by:

$$erf(x) = \frac{2}{\sqrt{\pi}} \int_0^x e^{-t^2} dt. \tag{57}$$

Due to its complexity, many programming code libraries use approximations to calculate these functions. These approximations, in turn, result in errors. The authors then used the Faddeeva Package [28] to minimise these, especially for the tails (far from the resonance peaks) regions.

5. Kaniadakis' Entropy Validation in the Nuclear Reactor Physics

Other areas that employed the Kaniadakis statistics can validate its results through observational data (stellar clusters, for instance). For an example of this validation, one can present the high-quality agreement between the theoretical curves for cosmic rays' flux using the Kaniadakis distribution and observational data. As one can see in reference [2], the Kaniadakis statistics represented the observational data better than the standard Maxwell-Boltzmann one.

However, in nuclear physics reactors, this is not possible. Thus, one needs to find ways to validate the applicability of this generalised entropy in the nuclear engineering area.

From this perspective, it is crucial—from a validation point of view—to verify the existence of a pertinency in applying the Kaniadakis method in the reactor physics area. Hence, one of the first necessary steps is to generate deformed nuclear data and compare it to the Maxwell–Boltzmann results.

Nevertheless, before this step, it was necessary to validate the analytical solution itself. Hence, in a 2020 work, de Abreu [28] generated the first deformed data using the following equations for the deformed radiative capture cross sections obtained through the Doppler broadening function considering the Kaniadakis distribution:

$$\sigma_\gamma^\kappa = \sigma_0 \left(\frac{\Gamma_\gamma}{\Gamma}\right) \left(\frac{E_0}{E}\right)^{\frac{1}{2}} \psi_\kappa(\xi, x) \tag{58}$$

To calculate these cross sections, the authors used the parameters — illustrated in Table 1—of the 238 isotope of uranium in the 6.67 eV peak resonance:

Table 1. Resonance data for the 6.67 eV line.

E_0(eV)	Γ_n(eV)	Γ_γ(eV)	σ_0 (b)	Γ_p(eV)
6.67	0.00152	0.026	2.16×10^5	1.26

Source: Duderstadt and Hamilton [24] (1976, p. 335).

Using the above-mentioned data and Equation (58), the authors generated the deformed analytical data by using the numerical and analytical expressions for $\psi_\kappa(\xi, x)$ to compare them with each other, as shown in Figure 4.

The results show a robust agreement between both solutions, which gives conditions to proceed in the validation steps.

However, even with the results in conformity with the predicted behaviour concerning the numerical results, applying the Kaniadakis distribution in a proper nuclear data generator tool is still necessary for certification purposes. In this sense, one needs to insert the analytical solution into nuclear data processing codes, preferably into ones recognised and used by peers around the world, for instance, FRENDY [28,29], NJOY [40], PREPRO [41], or NECP-Atlas [42]. The advantage of using these systems is that they can process official evaluated nuclear data libraries, such as ENDF, CENDL, JEFF, or JENDL.

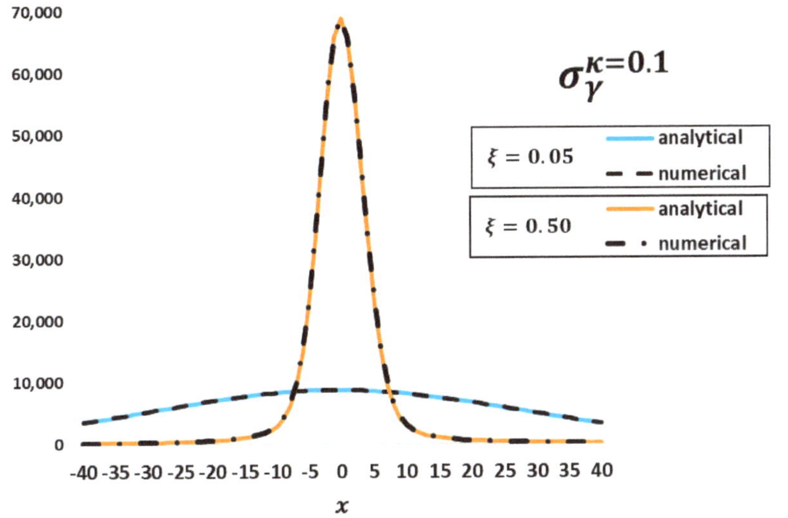

Figure 4. A comparison between the analytical and numerical results for the radiative capture cross-sections using the Kaniadakis distribution [28]. Source: de Abreu (2020, p. 90, our translation).

Therefore, a paper in this twenty-year commemorative issue was published, generating, for the first time, deformed data using the Kaniadakis distribution for two important nuclides: technetium 99 and plutonium 238, as one can see in Figures 5 and 6. To do that, they chose the FRENDY nuclear data processing code [28,29].

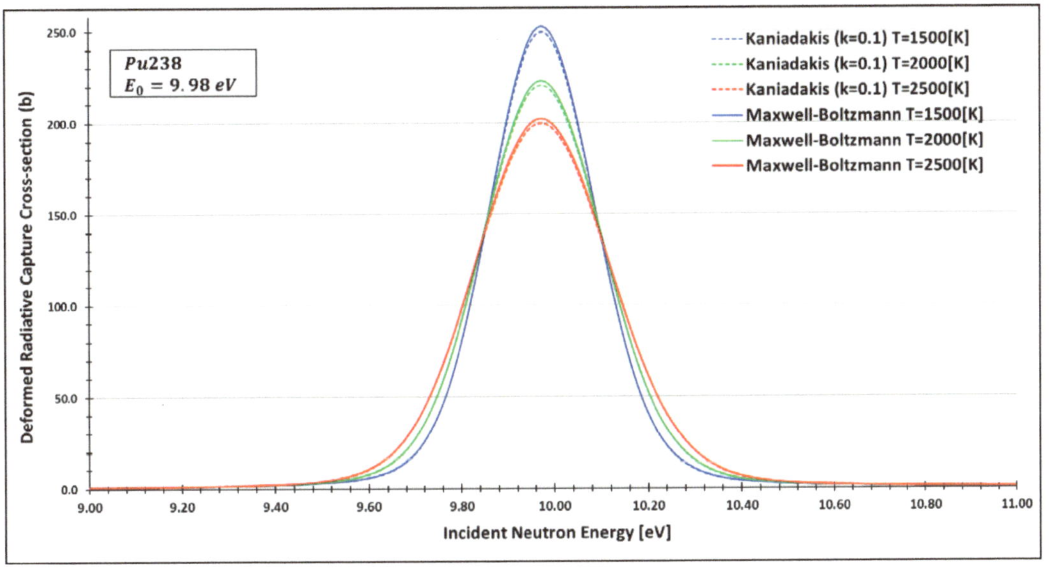

Figure 5. A plot of the deformed radiative capture cross section for plutonium 238 considering $k = 0.1$ and the 9.98 eV peak generated by the FRENDY data processing code [23]. Source: de Abreu et al. (2022, p. 10).

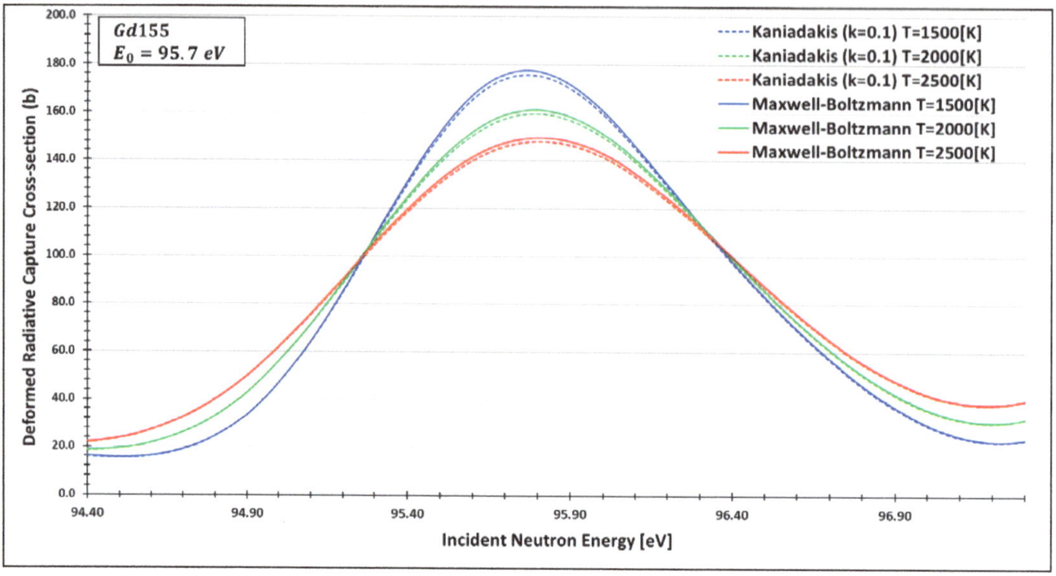

Figure 6. A plot of the deformed radiative capture cross section for gadolinium 155 considering $k = 0.1$ and the 95.7 eV peak generated by the FRENDY data processing code [23]. Source: de Abreu et al. (2022, p. 12).

By analysing the data, the authors found compatibility between the expected peak attenuation behaviour of the deformed curves compared to the Maxwell–Boltzmann data for different resonance peaks.

These results represent a significant step in validating the Kaniadakis statistics in the nuclear fission area.

6. Next Steps and Perspectives of Kaniadakis Entropy in the Nuclear Reactor Physics

Much work has been conducted on nuclear reactor physics with the Kaniadakis deformed statistics. However, to proceed with this investigation, there is still a question to answer: which value (or range) of κ fits with current and future nuclear reactors?

In this sense, future works will need to proceed with the work inside data generation codes and use its results of deformed nuclear cross sections in other nuclear reactor simulations to find the closest values of the deformation factor, κ, in order to fit the standard nuclear reactor behaviour.

Another possible and relevant step is implementing the Kaniadakis methodology in modern theoretical methods, such as the multi-level Breit–Wigner and Reich–Moore methods. With this implementation, it would be possible, for instance, to generate a deformed nuclear cross section for other nuclides, such as the isotopes 235 and 238 of uranium.

Author Contributions: Conceptualization, A.S.M., W.V.d.A.; methodology, A.S.M., W.V.d.A.; writing—original draft preparation, W.V.d.A.; writing—review and editing, A.S.M., W.V.d.A.; supervision, A.S.M.; project administration, W.V.d.A. All authors have read and agreed to the published version of the manuscript.

Funding: This research was funded by Fundação Carlos Chagas Filho (FAPERJ/Pós-doutorado Nota 10 e Cientistas do Nosso Estado), grants numbers E-26/204.444/2021 (registration number 2021.00898.8) and E-26/201.090/2022 and by the Conselho Nacional de Desenvolvimento Científico e Tecnológico, grant number CNPq/304.580/2019-8.

Institutional Review Board Statement: Not applicable.

Data Availability Statement: Not applicable.

Acknowledgments: The first author thanks CNPq for the Productivity Grant and FAPERJ for the Scientist of Our State Grant. The second author thanks FAPERJ for his post-doctoral grant.

Conflicts of Interest: The authors declare no conflict of interest.

References

1. Kaniadakis, G. Non-Linear Kinetics Underlying Generalized Statistics. *Phys. A Stat. Mech. Its Appl.* **2001**, *296*, 405–425. [CrossRef]
2. Kaniadakis, G. Statistical Mechanics in the Context of Special Relativity. *Phys. Rev. E* **2002**, *66*, 056125. [CrossRef]
3. Guedes, G.; Gonçalves, A.C.; Palma, D.A. The Doppler Broadening Function Using the Kaniadakis Distribution. *Ann. Nucl. Energy* **2017**, *110*, 453–458. [CrossRef]
4. Tsallis, C. Possible Generalization of Boltzmann-Gibbs Statistics. *J. Stat. Phys.* **1988**, *52*, 479–487. [CrossRef]
5. Chavanis, P.-H. Generalized Thermodynamics and Fokker-Planck Equations: Applications to Stellar Dynamics and Two-Dimensional Turbulence. *Phys. Rev. E* **2003**, *68*, 036108. [CrossRef] [PubMed]
6. Carvalho, J.C.; Silva, R.; do Nascimento, J.D., Jr.; Soares, B.B.; De Medeiros, J.R. Observational Measurement of Open Stellar Clusters: A Test of Kaniadakis and Tsallis Statistics. *EPL Europhys. Lett.* **2010**, *91*, 69002. [CrossRef]
7. Luciano, G.G. Gravity and Cosmology in Kaniadakis Statistics: Current Status and Future Challenges. *Entropy* **2022**, *24*, 1712. [CrossRef]
8. de Lima, M.M.F.; Anselmo, D.H.A.L.; Silva, R.; Nunes, G.H.S.; Fulco, U.L.; Vasconcelos, M.S.; Mello, V.D. A Bayesian Analysis of Plant DNA Length Distribution via κ-Statistics. *Entropy* **2022**, *24*, 1225. [CrossRef]
9. Teweldeberhan, A.M.; Miller, H.G.; Tegen, R. κ-deformed statistics and the formation of a quark-gluon plasma. *Int. J. Mod. Phys. E* **2003**, *12*, 669–673. [CrossRef]
10. Topsøe, F. Entropy and Equilibrium via Games of Complexity. *Phys. A Stat. Mech. Its Appl.* **2004**, *340*, 11–31. [CrossRef]
11. Wada, T.; Suyari, H. κ-Generalization of Gauss' Law of Error. *Phys. Lett. A* **2006**, *348*, 89–93. [CrossRef]
12. Wada, T.; Suyari, H. A Two-Parameter Generalization of Shannon–Khinchin Axioms and the Uniqueness Theorem. *Phys. Lett. A* **2007**, *368*, 199–205. [CrossRef]
13. Abul-Magd, A.Y. Nonextensive Random-Matrix Theory Based on Kaniadakis Entropy. *Phys. Lett. A* **2007**, *361*, 450–454. [CrossRef]
14. Olemskoi, A.I.; Kharchenko, V.O.; Borisyuk, V.N. Multifractal Spectrum of Phase Space Related to Generalized Thermostatistics. *Phys. A Stat. Mech. Its Appl.* **2008**, *387*, 1895–1906. [CrossRef]
15. Lourek, I.; Tribeche, M. Dust Charging Current in Non Equilibrium Dusty Plasma in the Context of Kaniadakis Generalization. *Phys. A Stat. Mech. Its Appl.* **2019**, *517*, 522–529. [CrossRef]
16. Abreu, E.M.C.; Neto, J.A.; Barboza, E.M.; Nunes, R.C. Tsallis and Kaniadakis Statistics from the Viewpoint of Entropic Gravity Formalism. *Int. J. Mod. Phys. A* **2017**, *32*, 1750028. [CrossRef]
17. Abreu, E.M.C.; Neto, J.A.; Mendes, A.C.R.; Bonilla, A.; de Paula, R.M. Cosmological Considerations in Kaniadakis Statistics. *EPL Europhys. Lett.* **2018**, *124*, 30003. [CrossRef]
18. Kaniadakis, G.; Baldi, M.M.; Deisboeck, T.S.; Grisolia, G.; Hristopulos, D.T.; Scarfone, A.M.; Sparavigna, A.; Wada, T.; Lucia, U. The κ-Statistics Approach to Epidemiology. *Sci. Rep.* **2020**, *10*, 19949. [CrossRef] [PubMed]
19. He, K.-R. Jeans Analysis with κ-Deformed Kaniadakis Distribution in f (R) Gravity. *Phys. Scr.* **2022**, *97*, 025601. [CrossRef]
20. de Abreu, W.V.; Gonçalves, A.C.; Martinez, A.S. An Analytical Approximation for the Generalized Interference Term Using the Kaniadakis Distribution. *Proc. Int. Conf. Nucl. Eng.* **2019**, *2019*, 1912. [CrossRef]
21. de Abreu, W.V.; Gonçalves, A.C.; Martinez, A.S. Analytical Solution for the Doppler Broadening Function Using the Kaniadakis Distribution. *Ann. Nucl. Energy* **2019**, *126*, 262–268. [CrossRef]
22. de Abreu, W.V.; Gonçalves, A.C.; Martinez, A.S. New Analytical Formulations for the Doppler Broadening Function and Interference Term Based on Kaniadakis Distributions. *Ann. Nucl. Energy* **2020**, *135*, 106960. [CrossRef]
23. de Abreu, W.V.; Maciel, J.M.; Martinez, A.S.; da Cruz Gonçalves, A.; Schmidt, L. Doppler Broadening of Neutron Cross-Sections Using Kaniadakis Entropy. *Entropy* **2022**, *24*, 1437. [CrossRef]
24. Duderstadt, J.J.; Hamilton, L.J. *Nuclear Reactor Analysis*, 1st ed.; John Wiley & Sons: New York, NY, USA, 1976; Volume 1, ISBN 0-471-22363-8.
25. de Abreu, W.V.; Martinez, A.S.; do Carmo, E.D.; Gonçalves, A.C. A Novel Analytical Solution of the Deformed Doppler Broadening Function Using the Kaniadakis Distribution and the Comparison of Computational Efficiencies with the Numerical Solution. *Nucl. Eng. Technol.* **2021**, *54*, 1471–1481. [CrossRef]
26. Pathria, R.K.; Beale, P.D. *Statistical Mechanics*; Elsevier: Amsterdam, The Netherlands, 2011; ISBN 9780123821881.
27. Bethe, H.A.; Placzek, G. Resonance Effects in Nuclear Processes. *Phys. Rev.* **1937**, *51*, 450–484. [CrossRef]
28. de Abreu, W.V. de Solução Analítica Da Função de Alargamento Doppler Usando a Distribuição de Kaniadakis. Ph.D. Thesis (Nuclear Engineering), Universidade Federal do Rio de Janeiro, Rio de Janeiro, RJ, Brazil, 2020. [CrossRef]
29. Stacey, W.M. *Nuclear Reactor Physics*; Wiley-VCH Verlag GmbH & Co. KGaA: Weinheim, Germany, 2007; ISBN 9783527611041.
30. Beynon, T.D.; Grant, I.S. Evaluation of the Doppler-Broadened Single-Level and Interference Functions. *Nucl. Sci. Eng.* **1963**, *17*, 547–550. [CrossRef]
31. Campos, T.P.R.; Martinez, A.S. The Dependence of Practical Width on Temperature. *Ann. Nucl. Energy* **1987**, *14*, 241–247. [CrossRef]

32. Arfken, G.B.; Weber, H.J.; Harris, F.H. *Mathematical Methods for Physicists*, 7th ed.; Academic Press; Elesvier: Waltham, MA, USA, 2013; ISBN 978-0-12-384654-9.
33. Melkonian, E.; Havens, W.W.; Rainwater, L.J. Slow Neutron Velocity Spectrometer Studies. V. Re, Ta, Ru, Cr, Ga. *Phys. Rev.* **1953**, *92*, 702–715. [CrossRef]
34. Gonçalves, A. da C. Cálculo Da Função de Alargamento Doppler Utilizando Análise de Fourier. Ph.D. Thesis (Nuclear Engineering), Universidade Federal do Rio de Janeiro, Rio de Janeiro, RJ, Brazil, 2010. Available online: http://antigo.nuclear.ufrj.br/DScTeses/teses2010/Tese_Alessandro.pdf (accessed on 25 February 2023).
35. Marguet, S. *The Physics of Nuclear Reactors*; Springer International Publishing: Cham, Switzerland, 2017; ISBN 978-3-319-59559-7.
36. Palma, D.A.; Martinez, A.S.; Silva, F.C. The Derivation of the Doppler Broadening Function Using Frobenius Method. *J. Nucl. Sci. Technol.* **2006**, *43*, 617–622. [CrossRef]
37. Keshavamurthy, R.S.; Harish, R. Use of Padé Approximations in the Analytical Evaluation of the J (θ,β) Function and Its Temperature Derivative. *Nucl. Sci. Eng.* **1993**, *115*, 81–88. [CrossRef]
38. Shcherbakov, O.; Harada, H. Resonance Self-Shielding Corrections for Activation Cross Section Measurements. *J. Nucl. Sci. Technol.* **2002**, *39*, 548–553. [CrossRef]
39. da Silva, M.V.; Martinez, A.S.; Gonçalves, A.C. Effective Medium Temperature for Calculating the Doppler Broadening Function Using Kaniadakis Distribution. *Ann. Nucl. Energy* **2021**, *161*, 108500. [CrossRef]
40. Macfarlane, R.; Muir, D.W.; Boicourt, R.M.; Kahler, I.A.C.; Conlin, J.L. *The NJOY Nuclear Data Process. System, Version 2016*; U.S. Department of Energy Office of Scientific and Technical Information: Los Alamos, NM, USA,, 2017.
41. International Atomic Energy Agency PREPRO. 2010. Available online: https://www-nds.iaea.org/public/endf/prepro/ (accessed on 25 February 2023).
42. Zu, T.; Xu, J.; Tang, Y.; Bi, H.; Zhao, F.; Cao, L.; Wu, H. NECP-Atlas: A New Nuclear Data Processing Code. *Ann. Nucl. Energy* **2019**, *123*, 153–161. [CrossRef]

entropy

MDPI

Article

A Graph-Space Optimal Transport Approach Based on Kaniadakis κ-Gaussian Distribution for Inverse Problems Related to Wave Propagation

Sérgio Luiz E. F. da Silva [1,2,*,†], João M. de Araújo [3,†], Erick de la Barra [4,†] and Gilberto Corso [3,†]

1 Department of Applied Science and Technology, Politecnico di Torino, 10129 Torino, Italy
2 Geoscience Institute, Fluminense Federal University, Niterói 24210-346, RJ, Brazil
3 Department of Theoretical and Experimental Physics, Federal University of Rio Grande do Norte, Natal 59072-970, RN, Brazil; gfcorso@gmail.com (G.C.)
4 School of Business, Universidad Católica del Norte, Coquimbo 1780000, CO, Chile
* Correspondence: sergio.dasilva@polito.it
† These authors contributed equally to this work.

Abstract: Data-centric inverse problems are a process of inferring physical attributes from indirect measurements. Full-waveform inversion (FWI) is a non-linear inverse problem that attempts to obtain a quantitative physical model by comparing the wave equation solution with observed data, optimizing an objective function. However, the FWI is strenuously dependent on a robust objective function, especially for dealing with cycle-skipping issues and non-Gaussian noises in the dataset. In this work, we present an objective function based on the Kaniadakis κ-Gaussian distribution and the optimal transport (OT) theory to mitigate non-Gaussian noise effects and phase ambiguity concerns that cause cycle skipping. We construct the κ-objective function using the probabilistic maximum likelihood procedure and include it within a well-posed version of the original OT formulation, known as the Kantorovich–Rubinstein metric. We represent the data in the graph space to satisfy the probability axioms required by the Kantorovich–Rubinstein framework. We call our proposal the κ-Graph-Space Optimal Transport FWI (κ-GSOT-FWI). The results suggest that the κ-GSOT-FWI is an effective procedure to circumvent the effects of non-Gaussian noise and cycle-skipping problems. They also show that the Kaniadakis κ-statistics significantly improve the FWI objective function convergence, resulting in higher-resolution models than classical techniques, especially when $\kappa = 0.6$.

Keywords: κ-Gaussian distribution; optimal transport; seismic imaging; cycle skipping; non-linear optimization; Wasserstein metric; inverse problems; wave propagation

Citation: da Silva, S.L.E.F.; de Araújo, J.M.; de la Barra, E.; Corso, G. A Graph-Space Optimal Transport Approach Based on Kaniadakis κ-Gaussian Distribution for Inverse Problems Related to Wave Propagation. *Entropy* **2023**, *25*, 990. https://doi.org/10.3390/e25070990

Academic Editor: Ali Mohammad-Djafari

Received: 3 May 2023
Revised: 15 June 2023
Accepted: 25 June 2023
Published: 28 June 2023

1. Introduction

The task of inferencing physical parameters from indirect observations arises in various practical problems. Determining parameters that cannot be directly observed remains a complex issue and involves a robust set of tools that compose the theoretical basis of the inverse problem theory [1]. The goal of an inverse problem consists of obtaining a quantitative model m that explains the observations (or observed data) by matching modeled data $d^{mod} = G(m)$ to observed data d^{obs}, in which G denotes the so-called forward operator. The forward operator maps the variables from the model space to the data space through a physical law [2]. For instance, we may want to determine the thermal diffusivity of a material (physical system) by analyzing the observed data: the temporal distribution of the diffusing material density at a determined location. In this regard, the model consists of the collective diffusion coefficient, and a diffusion equation represents the forward operator G. So, the diffusion coefficients are determined by optimizing an objective function, which measures the distance between modeled and observed data.

In this work, we consider a non-linear inverse problem that has attracted increasing interest in several fields, such as astrophysics [3], biomedicine [4], machine learning [5], and geophysics [6], named the Full-Waveform Inversion (FWI) [7]. FWI is a powerful imaging technique to obtain high-resolution quantitative physical models by analyzing the complete information of a collection of waveforms [8]. In particular, we consider the FWI technique in a geophysical context, in which the forward problem consists of simulating the propagation of acoustic waves by solving a wave equation. In this regard, the acoustic wave equation represents the forward operator G. At the same time, the data d and the model m are the pressure waveforms and the distribution of acoustic wave velocities (coefficients of the wave equation) of the subsurface medium. The inverse problem involves inferencing the coefficients of the wave equation (model parameters) by comparing the modeled data (wave equation solution) with the observed data by employing an objective function [9].

The objective function based on the least-squares method (or squared l_2-norm) is the most employed for handling FWI issues [7]. The least-squares objective function (from now on, classical objective function) computes the square root of the sum of the absolute squares of the residual data (or errors), the difference between the modeled and the observed data. Each objective function is closely connected to a statistical interpretation of the errors [2]. The classical objective function bears a relationship to Gaussian statistics. Indeed, in this classical framework, the errors are assumed to be independent and identically distributed according to a Gaussian probability distribution [10]. However, this assumption is sometimes adequate since the errors seldom are Gaussian in non-linear problems [11,12]. Let us remind the reader that errors arise from different natures, comprising the noise in the observations and uncertainties related to the physical rule employed in the forward problem. In fact, non-Gaussian noises are present in geophysical datasets and are caused by several elements, such as weather-related mechanisms [13] and instrument noise [14]. Several objective functions based on non-Gaussian statistics have been presented in the literature as alternative criteria. Non-Gaussian distributions exhibit much longer tails than the Gaussian ones, a crucial feature for dealing with erratic data (outliers) [15]. Several works have shown the effectiveness of non-Gaussian criteria in geophysical data inversions, such as objective functions based on Laplace distribution [16], Student's t distribution [17], generalized approaches [18,19], and hybrid criteria [20,21].

Recently, ref. [22] introduced a new non-Gaussian criterion, namely ,the κ-objective function, based on the Kaniadakis statistics (or κ-statistics) [23–27], which is robust to erratic data. The κ-objective function assumes that the errors are independent and identically distributed according to the κ-deformation of a Gaussian distribution (or κ-Gaussian distribution), in which the classical approach is a particular case [28]. The κ-Gaussian distribution arises from optimizing the Kaniadakis κ-entropy as a generalization of the well-known Gaussian distribution [29]. The κ-criterion exhibits robust characteristics thanks to the much longer tail of the κ-Gaussian distribution than the classical Gaussian probability function, which is crucial to mitigate the effects of non-Gaussian errors in FWI problems [30].

Due to the high computational efforts to solve the wave equation several times during the FWI process, the minimization process of the objective function is usually solved by local optimization methods [7]. Thus, FWI is prone to trapping into a non-informative local minimum if the initial background velocity model is not kinematically accurate [31]. Such an intrinsic limitation of FWI is associated with the absence of low-frequency contents, causing cycle-skipping issues [32,33]. Cycle-skipping is a phase ambiguity problem when the phase correspondence between two waveforms is greater than half a wavelength [34]. Although the approaches mentioned above are robust to non-Gaussian errors, they measure sample-by-sample the data misfit, making them sensitive to cycle skipping. Hence, a vast body of objective functions has been introduced for mitigating the cycle-skipping effects, such as those based on the waveform envelopes [35], convolutional filters [36], non-parametric techniques [37], and optimal transport metrics [38]; this is the methodology employed in the present study.

The theory of optimal transport (OT) was formally introduced by Gaspard Monge [39], who sought to understand the most effective allocation of resources by redistributing materials (mass) from sources to sinks. In recent years, OT theory has received much attention in broad literature (e.g., refs. [40–43]), such as geophysics issues [44–46]. However, the OT-based objective function is suitable for comparing probability distributions, that is, positive and normalized quantities, two requirements that seismic signals do not greet due to their oscillatory nature. In this way, waveforms are commonly distorted through transformations to satisfy the probability axioms, which may manufacture unwanted information. Indeed, several applications have demonstrated the effectiveness of OT-based objective functions to mitigate the effects of phase ambiguity; however, all assume that the errors obey Gaussian statistics.

In this work, we explore the κ-objective function in the context of OT theory to introduce an objective function resistant to non-Gaussian noise and less sensitive to cycle-skipping issues. In this regard, we propose a robust framework for matching seismic waveforms using the Wasserstein distance, a well-posed relaxation of the OT formulation. Furthermore, inspired by ref. [47], we consider the representation of the waveforms in the graph space suitable for real large-scale problems [33]. In this approach, the waveforms are represented by Dirac delta functions in a two-dimensional space (amplitude *versus* time).

We organize the present work as follows. In Section 2, we briefly introduce the theoretical basis of inverse problems in the context of κ-Gaussian statistics and their robustness properties. In Section 3, we present a well-posed relaxation of the original optimal transport formulation using the Kaniadakis κ-objective function. Then, in Section 4 we present FWI based on optimal transport and κ-Gaussian statistics in the context of the adjoint state method. In Section 5, we demonstrate how the proposed objective function deals with cycle-skipping issues and non-Gaussian noise by considering a Brazilian pre-salt case study. Finally, we devote Section 6 to the final remarks and future applications.

2. Inverse Problems in the Context of Kaniadakis κ-Statistics

In science issues, several practical problems are data-centric. Indeed, determining a quantitative physical model that explains the observations is crucial to more accurately model and describe a wide variety of existing physical systems. In this context, the inverse problem theory is an excellent tool.

From a practical point of view, an inverse problem is formulated as an optimization task for obtaining a quantitative model by comparing modeled data to observed data. Modeled data are calculated using an appropriate physical law. The comparison between modeled and observed data is performed through an objective function. In the classical approach, the objective function is constructed from the assumption that the errors (the difference between modeled and observed data) obey Gaussian statistics. Let $\vec{\varepsilon} = \{\varepsilon_1, \varepsilon_2, \cdots, \varepsilon_N\}$ be the errors. From the assumption that the errors are independent and identically distributed according to a standard Gaussian distribution,

$$p_0(\varepsilon_i) = \frac{1}{\sqrt{2\pi}} \exp\left(-\frac{1}{2}\varepsilon_i^2\right), \tag{1}$$

we can determine the associated likelihood function as follows [11]:

$$\mathcal{L}_0 = \prod_{i=1}^{N} p_0(\varepsilon_i) = \left(\frac{1}{\sqrt{2\pi}}\right)^N \exp\left(-\frac{1}{2}\sum_{i=1}^{N}\varepsilon_i^2\right), \tag{2}$$

where $\mathcal{L}_0$ is the Gaussian likelihood. The use of index 0 will become clear later on. It is worth remembering that the standard Gaussian distribution can be determined from the maximization of the Boltzmann–Gibbs–Shannon entropy subject to the normalization condition

$$\int_{-\infty}^{+\infty} p(\varepsilon)d\varepsilon = 1 \tag{3}$$

and the unit variance constraint,

$$\int_{-\infty}^{+\infty} \varepsilon^2 \, p(\varepsilon) d\varepsilon = 1. \tag{4}$$

In inverse problems, the errors $\vec{\varepsilon}$ depend on the model parameter m and are computed through the difference between modeled ($d^{mod} = G(m)$) and observed (d^{obs}) data, i.e., $\varepsilon_i(m) = d_i^{mod}(m) - d_i^{obs}$, where G represents the forward operator. In this way, obtaining the model parameter can be performed by employing the maximum likelihood estimation (MLE) method, which is achieved by maximizing the likelihood function as follows:

$$\hat{m} = \max_m \mathcal{L}_0(m|d^{obs}) \tag{5}$$

where $\hat{m}$ represents the estimated model. The MLE estimates an unknown model parameter by considering that its optimal value maximizes the probability that the observed data are measured. Since the minimum of the negative log-likelihood coincides with the maximum of the likelihood function, maximizing $\mathcal{L}_0$ (5) is equivalent to minimizing the negative log-likelihood, i.e.,

$$\max_m \mathcal{L}_0(m|d^{obs}) \equiv \min_m - \ln\left(\mathcal{L}_0(m|d^{obs})\right). \tag{6}$$

From the principle of maximum likelihood, an objective function ϕ_0 can be obtained from [11]:

$$\phi_0(m) \propto - \ln\left(\mathcal{L}_0(m|d^{obs})\right), \tag{7}$$

which can be rewritten as:

$$\phi_0(m) \propto \frac{N}{2} \ln(2\,\pi) + \frac{1}{2} \sum_{i=1}^{N} \varepsilon_i^2 \tag{8}$$

$$\phi_0(m) = \frac{1}{2} \sum_{i=1}^{N} \varepsilon_i^2. \tag{9}$$

We notice that minimizing Equation (8) or (9) is the same since the term $\frac{N}{2} \ln(2\,\pi)$ is constant. The latter equation is well-known and used in solving problems via the least squares method. Please see Section 2 of ref. [48] for more detail.

However, due to the non-Gaussianity of the errors, it is reasonable to assume that the errors are non-Gaussian. In this study, we consider that the errors are distributed according to a Kaniadakis κ-Gaussian distribution of the form [22]:

$$p_\kappa(\varepsilon_i) = \frac{1}{Z_\kappa} \exp_\kappa\left(-\beta_\kappa \, \varepsilon_i^2\right), \tag{10}$$

where Z_κ is a normalizing constant, β_κ is a scale parameter, and

$$\exp_\kappa(y) = \exp\left(\frac{1}{\kappa} \text{arcsinh}(\kappa y)\right) = \left(\sqrt{1 + \kappa^2 y^2} + \kappa y\right)^{\frac{1}{\kappa}} \tag{11}$$

with $0 \leq |\kappa| < 1$, is the κ-exponential function [26], a generalization of the exponential function. The κ-exponential becomes the ordinary exponential function in the limit $\kappa \to 0$: $\exp_0(y) = \exp(y)$.

Considering the normalization (3) and unitary variance (4) conditions, we obtain

$$Z_\kappa = \sqrt{\frac{\pi}{\beta_\kappa}} \frac{|2\kappa|^{-1/2}}{1 + \frac{1}{2}|\kappa|} \frac{\Gamma\left(\frac{1}{|2\kappa|} - \frac{1}{4}\right)}{\Gamma\left(\frac{1}{|2\kappa|} + \frac{1}{4}\right)} \tag{12}$$

and

$$\beta_\kappa = \frac{|2\kappa|^{-1}}{2} \frac{1+\frac{1}{2}|\kappa|}{1+\frac{3}{2}|\kappa|} \frac{\Gamma\left(\frac{1}{|2\kappa|}-\frac{3}{4}\right)\Gamma\left(\frac{1}{|2\kappa|}+\frac{1}{4}\right)}{\Gamma\left(\frac{1}{|2\kappa|}+\frac{3}{4}\right)\Gamma\left(\frac{1}{|2\kappa|}-\frac{1}{4}\right)} \tag{13}$$

holding for $|\kappa| < 2/3$. The standard Gaussian distribution (1) is a particular case of the κ-Gaussian distribution (10) in the classical limit $\kappa \to 0$ since $\lim_{\kappa\to 0} \beta_\kappa = \frac{1}{2}$ and $\lim_{\kappa\to 0} Z_\kappa = \sqrt{2\pi}$. Figure 1 depicts the plots of the κ-Gaussian distribution (10) for typical κ-values, with the solid black curve referring to the standard Gaussian distribution ($\kappa \to 0$).

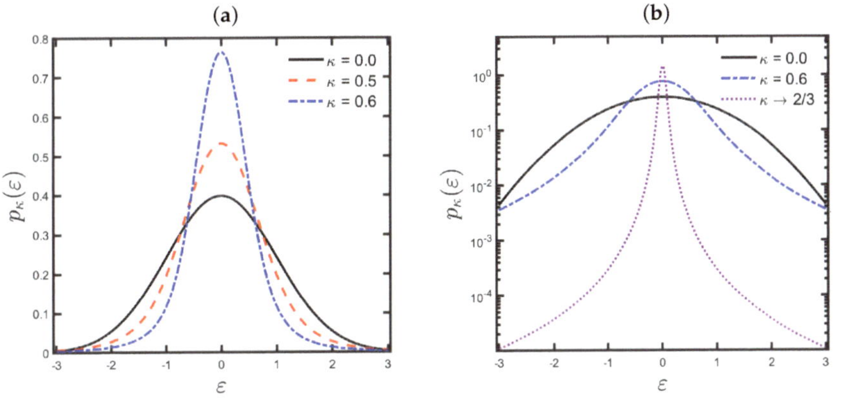

Figure 1. Probability plots of the κ-Gaussian distribution (10) for some κ-values using (**a**) a linear scale and (**b**) a linear scale on the axis of ordinates, and a logarithmic scale on the axis of abscissas. The solid black line represents the standard Gaussian distribution ($\kappa \to 0$).

Because we assume that the errors are independent and identically distributed by the power law distribution represented in (10), we can calculate the corresponding objective function by estimating the most likely state using the probabilistic maximum likelihood method:

$$\min_{m} \; \phi_\kappa(m) \equiv \max_{m} \mathcal{L}_\kappa(m|d^{obs}), \tag{14}$$

where $\mathcal{L}_\kappa(m|d^{obs}) := \prod_{i=1}^{N} p_\kappa(\varepsilon_i(m))$ represents the likelihood function. It is crucial to remember that minimizing the negative log-likelihood is the same as maximizing the likelihood function. In this way, the objective function ϕ_κ can be obtained from (14):

$$\phi_\kappa(m) \propto N \ln\left(Z_\kappa\right) - \sum_{i=1}^{N} \ln\left[\exp_\kappa\left(-\beta_\kappa \varepsilon_i^2(m)\right)\right] \tag{15}$$

$$\phi_\kappa(m) = -\sum_{i=1}^{N} \ln\left[\exp_\kappa\left(-\beta_\kappa \varepsilon_i^2(m)\right)\right], \tag{16}$$

where ϕ_κ is the κ-objective function, which converges to the classical objective function (9) in the limit $\kappa \to 0$.

The κ-objective function is not easily influenced by aberrant measurements (outliers), as it is based on κ-Gaussian criteria [49]. To demonstrate this, we compute the influence function Υ related to the objective function. According to ref. [50], a statistical criterion is not robust if $\Upsilon \to \pm\infty$ under $|\varepsilon| \to \infty$, and robust (outlier-resistant) if $\Upsilon \to 0$ under $\varepsilon \to \pm\infty$. Given a model m_i, the influence function is defined by [50]:

$$\Upsilon_\kappa(m_i) := \frac{\partial \phi_\kappa(\varepsilon|m_i)}{\partial \varepsilon}, \tag{17}$$

where $\phi_\kappa(\varepsilon|m_i)$ is the κ-objective function computed from the errors ε given the model m_i. Thus, the κ-objective function generates the following influence function:

$$\Upsilon_\kappa = \frac{2\beta_\kappa \varepsilon}{\sqrt{1 + \kappa^2 \beta_\kappa^2 \varepsilon^4}} \tag{18}$$

for $0 < \kappa < 2/3$, with $\Upsilon_\kappa = \Upsilon_\kappa(m_i)$ and $\varepsilon = \varepsilon(m_i)$. We notice that, as ε tends to $\pm\infty$, the influence function associated with the κ-criterion (Υ_κ) approaches to 0; the κ-objective function is then robust (outlier-resistant). Indeed, the influence function in its valid domain ($0 < \kappa < 2/3$) is proportional to $1/\varepsilon$ for large errors (suppressing these) and to ε for small errors (magnifying these). Figure 2 depicts the behavior of the κ-objective function and the associated influence function.

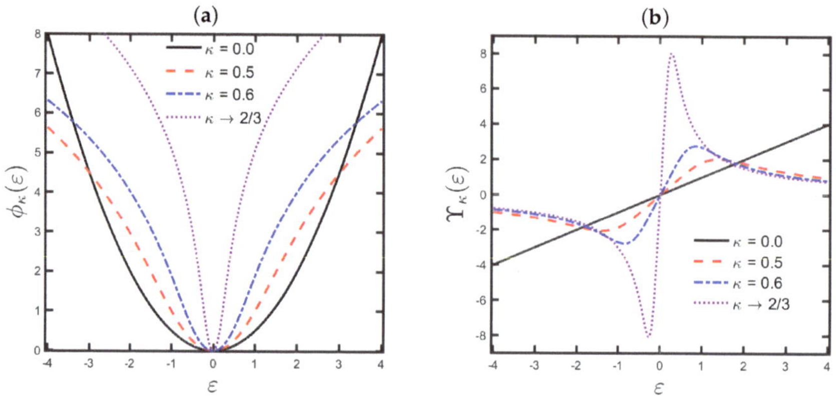

Figure 2. (**a**) Graphical representation of the κ-objective function (16), and (**b**) the associated influence function (18) for some κ-values. The solid black line represents the classical criterion ($\kappa \to 0$).

3. Optimal Transport Metric Based on Kaniadakis κ-Statistics

In 1781, Gaspard Monge first raised a challenger transportation problem [39], which consisted of moving a pile of sand from one place to another optimally and without losing mass. As formulated by Monge, the optimal transport (OT) issue is an ill-posed problem; hence the solution, if it exists, is not unique. Nearly 200 years later, Leonid Kantorovitch proposed a well-posed relaxation of Monge's OT problem in the context of optimal economic resource allocation [51]. In this regard, Kantorovich proposed what is now known as the Kantorovich–Rubinstein metric (also referred to as Wasserstein distance), which earned him the 1975 Nobel Memorial Prize in Economic Science.

The Wasserstein criterion is a metric that defines a distance between two probability distributions. Let us consider two sets of points $\Omega_1 = \{x_i; i = 1, 2, \cdots, N_1\}$ and $\Omega_2 = \{y_j; j = 1, 2, \cdots, N_2\}$, in which each point x_i and y_j are represented by "mass" functions, namely $\mu(x_i)$ and $v(y_j)$, respectively. Considering the mass conservation constraint ($\sum_i \mu(x_i) = \sum_j v(y_j) = 1$), we can define the κ-optimal total transport cost $\mathcal{W}_\kappa$ as [30,40]:

$$\mathcal{W}_\kappa(\mu, v) = \min_{\mathcal{T} \in \Lambda(\mu, v)} \sum_{i,j} \mathcal{T}_{i,j}\, \phi_{\kappa,i,j}, \tag{19}$$

where $\Lambda(\mu, v)$ denotes the set of transport maps $\mathcal{T}$ defined in

$$\Lambda(\mu, v) = \left\{ \mathcal{T}_{i,j} \geq 0,\ \forall (i,j);\quad \sum_{j=1}^{N_2} \mathcal{T}_{i,j} = \mu(x_i),\ \forall\ i;\quad \sum_{i=1}^{N_1} \mathcal{T}_{i,j} = v(y_j),\ \forall\ j \right\}. \tag{20}$$

The transport map $\mathcal{T}$ assigns how many "sand particles" from $\mu(x_i)$ should be transported to $v(y_j)$ for each pair (x_i, y_j), while the κ-objective function maps each pair (x_i, y_j)

to $[0; +\infty]$. The Monge–Kantorovich transportation relaxed problem (19), using the Kaniadakis κ-statistics, can therefore be solved by determining an optimal transport plan $\mathcal{T}$ that minimizes the κ-optimal total transport cost $\mathcal{W}_\kappa$ from μ to v, given an κ-objective function ϕ_κ.

Let us consider a metric space $\mathcal{P}(\mathcal{X} \times \mathcal{Y})$ formed by a set of probability measures, in which $\mathcal{X}$ and $\mathcal{Y}$ are two separable and complete metric spaces with $\mu \in \mathcal{X}$ and $v \in \mathcal{Y}$. From this point forward, for practical reasons, we assume that $\mathcal{X} = \mathcal{Y} \subset \mathbb{R}^N$ (with $N_1 = N_2 = N \in \mathbb{N}$). In addition, let us consider the mass distributions μ and v represented in terms of the Dirac delta function as follows: $\mu(x) = \frac{1}{N}\sum_{l=1}^{N}\delta(x - u_l)$ and $v(y) = \frac{1}{N}\sum_{l=1}^{N}\delta(y - w_l)$, in which $u_l \in \Omega_1$ and $w_l \in \Omega_2$ point out the data points describing $\mu(x)$ and $v(x)$. In this context, we can reformulate the optimization problem in Equation (19) as follows:

$$\mathcal{W}_\kappa(\mu, v) = \min_{\mathcal{T}_{i,j}} \quad -\frac{1}{N}\sum_{i,j=1}^{N} \mathcal{T}_{i,j} \ln\left\{ \exp_\kappa\left[-\beta_\kappa\Big(\mu(x_i) - v(y_j)\Big)^2 \right] \right\} \tag{21}$$

subject to

$$\mathcal{T}_{i,j} \geq 0, \quad \sum_{j=1}^{N} \mathcal{T}_{i,j} = 1, \quad \sum_{i=1}^{N} \mathcal{T}_{i,j} = 1. \tag{22}$$

From a practical viewpoint, we notice that solving (21) consists of obtaining an optimal transport plan that links data points from $\mathcal{P}(\mathcal{X})$ to the corresponding data points in $\mathcal{P}(\mathcal{Y})$ that minimizes the κ-optimal total transport cost $\mathcal{W}_\kappa$. Although each element of the optimal transport plan $\mathcal{T}_{i,j}$ can assume fractional values, a classic result states that the optimal solution values are integer values, specifically 0 or 1 when the constraints described in Equation (22) are considered [52,53]. Indeed, obtaining the minimum of $\mathcal{W}_\kappa$ implicates solving a combinatorial optimization issue, which can be defined as:

$$\mathcal{W}_\kappa(\mu, v) = \min_{\sigma \in \mathcal{S}} \quad -\frac{1}{N}\sum_{i=1}^{N} \ln\left\{ \exp_\kappa\left[-\beta_\kappa\Big(\mu(x_{\sigma(i)}) - v(y_i)\Big)^2 \right] \right\}, \tag{23}$$

where σ represents a permutation solution for the linear sum assignment problem in (21) related with $\mathcal{T}$, and $\mathcal{S}(N) = \{1, 2, \cdots, N\}$ is a set of permutations. Equation (23) represents the Wasserstein metric in the context of κ-Gaussian statistics.

Naturally, the Wasserstein metric based on Kaniadakis κ-statistics appreciates the advantages provided by κ-Gaussian statistics. However, this approach in this format is only valid for comparing probability distributions, which is not interesting for geophysical applications like the FWI case. This incompatibility is because seismic signals are not normalized and positive-definite quantities like probability functions.

4. Kaniadakis κ-Graph-Space Optimal Transport FWI

4.1. FWI Based on Kaniadakis κ-Gaussian Distribution

In this section, we present the main elements of FWI based on the Kaniadakis κ-Gaussian distribution, the metric explained in Section 2. The FWI is a non-linear inverse problem whose main goal consists of inferring a quantitative physical model by comparing modeled waveforms (modeled data) with measured waveforms (observed data) [7]. FWI is often formulated as a gradient-based minimization due to the computational costs, in which the model parameters are iteratively updated, from an initial model m_0, as follows [8]:

$$m_{i+1} = m_i - \alpha_i \, h_\kappa(m_i) \quad \text{for} \quad i = 0, 1, 2, \cdots, N_{iter}, \tag{24}$$

where m represents the model parameter, $\alpha_i > 0$ is the so-called step length [54], N_{iter} represents the number of FWI iterations, and h_κ denotes the descent direction at the i-th iteration.

In this work, we employ a non-linear conjugate gradient optimization method based on the so-called Polak–Ribière–Polyak algorithm. In this regard, the descent direction is defined by [55,56]:

$$h_\kappa(m_i) = \begin{cases} \nabla_m \phi_\kappa(m_0) & , \text{ if } \quad i = 0 \\ \nabla_m \phi_\kappa(m_i) + \zeta_\kappa(m_i) h_\kappa(m_{i-1}), & \text{for } i = 1, 2, \cdots, N_{iter} \end{cases} \tag{25}$$

with

$$\zeta_\kappa(m_i) = \frac{\nabla_m \phi_\kappa(m_i)(\nabla_m \phi_\kappa(m_i) - \nabla_m \phi_\kappa(m_{i-1}))}{\nabla_m \phi_\kappa(m_{i-1}) \nabla_m \phi_\kappa(m_{i-1})}, \tag{26}$$

where $\nabla_m \phi_\kappa(m)$ is the gradient of the κ-objective function.

Thus, it is remarkable that the objective function plays a crucial role in obtaining models via FWI, which is defined for our problem as (10):

$$\min_m \quad \phi_\kappa(m) := -\sum_{s,r} \int_0^T \ln\left\{ \exp_\kappa\left[-\beta_\kappa \left(\Gamma_{s,r}\psi_s(\vec{x}, m, t) - d_{s,r}(\vec{x}_{s,r}, t) \right)^2 \right] \right\} dt, \tag{27}$$

where $\Gamma_{s,r}\psi_s = d^{mod}_{s,r}$ and $d_{s,r} = d^{obs}_{s,r}$ represent the modeled and observed data generated by the seismic source s and recorded in the receiver r, while $\vec{x} \in \mathbb{R}^2$ and $t \in [0, T]$ denote the spatial coordinates and the seismic acquisition time.

It is worth mentioning that the observed data $d_{s,r}$ are registered only in the receiver positions $\vec{x} = \vec{x}_{s,r}$, the available and chosen positions during a seismic survey. The seismic wavefield ψ_s is computed in the entire physical domain for each seismic source s by solving a wave equation. Thus, $\Gamma_{s,r}$ represents a sampling operator that acts as a measurement processor onto the receiver r from the source s. In this work, we consider the acoustic case; therefore, ψ_s are the pressure wavefields that satisfy the following model:

$$\frac{1}{c^2(\vec{x})} \frac{\partial^2 \psi_s(\vec{x}, t)}{\partial t^2} - \nabla^2 \psi_s(\vec{x}, t) = g_s(t)\delta(\vec{x} - \vec{x}_s) \tag{28}$$

where g_s represents a seismic source signature at the fixed position $\vec{x} = \vec{x}_s$, c is the P-wave velocity model of the medium, and ∇^2 denotes de Laplacian operator.

Thus, the gradient of the κ-objective function (27) with respect to the model parameters is given by:

$$\nabla_m \phi_\kappa(m) = \frac{\partial \phi_\kappa(m)}{\partial m_l} = 2\beta_\kappa \sum_{s,r} \int_0^T \frac{\mathcal{J}_{s,r}(m, t) \Delta d_{s,r}(m, t)}{\sqrt{1 + \kappa^2 \beta_\kappa^2 \Delta d^4_{s,r}(m, t)}} dt, \tag{29}$$

where $\Delta d_{s,r}(m, t) = \Gamma_{s,r}\psi_s(\vec{x}, m, t) - d_{s,r}(\vec{x}_{s,r}, t)$ represents the error (or residual data) and

$$\mathcal{J}_{s,r}(m, t) = \frac{\partial}{\partial m_l} \left(\Gamma_{s,r}\psi_s(\vec{x}, m, t) \right) \tag{30}$$

is known as the Fréchet derivative. It is worth emphasizing that FWI problems involve many elements from the model parameters that typically comprise 10^6 to 10^{12} variables (coefficients of the wave equation). In this context, we need to solve the wave equation once in the forward modeling process plus at least 10^6 times in calculating the gradient of the κ-objective function through Fréchet derivatives, being unfeasible in industrial problems.

4.2. Adjoint-State Method

Since calculating Fréchet derivatives can be computationally prohibitive, we compute the gradient of the κ-objective function using the adjoint-state method, which was developed in the 1970s [57]. There are several ways to formulate the state-adjoint approach, such as in techniques based on the augmented Lagrangian method or Green's functions. However, in this work we consider the perturbation theory to calculate the gradient efficiently. We notice that the κ-objective function can be written as:

$$\phi_\kappa(m) = f\Big(\psi(m), m\Big), \tag{31}$$

where ψ is a state variable that belongs to the complex space $\mathcal{Q}$; ψ satisfies the following equation of state:

$$F\Big(\psi(m), m\Big) = \mathcal{A}(m,t)\psi(m,t) - g(t) = 0, \tag{32}$$

in which we suppress the subscript s for the sake of a simplified notation. In the latter equation, $\mathcal{A}(m,t)\psi(t) = q(t)$ represents the wave equation written in a compact form, where $\mathcal{A}(m,t) = m\frac{\partial^2}{\partial t^2} - \nabla^2$ is the d'Alembert wave operator with $m = \frac{1}{c^2(\vec{x})}$ belonging to the real space $\mathcal{M}$, whilst $g(t) = g_s(t)\delta(\vec{x} - \vec{x}_s)$.

Suppose we consider an arbitrary variation δm concerning the model parameter m. In that case, the state variable ψ will be disturbed by a variation $\delta\psi$; consequently, the κ-objective function in Equation (31) will also be disturbed. In this way, we have to:

$$\delta\phi_\kappa = \frac{\partial f(\psi,m)}{\partial m}\delta m + \left\langle \frac{\partial f(\psi,m)}{\partial \psi_j}, \delta\psi \right\rangle_{\mathcal{Q}}, \tag{33}$$

where we only consider the first-order terms in δm and $\delta\psi$. Furthermore, ψ_j is any element of the space $\mathcal{Q}$, and $\langle , \rangle_{\mathcal{Q}}$ is the inner product in $\mathcal{Q}$.

It is worth emphasizing that the perturbations δm and $\delta\psi$ also induce variations in the equation of state (32). Moreover, assuming that there is a unique solution ψ for any model parameter m, we can state that $\psi + \delta\psi$ is the unique solution of $F(\psi + \delta\psi, m + \delta m) = 0$. In other words, for a physical realization ψ (that is, $F(\psi, m) = 0$), we have the following first-order development in δm and $\delta\psi$:

$$F(\psi + \delta\psi, m + \delta m) = F(\psi, m) + \left(\frac{\partial F(\psi,m)}{\partial m}\right)\delta m + \left(\frac{\partial F(\psi,m)}{\partial \psi_j}\right)\delta\psi = 0. \tag{34}$$

From the latter equation, we have that the perturbation in the state variable ψ is given by:

$$\delta\psi = -\left(\frac{\partial F(\psi,m)}{\partial \psi_j}\right)^{-1}\left(\frac{\partial F(\psi,m)}{\partial m}\right)\delta m, \tag{35}$$

where a^{-1} denotes the inverse of a. So, replacing the resulting from Equation (35) in Equation (33), we have an efficient way to compute the gradient of the κ-objective function without the Fréchet derivatives:

$$\delta\phi_\kappa = \frac{\partial f(\psi,m)}{\partial m}\delta m - \left\langle \frac{\partial f(\psi,m)}{\partial \psi_j}, \left(\frac{\partial F(\psi,m)}{\partial \psi_j}\right)^{-1}\left(\frac{\partial F(\psi,m)}{\partial m}\right)\delta m \right\rangle_{\mathcal{Q}}. \tag{36}$$

On the other hand, to obtain an intuitive way to calculate the gradient, Equation (36) can be rewritten so that in the inner product in $\mathcal{Q}$, one of the terms varies only with ψ and the other with m. For this, we consider the following adjoint operator property for any x and y variables:

$$\langle x, \mathcal{R}y\rangle = \langle \mathcal{R}^\dagger x, y\rangle \tag{37}$$

where $\mathcal{R}^\dagger$ is the adjoint operator of $\mathcal{R}$, while the superscript $\dagger$ represents the adjoint operation (complex-conjugate transpose). Applying the property (37) to the second term of Equation (36), we obtain:

$$\delta\phi_\kappa = \frac{\partial f(\psi,m)}{\partial m}\delta m - \left\langle \left[\left(\frac{\partial F(\psi,m)}{\partial \psi_j} \right)^{-1} \right]^\dagger \frac{\partial f(\psi,m)}{\partial u_j}, \left(\frac{\partial F(\psi,m)}{\partial m} \right) \delta m \right\rangle_{\mathcal{Q}}. \tag{38}$$

Furthermore, if we consider a new state variable v belonging to the complex space $\mathcal{V}$, given by:

$$v = \left[\left(\frac{\partial F(\psi,m)}{\partial \psi_j} \right)^{-1} \right]^\dagger \frac{\partial f(\psi,m)}{\partial \psi_j}, \tag{39}$$

where v is the first term of the inner product in Equation (38), we have the following equation of state:

$$\left(\frac{\partial F(\psi,m)}{\partial \psi_j} \right)^\dagger v = \frac{\partial f(\psi,m)}{\partial \psi_j}, \tag{40}$$

which is known as the adjoint-state equation [58,59], and therefore v is called the adjoint-state variable.

In summary, the calculation of the gradient of the κ-objective function through the state-adjoint method is given by:

$$\nabla_m \phi_\kappa(m) = \frac{\partial \phi_\kappa(m)}{\partial m} = \frac{\partial f(\psi,m)}{\partial m} - \left\langle v, \frac{\partial F(\psi,m)}{\partial m} \right\rangle_{\mathcal{Q}}, \tag{41}$$

where the state-adjoint variable v is calculated from the state-adjoint equation in (40). In this way, for our problem we have:

$$f\Big(\psi_s(m,t), m\Big) = -\sum_r \ln\left\{ \exp_\kappa\left[-\beta_\kappa \Big(\Gamma_{s,r}\psi_s(m,t) - d_{s,r}(t) \Big)^2 \right] \right\}, \tag{42}$$

where $\phi_\kappa(m) = \sum_s f\Big(\psi_s(m,t), m\Big)$. In addition, for any model parameter m, let ψ_s be a solution of the equation of state given in (32), that is, a physical realization. We obtain:

$$F(\psi_s(t), m) = \mathcal{A}(m,t)\psi_s(t) - g_s(t) = 0 \tag{43}$$

and

$$f\Big(\psi_s(t), m\Big) = -\sum_r \ln\left\{ \exp_\kappa\left[-\beta_\kappa \Big(\Gamma_{s,r}\psi_s(t) - d_{s,r}(t) \Big)^2 \right] \right\} = f\Big(\psi_s(t)\Big). \tag{44}$$

Therefore, we obtain the following derivatives of the equation of state:

$$\frac{\partial F(\psi_s(t), m)}{\partial m} = \frac{\partial \mathcal{A}(m,t)}{\partial m}\psi_s(t) \quad \text{and} \quad \frac{\partial F(\psi_s(t), m)}{\partial \psi_{s_j}} = \mathcal{A}(m,t), \tag{45}$$

and the following for the κ-objective function:

$$\frac{\partial f(\psi_s(t), m)}{\partial m} = 0 \quad \text{and} \quad \frac{\partial f(\psi_s(t), m)}{\partial \psi_{s_j}} = \sum_r \frac{2\beta_\kappa \Gamma_{s,r}^\dagger \Big(\Gamma_{s,r}\psi_s(t) - d_{s,r}(t) \Big)}{\sqrt{1 + \kappa^2 \beta_\kappa^2 \big(\Gamma_{s,r}\psi_s(t) - d_{s,r}(t) \big)^4}}. \tag{46}$$

Thus, the gradient of the κ-objective function via the state-adjoint method is given by substituting the derivatives calculated in Equations (45) and (46) in Equations (40) and (41), as follows:

$$\nabla_m \phi_\kappa(m) = -\sum_s \int_0^T \left\langle v_s(\vec{x}, t; \kappa), \frac{\partial^2 \psi_s(\vec{x}, t)}{\partial t^2} \right\rangle_{\vec{x}} dt \tag{47}$$

with v_s being the solution of the adjoint-wave equation given by

$$m(\vec{x}) \frac{\partial^2 v_s(\vec{x}, t)}{\partial t^2} - \nabla^2 v_s(\vec{x}, t) = \sum_r \frac{2\beta_\kappa \Gamma_{s,r}^\dagger \left(\Gamma_{s,r}\psi_s(t) - d_{s,r}(t)\right)}{\sqrt{1 + \kappa^2 \beta_\kappa^2 \left(\Gamma_{s,r}\psi_s(t) - d_{s,r}(t)\right)^4}} \tag{48}$$

where $m(\vec{x}) = \frac{1}{c^2(\vec{x})}$.

In search of a physical meaning for Equations (47) and (48), let us consider a new state variable given by $\lambda_s(\vec{x}, t) = v_s(\vec{x}, T - t)$. So, the latter equation becomes:

$$m(\vec{x}) \frac{\partial^2 \lambda_s(\vec{x}, t)}{\partial t^2} - \nabla^2 \lambda_s(\vec{x}, t) = \sum_r \frac{2\beta_\kappa \Gamma_{s,r}^\dagger \left(\Gamma_{s,r}\psi_s(T-t) - d_{s,r}(T-t)\right)}{\sqrt{1 + \kappa^2 \beta_\kappa^2 \left(\Gamma_{s,r}\psi_s(T-t) - d_{s,r}(T-t)\right)^4}}. \tag{49}$$

We notice that the adjoint-state variable λ_s is calculated in reverse time from Equation (49), i.e., starting the wave propagation from the final time T to the initial time 0. For this reason, this state-adjoint variable is commonly called the backpropagated wavefield, while Equation (49) is called the adjoint-wave equation, in which the right-hand term is named the adjoint source [59]. In this way, the κ-objective function gradient is calculated efficiently from the cross-correlation of the forward wavefield with the backpropagated wavefield.

In this context, computing the gradient via the state-adjoint method for each seismic source requires solving the wave equation only twice, first in the forward modeling and second in backpropagation modeling. We also point out that the robustness properties of the objective function discussed in Section 2 are indispensable in calculating the gradient. Indeed, the influence function (18) gives the adjoint source used in the inversion process. The particular classical case $\kappa \to 0$ provides the residual data as the adjoint source.

4.3. κ-Graph-Space Optimal Transport FWI

From a statistical point of view, non-Gaussian criteria are critical to handle noisy datasets in FWI analysis [9]. In this sense, we also consider the κ-Gaussian-based metric to deal with a challenging issue in FWI called cycle skipping [7]. Cycle skipping occurs when the initial model used in the FWI process is not kinematically accurate or lacks low-frequency contents in the analyzed dataset [34]. So, we consider the criterion based on κ-Gaussian statistics in the context of OT metric to mitigate the effects of non-Gaussian errors and cycle-skipping issues. However, let us remind the reader that the OT metric measures the distance between probability distributions, incompatible with a comparison between seismic signals, as discussed earlier. Thus, to work around this incompatibility, in this work we represent the non-normalized and oscillatory waveforms in the graph space [47]. Graphs are mathematical structures formed by ordered pairs of disjoint sets (V, E), where V denotes the so-called vertices and E represents an edge that connects paired vertices [60].

Hence, we discretize the waveforms $d(t)$ as an ensemble of ordered pairs of the form $\{(t_i, d_i) \in \mathbb{R}^2; i = 1, 2, \cdots, N\}$ with $d_i = d(t_i)$. So, the graph-transformed representation of a discretized waveform $\mathbf{d} = \{d_i; i = 1, 2, \cdots, N\}$ is defined as:

$$\begin{aligned} \mathcal{G} : \mathbf{d} &\to \mathcal{G}(\mathbf{d}) = d^{\mathcal{G}}(y, t) \\ \mathbb{R}^N &\to \mathcal{D}(\mathbb{R}^2), \end{aligned} \tag{50}$$

where $\mathcal{G}$ denotes the graph transformation, $d^{\mathcal{G}}(y,t)$ is the graph-transformed waveform, and $\mathcal{D}(\mathbb{R}^2)$ is a probability space on $\mathbb{R}^2$. The graph-transformed waveform is defined as:

$$d^{\mathcal{G}}(y,t) = \frac{1}{N}\sum_{i=1}^{N}\delta(t-t_i)\delta(y-d_i), \tag{51}$$

where y is associated with the waveform amplitude. In this way, waveforms are represented by normalized and positive quantities.

However, in many contexts like FWI, one needs to calculate the derivative of waveforms on some occasions (as explained in the previous sections); the Dirac delta function is not differentiable. Due to this, we consider a smoothed graph transformation by representing Dirac functions using κ-Gaussian distributions (10). Thus, the κ-graph-transformed representation of a discretized waveform is given by [61]:

$$\begin{aligned} \mathcal{G}_\kappa : \mathbf{d} \quad &\to \mathcal{G}_\kappa(\mathbf{d}) = d^{\mathcal{G}_\kappa}(y,t) \\ \mathbb{R}^N &\to \mathbb{C}^\infty(\mathbb{R},\mathbb{R}_*^+) \end{aligned} \tag{52}$$

with

$$d^{\mathcal{G}_\kappa}(y,t) = \frac{1}{Z_\kappa}\sum_{i=1}^{N}\exp_\kappa\Big(-\beta_\kappa(t-t_i)^2\Big)\exp_\kappa\Big(-\beta_\kappa(y-d_i)^2\Big), \tag{53}$$

where $\mathbb{C}^\infty(\mathbb{R},\mathbb{R}_*^+)$ represents a set of strictly positive and infinitely differentiable functions. In this context, the graph-space κ-OT objective function is defined as:

$$\phi_{\mathcal{W}_\kappa^{\mathcal{G}_\kappa}}(m) := \sum_{s,r}\mathcal{C}_\kappa\left(d_{s,r}^{mod}(m), d_{s,r}^{obs}\right), \tag{54}$$

where $d_{mod,i}^{\mathcal{G}_\kappa} = (t_i, d_{mod,i})$ and $d_{obs,i}^{\mathcal{G}_\kappa} = (t_i, d_{obs,i})$, and $\mathcal{C}_\kappa\left(d^{mod}, d^{obs}\right) = \mathcal{W}_\kappa^{\mathcal{G}_\kappa}\left(d_{mod}^{\mathcal{G}_\kappa}, d_{obs}^{\mathcal{G}_\kappa}\right)$ represents the κ-Wasserstein criterion applied to the graph-transformed seismic data. The κ-Wasserstein distance $\mathcal{W}_\kappa^{\mathcal{G}_\kappa}\left(d_{mod}^{\mathcal{G}_\kappa}, d_{obs}^{\mathcal{G}_\kappa}\right) = \mathcal{W}_\kappa^{\mathcal{G}_\kappa}$ is then computed via the following minimization task:

$$\mathcal{W}_\kappa^{\mathcal{G}_\kappa} = \min_{\sigma\in\mathcal{S}(N)} -\sum_{i=1}^{N_t}\ln\left\{\exp_\kappa\left[-\beta_\kappa\left(t_{\sigma(i)}-t_i\right)^2\right]\exp_\kappa\left[-\beta_\kappa\left(d_{\sigma(i)}^{mod}-d_i^{obs}\right)^2\right]\right\}. \tag{55}$$

where N_t denotes the number of time samples for each waveform, while σ is the permutation solution for the linear sum assignment problem in (21) related with a transport map $\mathcal{T}$, and $\mathcal{S}(N) = \{1,2,\cdots,N\}$ is an ensemble of permutations. For simplicity, we multiply the κ-Wasserstein distance (23) by the scalar N in the latter equation. Indeed, optimizing $\mathcal{W}_\kappa$ is equivalent to optimizing the product $N \times \mathcal{W}_\kappa$. Equation (55) represents the FWI objective function based on κ-OT, namely, the κ-GSOT-FWI, for short, in reference to κ-Graph-Space Optimal Transport FWI.

The gradient of the κ-GSOT-objective function (55), that is, the derivative of $\mathcal{W}_\kappa^{\mathcal{G}_\kappa}$ with respect to the model parameters, is given by:

$$\nabla_m\mathcal{W}_\kappa^{\mathcal{G}_\kappa}(m) = \frac{\partial\mathcal{W}_\kappa^{\mathcal{G}_\kappa}(m)}{\partial m} = \sum_{s=1}^{N_s}\sum_{r=1}^{N_r}\sum_{i=1}^{N_t}\mathcal{J}_{s,r,i}(m)\mathcal{U}_{s,r,i}(m;\kappa), \tag{56}$$

in which

$$\mathcal{U}_{s,r,i}(m;\kappa) = \frac{2\beta_\kappa\left(d_{s,r,\sigma(i)}^{mod}(m) - d_{s,r,i}^{obs}\right)}{\sqrt{1+\kappa^2\beta_\kappa^2\left(d_{s,r,\sigma(i)}^{mod}(m) - d_{s,r,i}^{obs}\right)^4}}. \tag{57}$$

is the adjoint-source related with the κ-GSOT-FWI framework, while N_s, N_r and N_t represent the number of seismic sources, receivers, and time samples used in the acquisition of seismic data.

The statistical interpretation of the residual data (error) associated with the κ-Gaussian statistics is preserved in the κ-GSOT-FWI case. The critical difference is that in the approach without OT, the waveforms are compared sample by sample. In contrast, in the κ-GSOT-FWI approach, the waveforms are analyzed more completely, comparing each time sample of the observed data with all the time samples of the modeled data in according to an optimal assignment using the permutation solution σ.

Figure 3 shows a flow chart of the FWI algorithm, which is an iterative process, which means that model updates are computed concerning the previous model as described by Equation (24). The first step, called Initial Setup, consists of introducing the input variables, i.e., the initial model, the parameters of the seismic acquisition (the positions of the sources and receivers, the seismic source signature, acquisition time). After configuring and organizing all the input variables of the FWI algorithm, modeled wavefields are obtained in the forward problem through the numerical solution of the wave Equation (28) by employing the finite difference method [62]. Then, a sampling operation ($\Gamma_{s,r}$) is carried out from the modeled wavefields ψ_s to obtain the modeled data ($d^{mod}_{s,r} = \Gamma_{s,r}\psi_s$), extracting the wavefields in the positions of the seismic acquisition receivers. After, the objective function gradient is obtained through the adjoint-state method described in Section 4.2 and used to update the model following Equation (24). Finally, the FWI algorithm checks whether the optimization process reached the pre-defined stopping criteria (which, in our case, was the maximum number of iterations equal to 50). As long as the criteria are not met, the cycle is repeated. If so, the iterative process is interrupted, and the resulting model is the one that minimizes the difference between modeled and observed data.

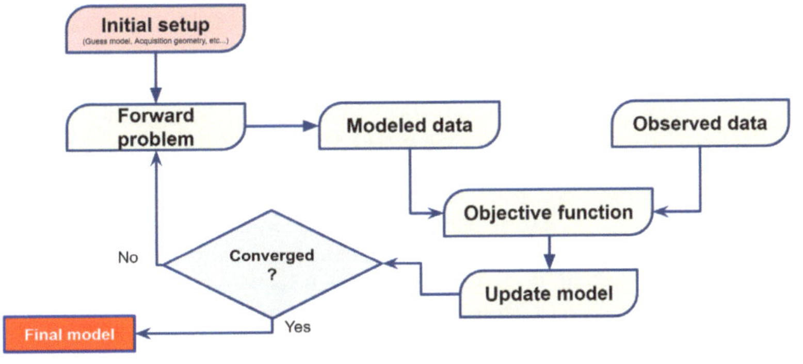

Figure 3. Flow chart of the full-waveform inversion (FWI) process.

5. Numerical Experiments

To demonstrate how the κ-GSOT-FWI deals with non-Gaussian noise and cycle-skipping issues, we carried out numerical examples involving a 2D acoustic time-domain FWI to estimate a P-wave velocity model in a typical Brazilian pre-salt oil region. Such an Earth model, namely, Chalda, represents a region with approximate dimensions 16 by 7 km in lateral distance and depth, respectively, as depicted in Figure 4a. Our problem has 720,484 unknown variables because we discretize the Chalda model in a regular grid with 12.5 m spacing, generating 562 and 1282 grid cells in the vertical and horizontal directions, respectively.

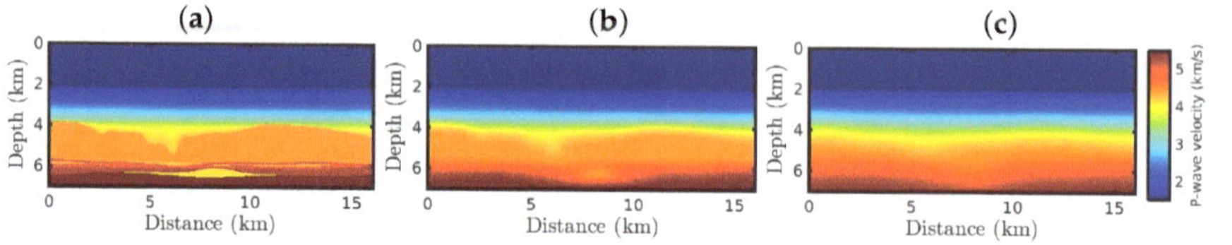

Figure 4. (**a**) Chalda model representing the Brazilian pre-salt oil region, used as the true model. Initial models used in the (**b**) first and (**c**) second scenarios.

In all numerical experiments, we consider a seismic survey comprising 161 seismic sources equally spaced every 75 m at 12.5 m in-depth. We employ a Ricker wavelet as a seismic source, which is mathematically described by: $f(t) = \left(1 - 2\pi^2\mu_p^2 t^2\right) \exp\left(-\pi^2\mu_p^2 t^2\right)$, in which μ_p represents the peak frequency (maximum energy in the spectrum of frequencies). Moreover, to simulate a sparse node acquisition, named the ocean bottom nodes survey, we take into account 21 receivers implanted on the ocean floor at 400 m intervals. We consider the Chalda model depicted in Figure 4a as a benchmark (or true model). Thus, we generate a seismic dataset by considering the true model, the acquisition geometry, and the finite difference method with second and eighth order approximations for time and space. In order to simulate an infinite medium, we implement the perfectly matched layer [63] absorbing boundaries for spatial discretization. We consider 7 s as the seismic acquisition time at a sampling rate of 2 ms. In addition, to simulate a realistic case, we also employ a high-pass filter on the seismic dataset to remove energy less than 2.5 Hz.

In the FWI experiments, we consider two scenarios involving different initial models to confirm the significance of our proposal. In the first one, we consider an initial model similar to the true model, which is depicted in Figure 4b. We produce such a velocity model by weakly smoothing the true model by applying a Gaussian filter with a standard deviation of 250 m. This scenario's idea is to simulate a seismic imaging process starting from a kinetically accurate model. We call this model the Good Model. In contrast, we produce the second initial model, referring to the second scenario, by applying a more severe Gaussian filter with a standard deviation of 750 m. We call this model the Bad Model. We notice that the Bad Model lacks the main structures of the true model, particularly in the pre-salt oil region, as depicted in Figure 4c. Since the Bad Model is kinematically inaccurate, it generates cycle-skipped data [34].

For each initial-model scenario, we conduct time-domain FWI by applying the classical FWI based on Gaussian statistics, and the κ-GSOT objective function (55) in the classical limit $\kappa \to 0$ and for $\kappa = 0.1, 0.3, 0.5$ and 0.6. We consider 50 FWI iterations in all numerical experiments. To evade the so-called inversion crime, we perform the forward modeling using a different algorithm than the one used to generate the observed dataset. In this regard, our algorithm solves the forward problem using a finite difference scheme with second and fourth order approximations for time and space in a regular grid with 25 m spacing. In addition, we consider two different circumstances concerning the type of noise in the seismic dataset. First, we consider a dataset contaminated by Gaussian noise with a signal-to-noise ratio (SNR) of 20. In contrast, in the second circumstance, we consider a non-Gaussian noise from which the dataset is polluted by Gaussian noise with an SNR of 20 and a collection of spikes (erratic data or outliers) with different amplitudes. In this regard, 10% of the time samples were contaminated by outliers, where locations were randomly drawn. The spikes have intensities that range from 5 P to 15 P multiplied by the original waveform amplitude, where P is a standard normal random variable.

We perform our numerical simulations using a computer hosting a quad-core processor at 3.50 GHz and 256 GB RAM. Each FWI iteration takes approximately 6 min; 71.4% is associated with calculating the gradient of the objective function using the state-adjoint method described in Section 4.2, 26.7% is related to the forward modeling process (i.e., in generating the modeled data by numerically solving Equation (28)), 1.3% of this time is dedicated to solving the combinatorial optimization problem in (55), and 0.6% is spent on the rest of the algorithm in I/O initialization and initial set-ups loading.

Figures 5 and 6 show the FWI resulting P-wave models starting from the Good Model for the Gaussian and non-Gaussian noise cases, respectively. From a visual inspection, when only Gaussian noise is considered, all resulting models are satisfactory (Figure 5) since they are very similar to the true model (Figure 4a), regardless of the κ-value. Such successful results are due to the weak Gaussian noise in the observed data simultaneously with a kinetically accurate initial model (Figure 4b).

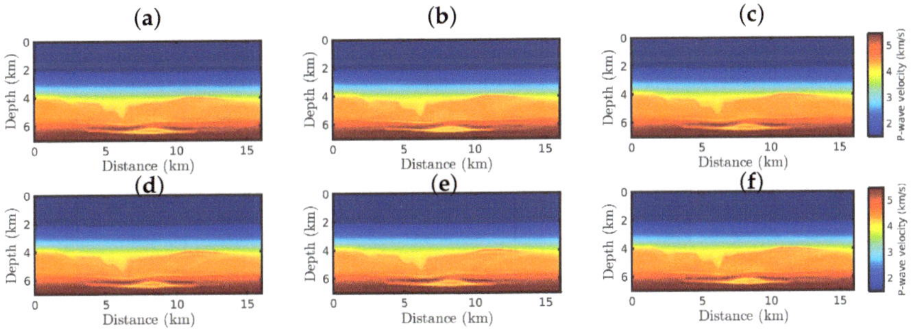

Figure 5. The resulting models starting from the Good Model for the Gaussian noise case, by employing the (**a**) classical FWI approach, and the κ-GSOT-FWI framework with (**b**) $\kappa \to 0$, (**c**) $\kappa = 0.1$, (**d**) $\kappa = 0.3$, (**e**) $\kappa = 0.5$, and (**f**) $\kappa = 0.6$.

Furthermore, we quantitatively compare our FWI resulting models with the true model by employing Pearson's correlation coefficient ($\mathtt{R}$) and the normalized root-mean-square ($\mathtt{NRMS}$), defined as

$$\mathtt{R} = \frac{\mathrm{cov}\left(c^{true}, c^{inv}\right)}{\mathrm{std}\left(c^{true}\right)\mathrm{std}\left(c^{inv}\right)} \quad \text{and} \quad \mathtt{NRMS} = \left[\frac{\sum_i \left(c_i^{true} - c_i^{inv}\right)^2}{\sum_i \left(c_i^{true}\right)^2}\right]^{1/2}, \tag{58}$$

where c^{true} and c^{inv} are the true and the resulting models, while $\mathrm{cov}(\cdot)$ and $\mathrm{std}(\cdot)$ denote covariance and standard deviation, respectively. The $\mathtt{R}$-value ranges from -1 to 1, with -1 representing, in this context, a wrong resulting model, while 1 represents a perfect resulting model. The $\mathtt{NRMS}$-value range from 0 (perfect resulting model) to ∞ (wrong resulting model).

Table 1 summarizes the comparative metrics between the true model and the P-wave velocity models resulting from the first scenario by analyzing data contaminated only by Gaussian noise. In this table, we can see that all resulting models have a low error and are strongly correlated with the true model ($\mathtt{R} \geq 0.8$, following the strength-scale suggested by ref. [64]).

Table 1. The comparative metrics between the true model and the resulting models, depicted in Figure 1, from the first scenario in the Gaussian noise case. $\mathtt{R}$ represents the Pearson's correlation coefficient, while $\mathtt{NRMS}$ represents the normalized root-mean-square.

Strategy	κ	NRMS	R
Classical FWI	-	0.0256	0.9982
κ-GSOT-FWI	0.0	0.0272	0.9979
	0.1	0.0278	0.9979
	0.3	0.0288	0.9977
	0.5	0.0293	0.9976
	0.6	0.0277	0.9979

However, when non-Gaussian noise is considered, the classical approach fails as expected (Figure 6a). Such a wrong model is due to the classical approach being based on Gaussian statistics and sensitive to cycle-skipping issues. Figure 6b shows the resulting model from the classical GSOT-FWI, which is also based on Gaussian statistics. However, the Wasserstein metric was able to mitigate the effects of the outliers, building a satisfactory model. Nevertheless, as the κ-value increases (which means a more significant deviation from Gaussian behaviors), the κ-GSOT-FWI models present a better resolution (Figure 6c–f), especially in the deeper regions of the analyzed area. Although all κ-GSOT-FWI models are strongly correlated with the true model, the case $\kappa = 0.6$ has a higher Pearson's coefficient and a smaller NRMS error, as summarized in Table 2.

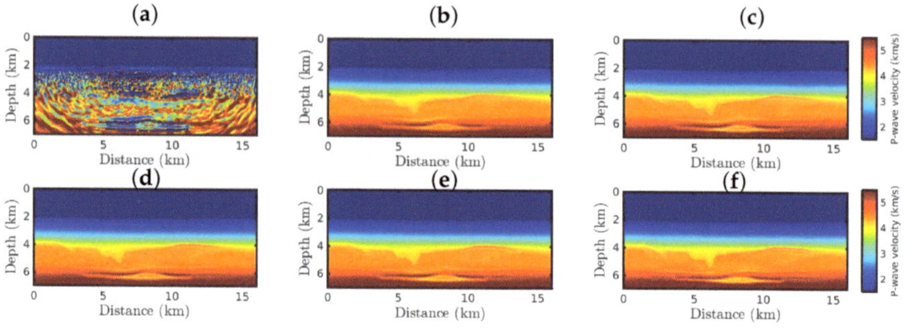

Figure 6. The resulting models starting from the Good Model for the non-Gaussian noise case, by employing the (**a**) classical FWI approach, and the κ-GSOT-FWI framework with (**b**) $\kappa \to 0$, (**c**) $\kappa = 0.1$, (**d**) $\kappa = 0.3$, (**e**) $\kappa = 0.5$, and (**f**) $\kappa = 0.6$.

Table 2. The comparative metrics between the true model and the resulting models, depicted in Figure 6, from the first scenario in the non-Gaussian noise case. R represents the Pearson's correlation coefficient, while NRMS represents the normalized root-mean-square.

Strategy	κ	NRMS	R
Classical FWI	-	0.3009	0.7627
κ-GSOT-FWI	0.0	0.0306	0.9974
	0.1	0.0296	0.9976
	0.3	0.0293	0.9976
	0.5	0.0293	0.9976
	0.6	0.0277	0.9979

Figures 7 and 8 show the FWI resulting P-wave models starting from the Bad Model for the Gaussian and non-Gaussian noise cases, respectively. From a visual inspection, it is noticeable that the classical FWI approach fails when the initial model is kinetically inaccurate, regardless of whether the data are polluted by Gaussian or non-Gaussian noise, as depicted in Figures 7a and 8a. In contrast, the FWI based on the κ-GSOT approach generates satisfactory models when Gaussian noise is considered, regardless of the κ-value (Figure 7b–f). Again, as the κ-value increases, the resulting models (Figure 7) are closer to the true model (Figure 4a), as endorsed by the statistical metrics summarized in Table 3.

Finally, in the second scenario with non-Gaussian noise, the resulting models are drastically affected by the outliers and poverty from the initial model, as depicted in Figure 8. However, when the κ-GSOT-based objective function is applied, the large geological structures of the true model are reconstructed regardless of the κ-value. However, the case $\kappa = 0.6$ reveals a P-wave velocity model (Figure 8f) that is quite accurate and comparable to the true model (Figure 4a). Likewise, the case $\kappa = 0.6$ generated a model closer to the true model, as summarized in Table 4. Indeed, in all numerical tests, the κ-GSOT-FWI for $\kappa = 0.6$ generated accurate velocity models, leading to accurate parameter estimations.

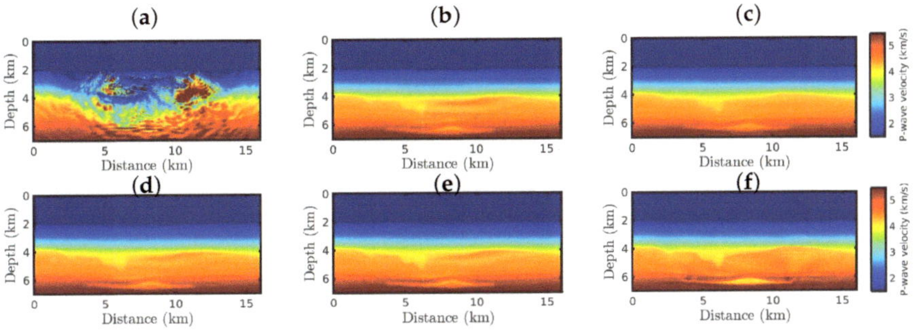

Figure 7. The resulting models starting from the Bad Model for the Gaussian noise case, by employing the (**a**) classical FWI approach, and the κ-GSOT-FWI framework with (**b**) $\kappa \to 0$, (**c**) $\kappa = 0.1$, (**d**) $\kappa = 0.3$, (**e**) $\kappa = 0.5$, and (**f**) $\kappa = 0.6$.

Table 3. The comparative metrics between the true model and the resulting models, depicted in Figure 7, from the second scenario in the Gaussian noise case. R represents the Pearson's correlation coefficient, while NRMS represents the normalized root-mean-square.

Strategy	κ	NRMS	R
Classical FWI	-	0.2947	0.7982
κ-GSOT-FWI	0.0	0.0362	0.9964
	0.1	0.0333	0.9970
	0.3	0.0363	0.9964
	0.5	0.0372	0.9962
	0.6	0.0341	0.9968

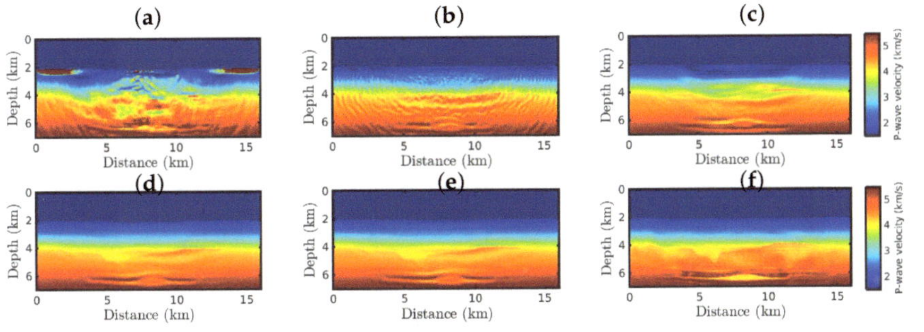

Figure 8. The resulting models starting from the Bad Model for the non-Gaussian noise case, by employing the (**a**) classical FWI approach, and the κ-GSOT-FWI framework with (**b**) $\kappa \to 0$, (**c**) $\kappa = 0.1$, (**d**) $\kappa = 0.3$, (**e**) $\kappa = 0.5$, and (**f**) $\kappa = 0.6$.

Table 4. The comparative metrics between the true model and the resulting models, depicted in Figure 8, from the second scenario in the non-Gaussian noise case. R represents the Pearson's correlation coefficient, while NRMS represents the normalized root mean square.

Strategy	κ	NRMS	R
Classical FWI	-	0.2715	0.7192
κ-GSOT-FWI	0.0	0.0610	0.9899
	0.1	0.0673	0.9874
	0.3	0.0564	0.9913
	0.5	0.0578	0.9908
	0.6	0.0509	0.9928

Figure 9 shows the normalized κ-GSOT-objective function decay for all numerical tests, in which panels (a) and (b) refer to the first scenario, while panels (c) and (d) correspond to the second scenario. In this regard, the left column refers to the case in which Gaussian noise is considered, and the right column is the non-Gaussian noise case. The convergence curve of the classical objective is represented by the solid black line in Figure 9. We notice that the classical objective function monotonically decays only in the most straightforward situation, where the initial model is the Good Model, and the data is contaminated by Gaussian noise (Figure 9a). In this case, the classical approach is the most indicated because the convergence rate is higher than our proposal, in addition to generating more accurate models (as summarized in Table 1). In cases where the noise is non-Gaussian or when the inversion process starts from the Bad Model, our proposal with $\kappa = 0.6$ exhibits a higher objective function decay rate (see red curves in Figure 9b–d), reconstructing P-wave velocity models closer to the true model, as summarized in Tables 2–4.

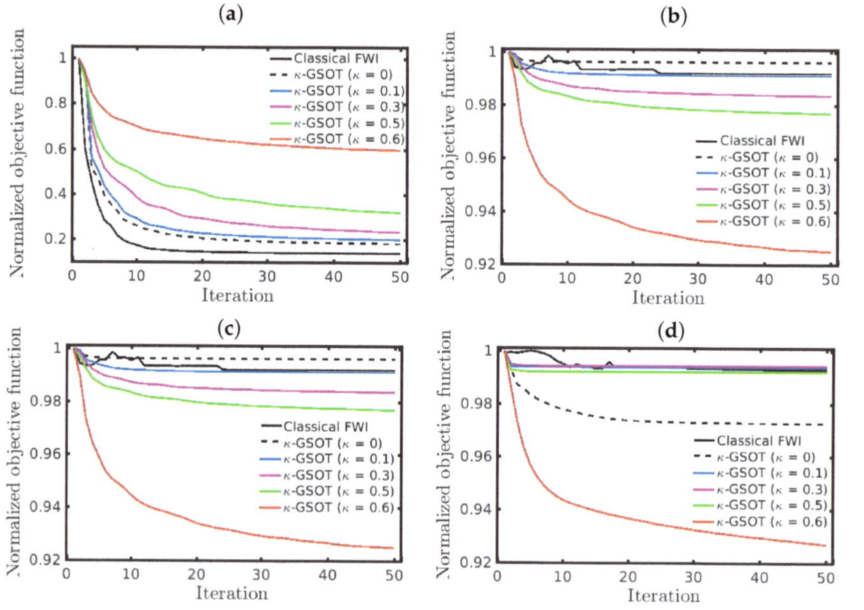

Figure 9. Convergence curves for the first scenario with (**a**) Gaussian noise, (**b**) non-Gaussian noise, and for the second scenario with (**c**) Gaussian noise, (**d**) non-Gaussian noise.

6. Final Remarks

In this work, we have examined the portability of the objective function based on the graph-space optimal transport and Kaniadakis κ-Gaussian statistics in the FWI context. In particular, we have analyzed the robustness of our proposal in mitigating two critical problems in seismic imaging via FWI, which are associated with cycle-skipping issues and the non-Gaussian nature of the errors. We have set up an objective function by employing the probabilistic maximum likelihood method for computing the most probable state using a κ-Gaussian distribution. Furthermore, we have formulated the FWI in a relaxed version of the optimal transport problem, known as the Kantorovich–Rubinstein metric or Wasserstein distance. So, we have considered the graph of the seismic data rather than the original data because the optimal transport framework is predicated on the idea that the compared entities adhere to the probability axioms. We named our proposal the κ-Graph-Space Optimal Transport FWI (or κ-GSOT-FWI, for short).

The Brazilian pre-salt case study disclosed how the κ-GSOT-FWI could be employed to deal with flawed initial models and non-Gaussian noise. The findings have demonstrated that the classical approach is ineffective in producing accurate physical models when the initial model is crude or if the observed waveforms are contaminated by non-Gaussian errors. However, when the initial model is kinetically precise and the data well-behaved, the classical approach is the best alternative in terms of computational cost. The results also revealed that the κ-GSOT-FWI lessens the impact of phase ambiguity and non-Gaussian errors on the waveform inversion, demonstrating that our proposal is a powerful way to deal with non-linear inverse problems related to wave propagation. Moreover, we notice that the κ-GSOT-FWI produces more accurate models than those produced by classical approaches, leading to a notable improvement in objective function convergence. Additionally, our numerical experiments demonstrated that a more significant deviation from a Gaussian behavior (which in our applications was typified by the $\kappa = 0.6$ case) results in a more authentic P-wave velocity model. However, our proposal depends on the choice of a hyperparameter, which demands special investigations on how to obtain it in a real setting application. This issue should be examined in future applications.

From a practical point of view, extensive and arduous data processing is required to engineer a good initial model to alleviate phase-ambiguity issues and eliminate erratic data points. In this context, the κ-GSOT-FWI decreases the requirement of human subjectivity, which is appealing for automated techniques to analyze, for instance, recent big datasets. Thus, the κ-OT-based approach has enormous potential for dealing with modern data-centric problems. As a perspective, we intend to test our proposed methodology to analyze field data and evaluate its robustness from several initial conditions. Finally, we underline how readily our concept may be applied to a wide variety of inverse problems, ranging from estimating critical exponents of power-law distributions to modern artificial intelligence applications.

Author Contributions: Conceptualization, S.L.E.F.d.S., J.M.d.A., E.d.l.B. and G.C.; Methodology, S.L.E.F.d.S., J.M.d.A., E.d.l.B. and G.C.; Software, S.L.E.F.d.S.; Validation, S.L.E.F.d.S., J.M.d.A., E.d.l.B. and G.C.; Formal analysis, S.L.E.F.d.S., J.M.d.A., E.d.l.B. and G.C.; Investigation, S.L.E.F.d.S.; Resources, J.M.d.A. and G.C.; Data curation, S.L.E.F.d.S.; Writing—original draft, S.L.E.F.d.S., J.M.d.A., E.d.l.B. and G.C.; Supervision, G.C.; Project administration, J.M.d.A. All authors have read and agreed to the published version of the manuscript.

Funding: This research received no external funding.

Data Availability Statement: Not applicable.

Conflicts of Interest: The authors declare no conflict of interest.

Abbreviations

The following abbreviations are used in this manuscript:

FWI	Full-Waveform Inversion
GSOT	Graph-Space Optimal Transport
OT	Optimal Transport
MLE	Maximum Likelihood Estimation
NRMS	Normalized Root-Mean-Square

References

1. Razavy, M. *An Introduction to Inverse Problems in Physics*, 1st ed.; World Scientific: Hackensack, NJ, USA, 2020; pp. 1–388.
2. Menke, W. *Geophysical Data Analysis: Discrete Inverse Theory*, 4th ed.; Academic Press: London, UK, 2018; pp. 1–352.
3. Hanasoge, S.M.; Tromp, J. Full waveform inversion for time-distance helioseismology. *APJ* **2014**, *784*, 69. [CrossRef]
4. Guasch, L.; Calderón Agudo, O.; Tang, M.X.; Nachev, P.; Warner, M. Full-waveform inversion imaging of the human brain. *NPJ Digit. Med.* **2020**, *3*, 28. [CrossRef] [PubMed]
5. Robins, T.; Camacho, J.; Calderón Agudo, O.; Herraiz, J.L.; Guasch, L. Deep-Learning-Driven Full-Waveform Inversion for Ultrasound Breast Imaging. *Sensors* **2021**, *21*, 4570. [CrossRef] [PubMed]
6. Cao, J.; Brossier, R.; Górszczyk, A.; Métivier, L.; Virieux, J. 3-D multiparameter full-waveform inversion for ocean-bottom seismic data using an efficient fluid–solid coupled spectral-element solver. *Geophys. J. Int.* **2022**, *229*, 671–703. [CrossRef]

7. Virieux, J.; Operto, S. An overview of full-waveform inversion in exploration geophysics. *Geophysics* **2009**, *74*, WCC1–WCC26. [CrossRef]
8. Fichtner, A. *Full Seismic Waveform Modelling and Inversion*, 1st ed.; Springer: Berlin/Heidelberg, Germay, 2011; pp. 1–343.
9. Brossier, R.; Operto, S.; Virieux, J. Which data residual norm for robust elastic frequency-domain full waveform inversion? *Geophysics* **2010**, *75*, R37–R46. [CrossRef]
10. de Lima, I.P.; da Silva, S.L.E.F.; Corso, G.; de Araújo, J.M. Tsallis Entropy, Likelihood, and the Robust Seismic Inversion. *Entropy* **2020**, *22*, 464. [CrossRef]
11. Tarantola, A. *Inverse Problem Theory and Methods for Model Parameter Estimation*, 1st ed.; Society for Industrial and Applied Mathematics: Philadelphia, PA, USA, 2005; pp. 1–352.
12. Khokhlov, A.; Hulot, G. On the cause of the non-Gaussian distribution of residuals in geomagnetism. *Geophys. J. Int.* **2017**, *209*, 1036–1047. [CrossRef]
13. Elboth, T.; Reif, B.A.; Andreassen, Ø. Flow and swell noise in marine seismic data. *Geophysics* **2009**, *74*, Q17–Q25. [CrossRef]
14. Hlebnikov, V.; Elboth, T.; Vinje, V.; Gelius, L.-J. Noise types and their attenuation in towed marine seismic: A tutorial. *Geophysics* **2021**, *86*, W1–W19. [CrossRef]
15. Claerbout, J.F.; Muir, F. Robust modeling with erratic data. *Geophysics* **1973**, *38*, 826–844. [CrossRef]
16. Crase, E.; Pica, A.; Noble, M.; McDonald, J.; Tarantola, A. Robust elastic nonlinear waveform inversion: Application to real data. *Geophysics* **1990**, *55*, 1942–2156. [CrossRef]
17. Aravkin, A.Y.; Friedlander, M.P.; Herrmann, F.J.; van Leeuwen, T. Robust inversion, dimensionality reduction and randomized sampling. *Math. Program.* **2012**, *135*, 101–125. [CrossRef]
18. Da Silva, S.L.E.F.; da Costa, C.A.N.; Carvalho, P.T.C.; de Araújo, J.M.; dos Santos Lucena, L.; Corso, G. Robust full-waveform inversion using *Q*-Statistics. *Phys. A* **2020**, *548*, 124473. [CrossRef]
19. Silva, S.A.; da Silva, S.L.E.F.; de Souza, R.F.; Marinho, A.A.; de Araújo, J.M.; Bezerra, C.G. Improving Seismic Inversion Robustness via Deformed Jackson Gaussian. *Entropy* **2021**, *23*, 1081. [CrossRef] [PubMed]
20. Bube, K.P.; Langan, R.T. Hybrid *L*1/*l*2 Minimization Appl. Tomography. *Geophysics* **1997**, *62*, 1045–1346. [CrossRef]
21. Guitton, A.; Symes, W.W. Robust inversion of seismic data using the Huber norm. *Geophysics* **2003**, *68*, 1126–1422. [CrossRef]
22. Da Silva, S.L.E.F.; Silva, R.; dos Santos Lima, G.Z.; de Araújo, J.M.; Corso, G. An outlier-resistant κ-generalized approach for robust physical parameter estimation. *Physica A* **2022**, *600*, 127554. [CrossRef]
23. Kaniadakis, G. Non-linear kinetics underlying generalized statistics. *Phys. A* **2001**, *296*, 405–425. [CrossRef]
24. Kaniadakis, G. Statistical mechanics in the context of special relativity. *Phys. Rev. E* **2002**, *66*, 056125. [CrossRef]
25. Kaniadakis, G.; Scarfone, A.M. A new one-parameter deformation of the exponential function. *Phys. A* **2002**, *305*, 69–75. [CrossRef]
26. Kaniadakis, G. Statistical mechanics in the context of special relativity. II. *Phys. Rev. E* **2005**, *72*, 036108. [CrossRef] [PubMed]
27. Kaniadakis, G. Theoretical Foundations and Mathematical Formalism of the Power-Law Tailed Statistical Distributions. *Entropy* **2013**, *15*, 3983–4010. [CrossRef]
28. Da Silva, S.L.E.F.; dos Santos Lima, G.Z.; de Araújo, J.M.; Corso, G. Extensive and nonextensive statistics in seismic inversion. *Phys. A* **2021**, *563*, 125496. [CrossRef]
29. Wada, T.; Suyari, H. κ-generalization of Gauss' law of error. *Phys. Lett. A* **2006**, *348*, 89–93. [CrossRef]
30. Da Silva, S.L.E.F.; Carvalho, P.T.C.; de Araújo, J.M.; Corso, G. Full-waveform inversion based on Kaniadakis statistics. *Phys. Rev. E* **2020**, *101*, 053311. [CrossRef]
31. Bunks, C.; Saleck, F.M.; Zaleski, S.; Chavent, G. Multiscale seismic waveform inversion. *Geophysics* **1995**, *60*, 1457–1473. [CrossRef]
32. Liu, X.; Zhu, T.; Hayes, J. Critical zone structure by elastic full waveform inversion of seismic refractions in a sandstone catchment, central Pennsylvania, USA. *J. Geophys. Res. Solid Earth* **2022**, *127*, e2021JB023321. [CrossRef]
33. Górszczyk, A.; Brossier, R.; Métivier, L. Graph-space optimal transport concept for time-domain full-waveform inversion of ocean-bottom seismometer data: Nankai Trough velocity structure reconstructed from a 1D model. *J. Geophys. Res. Solid Earth* **2021**, *126*, e2020JB021504. [CrossRef]
34. Hu, W.; Chen, J.; Liu, J.; Abubakar, A. Retrieving Low Wavenumber Information in FWI: An Overview of the Cycle-Skipping Phenomenon and Solutions. *IEEE Signal Process. Mag.* **2018**, *35*, 132–141. [CrossRef]
35. Bozdağ, E.; Trampert, J.; Tromp, J. Misfit functions for full waveform inversion based on instantaneous phase and envelope measurement. *Geophys. J. Int.* **2011**, *185*, 845–870. [CrossRef]
36. Warner, M.; Guasch, L. Adaptive waveform inversion: Theory. *Geophysics* **2016**, *81*, R429–R445. [CrossRef]
37. Carvalho, P.T.C.; da Silva, S.E.F.; Duarte, E.F.; Brossier, R.; Corso, G.; de Araújo, J.M. Full waveform inversion based on the non-parametric estimate of the probability distribution of the residuals. *Geophys. J. Int.* **2022**, *229*, 35–55. [CrossRef]
38. Métivier, L.; Brossier, R.; Mérigot, Q.; Oudet, E.; Virieux, J. Measuring the misfit between seismograms using an optimal transport distance: Application to full waveform inversion. *Geophys. J. Int.* **2016**, *205*, 345–377. [CrossRef]
39. Monge, G. *Mémoire sur la Théorie des Déblais et des Remblais*, 1st ed.; Histoire de l'Académie Royale des Sciences de Paris: Paris, France, 1781; pp. 666–704.
40. Villani, C. *Optimal Transport: Old and New*, 1st ed.; Springer: Berlin/Heidelberg, Germany, 2008; pp. 1–976.
41. Figalli, A. The Optimal Partial Transport Problem. *Arch. Ration. Mech. Anal.* **2010**, *195*, 533–560. [CrossRef]
42. Ambrosio, L.; Gigli, N. User's Guide to Optimal Transport. In *Modelling and Optimisation of Flows on Networks. Lecture Notes in Mathematics*; Springer: Berlin/Heidelberg, Germany, 2012; Volume 2062, pp. 1–155.

43. Wang, S.; Stavrou, P.A.; Skoglund, M. Generalizations of Talagrand Inequality for Sinkhorn Distance Using Entropy Power Inequality. *Entropy* **2022**, *24*, 306. [CrossRef]
44. Messud, J.; Poncet, R.; Lambaré, G. Optimal transport in full-waveform inversion: Analysis and practice of the multidimensional Kantorovich–Rubinstein norm. *Inverse Probl.* **2021**, *36*, 065012. [CrossRef]
45. Sambridge, M.; Jackson, A.; Valentine, A.P. Geophysical inversion and optimal transport. *Geophys. J. Int.* **2022**, *231*, 172–198. [CrossRef]
46. Da Silva, S.L.E.F.; Karsou, A.; de Souza, A.; Capuzzo, F.; Costa, F.; Moreira, R.; Cetale, M. A graph-space optimal transport objective function based on q-statistics to mitigate cycle-skipping issues in FWI. *Geophys. J. Int.* **2022**, *231*, 1363–1385. [CrossRef]
47. Métivier, L.; Allain, A.; Brossier, R.; Mérigot, Q.; Oudet, E.; Virieux, J. Optimal transport for mitigating cycle skipping in full-waveform inversion: A graph-space transform approach. *Geophysics* **2018**, *83*, R515–R540. [CrossRef]
48. De Lima, J.V.T.; da Silva, S.L.E.F.; de Araújo, J.M.; Corso, G.; dos Santos Lima, G.Z. Nonextensive statistical mechanics for robust physical parameter estimation: The role of entropic index. *Eur. Phys. J. Plus* **2021**, *136*, 269. [CrossRef]
49. Dos Santos Lima, G.Z.; de Lima, J.V.T.; de Araújo, J.M.; Corso, G.; Da Silva, S.L.E.F. Generalized statistics: Applications to data inverse problems with outlier-resistance. *PLoS ONE* **2023**, *18*, e0282578. [CrossRef]
50. Hampel, F.R.; Ronchetti, E.M.; Rousseeuw, P.J.; Stahel, W.A. *Robust Statistics: The Approach Based on Influence Functions*, 1st ed.; Wiley-Intersciencec: Hoboken, NJ, USA, 2005; pp. 1–502.
51. Kantorovich, L. On the Translocation of Masses. *J. Manag. Sci.* **1958**, *5*, 1381–1382. [CrossRef]
52. Birkhoff, G. Three observations on linear algebra. *Univ. Nac. Tucumán. Rev. A* **1946**, *5*, 147–151.
53. Burkard, R.; Dell'Amico, M.; Martello, S. *Assignment Problems*; Society for Industrial and Applied Mathematics: Philadelphia, PA, USA, 2012; pp. 1–393.
54. Nocedal, J.; Wright, S.J. *Numerical Optimization*; Springer: New York, NY, USA, 2006; pp. 1–597.
55. Polak, B.; Ribière, G. Note surla convergence des mèthodes de directions conjuguèes. *Rev. Fr. Inform. Rech. Oper.* **1969**, *3*, 35–43.
56. Polyak, B.T. The conjugate gradient method in extreme problems. *USSR Comput. Math. Math. Phys.* **1969**, *9*, 94–112. [CrossRef]
57. Chavent, G. Identification of function parameters in partial differential equations. In *Identification of Parameter Distributed Systems*; R. E. Goodson and M. Polis: New York, NY, USA, 1974; pp. 155–156.
58. Haber, E.; Ascher, U.M. On optimization techniques for solving nonlinear inverse problems. *Inverse Probl.* **2000**, *16*, 1263–1280. [CrossRef]
59. Plessix, R.-E. A review of the adjoint-state method for computing the gradient of a functional with geophysical applications. *Geophys. J. Int.* **2006**, *167*, 495–503. [CrossRef]
60. Bollobás, B. *Modern Graph Theory, Graduate Texts in Mathematics*, 1st ed.; Springer: New York, NY, USA, 1998; pp. 1–408.
61. da Silva, S.L.E.F.; Kaniadakis, G. κ-statistics approach to optimal transport waveform inversion. *Phys. Rev. E* **2022**, *106*, 034113. [CrossRef]
62. Grossmann, C.; Roos, H.-G.; Stynes, M. *Numerical Treatment of Partial Differential Equations*; Springer Science & Business Media: Berlin/Heidelberg, Germany, 2007; p. 23.
63. Berenger, J.-P. A perfectly matched layer for the absorption of electromagnetic waves. *J. Comput. Phys.* **1994**, *114*, 185–200. [CrossRef]
64. Evans, J.D. *Straightforward Statistics for the Behavioral Sciences*, 1st ed.; Brooks/Cole Publishing Company: Monterey, CA, USA, 1996; pp. 1–600.

entropy

MDPI

Article

Kaniadakis's Information Geometry of Compositional Data

Giovanni Pistone [1,*] and Muhammad Shoaib [2]

[1] De Castro Statistics, Collegio Carlo Alberto, 10122 Torino, Italy
[2] Department of Mathematics, University of Genoa, 16144 Genova, Italy; muhammad.shoaib@dima.unige.it
* Correspondence: giovanni.pistone@carloalberto.org

Abstract: We propose to use a particular case of Kaniadakis' logarithm for the exploratory analysis of compositional data following the Aitchison approach. The affine information geometry derived from Kaniadakis' logarithm provides a consistent setup for the geometric analysis of compositional data. Moreover, the affine setup suggests a rationale for choosing a specific divergence, which we name the Kaniadakis divergence.

Keywords: Kaniadakis logarithm; information geometry; compositional data; affine displacement; affine statistical bundle; barycenter; Kaniadakis divergence

Citation: Pistone, G.; Shoaib, M. Kaniadakis's Information Geometry of Compositional Data. *Entropy* **2023**, *25*, 1107. https://doi.org/10.3390/e25071107

Academic Editors: Antonio M. Scarfone, Dionissios T. Hristopulos and Sergio Luiz E. F. da Silva

Received: 29 June 2023
Revised: 20 July 2023
Accepted: 21 July 2023
Published: 24 July 2023

1. Introduction

This paper describes Kaniadakis' statistics as a methodology in data science. Precisely, we discuss Kaniadakis' formalism for defining an affine structure on the open probability simplex. We present the methods in some generality and use them for the exploratory analysis of compositional data. The illustrating example is a small dataset, and we do not discuss any scaling issues of our methods. However, the dataset has an independent interest in financial risk analysis.

1.1. Why a Geometric Methodology

Kaniadakis' logarithm [1,2] generalises the ordinary logarithm in a way that supports the development of deformed exponential families, deformed statistical divergences, and deformed entropy. Kaniadakis was originally motivated by the applications to non-extensive statistical physics in the sense of [3,4]. In this paper, we present the geometry of the probability simplex as a system of two affine spaces in duality from the perspective of information geometry (IG) [5]. The affine setup was first applied to deformed statistical models in [6].

The systematic use of this formal geometric perspective provides a robust and unified rationale for discussing key descriptive concepts. Defining geometry is much more than providing a topology or a distance. We provide a definition of affine geodesics and a natural duality so that the orthogonal surfaces of the geodesics are well-defined by a specific divergence function. The divergence level sets form a neighbourhood system and, eventually, a topology. In this setup, we define the barycentre, the displacement from the barycentre, and dimensionality reduction. For the standard affine geometry of the probability simplex, see, for example, the tutorial reference [7]. We use a special kind of Kaniadakis' logarithm that appears with a different name in compositional data (CoDa) ([8] Example 4.20).

1.2. CoDa

Compositional data (or CoDa) are the data where all of a (row) vector's (i.e., $[x_1, x_2, \ldots, x_D]$) components are strictly positive real values, can also have zero values, and thus contain solely relative information; the composition is called a D-part composition. Compositional data are often expressed in closed form and totalled a fixed value, such as 1 for parts per unit or 100 for percentage measurements ([9] Chapter 2).

Compositional data are often found in geosciences and other scientific disciplines, and classification, discrimination, and categorization need to be adapted to the case of CoDa. CoDa analysis is closely related to geosciences and biology, where the data are mostly expressed as proportions or concentrations without mentioning the total size or amount explicitly [10].

Significant advancement has been accomplished during the last thirty to forty years. Recently, the term CoDa analysis has been employed to "Insist on the idea that the study goals or hypotheses, which place more of an emphasis on relative than absolute values, are what ultimately determine composition rather than the data, which may not be pieces of a whole or may not have a fixed sum" [11]. These qualities make CoDa analysis the most powerful tool for applications outside the tradition of hard sciences [12]. Current studies in management, economics, and social sciences have shown in practice the benefits of compositional methods in handling a wide variety of problems, which range from market shares and customer segmentation to tourism, transport systems, financial ratios, and many more (see [11,13–15]).

1.3. CoDa and Systemic Financial Risk

The Center for Risk Management at the University of Lausanne (http://www.crml.ch, accessed on 28 June 2023) provides systemic risk assessments for European financial institutions, which we used in our empirical study using the above Kaniadakis methods. The dataset enables the determination of SRISK country-level values, a market-based systemic risk indicator first proposed in [16,17] and most recently examined in [18].

The characteristics of SRISK are popular in the literature, and SRISK is mainly used to recognize weak institutions and countries with a system-wide impact before a crisis occurs [19] and can help forecast actual sector performance [20].
Most of the previous literature has mainly focused on the absolute values of SRISK. In this work, we focus on implementing the Kaniadakis methods to see the different European countries as a part of compositional data. We developed work started in [12], where they first introduced compositional data analysis to examine the distribution of relative contributions to SRISK connected with key European nations from 2008 to 2021.

Atchison [21] first introduced CoDa analysis. The research conducted by the [12] on financial data used the Atchison methods to examine how European nations contribute to the total amount of systemic risk (SRISK). They find that the distinctive quality of CoDa analysis, especially the Atchison geometry, is very effective in determining the threats of possible instability offered by smaller institutions and nations that might not completely emerge from the scale of their systemic risk.

1.4. Data and Methods

This paper first establishes a novel theoretical framework for compositional data using Kaniadakis' logarithm. Second, we implement the Kaniadakis divergence on the compositional data and calculate the exponential and mixture displacements on compositional data. Next, we calculate the barycenter and deviation. The purpose of calculating the barycenter is to check how far the values of SRISK are from their centre value.

We consider ten European economies (Belgium, Denmark, France, Germany, Greece, Italy, Netherlands, Spain, Switzerland, and the UK) with annual SRISK measurements collected at the end of December for 2008–2021. Every number is stated in billions of Euros. Like most CoDa method applications, the sample does not cover Europe. Therefore, the ten components that make up our SRISK compositions are just a portion of all those that may be used. CoDa analysis, however, is predicated on the basic notion of sub-compositional coherence, which ensures that a compositional study conducted on a subset of components is consistent with the same analysis performed on the entire composition.

1.5. Kaniadakis' Logarithm

We summarize the particular case of Kaniadakis' logarithm with a purely algebraic form. In the suggestive formalism introduced by [22], the generalised logarithms are associated with the reciprocal derivative function A

$$\log_\kappa(x) = \frac{1}{2}\left(x - \frac{1}{x}\right) = \int_1^x \frac{du}{A(u)}, \quad \textit{with} \quad A(u) = \frac{2u^2}{1+u^2} \tag{1}$$

$$\exp_\kappa(y) = \log_\kappa^{-1} y = y + \sqrt{1+y^2} = \exp \int_0^y \frac{dv}{\sqrt{1+v^2}}, \quad x^2 - 2xy - 1 = 0\,, \quad x > 0\,.$$

Notice that the growth is linear in both directions.

Notice that the above equation reduces any polynomial in y and $x = \exp_\kappa(y)$, for example:

$$\exp_\kappa(y)^2 = 2y\exp_\kappa(y) + 1\,,$$
$$\exp_\kappa(y)^3 = \exp_\kappa(y)(2y\exp_\kappa(y) + 1) = \ldots,$$

and so on. This is an algebraic feature, and this theory is a case of algebraic statistics [23].

The main known properties of the $\kappa-$logarithm and $\kappa-$exponential are

$$\log_\kappa\left(\frac{1}{x}\right) = -\log_\kappa(x)\,, \quad \exp_\kappa(y)\exp_\kappa(-y) = 1$$

$$\frac{d}{dy}\exp_\kappa(y) = A(\exp_\kappa(y)) = \frac{1}{\sqrt{1+y^2}}\exp_\kappa(y)$$

$$\exp_\kappa(x)\exp_\kappa(y) = \exp_\kappa\left(x\sqrt{1+y^2} + y\sqrt{1+x^2}\right) \tag{2}$$

$$\log_\kappa(x_2) - \log_\kappa(x_1) < \frac{1}{A(x_1)}(x_2 - x_1)\,, \quad x_1 \neq x_2 \tag{3}$$

$$\exp_\kappa(y_2) - \exp_\kappa(y_1) > A(\exp_\kappa(y_1))(y_2 - y_1)\,, \quad y_1 \neq y_2 \tag{4}$$

1.6. Kaniadakis' Exponential form of a Positive Probability Function

If the sample space Ω is a finite set, then the probability simplex on Ω is $\mathcal{P}(\Omega)$, and the open probability simplex is $\mathcal{E}(\Omega)$.

For all $p \in \mathcal{E}(\Omega)$, the function $A \circ p$, A as in Equation (1), is strictly positive and provides a positive weight on Ω. It is proportional, but usually not equal, to a probability function,

$$\widetilde{p} \propto A \circ p\,, \quad \text{namely,} \quad \widetilde{p}(x) = \frac{A(p(x))}{\sum_{x\in\Omega} A(p(x))}\,.$$

We will also write $\widetilde{p} = A \circ p / \overline{A \circ p}$. The mapping $\mathcal{E}(\Omega) \ni p \mapsto \widetilde{p} \in \mathcal{E}(\Omega)$ is called the *escort* mapping; see [22]. See ([24] §3.1) for a discussion of its injectivity and surjectivity. We introduce a notation for the escort expectation, $\widetilde{\mathbb{E}}_p[u] = \sum_{x\in\Omega} u(x)\widetilde{p}(x)$.

For $p, q \in \mathcal{E}(\Omega)$, the *Kaniadakis divergence* can be defined by changes in the usual definition of the logarithm to the Kaniadakis logarithm and the probability function p with the escort $\widetilde{p}$:

$$\begin{aligned}\widetilde{D}(p|q) &= \widetilde{\mathbb{E}}_p[\log_\kappa(p) - \log_\kappa(q)] \\ &= \sum_{x\in\Omega} (\log_\kappa(p) - \log_\kappa(q))\, A \circ p / \overline{A \circ p} \\ &= \sum_{x\in\Omega} (\log_\kappa(p) - \log_\kappa(q))\widetilde{p}(x)\end{aligned} \tag{5}$$

Clearly, $\widetilde{D}(p|p) = 0$. If $p \neq q$, from the concavity in Equation (3),

$$\begin{aligned}
\widetilde{D}(p|q) &= \left(\overline{A \circ p}\right)^{-1} \sum_{x\in\Omega} A(p(x))(\log_\kappa(p(x)) - \log_\kappa(q(x))) \\
&> \left(\overline{A \circ p}\right)^{-1} \sum_{x\in\Omega} (p(x) - q(x)) \\
&= 0\,.
\end{aligned}$$

Fix $p \in \mathcal{E}(\Omega)$. For all $q \in \mathcal{E}(\Omega)$, define

$$s_p(q) = (\log_\kappa(q) - \log_\kappa(p)) + \widetilde{D}(p|q) = (\log_\kappa(q) - \log_\kappa(p)) - \widetilde{\mathbb{E}}_p[(\log_\kappa(q) - \log_\kappa(p))]\,,$$

then, for $u = s_p(q)$,

$$q = \exp_\kappa\left(u - \widetilde{D}(p|q) + \log_\kappa(p)\right), \quad \widetilde{\mathbb{E}}_p[u] = 0\,. \tag{6}$$

Conversely, for all $p \in \mathcal{E}(\Omega)$, if u is a random variable such that $\widetilde{\mathbb{E}}_p[u] = 0$, the real function

$$\mathbb{R}_+ \ni \psi \mapsto \sum_{x\in\Omega} \exp_\kappa(u(x) - \psi + \log_\kappa(p(x)))$$

is continuous, goes to 0 as $\psi \to \infty$, and, for $\psi = 0$, takes a value larger than 1 because of Equation (4):

$$\begin{aligned}
\sum_{x\in\Omega} \exp_\kappa(u(x) + \log_\kappa(p(x))) &> \sum_{x\in\Omega} \exp_\kappa(\log_\kappa(p(x))) + \sum_{x\in\Omega} A(\exp_\kappa(\log_\kappa(p(x)))u(x) \\
&= \sum_{x\in\Omega} p(x) + \sum_{x\in\Omega} A(p(x))u(x) \\
&= 1\,.
\end{aligned}$$

In conclusion, there exists a function,

$$K_p \colon S_p = \left\{u \in L(\Omega) \,\middle|\, \widetilde{\mathbb{E}}_p[u] = 0\right\} \ni u \mapsto K_p(u) \geq 0\,,$$

and $K_p(u) > 0$ provided $u \neq 0$, such that

$$q = \exp_\kappa(u - K_p(u) + \log_\kappa(p)) \in \mathcal{E}(\Omega)\,.$$

Hence, we have

$$K_p(u) = \widetilde{D}(p|q)$$

and the mapping

$$s_p \colon \mathcal{E}(\Omega) \to S_p$$

is a bijection with inverse

$$e_p \colon S_p \ni u \mapsto \exp_\kappa(u - K_p(u) + \log_\kappa(p)) \in \mathcal{E}(\Omega)\,. \tag{7}$$

1.7. Properties of the Cumulant Function K_p

Let us compute the derivatives of the function K_p. We use a square bracket notation for the direction

$$dK_p(u)[h] = \lim_{\theta\to 0} \theta^{-1}(K_p(u + \theta h) - K_p(u)) = \left.\frac{d}{d\theta} K_p(u + \theta h)\right|_{\theta=0}\,.$$

From Equation (7),

$$\begin{aligned}
0 &= \frac{d}{d\theta} \sum_{x \in \Omega} \exp_\kappa(u(x) + \theta h(x) - K_p(u+\theta h) + \log_\kappa(p(x)))\Big|_{\theta=0} \\
&= \sum_{x\in\Omega} A(\exp_\kappa(u(x) + \theta h(x) - K_p(u+\theta h) + \log_\kappa(p(x))))(h(x) - \frac{d}{dt}K_p(u+\theta h))\Big|_{\theta=0} \\
&= \sum_{x\in\Omega} A(\exp_\kappa(u(x) - K_p(u) + \log_\kappa(p(x))))(h(x) - dK_p(u)[h]) \\
&= \sum_{x\in\Omega} A(e_p(u))(h - dK_p(u)[h])\,.
\end{aligned}$$

It follows that, for each $p \in \mathcal{E}(\Omega)$ and $u, h \in S_p$, it holds

$$dK_p(u)[h] = \widetilde{\mathbb{E}}_q[h]\,, \tag{8}$$

where $q = e_p(u)$; see Equation (7).

If the curve $t \mapsto q(t)$ has constant divergence, that is, $\widetilde{D}(p|q(t)) = \widetilde{D}(p|q(0))$, derivation provides

$$0 = \frac{d}{dt}\widetilde{D}(p|q(t)) = \frac{d}{dt}K_p(u(t)) = dK_p(u(t))[\dot{u}(t)] = \widetilde{\mathbb{E}}_{q(t)}[\dot{u}(t)]\,.$$

Notice that $\widetilde{\mathbb{E}}_{q(0)}[u(t)] = 0$, but this does not imply $\widetilde{\mathbb{E}}_{q(t)}[\dot{u}(t)]$ unless the previous conditions hold true.

1.8. Bibliographical Notes

Similarly, d^2K_p and the convex conjugate of K_p can be computed. See below for the duality and see also [7,25]. Kaniadakis logarithm and exponential were first introduced in [26,27]. The application to IG used here appeared in [6,24,28]. These papers discuss both the finite state space and the general state space.

2. Affine Space

The Kaniadakis non-parametric affine geometry of the open probability simplex is a variation of the standard case [7]. The main difference is the substitution of the expectation with the escort expectation.

2.1. Statistical Bundle

The statistical bundle is an expression of the tangent space of $\mathcal{E}(\Omega)$ as a dually flat affine statistical manifold in the sense of [5]. The statistical bundle $S\,\mathcal{E}(\Omega)$ and each fiber $S_q\,\mathcal{E}(\Omega)$ are defined by

$$S\,\mathcal{E}(\Omega) = \left\{(q,v)\,\middle|\, q \in \mathcal{E}(\Omega), \widetilde{\mathbb{E}}_q[v] = 0\right\}, \tag{9}$$

$$S_q\,\mathcal{E}(\Omega) = \left\{v \in L(\Omega)\,\middle|\, \widetilde{\mathbb{E}}_q[u] = 0\right\}, \quad q \in \mathcal{E}(\Omega)\,. \tag{10}$$

In our setup, each fibre is a finite-dimensional vector space and can be identified with its dual. However, it is convenient to distinguish the two statistical bundles. The previous one is called *exponential* statistical bundle, while the *mixture* statistical bundle is

$$^*S\,\mathcal{E}(\Omega) = \left\{(q,v)\,\middle|\, q \in \mathcal{E}(\Omega), \widetilde{\mathbb{E}}_q[v] = 0\right\}, \tag{11}$$

$$^*S_q\,\mathcal{E}(\Omega) = \left\{v \in L(\Omega)\,\middle|\, \widetilde{\mathbb{E}}_q[u] = 0\right\}, \quad q \in \mathcal{E}(\Omega)\,. \tag{12}$$

For each couple $p, q \in \mathcal{E}(\Omega)$, the mapping

$$^{e}\mathbb{U}_p^q \colon S_p\,\mathcal{E}(\Omega) \ni v \mapsto v - \widetilde{\mathbb{E}}_q[v] \in S_q\,\mathcal{E}(\Omega) \tag{13}$$

is a bijection. The ${}^{\mathrm{e}}\mathbb{U}_p^p$ is the identity mapping, and

$$ {}^{\mathrm{e}}\mathbb{U}_q^r \, {}^{\mathrm{e}}\mathbb{U}_p^q = {}^{\mathrm{e}}\mathbb{U}_p^r \, . $$

The co-cycle of mappings $({}^{\mathrm{e}}\mathbb{U}_p^q)_{p,q}$ is the *exponential parallel transport* of the exponential statistical bundle.

The mapping defined for all $p \in \mathcal{E}(\Omega)$, $v \in {}^*S_p\,\mathcal{E}(\Omega)$, and $w \in S_p\,\mathcal{E}(\Omega)$ by

$$ g\colon (p,v,w) \mapsto g_p(v,w) = \langle v,w\rangle_{A\circ p} = \sum_{x\in\Omega} A(p(x))u(x)v(x) = \overline{A\circ p}\,\widetilde{\mathbb{E}}_p[vw] $$

provides a duality between the fibres of $S\,\mathcal{E}(\Omega)$ and ${}^*S\,\mathcal{E}(\Omega)$.

The dual of the exponential transport can be computed as follows. For $p,q \in \mathcal{E}(\Omega)$, $v \in {}^*S_q\,\mathcal{E}(\Omega)$, and $w \in S_p\,\mathcal{E}(\Omega)$,

$$ \begin{aligned} \left\langle v, {}^{\mathrm{e}}\mathbb{U}_p^q w \right\rangle_{A\circ q} &= \sum_{x\in\Omega} A\circ q\, v(w - \widetilde{\mathbb{E}}_q[w]) \\ &= \sum_{x\in\Omega} A\circ q\, vw - \widetilde{\mathbb{E}}_q[w] \sum A\circ qv \\ &= \sum_{x\in\Omega} A\circ q\, vw \\ &= \sum_{x\in\Omega} A\circ p \left(\frac{A\circ q}{A\circ p}u\right) w \\ &= \left\langle \frac{A\circ q}{A\circ p}u, w \right\rangle_{A\circ p} . \end{aligned} $$

Now, $\frac{A\circ q}{A\circ p}v \in {}^*S_p\,\mathcal{E}(\Omega)$; hence, the dual of the exponential transport is the *mixture transport*,

$$ {}^*\left({}^{\mathrm{e}}\mathbb{U}_q^p\right)v = {}^{\mathrm{m}}\mathbb{U}_q^p v = \frac{A\circ q}{A\circ p}v \, . \tag{14} $$

2.2. Velocity and Auto-Parallel Curves

The following computation is a version of the original argument about Fisher's score. Let $t \mapsto q(t) \in \mathcal{E}(\Omega)$ be a one-dimensional parametric statistical model, namely a curve in geometric language. We assume the curve is smooth and twice differentiable as a mapping in the vector space $L(\Omega)$. For each random variable $f \in L(\Omega)$,

$$ \begin{aligned} \frac{d}{dt}\mathbb{E}_{q(t)}[f] &= \sum_{x\in\Omega} f(x)\dot{q}(x;t) \\ &= \sum_{x\in\Omega} f(x)\frac{\dot{q}(x;t)}{A(q(x;t))}A(q(x;t)) \\ &= \sum_{x\in\Omega} f(x)\frac{d}{dt}\log_\kappa(q(x;t)A(q(x;t)) \\ &= \sum_{x\in\Omega} A\circ q(t)(f - \mathbb{E}_{\widetilde{q}(t)}[f])\frac{d}{dt}\log_\kappa(q(t)) \\ &= \left\langle f - \mathbb{E}_{\widetilde{q}(t)}[f], \frac{d}{dt}\log_\kappa(q(t)) \right\rangle_{A\circ q(t)} \\ &= \sum_{x\in\Omega} f(x)\dot{q}(x;t) \\ &= \left\langle f - \mathbb{E}_{\widetilde{q}(t)}[f], \frac{d}{dt}\log_\kappa(q(t)) \right\rangle_{A\circ q(t)} \end{aligned} \tag{15} $$

The *velocity* of the curve is defined as

$$\mathring{q}(t) = \frac{d}{dt}\log_\kappa(q(t)) = \frac{\dot{q}(t)}{A(q(t))} . \tag{16}$$

We can check that $\mathring{q}(t) \in S_{q(t)}\mathcal{E}(\Omega)$ and $(f - \mathbb{E}_{\widetilde{q}(t)}[f]) \in {}^*S_{q(t)}\mathcal{E}(\Omega)$. The *Cramer–Rao bound* is

$$\begin{aligned}\left(\frac{d}{dt}\mathbb{E}_{q(t)}[f]\right)^2 &= \overline{A \circ q(t)}^2 \left\langle f - \mathbb{E}_{\widetilde{q}(t)}[f], \frac{d}{dt}\log_\kappa(q(t))\right\rangle_{\widetilde{q}(t)} \\ &\leq \overline{A \circ q(t)}^2\, \mathbb{E}_{\widetilde{q}(t)}\left[(f - \mathbb{E}_{\widetilde{q}(t)}[f])^2\right] \mathbb{E}_{\widetilde{q}(t)}\left[\left(\frac{\dot{q}(t)}{A \circ q(t)}\right)^2\right] \\ &= \sum_{x\in\Omega} A \circ q(t)(f - \widetilde{\mathbb{E}}_q[f])^2 \sum \frac{\dot{q}(t)^2}{A \circ q(t)} \\ &= \sum_{x\in\Omega} A \circ q(t)(f - \widetilde{\mathbb{E}}_q[f])^2 \sum \frac{\dot{q}(t)^2}{A \circ q(t)} \end{aligned} \tag{17}$$

The variation computed with the escort probability function, namely $f - \widetilde{\mathbb{E}}_q[f]$, appears in Equation (15) as a gradient of the expectation $f \mapsto \mathbb{E}_q[f]$.

A curve $t \mapsto q(t)$ is *auto-parallel* for the mixture trasport if

$${}^{\mathrm{m}}\mathbb{U}_{q(t)}^{q(s)}\mathring{q}(t) = \mathring{q}(s) .$$

For $q(0) = q_0$ and $q(1) = q_1$,

$$\frac{\dot{q}(0)}{A \circ q_0} = \frac{A \circ q(t)}{A \circ q_0}\frac{\dot{q}(t)}{A \circ q(t)} = \frac{1}{A \circ q_0}\dot{q}(t) ,$$

so that

$$q(t) = q_0 + t(q_1 - q_0) .$$

Let us compute the auto-parallel curves for the exponential transport,

$${}^{\mathrm{e}}\mathbb{U}_{q(t)}^{q(s)}\mathring{q}(t) = \mathring{q}(s) .$$

For $q(0) = q_0$ and $q(1) = q_1$,

$$\mathring{q}(0) = \mathring{q}(t) - \widetilde{\mathbb{E}}_{q_0}\left[\mathring{q}(t)\right] = \frac{d}{dt}\log_\kappa(q(t)) - \frac{1}{\overline{A \circ q_0}}\sum A \circ q_0 \frac{d}{dt}\log_\kappa(q(t)) ,$$

so that, for some function ψ,

$$\log_\kappa(q(t)) = \log_\kappa(q_0) + t\mathring{q}(0) - \psi(t) .$$

Comparing with Equation (7), we have that *the auto-parallel curve for the exponential transport is*

$$q(t) = \exp_\kappa(t\mathring{q}(0) - K_{q_0}(t\mathring{q}(0)) + \log_\kappa(q_0)) . \tag{18}$$

As observed above,

$$\mathring{q}(0) = \log_\kappa(q_1) - \log_\kappa(q_0) - \widetilde{\mathbb{E}}_{q_0}[\log_\kappa(q_1) - \log_\kappa(q_0)] = \log_\kappa(q_0) + \widetilde{D}(q_0|q_1) .$$

2.3. Surfaces of Constant Divergence

We have observed that an auto-parallel curve starting at $q(0)$ with velocity $\mathring{q}(0)$ has the form of Equation (18). For given extreme points $q_0 = q(0)$ and $q_1 = q(1)$, it holds that

$$q_1 = \exp_\kappa(u - K_{q_0}(u) + \log_\kappa(q_0)) \quad \text{hence} \quad u = \log_\kappa(q(0)) - \log_\kappa(q(1)) + \widetilde{D}(q_0|q_1)\,,$$

in particular, $\widetilde{\mathbb{E}}_{q_1}[u] = 0$.

The velocity of the auto-parallel curve at q_1 is constant,

$$\frac{d}{dt}(tu - K_{q_0}(tu) + \log_\kappa(q_0))\bigg|_{t=1} = u - \widetilde{\mathbb{E}}_{q_1}[u] = u.$$

Consider a curve γ starting at $\gamma(0) = q_1$ of the form,

$$t \mapsto \gamma(t) = \exp_\kappa(u + v(t) - K_{q_0}(u + v(t)) + \log_\kappa(q_0)) \quad \text{with } v(0) = 0$$

and assume a divergence is constant, precisely

$$\widetilde{D}(\gamma(t)|q_0) = \widetilde{D}(\gamma(0)|q_0) = \widetilde{D}(q_1|q_0)\,.$$

It holds

$$\begin{aligned}\widetilde{\mathbb{E}}_{\gamma(t)}[\log_\kappa(\gamma(t)) - \log_\kappa(q_0)] &= \widetilde{\mathbb{E}}_{\gamma(t)}\left[u + v(t) - K_{q_0}(u + v(t)) + \log_\kappa(q_0) - \log_\kappa(q_0)\right] \\ &= \widetilde{\mathbb{E}}_{\gamma(t)}[u + v(t)] - K_{q_0}(u + v(t)) \\ &= dK_{q(0)}(u + v(t))[u + v(t)] - K_{q_0}(u + v(t))\end{aligned}$$

is constant so that the derivative is zero. In particular, it is zero at $t = 0$

$$0 = d^2K_{q_0}[u + v(t)][u + v(t), \dot{v}(t)]\bigg|_{t=0} = d^2K_{q_0}[u, \dot{v}(0)]\,.$$

That is, this surface of equi-divergence is orthogonal to the auto-parallel curves in the sense of the quadratic form d^2K. This is actually the generalization of a well-known result in IG, where the Hessian of the cumulant function is the Fisher's information matrix. See, for example, [5].

2.4. Displacement

The machinery introduced above allows for explicitly defining the affine structure as originally defined by [29]. A textbook on affine geometry is ([30] Ch. 2,3,9). Below, we call the following two (dual) displacements on the statistical bundle. The mixture displacement is

$$\eta_p(q) = \frac{q - p}{A \circ p}\,. \tag{19}$$

The exponential displacement is

$$s_p(q) = (\log_\kappa(q) - \log_\kappa(p)) - \widetilde{\mathbb{E}}_p[(\log_\kappa(q) - \log_\kappa(p))] \tag{20}$$

Both displacements define affine coordinates in the statistical bundle. The easy proofs are the same as in the standard cases [7]. Each displacement defines an atlas of charts on the affine bundle.

The orthogonal surfaces of the affine exponential auto-parallel curves are discussed in the section above. The orthogonal surfaces to the affine mixture auto-parallel curves are easily observed to be associated with the other divergence. In fact, it is the classical result of the duality between the two divergences. See, for example, [5].

The availability of an affine bundle would allow for a coherent and straightforward definition of mechanical concepts such as velocity, acceleration, Lagrangian, and Hamiltonian. See [31,32] for the standard case. In the present paper, we develop the application to CoDa, and we stress the notion of affine barycenter and the fact that a system of charts can be observed as a preprocessing of data to be followed by any method adapted to actual vector data.

2.5. Barycenter and Deviation

Let $f_1, \ldots, f_n$ be a sequence of CoDa points with strictly positive components and normalized to one. Each data point is a point in the open probability simplex. The affine coordinates (20) centered at p are

$$\begin{aligned} s_p(f_1) &= (\log_\kappa(f_1) - \log_\kappa(p)) - \widetilde{\mathbb{E}}_p[(\log_\kappa(f_1) - \log_\kappa(p))] \\ &\vdots \\ s_p(f_n) &= (\log_\kappa(f_n) - \log_\kappa(p)) - \widetilde{\mathbb{E}}_p[(\log_\kappa(f_n) - \log_\kappa(p))] \end{aligned}$$

The *mean value* of the affine coordinates is

$$\overline{s_p} = \frac{1}{n}\sum_{j=1}^{n} s_p(f_j) = \frac{1}{n}\sum_{j=1}^{n}\Big(\log_\kappa(f_j) - \widetilde{\mathbb{E}}_p\left[(\log_\kappa(f_j))\right]\Big) - \Big(\log_\kappa(p) - \widetilde{\mathbb{E}}_p[\log_\kappa(p)]\Big) \tag{21}$$

If the mean value computed in the centering q is $\overline{s_q}$, the difference is

$$\overline{s_p} - \overline{s_q} = \log_\kappa(q) - \log_\kappa(p) - \frac{1}{n}\sum_{j=1}^{n}\widetilde{\mathbb{E}}_p\left[\log_\kappa(f_j)\right] + \frac{1}{n}\sum_{j=1}^{n}\widetilde{\mathbb{E}}_q\left[\log_\kappa(f_j)\right].$$

Hence,

$$\overline{s_p} + \log_\kappa(p) = \overline{s_q} + \log_\kappa(q) + \text{constant}.$$

The probability function is the same in both cases. In fact,

$$\begin{aligned} \exp_\kappa(\overline{s_p} - K_p(\overline{s_p}) + \log_\kappa(p)) &= \exp_\kappa(\overline{s_q} - K_p(\overline{s_p}) + \log_\kappa(q) + \text{constant}) \\ &= \exp_\kappa(\overline{s_q} - K_q(\overline{s_q}) + \log_\kappa(q)) \end{aligned}$$

because of the uniqueness of the normalizing constant.

In conclusion, the probability function

$$\begin{aligned} \overline{f} &= \exp_\kappa(\overline{s_p} - K_p(\overline{s_p}) + \log_\kappa(p)) \\ &= \exp_\kappa\left(\overline{s_p} + \widetilde{D}\left(p \middle| \overline{f}\right) + \log_\kappa(p)\right), \end{aligned}$$

with $\overline{s_p}$ as Equation (21) *does not depend on the reference p*. It is the *barycentre* of the given data points.

The displacement of each data point f_j from the barycentre $\overline{f}$ is

$$s_{\overline{f}}(f_j) = \overline{s_p} - s_p(f_j) - \widetilde{\mathbb{E}}_{\overline{f}}[\overline{s_p} - s_p(f_j)] = \frac{1}{n}\sum_k s_p(f_k) - s_p(f_j) - \widetilde{\mathbb{E}}_{\overline{f}}\left[\frac{1}{n}\sum_k s_p(f_k) - s_p(f_j)\right]$$

and the expression of each point f_j in the barycentre $\overline{f}$ is

$$f_j = \exp_\kappa\left(s_{\overline{f}}(f_j) + \widetilde{D}\left(\overline{f} \middle| f_j\right) + \log_\kappa\left(\overline{f}\right)\right).$$

A one-dimensional summary consistent with our formalism of the divergence of each point from the barycentre is the Kaniadakis' divergence $\widetilde{D}\left(\overline{f} \middle| f_j\right)$, which is the normalising constant in the equation above. Another option is the Kaniadakis' divergence $\widetilde{D}\left(f_j \middle| \overline{f}\right)$ that appears in the representation of the barycentre in the data point f_j.

3. Data Analysis

This section will use some geometric concepts derived from Kaniadakis' IG. It should be noted that our formalism is, in principle, affine and does not include any properly defined distance.

3.1. Kaniadakis Divergence

First, we compute the Kaniadakis divergence defined in Equation (5). Each point in (i, j) in Figure 1 is the Kaniadakis divergence of the CoDa point corresponding to the year in the i-th row with respect to the CoDa point for the year in the j-th column. For example, the Kaniadakis divergence between 2008 and 2009 is $\tilde{D}(2008|2009) = 0.14$. Most values are smaller than one, except when the reference distribution corresponds to 2008 or 2009 for the most recent years. The year 2009 deviates significantly from the other years.

3.2. Mixture Displacement

Equation (19) provides instructions for computing the mixture displacement. From Figure 2, the mixture displacement for Greece and Spain is very high for all the years. The value for Spain in 2009 was less than zero—the only negative value for Spain. On the contrary, all other countries do not have too many high values.

Equation (21) provides the mean value. After determining the mean, we compute the mixture displacement using the mean as a reference. We check that our values abruptly go from −10 to 10. However, the results for Greece and Spain decrease when the mixture displacement from the mean is calculated.

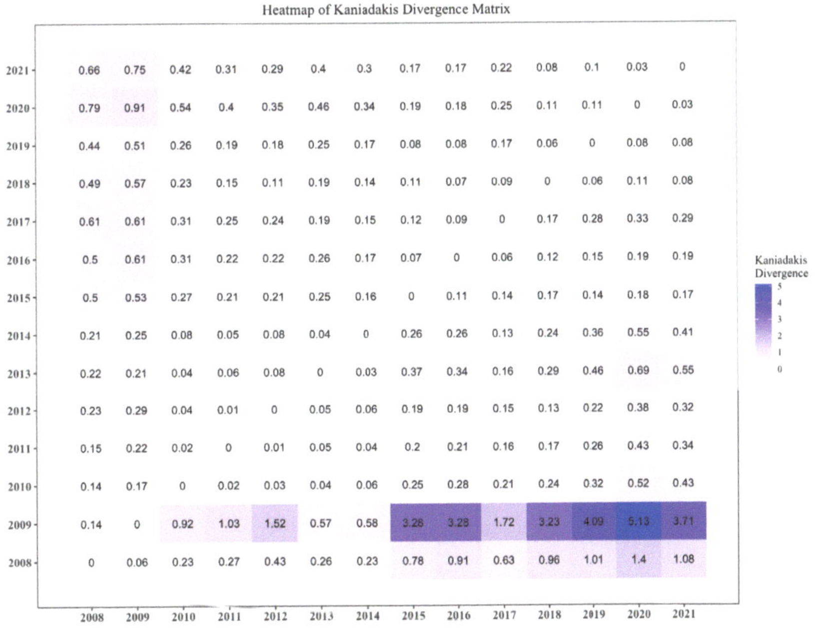

Figure 1. Kaniadakis divergence on compositional data.

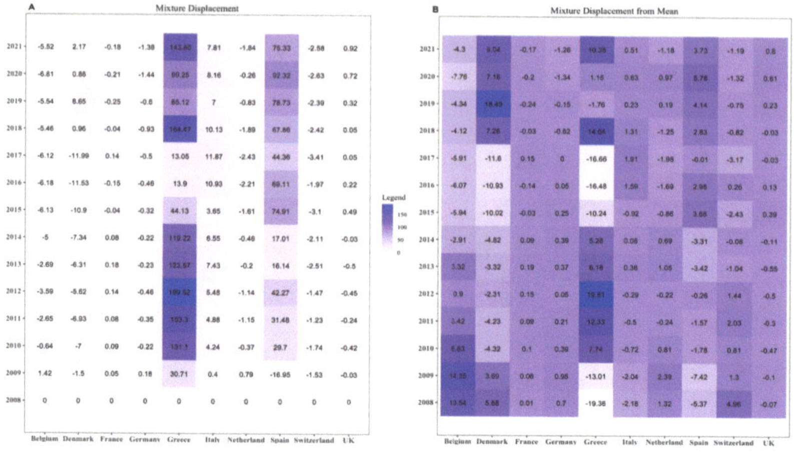

Figure 2. (**A**) Mixture displacement on compositional data by taking 2008 as a reference and (**B**) mixture displacement on compositional data by taking mean as reference.

3.3. *Exponential Displacement*

As above, Equation (20) returns the exponential displacement. Further, Figure 3 is the empirical result of Equation (20). As for the mixture displacement, we can see that Spain and Greece have higher displacement than other European countries. The value for Spain in the year 2009 is meagre.

If the mean is the reference point, the exponential displacement ranges from 0 to -60. The only significant changes are for the nations of Spain and Greece, where our values for 2009 for Spain decreased by about 18 times, and, for Greece, our values decreased significantly.

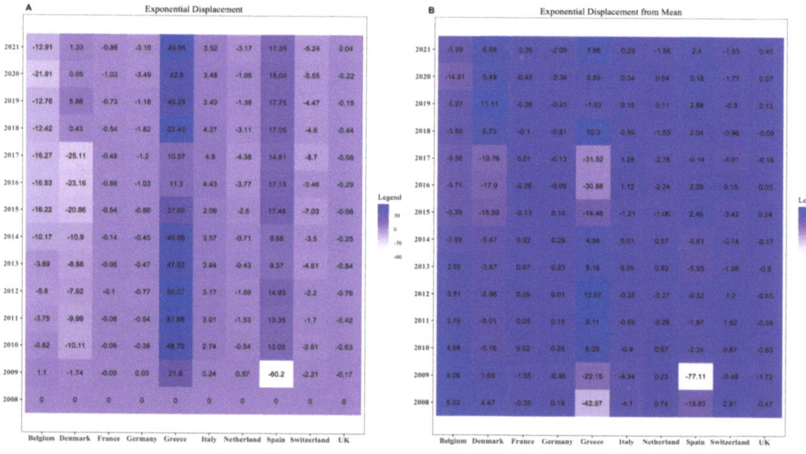

Figure 3. (**A**) Exponential displacement on compositional data by taking 2008 as a reference and (**B**) exponential displacement on compositional data by taking mean as a reference.

4. Conclusions and Discussion

In this research, we applied a particular type of divergence, Kaniadakis divergence, to compositional data, aligned with the symmetrised ratio transformation in ([8] Example 4.20). The dataset being examined spans the years 2008 through 2021. First, we built a theoretical framework for Kaniadakis divergence, mixture displacement, and exponential displacement.

Section 1 provided the mathematical framework for determining divergence and displacement, while Section 2 demonstrated how to apply those mathematical algorithms to compositional data. In the application, we found that Spain and Greece have more fluctuations when compared to the other European countries. The values of the mixture and exponential displacement confirm that Spain and Greece faced some financial crises compared to other countries.

This simple application shows the potential of IG for application to compositional data analysis. We suggest that Kaniadakis' logarithm can reduce the computations for monitoring systemic risk to algebraic computations. The Kaniadakis logarithm, mixture, and exponential displacement on compositional data can be considered to broaden traditional research methods for compositional data analysis.

We would like to add a few words regarding the specific tools and formalism we used here. First, we mimicked one of the possible presentations of non-parametric IG by following the basic dually-flat setup step by step. Another successful presentation of non-parametric IG starts with properly defining the divergences and deriving the geometry; see, for example, [33]. A popular approach, not equivalent to the affine one, defines the geometry of the probability simplex by introducing a metric tensor. As in other geometric theories, one should carefully distinguish between choosing charts and introducing a topology.

In the present approach, we define the charts so that the associated manifold is affine; in this setup, some specific divergences appear as naturally associated with the geometry and the basic statistical notion, namely the pairing between measures and random variables. Everything is applied to simple data manipulation in the spirit of Aitchison's methods.

No claim of optimality is made. The existence of many different but topologically equivalent divergences is only natural in our setup, where the topology actually depends on the geometry and not the other way around. Whenever needed, a choice must be based on some additional assumption. We carefully check the simple, useful operations on data, such that the geodesic connecting two given points, the velocity of variation, the barycentre, and the deviation from the barycentre are all defined correctly.

Author Contributions: Writing—review & editing, G.P. and M.S. All authors contributed equally to this work. All authors have read and agreed to the published version of the manuscript.

Funding: This project has received funding from the European Union's Horizon 2020 research and innovation programme under Marie Skłodowska-Curie GA No. 101034449.

Institutional Review Board Statement: Not applicable.

Informed Consent Statement: Not applicable.

Data Availability Statement: Data are publicly available on the website (http://www.crml.ch, accessed on 1 May 2023).

Acknowledgments: G.P. enjoys the support of de Castro Statistics, Collegio Carlo Alberto, and is a member of INDAM/GNAFA. Both G.P. and M.S. thank Francesco Porro (DIMA U. Genova) for providing and explaining the datasets. They also thank Eva Riccomagno (DIMA U. Genova) for carefully reading and providing advice on the final version of this paper.

Conflicts of Interest: The authors declare no conflict of interest.

Abbreviations

CoDa	Compositional Data
SRISK	Systemic Risk
IG	Information Geometry

References

1. Kaniadakis, G. Non-linear kinetics underlying generalized statistics. *Phys. A* **2001**, *296*, 405–425. [CrossRef]
2. Kaniadakis, G. H-theorem and generalized entropies within the framework of nonlinear kinetics. *Phys. Lett. A* **2001**, *288*, 283–291. [CrossRef]
3. Tsallis, C. Possible generalization of Boltzmann-Gibbs statistics. *J. Statist. Phys.* **1988**, *52*, 479–487. [CrossRef]
4. Tsallis, C. Introduction to nonextensive statistical mechanics. In *Approaching a Complex World*; Springer: New York, NY, USA, 2009; p. xvii+382.

5. Amari, S.; Nagaoka, H. *Methods of Information Geometry*; American Mathematical Society: Providence, RI, USA, 2000; p. x+206. (Translated from the 1993 Japanese original by Daishi Harada)
6. Pistone, G. κ-exponential models from the geometrical viewpoint. *Eur. Phys. J. B Condens. Matter Phys.* **2009**, *71*, 29–37. [CrossRef]
7. Chirco, G.; Pistone, G. Dually affine Information Geometry modeled on a Banach space. *arXiv* **2022**, arXiv:2204.00917.
8. Pawlowsky-Glahn, V.; Egozcue, J.J.; Tolosana-Delgado, R. *Modelling and Analysis of Compositional Data*; John Wiley & Sons, Ltd.: Hoboken, NJ, USA, 2015. [CrossRef]
9. Pawlowsky-Glahn, V.; Egozcue, J.J. Compositional data and their analysis: An introduction. *Geol. Soc. Lond. Spec. Publ.* **2006**, *264*, 1–10. [CrossRef]
10. Egozcue, J.J.; Pawlowsky-Glahn, V. Compositional data: The sample space and its structure. *Test* **2019**, *28*, 599–638. [CrossRef]
11. Coenders, G.; Ferrer-Rosell, B. Compositional data analysis in tourism: Review and future directions. *Tour. Anal.* **2020**, *25*, 153–168. [CrossRef]
12. Fiori, A.M.; Porro, F. A compositional analysis of systemic risk in European financial institutions. *Ann. Financ.* **2023**, *19*, 1–30. [CrossRef]
13. Boonen, T.J.; Guillen, M.; Santolino, M. Forecasting compositional risk allocations. *Insur. Math. Econ.* **2019**, *84*, 79–86. [CrossRef]
14. Grifoll, M.; Ortego, M.; Egozcue, J. Compositional data techniques for the analysis of the container traffic share in a multi-port region. *Eur. Transp. Res. Rev.* **2019**, *11*, 1–15. [CrossRef]
15. Linares-Mustarós, S.; Coenders, G.; Vives-Mestres, M. Financial performance and distress profiles. From classification according to financial ratios to compositional classification. *Adv. Account.* **2018**, *40*, 1–10. [CrossRef]
16. Acharya, V.; Engle, R.; Richardson, M. Capital shortfall: A new approach to ranking and regulating systemic risks. *Am. Econ. Rev.* **2012**, *102*, 59–64. [CrossRef]
17. Acharya, V.V.; Richardson, M.P. (Eds.) *Restoring Financial Stability: How to Repair a Failed System*; John Wiley & Sons, Ltd.: Hoboken, NJ, USA, 2009; Volume 542.
18. Engle, R. Systemic risk 10 years later. *Annu. Rev. Financ. Econ.* **2018**, *10*, 125–152. [CrossRef]
19. Stolbov, M.; Shchepeleva, M. Systemic risk in Europe: Deciphering leading measures, common patterns and real effects. *Ann. Financ.* **2018**, *14*, 49–91. [CrossRef]
20. Engle, R.; Jondeau, E.; Rockinger, M. Systemic risk in Europe. *Rev. Financ.* **2015**, *19*, 145–190. [CrossRef]
21. Aitchison, J. *The Statistical Analysis of Compositional Data*; Monographs on Statistics and Applied Probability; Chapman & Hall: London, UK, 1986; p. xvi+416. [CrossRef]
22. Naudts, J. Generalised exponential families and associated entropy functions. *Entropy* **2008**, *10*, 131–149. [CrossRef]
23. Pistone, G.; Riccomagno, E.; Wynn, H.P. *Algebraic Statistics: Computational Commutative Algebra in Statistics*; Monographs on Statistics and Applied Probability; Chapman & Hall/CRC: Boca Raton, FL, USA, 2001; Volume 89, p. xvii+160.
24. Montrucchio, L.; Pistone, G. A Class of Non-parametric Deformed Exponential Statistical Models. In *Geometric Structures of Information*; Springer International Publishing: Berlin/Heidelberg, Germany, 2018; pp. 15–35. [CrossRef]
25. Pistone, G. Information Geometry of the Probability Simplex: A Short Course. *Nonlinear Phenom. Complex Syst.* **2020**, *23*, 221–242. [CrossRef]
26. Kaniadakis, G. Statistical mechanics in the context of special relativity. II. *Phys. Rev. E* **2005**, *72*, 036108. [CrossRef]
27. Kaniadakis, G. Statistical mechanics in the context of special relativity. *Phys. Rev. E* **2002**, *66*, 056125. [CrossRef]
28. Montrucchio, L.; Pistone, G. Deformed Exponential Bundle: The Linear Growth Case. In *Lecture Notes in Computer Science*; Springer International Publishing: Berlin/Heidelberg, Germany, 2017; pp. 239–246. [CrossRef]
29. Weyl, H. *Space—Time—Matter*; Dover: New York, NY, USA, 1952. (Translation of the 1921 RAUM ZEIT MATERIE)
30. Berger, M. *Geometry I*; Universitext, Springer: Berlin, Germany, 1994; p. xiv+427. (Translated from the 1977 French original by M. Cole and S. Levy, Corrected reprint of the 1987 translation)
31. Pistone, G. Lagrangian Function on the Finite State Space Statistical Bundle. *Entropy* **2018**, *20*, 139. [CrossRef] [PubMed]
32. Chirco, G.; Malagò, L.; Pistone, G. Lagrangian and Hamiltonian dynamics for probabilities on the statistical bundle. *Int. J. Geom. Methods Mod. Phys.* **2022**, *19*, 2250214. [CrossRef]
33. Eguchi, S. Second order efficiency of minimum contrast estimators in a curved exponential family. *Ann. Statist.* **1983**, *11*, 793–803. [CrossRef]

entropy

MDPI

Review

The Kaniadakis Distribution for the Analysis of Income and Wealth Data

Fabio Clementi

Department of Political Science, Communication and International Relations, University of Macerata, Via Don Minzoni 22/A, 62100 Macerata, Italy; fabio.clementi@unimc.it; Tel.: +39-0733-258-2560

Abstract: The paper reviews the "κ-generalized distribution", a statistical model for the analysis of income data. Basic analytical properties, interrelationships with other distributions, and standard measures of inequality such as the Gini index and the Lorenz curve are covered. An extension of the basic model that best fits wealth data is also discussed. The new and old empirical evidence presented in the article shows that the κ-generalized model of income/wealth is often in very good agreement with the observed data.

Keywords: income and wealth distribution; parametric modeling; κ-generalized model

1. Introduction

The past two decades have seen a resurgence of interest in the study of income and wealth distribution in both the physics [1–4] and economics [5–9] communities. Scholars have focused particularly on the empirical analysis of large data sets to infer the shape of income and wealth distributions and to develop theoretical models that can reproduce them.

Pareto's observation that the number of people in a population whose income exceeds x is often well approximated by $Cx^{-\alpha}$ was a natural starting point for this field of analysis [10–13]. However, empirical research has shown that the Pareto distribution accurately models only high income levels, while it does a poor job of describing the lower end of distributions.

As research has continued, new models have been proposed to better describe the data, using either a combination of known statistical distributions [14–22] or parametric functional forms for the distribution as a whole. Among these, the two-parameter lognormal [23] and gamma [24] distributions were proposed as models for the size distributions of income and wealth, but later evidence showed that these models tend to exaggerate skewness and perform poorly at the upper end of the empirical distributions [25–28]. Three-parameter models such as the generalized gamma [29–32], Singh–Maddala [33], and Dagum Type I [34] provide better fits. These models converge to the Pareto model for large values of income/wealth and accurately describe lower and middle ranges.

Finally, models with more than three parameters have also been suggested to fit income and wealth data. For example, the generalized beta distribution of the second kind (GB2) is a four-parameter distribution that was first described by [35]. It fits the data very well and also includes some of the two- and three-parameter models mentioned above as special or limiting cases. (The generalized beta distribution of the first kind (GB1) [35] and the double Pareto-lognormal distribution [36] are other four-parameter models that fit the data well. Ref. [37] also developed the five-parameter generalized beta distribution family, which includes the GB1 and GB2 as special cases and all of the two- and three-parameter distributions nested inside them. In turn, the double Pareto-lognormal distribution has been generalized into a five-parameter family of distributions called the generalized double Pareto-lognormal distribution [38]. However, closed-form expressions for probability density and/or cumulative distribution functions do not always exist for

Citation: Clementi, F. The Kaniadakis Distribution for the Analysis of Income and Wealth Data. *Entropy* **2023**, *25*, 1141. https://doi.org/10.3390/e25081141

Academic Editors: Antonio M. Scarfone, Dionissios T. Hristopulos and Sergio Luiz E. F. da Silva

Received: 27 June 2023
Revised: 27 July 2023
Accepted: 28 July 2023
Published: 30 July 2023

these "super" models, making fitting them to data computationally difficult and slow due to the need to use numerical methods [39,40]).

Among models that seek to provide a unified framework for describing real-world data, including the power-law tails found in empirical distributions of income and wealth, the κ-generalized distribution has demonstrated exceptional performance and is often seen as a better alternative to other widely used parametric models. This model, which was initially introduced in 2007 and progressively expanded in the years that followed, has its origins in the framework of κ-generalized statistical mechanics [41–46]. It has a bulk very similar to the Weibull distribution and an upper tail that decays according to a Pareto power law for high values of income and wealth, providing a sort of middle ground between the two descriptions.

The purpose of this paper is to provide a comprehensive overview of the important results concerning the κ-generalized distribution. The desire to celebrate the 20th anniversary of Kaniadakis' notable contribution and the belief that an interdisciplinary approach integrating statistical mechanics and economics may give novel insights into economic relationships motivated this work. Giorgio Kaniadakis played a pivotal role in the development of the κ-generalized model, making valuable and direct contributions to its conception. The intention behind presenting information on the fundamental statistical properties and empirical plausibility of this distribution is to convince the reader of its importance and usefulness for future exploration.

The paper is structured as follows. Section 2 introduces the κ-generalized model, covering topics such as interrelations with other distributions, basic statistical properties, and inferential aspects. Section 3 presents recent results of fitting the κ-generalized distribution to empirical income data corresponding to the distribution of household incomes in Greece, and compares the relative merits of alternative income size distribution models using the same data. Section 4 reviews empirical applications showing that the κ-generalized model is often in excellent agreement with observed income data; the κ-generalized mixture model for net worth distribution, which best fits wealth data, is also discussed in this section. Section 5 concludes the paper with some remarks.

2. The κ-Generalized Model for Income Distribution

The κ-generalized statistical model, named after [47], is based on the use of κ-deformed exponential and logarithmic functions introduced by Kaniadakis [41–43] in the context of special relativity. Within this framework, the ordinary exponential function $\exp(x)$ deforms into the generalized exponential function $\exp_\kappa(x)$ given by:

$$\exp_\kappa(x) = \left(\sqrt{1+\kappa^2 x^2} + \kappa x\right)^{\frac{1}{\kappa}}, \quad x \in \mathbb{R}, \quad \kappa \in [0,1). \tag{1}$$

The deformed logarithmic function $\ln_\kappa(x)$, which is defined as the inverse of (1), can be written as:

$$\ln_\kappa(x) = \frac{x^\kappa - x^{-\kappa}}{2\kappa}, \quad x \in \mathbb{R}_+. \tag{2}$$

Kaniadakis' deformed functions have also been successfully used to analyze nonphysical systems. In economics, the κ-deformation has been used to study differentiated product markets [48,49], finance [50–55], and the distribution of income by size [47,56–62]. In the latter case, it is interesting to use such deformed functions because they can be used to statistically describe the entire spectrum of incomes, from the low to the middle range and up to the Pareto tail.

2.1. Definitions and Basic Properties

A random variable X is said to have a κ-generalized distribution, and we write $X \sim \kappa\text{-gen}(\alpha, \beta, \kappa)$, if it has a probability density function (PDF) given by:

$$f(x;\alpha,\beta,\kappa) = \frac{\alpha}{\beta}\left(\frac{x}{\beta}\right)^{\alpha-1} \frac{\exp_\kappa\left[-(x/\beta)^\alpha\right]}{\sqrt{1+\kappa^2(x/\beta)^{2\alpha}}}, \quad x > 0, \quad \alpha, \beta > 0, \quad \kappa \in [0,1). \tag{3}$$

Its cumulative distribution function (CDF) can be expressed as:

$$F(x;\alpha,\beta,\kappa) = 1 - \exp_\kappa\left[-(x/\beta)^\alpha\right]. \tag{4}$$

(For a complete description of the κ-generalized distributional properties, the reader is referred to [60] and the references cited therein. A heuristic derivation of the κ-generalized density, showing how this probability distribution emerges naturally within the field of κ-deformed analysis, is given in [61,63]).

Figure 1 illustrates the behavior of the κ-generalized PDF and the complementary CDF, $1 - F(x;\alpha,\beta,\kappa)$, for various parameter values.

Each of the three graph pairs holds two parameters constant and varies the remaining one.

The constant β is a characteristic scale that has the same dimension as income. For this reason, it takes into account the monetary unit and can be used to adjust for inflation and facilitate cross-country comparisons of income distributions expressed in different monetary units. Increases in the monetary unit result in a global increase in individual income and average income.

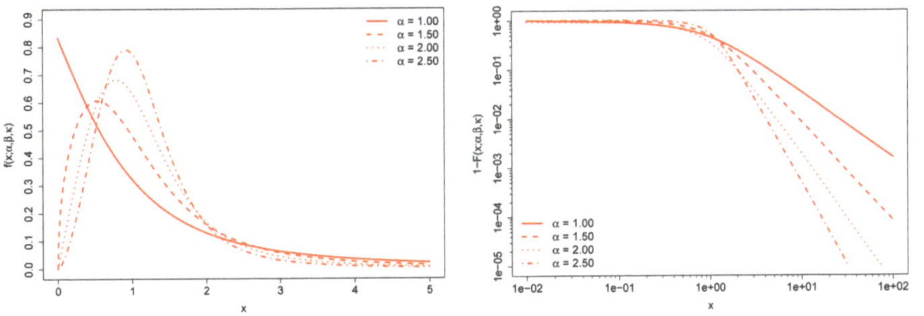

(a) $\beta = 1.20$, $\kappa = 0.75$ and $\alpha = 1.00, 1.50, 2.00, 2.50$

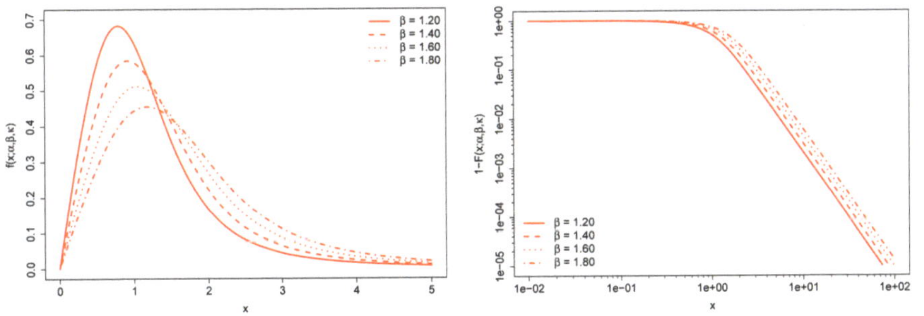

(b) $\alpha = 2.00$, $\kappa = 0.75$ and $\beta = 1.20, 1.40, 1.60, 1.80$

Figure 1. *Cont.*

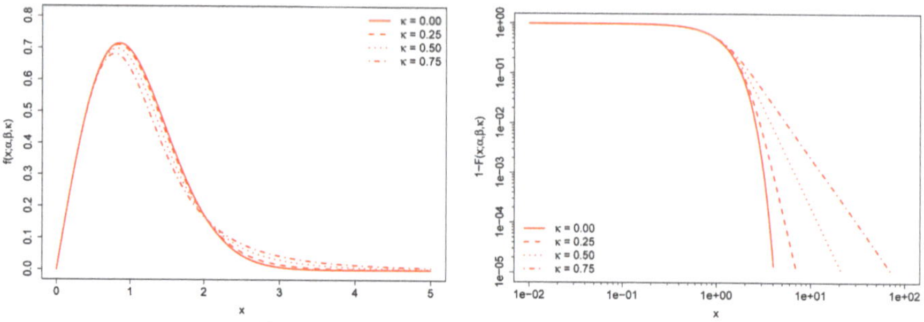

(**c**) $\alpha = 2.00, \beta = 1.20$ and $\kappa = 0.00, 0.25, 0.50, 0.75$

Figure 1. κ-generalized PDF (**left**) and complementary CDF (**right**) for different values of the parameters. The complementary CDF is plotted on double-log axes, which is the standard way to emphasize the right-tail behavior of a distribution.

The α and κ parameters are scale-free parameters that affect the distribution's shape. The region around the origin of the κ-generalized distribution is dominated by α, while the upper tail is dominated by both α and κ. Increasing κ leads to a thicker upper tail, while increasing α tapers both tails and increases the concentration of probability mass around the peak of the distribution.

As κ approaches 0, the distribution converges to the Weibull distribution; it is easy to verify that:

$$\lim_{\kappa \to 0} f(x; \alpha, \beta, \kappa) = \frac{\alpha}{\beta}\left(\frac{x}{\beta}\right)^{\alpha-1} \exp\left[-(x/\beta)^{\alpha}\right] \tag{5}$$

and:

$$\lim_{\kappa \to 0} F(x; \alpha, \beta, \kappa) = 1 - \exp\left[-(x/\beta)^{\alpha}\right]. \tag{6}$$

(The Weibull distribution is primarily studied in the engineering literature. In physics, it is known as the stretched exponential distribution when $\alpha < 1$. In economics, it has potential for income data, although it has only been used sporadically—some applications can be found in Refs. [29,35,64–69].) The distribution behaves similarly to the Weibull model for $x \to 0^+$, while for large x it approaches a Pareto distribution of the first kind with scale $k = \beta(2\kappa)^{-\frac{1}{\alpha}}$ and shape $a = \frac{\alpha}{\kappa}$, i.e.:

$$f(x; \alpha, \beta, \kappa) \underset{x \to +\infty}{\sim} \frac{a k^a}{x^{a+1}} \tag{7}$$

and:

$$F(x; \alpha, \beta, \kappa) \underset{x \to +\infty}{\sim} 1 - \left(\frac{k}{x}\right)^{a}, \tag{8}$$

thus satisfying the weak Pareto law [70]. (Additional versions of the Pareto law were introduced by [71], $\lim_{x \to +\infty} \frac{x f(x)}{1 - F(x)} = a$, and [30], $\lim_{x \to +\infty}\left[1 + \frac{x f'(x)}{f(x)}\right] = -a$. Since we have: $\lim_{x \to +\infty} \frac{x f(x;\alpha,\beta,\kappa)}{1 - F(x;\alpha,\beta,\kappa)} = \frac{\alpha}{\kappa} = a$ and $\lim_{x \to +\infty}\left[1 + \frac{x f'(x;\alpha,\beta,\kappa)}{f(x;\alpha,\beta,\kappa)}\right] = -\frac{\alpha}{\kappa} = -a$, the κ-generalized distribution also obeys these alternative versions of the weak Pareto law.)

Equation (4) implies that the quantile function is available in closed form:

$$F^{-1}(u; \alpha, \beta, \kappa) = x_u = \beta\left[\ln_{\kappa}\left(\frac{1}{1-u}\right)\right]^{\frac{1}{\alpha}}, \quad 0 < u < 1, \tag{9}$$

an attractive feature for generating random numbers from a κ-generalized distribution. The median of the distribution is:

$$x_{\text{med}} = \beta[\ln_\kappa(2)]^{\frac{1}{\alpha}}, \tag{10}$$

and the mode occurs at:

$$x_{\text{mode}} = \beta\left[\frac{\alpha^2 + 2\kappa^2(\alpha-1)}{2\kappa^2(\alpha^2-\kappa^2)}\right]^{\frac{1}{2\alpha}} \left\{\sqrt{1+\frac{4\kappa^2(\alpha^2-\kappa^2)(\alpha-1)^2}{[\alpha^2+2\kappa^2(\alpha-1)]^2}} - 1\right\}^{\frac{1}{2\alpha}} \tag{11}$$

if $\alpha > 1$; otherwise, the distribution is zero-modal with a pole at the origin.

Finally, the rth raw moment of the κ-generalized distribution is equal to:

$$\langle x^r \rangle = \int_0^\infty x^r f(x;\alpha,\beta,\kappa)\,\mathrm{d}\,x = \beta^r (2\kappa)^{-\frac{r}{\alpha}} \frac{\Gamma(1+\frac{r}{\alpha})}{1+\frac{r}{\alpha}\kappa} \frac{\Gamma\left(\frac{1}{2\kappa}-\frac{r}{2\alpha}\right)}{\Gamma\left(\frac{1}{2\kappa}+\frac{r}{2\alpha}\right)}, \tag{12}$$

where $\Gamma(\cdot)$ denotes the gamma function, and exists for $-\alpha < r < \frac{\alpha}{\kappa}$. Specifically:

$$\langle x \rangle = \beta(2\kappa)^{-\frac{1}{\alpha}} \frac{\Gamma\left(1+\frac{1}{\alpha}\right)}{1+\frac{1}{\alpha}\kappa} \frac{\Gamma\left(\frac{1}{2\kappa}-\frac{1}{2\alpha}\right)}{\Gamma\left(\frac{1}{2\kappa}+\frac{1}{2\alpha}\right)} \tag{13}$$

is the mean of the distribution and:

$$\left\langle x^2 \right\rangle - \langle x \rangle^2 = \beta^2(2\kappa)^{-\frac{2}{\alpha}} \left\{ \frac{\Gamma(1+\frac{2}{\alpha})}{1+2\frac{\kappa}{\alpha}} \frac{\Gamma\left(\frac{1}{2\kappa}-\frac{1}{\alpha}\right)}{\Gamma\left(\frac{1}{2\kappa}+\frac{1}{\alpha}\right)} - \left[\frac{\Gamma\left(1+\frac{1}{\alpha}\right)}{1+\frac{\kappa}{\alpha}} \frac{\Gamma\left(\frac{1}{2\kappa}-\frac{1}{2\alpha}\right)}{\Gamma\left(\frac{1}{2\kappa}+\frac{1}{2\alpha}\right)}\right]^2 \right\} \tag{14}$$

is the variance.

2.2. Measuring Income Inequality Using the κ-Generalized Distribution

The concept of inequality in economics dates back to Pareto's early work [10–13], which showed that the top 20% of population held about 80% of total income/wealth. Later, the American economist Lorenz [72] introduced the Lorenz curve, a widely used tool for measuring income/wealth inequality. This curve measures the difference in income or wealth distribution from an equal distribution. If there is perfect equality, the Lorenz curve coincides with the diagonal of a unit square, while worsening distribution (more inequality) moves the curve away from the diagonal.

The Lorenz curve for a random variable X with CDF $F(x)$ and finite mean $\langle x \rangle = \int x\,\mathrm{d}\,F(x)$ is defined as [73]:

$$L(u) = \frac{1}{\langle x \rangle} \int_0^u F^{-1}(t)\,\mathrm{d}\,t, \quad u \in [0,1]. \tag{15}$$

Using the closed form of the quantile function $F^{-1}(u)$ of the κ-generalized distribution, the Lorenz curve can be reformulated as follows [74]:

$$L(u) = I_x\left(1+\frac{1}{\alpha}, \frac{1}{2\kappa}-\frac{1}{2\alpha}\right), \quad x = 1-(1-u)^{2\kappa}, \tag{16}$$

where $I_x(\cdot,\cdot)$ is the regularized incomplete beta function defined in terms of the incomplete beta function and the complete beta function, that is, $I_x(\cdot,\cdot) = \frac{B_x(\cdot,\cdot)}{B(\cdot,\cdot)}$. The curve (16) exists if and only if $\frac{\alpha}{\kappa} > 1$. In particular, if $X_i \sim \kappa\text{-gen}(\alpha_i, \beta_i, \kappa_i)$, $i = 1,2$, the necessary and sufficient conditions for the Lorenz curves of X_1 and X_2 not to intersect (otherwise, it would be impossible to determine which distribution has more inequality) are [58]:

$$\alpha_1 \geq \alpha_2 \quad \text{and} \quad \frac{\alpha_1}{\kappa_1} \geq \frac{\alpha_2}{\kappa_2}. \tag{17}$$

The Lorenz curves of two κ-generalized distributions X_1 and X_2 with parameters chosen according to (17) are illustrated in Figure 2. The depicted curves indicate that X_1 exhibits lower inequality compared to X_2, as the Lorenz curve of X_1 does not intersect or fall below that of X_2.

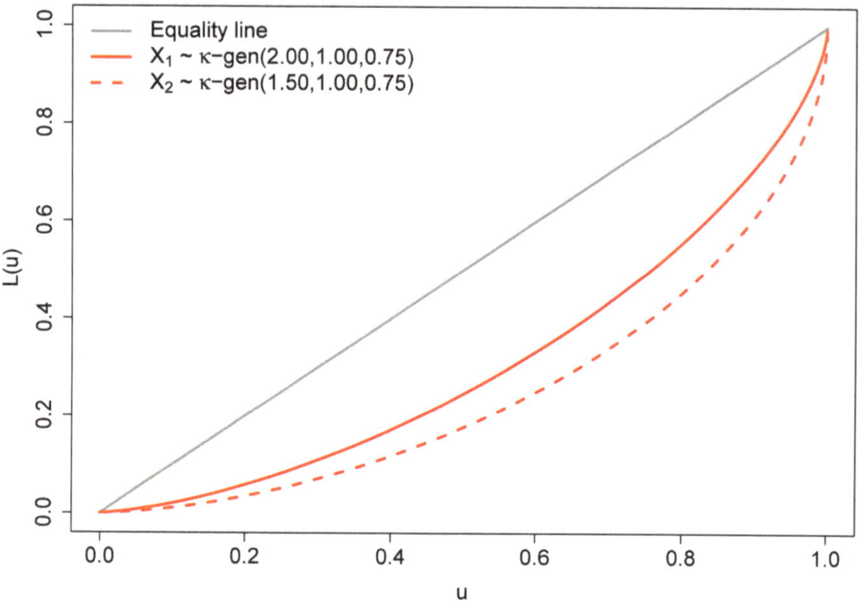

Figure 2. Lorenz curves for two κ-generalized distributions.

Economists have employed statistical metrics to quantify income and wealth inequality. The Gini coefficient, developed in 1914 by the Italian statistician Gini [75], is one of the best known. From the general definition $G = 1 - \frac{1}{\langle x \rangle} \int_0^\infty [1 - F(x)]^2 \, \mathrm{d}\, x$ due to [76], the Gini coefficient associated with the κ-generalized distribution is:

$$G = 1 - \frac{2\alpha + 2\kappa}{2\alpha + \kappa} \frac{\Gamma\left(\frac{1}{\kappa} - \frac{1}{2\alpha}\right) \Gamma\left(\frac{1}{2\kappa} + \frac{1}{2\alpha}\right)}{\Gamma\left(\frac{1}{\kappa} + \frac{1}{2\alpha}\right) \Gamma\left(\frac{1}{2\kappa} - \frac{1}{2\alpha}\right)}. \tag{18}$$

Using the Stirling approximation for the gamma function, $\Gamma(z) \approx \sqrt{2\pi} z^{z-\frac{1}{2}} \exp(-z)$, and taking the limit as $\kappa \to 0$ in Equation (18), after some simplification one arrives at $G = 1 - 2^{-\frac{1}{\alpha}}$, which is the explicit form of the Gini coefficient for the Weibull distribution (see e.g., [77], p. 177). Since the exponential distribution is a special case of the Weibull distribution with a shape parameter of 1, it follows directly that for $\kappa \to 0$ and $\alpha = 1$, the exponential law is also a special limiting case of the κ-generalized distribution with a true Gini coefficient of one half [16].

The Gini coefficient is a widely used measure of inequality, but it makes specific assumptions about income differences in different parts of the distribution. It is most sensitive to transfers around the middle of the income distribution and least sensitive to transfers among the very rich or very poor [78]. Differently, the generalized entropy class of inequality measures [79–83] provides a range of bottom-to-top sensitive indices used by analysts to assess inequality in different parts of the income distribution. The expression for this class of inequality indices in terms of the κ-generalized parameters is [57]:

$$GE(\theta) = \frac{1}{\theta^2 - \theta}\left\{\left(\frac{\beta}{m}\right)^{\theta}\left[\frac{(2\kappa)^{-\frac{\theta}{\alpha}}}{1+\frac{\theta}{\alpha}\kappa}\frac{\Gamma\left(\frac{1}{2\kappa}-\frac{\theta}{2\alpha}\right)}{\Gamma\left(\frac{1}{2\kappa}+\frac{\theta}{2\alpha}\right)}\Gamma\left(1+\frac{\theta}{\alpha}\right)\right]-1\right\}, \quad \theta \neq 0,1, \tag{19}$$

where $m = \langle x \rangle$ denotes the mean of the distribution given by Equation (13). Formula (19) defines a class because $GE(\theta)$ takes different forms depending on the value given to θ, the parameter that describes the sensitivity of the index to income differences in different parts of the income distribution—the more positive or negative θ is, the more sensitive $GE(\theta)$ is to income differences at the top or bottom of the distribution. Two limiting cases of (19), obtained when the parameter θ is set to 0 and 1, have gained attention in practical work for the purpose of measuring inequality; these are the mean logarithmic deviation index:

$$MLD = \lim_{\theta\to 0} GE(\theta) = \frac{1}{\alpha}\left[\gamma + \psi\left(\frac{1}{2\kappa}\right) + \ln(2\kappa) - \alpha\ln\left(\frac{\beta}{m}\right) + \kappa\right], \tag{20}$$

where $\gamma = -\psi(1)$ is the Euler–Mascheroni constant and $\psi(z) = \Gamma'(z)/\Gamma(z)$ is the digamma function, and the Theil index [84]:

$$\begin{aligned} T = \lim_{\theta\to 1} GE(\theta) =& \frac{1}{\alpha}\left[\psi\left(1+\frac{1}{\alpha}\right) - \frac{1}{2}\psi\left(\frac{1}{2\kappa}-\frac{1}{2\alpha}\right) - \frac{1}{2}\psi\left(\frac{1}{2\kappa}+\frac{1}{2\alpha}\right)\right. \\ &\left. - \ln(2\kappa) + \alpha\ln\left(\frac{\beta}{m}\right) - \frac{\alpha\kappa}{\alpha+\kappa}\right], \end{aligned} \tag{21}$$

where the former is more sensitive to variations in the lower tail, while the latter is more sensitive to variations in the upper tail [85]. (Equation (19) is not defined for $\theta = 0$ and $\theta = 1$, as $(\theta^2 - \theta) = 0$ in both cases. Expressions for these values of θ are therefore derived using l'Hôpital rule, which allows evaluating limits of indeterminate forms using derivatives. Expressions for any $GE(\theta)$ index other than the cases $\theta = 0, 1$ can be derived by simple substitution—see for example [60]).

Finally, the class of inequality measures introduced by Atkinson [86] can be derived from (19) by exploiting the relationship [87,88]:

$$A(\epsilon) = 1 - [\epsilon(\epsilon-1)GE(1-\epsilon)+1]^{\frac{1}{1-\epsilon}}, \quad \epsilon > 0, \quad \epsilon \neq 1, \tag{22}$$

where $\epsilon = 1 - \theta$ is the inequality aversion parameter. As ϵ increases, $A(\epsilon)$ becomes more sensitive to transfers among lower incomes and less sensitive to transfers among top incomes [78]. The limiting form of (22) is $A(1) = 1 - \exp(-MLD)$. (All measures considered here are functions of distributional moments, whose existence depends on conditions assuring the convergence of the appropriate integrals. The Gini coefficient (18) exists if and only if the mean of the distribution $\langle x \rangle = \int_0^\infty x f(x;\alpha,\beta,\kappa)\,\mathrm{d}\,x$ converges, which is true if and only if $\frac{\alpha}{\kappa} > 1$. According to [89], parametric income distribution models share the existence problem of popular inequality measures).

2.3. Estimation

The κ-generalized distribution's parameters can be estimated using the maximum likelihood technique, which produces estimators with good statistical properties [90,91]. If sample observations $\mathbf{x} = \{x_1, \ldots, x_n,\}$ are independent, the likelihood function is as follows:

$$L(\mathbf{x};\boldsymbol{\theta}) = \prod_{i=1}^{n} f(x_i;\boldsymbol{\theta})^{w_i} = \prod_{i=1}^{n}\left\{\frac{\alpha}{\beta}\left(\frac{x_i}{\beta}\right)^{\alpha-1}\frac{\exp_\kappa\left[-(x_i/\beta)^\alpha\right]}{\sqrt{1+\kappa^2(x_i/\beta)^{2\alpha}}}\right\}^{w_i}, \tag{23}$$

where $f(x_i;\boldsymbol{\theta})$ denotes the PDF, $\boldsymbol{\theta} = \{\alpha, \beta, \kappa\}$ the vector of unknown parameters, w_i the weight of the ith observation, and n the sample size. This leads to the problem of solving the partial derivatives with respect to α, β and κ for the log-likelihood function:

$$l(\mathbf{x};\boldsymbol{\theta}) = \ln[L(\mathbf{x};\boldsymbol{\theta})] = \sum_{i=1}^{n} w_i \ln[f(x_i;\boldsymbol{\theta})], \tag{24}$$

which is the same as finding the solution to the following nonlinear system of equations:

$$\sum_{i=1}^{n} w_i \frac{\partial}{\partial \alpha} \ln[f(x_i;\boldsymbol{\theta})] = 0, \tag{25}$$

$$\sum_{i=1}^{n} w_i \frac{\partial}{\partial \beta} \ln[f(x_i;\boldsymbol{\theta})] = 0, \tag{26}$$

$$\sum_{i=1}^{n} w_i \frac{\partial}{\partial \kappa} \ln[f(x_i;\boldsymbol{\theta})] = 0. \tag{27}$$

However, the derivation of explicit expressions for maximum likelihood estimators of the three κ-generalized parameters poses a challenge due to the absence of feasible analytical solutions. The utilization of numerical optimization algorithms becomes therefore imperative in order to solve the maximum likelihood estimation problem.

3. Application to the Income Distribution in Greece

To celebrate 20 years of Kaniadakis' contribution, it seems appropriate to consider the income distribution in his native Greece to demonstrate the κ-generalized model's capacity to fit real-world data. First, income data for parameter estimation are briefly described. Next, the κ-generalized distribution is fitted to Greek household income data. Finally, using the same income microdata, different income size distribution models are compared.

3.1. Description of the Income Data

Income distribution data for Greece were obtained from the Luxembourg Income Study (LIS) database, which provides public access to household-level data files for various countries, including both developed and developing economies. The data are remote-accessible, requiring program code to be sent to LIS rather than being run directly by the user. At the time of writing, LIS contains Greek income distribution data for the following years: 1995, 2000, 2004, 2007, 2010, 2013, and 2016. The data set used for this review is the 2016 data set based on the 2017 wave of the Greek EU-SILC survey conducted by the Hellenic Statistical Authority (ELSTAT). (EU-SILC is a cross-sectional and longitudinal sample survey coordinated by Eurostat, focusing on income, poverty, social exclusion, and living conditions in the European Union.) The sample size is 22,555 households.

The definition of income is "household disposable income", which is the income available to households to support consumption expenditure and saving during the reference period. The measure includes income from work, wealth, and direct government benefits, but subtracts direct taxes paid. It does not include sales taxes or noncash benefits, such as healthcare provided by a government or employer. Additionally, the income definition excludes income from capital gains, a significant source of nonwage income for wealthy individuals. As a result, many top incomes are likely to be underestimated.

Household disposable income is expressed in euro and "equivalized", i.e., divided by the square root of household size to adjust for differences in household demographics. Prior to equivalization, top and bottom coding is applied to set limits for extreme values. We also exclude all households with missing disposable income and use person-adjusted weights (the product of the household weights and the number of household members) when generating income indicators for the total population and estimating model parameters.

3.2. Results of Fitting

Figure 3 shows the results of fitting the κ-generalized distribution to empirical income data corresponding to the distribution of household income in Greece for the year 2016.

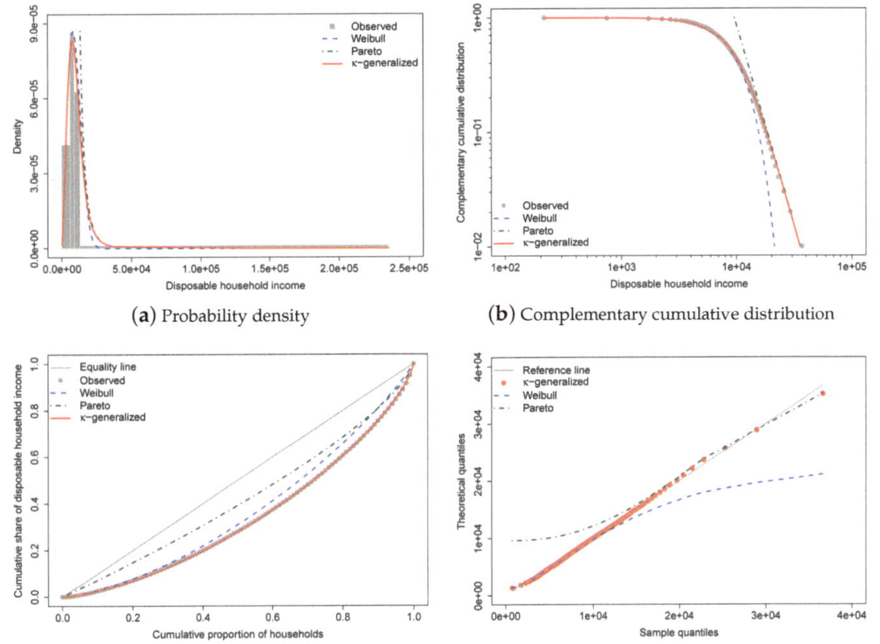

(**a**) Probability density (**b**) Complementary cumulative distribution

(**c**) Lorenz curve (**d**) Q-Q plot

Figure 3. κ-generalized distribution fitted to Greek household income data for 2016. The red solid line represents the κ-generalized model, which fits the data well over the whole range from low to high incomes, including the middle income region. It is compared to the Weibull (blue dashed line) and Pareto power-law (green dashed line) distributions. The complementary cumulative distribution is plotted on double-log axes, emphasizing the right-tail behavior of the distribution. The Lorenz curve plot compares the empirical and theoretical curves, with the gray solid line representing the Lorenz curve of a society with equal income distribution. The Q-Q plot of sample percentiles versus theoretical percentiles of the fitted κ-generalized shows excellent fit, with corresponding percentiles being close to the 45° line from the origin.

The best-fitting parameter values were determined using maximum likelihood estimation, resulting in estimates of $\alpha = 2.233 \pm 0.017$, $\beta = 10,667 \pm 46$, and $\kappa = 0.630 \pm 0.014$. The small errors indicate accurate estimations, and the comparison between the observed and fitted probabilities in panels (a) and (b) of Figure 3 suggests that the κ-generalized distribution has great potential for describing data across the range of low-to-middle-income to high-income power-law regimes, including the intermediate region where Weibull and Pareto distributions show clear departures. (In Figure 3, the curves for the Pareto and Weibull distributions have been drawn by expressing their parameters in terms of the estimated κ-generalized parameters—see Section 2.1).

Panel (c) of the same figure displays data points for the empirical Lorenz curve superimposed on the theoretical curve given by Equation (16) with estimates replacing α and κ as necessary. This formula, represented by the red solid line in the plot, matches the data exceptionally well. In addition, the plot contrasts the empirical Lorenz curve with the theoretical curves associated with the Weibull and Pareto distributions, respectively, given by:

$$\lim_{\kappa \to 0} L(u) = P\left(1 + \frac{1}{\alpha}, -\ln(1-u)\right), \tag{28}$$

where $P(\cdot, \cdot)$ is the lower regularized incomplete gamma function, and:

$$\lim_{x \to \infty} L(u) = 1 - (1-u)^{1-\frac{1}{a}}. \tag{29}$$

As one can easily see, these curves tell only a small part of the story.

To provide an indirect check on the validity of the parameter estimation, we have also computed predicted values for median and mean household disposable income, as well as the Gini and Atkinson coefficients—the latter with the inequality aversion parameter ϵ equal to 1. The results, obtained by substituting the estimated parameters into relevant expressions, are presented in Table 1, along with their empirical counterparts, corresponding to the LIS staff's "Inequality and Poverty Key Figures" for the considered country and year. (In this article, inequality measures are calculated using the most recent version of DASP, the Distributive Analysis Stata Package [92], which is available at http://dasp.ecn.ulaval.ca/—accessed on 29 June 2023. The complete set of corresponding "key figures" is available in an Excel workbook that can be downloaded from https://www.lisdatacenter.org/data-access/key-figures/—accessed on 29 June 2023.)

Table 1. Observed and predicted values of the median, the mean, the Gini index G and the Atkinson inequality measure $A(1)$.

Statistic	Observed			Predicted
	Value	LB [a]	UB [b]	
Median	9123	8983	9264	9181
Mean	10,548	10,292	10,805	10,488
G	0.323	0.312	0.334	0.322
$A(1)$	0.179	0.169	0.189	0.172

Notes: [a] lower bound of the 95% normal-based confidence interval obtained by adding −1.96 times the standard error to the sample indicator; [b] upper bound of the 95% normal-based confidence interval obtained by adding +1.96 times the standard error to the sample indicator. Source: author's calculations based on Greek LIS data for 2016.

The κ-generalized distribution predictions are fully covered by asymptotic normal 95% confidence intervals, confirming excellent agreement between the model and sample observations.

The linear behavior of the quantile-quantile (Q-Q) plot of sample percentiles against the fitted κ-generalized distribution and its limiting cases, shown in panel (d) of Figure 3, confirms the model's validity as well as the fact that the Weibull and Pareto distributions provide partial and incomplete data descriptions.

3.3. *Comparisons of Alternative Distributions*

This section compares the κ-generalized distribution's performance with other parametric models, including the three-parameter generalized gamma [93], Singh–Maddala [33], and Dagum type I [34] distributions, which have the following PDFs, respectively:

$$f(x; a, \beta, p) = \frac{a x^{ap-1} \exp\left[-(x/\beta)^a\right]}{\beta^{ap}\Gamma(a)}, \quad x > 0, \quad a, \beta, p > 0, \tag{30}$$

$$f(x; a, b, q) = \frac{a q x^{a-1}}{b^a \left[1 + (x/b)^a\right]^{1+q}}, \quad x > 0, \quad a, b, q > 0, \tag{31}$$

$$f(x;a,b,p)=\frac{apx^{ap-1}}{b^{ap}\left[1+(x/b)^{a}\right]^{p+1}},\quad x>0,\quad a,b,p>0. \tag{32}$$

Ref. [77] provides analytical expressions for distribution functions, moments, and tools for inequality measurement, including the Lorenz curve and Gini coefficient. Refs. [87,94] provide formulas for generalized entropy measures of the GB2 distribution, from which the Singh–Maddala and Dagum versions are easily obtained. For the generalized gamma distribution, closed expressions for the Theil entropy index and the mean logarithmic deviation are given in Refs. [85,95]. (Let X be a random variable following the generalized beta distribution of the second kind (GB2) with parameters a, b, p, and q, i.e., $X \sim \text{GB2}(a,b,p,q)$. The Singh–Maddala distribution is the special case of the GB2 distribution when $p=1$; the Dagum type I distribution is the special case when $q=1$. For a discussion of other special cases, see [35,77]).

Table 2 displays maximum likelihood estimates for the models under consideration.

Table 2. Maximum likelihood estimates for the generalized gamma, Singh–Maddala, Dagum type I and κ-generalized models of income distribution.

Model [a]	Parameters [b]			Goodness-of-Fit Criteria [c,d]			
	a (α)	b (β)	q, p, κ	*RMSE*	*MAE*	*LRMSE*	*LMAE*
GG	0.684 (0.018)	829 (115)	5.475 (0.270)	2.325	2.047	0.939	0.812
SM	2.441 (0.021)	12,531 (219)	1.835 (0.053)	0.716	0.574	0.319	0.187
D	3.705 (0.041)	11,705 (104)	0.560 (0.011)	0.539	0.437	0.281	0.211
κ-gen	2.233 (0.017)	10,667 (46)	0.630 (0.014)	0.530	0.418	0.188	0.139

Notes: [a] GG = generalized gamma, SM = Singh-Maddala, D = Dagum type I, κ-gen = κ-generalized; [b] numbers in parentheses: estimated standard errors; [c] *RMSE* = root mean square error, *MAE* = mean absolute error, *LRMSE* = root mean square error between the observed and estimated Lorenz curves, *LMAE* = mean absolute error between the observed and estimated Lorenz curves; [d] values multiplied by 100. Source: author's calculations based on Greek LIS data for 2016.

The κ-generalized model offers the best results, with parameter standard errors derived from the inverse Hessian matrix being the lowest among competing income distribution models.

The root mean square error and mean absolute error between observed and predicted probabilities were used to determine which distribution best fits the data. These goodness-of-fit measures are, respectively, defined by:

$$RMSE=\sqrt{\frac{1}{n}\sum_{i=1}^{n}\left[\hat{F}_W(t)-F(x_i;\hat{\boldsymbol{\theta}})\right]^2} \tag{33}$$

and:

$$MAE=\frac{1}{n}\sum_{i=1}^{n}\left|\hat{F}_W(t)-F(x_i;\hat{\boldsymbol{\theta}})\right|, \tag{34}$$

where $\hat{F}_W(t)=\frac{1}{W}\sum_{i=1}^{n}w_i\mathbb{1}\{x_i\le t\}$, with $W=\sum_{i=1}^{n}w_i$, denotes the weighted empirical cumulative distribution function—equal to the sum of the income weights where $x\le t$ divided by the total sum of weights—and $\hat{\boldsymbol{\theta}}$ is the vector of estimated parameters. (In the formulas above, $\mathbb{1}\{\cdot\}$ is an indicator function that takes the value 1 if the condition in $\{\cdot\}$ is true, 0 otherwise.) The *RMSE* and *MAE* between the observed and estimated Lorenz curves have also been used as goodness-of-fit criteria, as they are expected to better reflect the accuracy of the inequality estimates. These additional measures are given by:

$$LRMSE = \sqrt{\frac{1}{n}\sum_{i=1}^{n}\left[L_i - L(\lambda_i; \hat{\boldsymbol{\theta}})\right]^2} \tag{35}$$

and:

$$LMAE = \frac{1}{n}\sum_{i=1}^{n}\left|L_i - L(\lambda_i; \hat{\boldsymbol{\theta}})\right|, \tag{36}$$

where $\lambda_i = \hat{F}_W(t)$ and L_i denote the cumulative share of population and income, respectively, up to percentile i—i.e., (λ_i, L_i) is a point on the empirical Lorenz curve.

Based on the above goodness-of-fit criteria, the κ-generalized model is clearly the best fit. As shown in the last three columns of Table 2, the generalized gamma, Singh–Maddala, and Dagum type I have larger $RMSE$ and MAE values for both probabilities and Lorenz curves, suggesting that these models perform worse than the κ-generalized distribution.

The performance of the four models is further evaluated by considering the accuracy of selected distributional statistics implied by parameter estimates. Table 3 presents the predicted values for the median, mean, and several inequality measures derived from estimates in Table 2. (The Gini coefficient of the generalized gamma distribution is available in [35] as a long expression involving the Gaussian hypergeometric function ${}_2F_1$, which is not currently available in the online statistical evaluator provided by the LIS web-based interface. An estimate of the Gini index for the generalized gamma distribution was therefore obtained by numerically integrating the area between the predicted Lorenz curve and the line of hypothetical equality. Ref. [96] reviews various methods for numerically estimating the Gini.)

Table 3. Observed and predicted values of selected distributional statistics.

Statistic [a]	Observed			Predicted [d]			
	Value	LB [b]	UB [c]	GG	SM	D	κ-gen
Median	9123	8983	9264	9076	9108	9189	9181
Mean	10,548	10,292	10,805	10,532	10,463	10,447	10,488
G	0.323	0.312	0.334	0.333	0.321	0.320	0.322
MLD	0.197	0.185	0.209	0.196	0.183	0.188	0.188
T	0.191	0.175	0.207	0.180	0.174	0.178	0.181
$A(1)$	0.179	0.169	0.189	0.178	0.168	0.172	0.172

Notes: [a] G = Gini index, MLD = mean logarithmic deviation index, T = Theil index, $A(1)$ = Atkinson coefficient with inequality aversion parameter ϵ equal to 1; [b] lower bound of the 95% normal-based confidence interval obtained by adding -1.96 times the standard error to the sample indicator; [c] upper bound of the 95% normal-based confidence interval obtained by adding $+1.96$ times the standard error to the sample indicator; [d] GG = generalized gamma, SM = Singh–Maddala, D = Dagum type I, κ-gen = κ-generalized. Source: author's calculations based on Greek LIS data for 2016.

For each of the models examined, the accuracy of the implied statistics is evaluated by calculating the absolute percentage error:

$$APE = \frac{|P - A|}{A} \times 100 \tag{37}$$

between the predicted values (P) and the actual sample estimates (A) given in Table 3. The results are summarized in Figure 4.

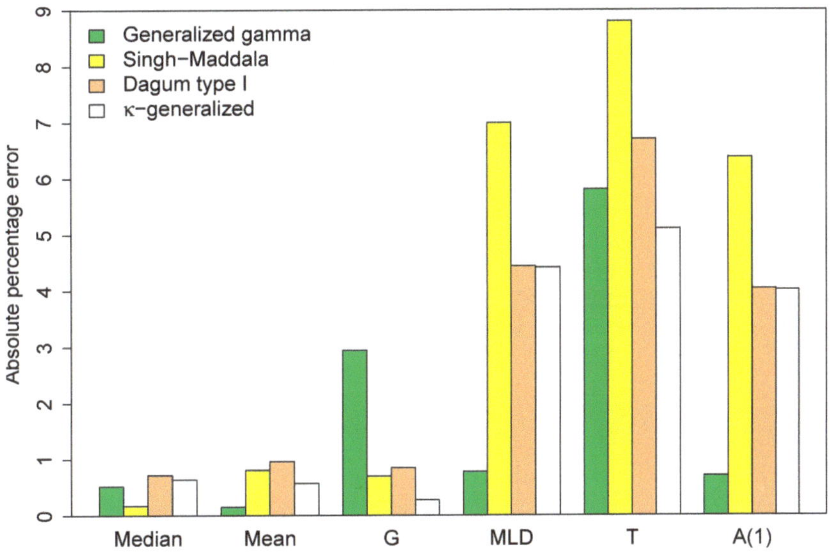

Figure 4. Absolute percentage error between the predicted values for key distributional summary measures and their sample counterparts.

Except for the median, the κ-generalized distribution has more accurate implied estimates of selected distributional statistics than the Singh–Maddala and Dagum type I models, with the Gini coefficient being significantly more accurate. This implies that the κ-generalized estimation procedure preserves the mean characteristic of the analyzed data and accurately models intra- and/or inter-group variation. Additionally, when considering income differences in different parts of the income distribution, the κ-generalized provides more accurate estimates than the two competitors of the MLD index, Theil index T and the Atkinson inequality measure $A(1)$. The Gini is an inequality index sensitive to the middle, while the other indices are more sensitive to the top and bottom of the income distribution. These results support the closest approximation to the income distribution found for the κ-generalized model.

The κ generalized distribution also outperforms the generalized gamma in predicting the Gini coefficient and Theil index, while the generalized gamma provides more accurate estimates for the MLD index, the $A(1)$ measure, the median, and the mean. This agreement is due to better fit in the lower part of the observed distribution, while disagreements arise from poorer fit in the upper-middle range, especially at the top end. This is demonstrated by the double-logarithmic plot in Figure 5, known as the Zipf plot, which shows the relationship between income and the complementary CDF of income for the data under study.

The Zipf plot is natural to use when looking at the upper part of the distribution because it puts more emphasis on the upper tail and makes it easier to detect deviations in that part of the distribution from what a model would predict [97]. The lines show the Zipf plots that were predicted by fitting the generalized gamma and κ-generalized models. As the graph shows, both are pretty close to the actual data in the lower part of the income distribution. However, the empirical observations of the upper tail are very different from what the generalized gamma says they should be, while the theoretical Zipf plot for the κ-generalized distribution is much closer to the empirical one in the same part of the observed income distribution.

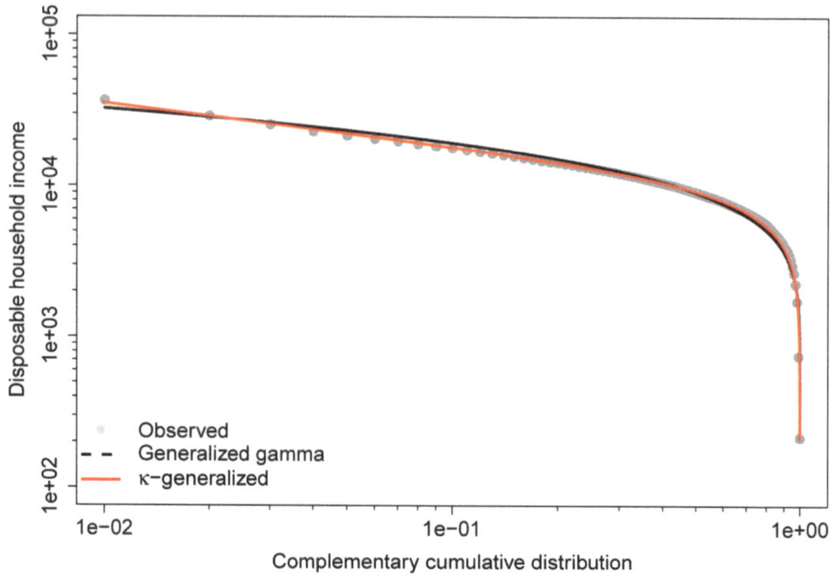

Figure 5. Zipf plot for the 2016 Greek household income data. The lines are the predicted Zipf plots obtained from the fit of the generalized gamma and κ-generalized models.

4. Applications of κ-Generalized Models to Income and Wealth Data

Apart from the one considered in this review, there have been numerous applications of the κ-generalized model to real-world income data over the past two decades.

The first study was conducted by [47], who analyzed 2001–2002 household incomes in Germany, Italy, and the United Kingdom. They found excellent agreement between the model and the empirical distributions across the full spectrum of incomes, including the intermediate income range where clear deviation was found when the Weibull model and pure Pareto law were used for interpolation.

The κ-generalized distribution was later applied to Australian household incomes in 2002–2003 [56] and US family incomes in 2003 [56,57]. The model again described the entire income range well and accurately estimated the inequality level in both countries using the Lorenz curve and Gini measure.

Comparative studies that fit multiple distributions to the same data are crucial for comparing performance. For example, Ref. [58], which examined the distribution of household income in Italy from 1989 to 2006, showed that the κ-generalized model outperforms three-parameter competitors such as the Singh–Maddala and Dagum type I distributions, except for the GB2, which has an extra parameter. The model has also also been used to analyze household income data for Germany between 1984 and 2007, the United Kingdom between 1991 and 2004, and the United States between 1980 and 2005. In many cases, the distribution of household income is observed to conform to the κ-generalized model, rather than the Singh–Maddala or Dagum type I distributions. In particular, the κ-generalized distribution is found to outperform competitors in the right tail of the data. The three-parameter κ-generalized model provides superior income inequality estimates even when the fit is worse than distributions belonging to the GB2 family, as obtained by [98] when comparing US and Italian income data for the 2000s. Finally, Ref. [60] finds that the κ-generalized distribution offers a superior fit to the data and, in many cases, estimates income inequality more accurately than alternatives using household income data for 45 countries from Wave IV to Wave IX of the LIS database. (Four-parameter extensions of the κ-generalized distribution, called *extended κ-generalized distributions of the first and second kind*—EκG1 and EκG2, respectively—were introduced by [74]. These two extensions are not discussed here, but Refs. [60,61,74] provide formulas for the moments, Lorenz curve, Gini index, coefficient of variation, mean logarithmic deviation, and Theil index for both the models. The new variants of the κ-generalized distribution outperform other four-parameter models in almost all cases, especially in estimating inequality indices with greater precision. In addition, a κ-deformation of the generalized gamma distribution with a power-law tail has recently been proposed by [99], to which the reader is referred for further details.)

The κ-generalized distribution has also been used to analyze the singularities of survey data on *net* wealth, which is gross wealth minus total debt [60,61,100]. These data show highly significant frequencies of households or individuals with wealth that is either null or negative. The κ-generalized model of wealth distribution is a mixture of an atomic and two continuous distributions. The atomic distribution accounts for economic units with no net worth, while a Weibull function accounts for negative net worth data. Positive net worth values, on the other hand, are represented by the κ-generalized model (3). The κ-generalized mixture model for wealth distribution was used to model US net worth data from 1984 to 2011 [100]. The model was generally accurate and its performance was superior to that of finite mixture models based on the Singh–Maddala and Dagum type I distributions for positive net worth values. Similar results were later obtained by Ref. [60] when analyzing net wealth data for nine countries selected from the Luxembourg Wealth Study (LWS) database. (The Luxembourg Wealth Study database—see https://www.lisdatacenter.org/our-data/lws-database/, accessed on 29 June 2023—is a collaborative project to assemble existing microdata on household wealth into a coherent database, aiming to do for wealth what the LIS database has achieved for income. The LWS was officially launched in 2004 and currently provides wealth data sets for several countries and years.)

5. Concluding Remarks

The κ-generalized distribution, a statistical model developed over several years of collaborative, multidisciplinary research, is a valuable tool for studying income and wealth distributions. This article discussed its basic properties, relationships with other distributions, and important extensions. It also discussed common inequality measures such as the Lorenz curve and Gini index, and how they can be computed from κ-generalized parameter estimates. A review of empirical applications showed excellent agreement with observed data. It is hoped that the collection of all these results in a single source will facilitate and promote the use of the κ-generalized distribution.

Funding: This research received no external funding.

Institutional Review Board Statement: Not applicable.

Informed Consent Statement: Not applicable.

Data Availability Statement: The Luxembourg Income Study Database (https://www.lisdatacenter.org/, accessed on 26 June 2023) provides remote access to the microdata through a web-based Job Submission Interface (LISSY). Users have to register to the platform and submit through the LISSY interface their statistical programs written in R, SAS, SPSS or Stata. Data analysis was performed using Stata software version 17 [101] while graphs were generated using R software version 4.3.1 [102]. To allow reproduction of the analysis, software code used in this article is available from the author on request.

Acknowledgments: The author would like to thank the anonymous referees whose comments and suggestions made it possible to improve this paper greatly.

Conflicts of Interest: The author declares no conflict of interest.

References

1. Chatterjee, A.; Yarlagadda, S.; Chakrabarti, B.K. *Econophysics of Wealth Distributions*; Springer: Milano, Italia, 2005.
2. Yakovenko, V.M. Econophysics, statistical mechanics approach to. In *Encyclopedia of Complexity and Systems Science*; Meyers, R.A., Ed.; Springer: New York, NY, USA, 2009; pp. 2800–2826.
3. Yakovenko, V.M.; Rosser, J.B. *Colloquium*: Statistical mechanics money, wealth, income. *Rev. Mod. Phys.* **2009**, *81*, 1703–1725. [CrossRef]
4. Chakrabarti, B.K.; Chakraborti, A.; Chakravarty, S.R.; Chatterjee, A. *Econophysics of Income and Wealth Distributions*; Cambridge University Press: New York, NY, USA, 2013.
5. Milanovic, B. *The Haves and the Have-Nots: A Brief and Idiosyncratic History of Global Inequality*; Basic Books: New York, NY, USA, 2011.
6. Stiglitz, J.E. *The Price of Inequality: How Today's Divided Society Endangers Our Future*; W. W. Norton & Company: New York, NY, USA, 2012.
7. Piketty, T. *Capital in the Twenty-First Century*; The Belknap Press of Harvard University Press: Cambridge, MA, USA, 2014.
8. Atkinson, A.B. *Inequality: What Can Be Done?* Harvard University Press: Cambridge, MA, USA, 2015.
9. Stiglitz, J.E. *The Great Divide: Unequal Societies and What We Can Do about Them*; W. W. Norton & Company: New York, NY, USA, 2015.
10. Pareto, V. La legge della domanda. *G. Degli Econ.* **1895**, *10*, 59–68.
11. Pareto, V. La courbe de la répartition de la richesse. In *Recueil Publié par la Faculté de Droit à l'Occasion de l'Exposition Nationale Suisse*; Viret-Genton, C., Ed.; Université de Lausanne: Lausanne, Switerland, 1896; pp. 373–387.
12. Pareto, V. *Cours d'Économie Politique*; Macmillan: London, UK, 1897.
13. Pareto, V. Aggiunta allo studio della curva delle entrate. *G. Degli Econ.* **1897**, *14*, 15–26.
14. Montroll, E.W.; Shlesinger, M.F. On $1/f$ noise and other distributions with long tails. *Proc. Natl. Acad. Sci. USA* **1982**, *79*, 3380–3383. [CrossRef] [PubMed]
15. Montroll, E.W.; Shlesinger, M.F. Maximum entropy formalism, fractals, scaling phenomena, and $1/f$ noise: A tale of tails. *J. Stat. Phys.* **1983**, *32*, 209–230. [CrossRef]
16. Drăgulescu, A.; Yakovenko, V.M. Evidence for the exponential distribution of income in the USA. *Eur. Phys. J. B* **2001**, *20*, 585–589. [CrossRef]
17. Drăgulescu, A.; Yakovenko, V.M. Exponential and power-law probability distributions of wealth and income in the United Kingdom and the United States. *Phys. A Stat. Mech. Its Appl.* **2001**, *299*, 213–221. [CrossRef]
18. Souma, W. Universal structure of the personal income distribution. *Fractals* **2001**, *9*, 463–470. [CrossRef]
19. Clementi, F.; Gallegati, M. Power law tails in the Italian personal income distribution. *Phys. A Stat. Mech. Its Appl.* **2005**, *350*, 427–438. [CrossRef]
20. Clementi, F.; Gallegati, M. Pareto's law of income distribution: Evidence for Germany, the United Kingdom, and the United States. In *Econophysics of Wealth Distributions*; Chatterjee, A., Yarlagadda, S., Chakrabarti, B.K., Eds.; Springer: Milan, Italy, 2005; pp. 3–14.
21. Silva, A.C.; Yakovenko, V.M. Temporal evolution of the "thermal" and "superthermal" income classes in the USA during 1983–2001. *Europhys. Lett.* **2005**, *69*, 304–310. [CrossRef]
22. Nirei, M.; Souma, W. A two factor model of income distribution dynamics. *Rev. Income Wealth* **2007**, *53*, 440–459. [CrossRef]
23. Gibrat, R. *Les Inégalités économiques. Applications: Aux Inégalités des Richesses, à la Concentration des Entreprises, aux Population des Villes, aux Statistiques des Familles, etc., d'une loi Nouvelle: La loi de l'Effet Proportionnel*; Librairie du Recueil Sirey: Paris, France, 1931.
24. Salem, A.B.Z.; Mount, T.D. A convenient descriptive model of income distribution: The gamma density. *Econometrica* **1974**, *42*, 1115–1127. [CrossRef]
25. Aitchison, J.; Brown, J.A.C. On criteria for descriptions of income distribuition. *Metroeconomica* **1954**, *6*, 88–107. [CrossRef]
26. Aitchison, J.; Brown, J.A.C. *The Lognormal Distribution with Special Reference to Its Use in Economics*; Cambridge University Press: New York, NY, USA, 1957.
27. McDonald, J.B.; Ransom, M.R. Functional forms, estimation techniques and the distribution of income. *Econometrica* **1979**, *47*, 1513–1525. [CrossRef]
28. Majumder, A.; Chakravarty, S.R. Distribution of personal income: Development of a new model and its application to U. S. income aata. *J. Appl. Econom.* **1990**, *5*, 189–196. [CrossRef]
29. Atoda, N.; Suruga, T.; Tachibanaki, T. Statistical inference of functional forms for income distribution. *Econ. Stud. Q.* **1988**, *39*, 14–40.
30. Esteban, J.M. Income-share elasticity and the size distribution of income. *Int. Econ. Rev.* **1986**, *27*, 439–444. [CrossRef]
31. Kloek, T.; van Dijk, H.K. Efficient estimation of income distribution parameters. *J. Econom.* **1978**, *8*, 61–74. [CrossRef]
32. Taillie, C. Lorenz ordering within the generalized gamma family of income distributions. In *Statistical Distributions in Scientific Work*; Taillie, C., Patil, G.P., Baldessari, B.A., Eds.; D. Reidel Publishing Company: Dordrecht, The Netherlands, 1981; Volume 6, pp. 181–192.
33. Singh, S.K.; Maddala, G.S. A function for size distribution of incomes. *Econometrica* **1976**, *44*, 963–970. [CrossRef]
34. Dagum, C. A new model of personal income distribution: Specification and estimation. *Econ. Appliquée* **1977**, *30*, 413–436.
35. McDonald, J.B. Some generalized functions for the size distribution of income. *Econometrica* **1984**, *52*, 647–665. [CrossRef]

36. Reed, W.J.; Jorgensen, M. The double Pareto-lognormal distribution—A new parametric model for size distributions. *Commun. Stat.-Theory Methods* **2004**, *33*, 1733–1753. [CrossRef]
37. McDonald, J.B.; Xu, Y.J. A generalization of the beta distribution with applications. *J. Econom.* **1995**, *66*, 133–152. [CrossRef]
38. Reed, W.J. Brownian-Laplace motion and its use in financial modelling. *Commun. Stat. Theory Methods* **2007**, *36*, 473–484. [CrossRef]
39. McDonald, J.B.; Ransom, M.R. The generalized beta distribution as a model for the distribution of income: Estimation of related measures of Inequality. In *Modeling Income Distributions and Lorenz Curves*; Chotikapanich, D., Ed.; Springer: New York, NY, USA, 2008; pp. 147–166.
40. Reed, W.J.; Wu, F. New four- and five-parameter models for income distributions. In *Modeling Income Distributions and Lorenz Curves*; Chotikapanich, D., Ed.; Springer: New York, NY, USA, 2008; pp. 211–223.
41. Kaniadakis, G. Non-linear kinetics underlying generalized statistics. *Phys. A Stat. Mech. Its Appl.* **2001**, *296*, 405–425. [CrossRef]
42. Kaniadakis, G. Statistical mechanics in the context of special relativity. *Phys. Rev. E* **2002**, *66*, 056125. [CrossRef] [PubMed]
43. Kaniadakis, G. Statistical mechanics in the context of special relativity. II. *Phys. Rev. E* **2005**, *72*, 036108. [CrossRef] [PubMed]
44. Kaniadakis, G. Maximum entropy principle and power-law tailed distributions. *Eur. Phys. J. B* **2009**, *70*, 3–13. [CrossRef]
45. Kaniadakis, G. Relativistic entropy and related Boltzmann kinetics. *Eur. Phys. J. A* **2009**, *40*, 275–287. [CrossRef]
46. Kaniadakis, G. Theoretical foundations and mathematical formalism of the power-law tailed statistical distributions. *Entropy* **2013**, *15*, 3983–4010. [CrossRef]
47. Clementi, F.; Gallegati, M.; Kaniadakis, G. κ-generalized statistics in personal income distribution. *Eur. Phys. J. B* **2007**, *57*, 187–193. [CrossRef]
48. Rajaonarison, D.; Bolduc, D.; Jayet, H. The *K*-deformed multinomial logit model. *Econ. Lett.* **2005**, *86*, 13–20. [CrossRef]
49. Rajaonarison, D. Deterministic heterogeneity in tastes and product differentiation in the *K*-logit model. *Econ. Lett.* **2008**, *100*, 396–398. [CrossRef]
50. Trivellato, B. Replication and shortfall risk in a binomial model with transaction costs. *Math. Methods Oper. Res.* **2009**, *69*, 1–26. [CrossRef]
51. Imparato, D.; Trivellato, B. Geometry of extended exponential models. In *Algebraic and Geometric Methods in Statistics*; Gibilisco, P., Riccomagno, E., Pistone, G., Wynn, H.P., Eds.; Cambridge University Press: Cambridge, UK, 2009; pp. 307–326.
52. Trivellato, B. The minimal κ-entropy martingale measure. *Int. J. Theor. Appl. Financ.* **2012**, *15*, 1250038. [CrossRef]
53. Trivellato, B. Deformed exponentials and applications to finance. *Entropy* **2013**, *15*, 3471–3489. [CrossRef]
54. Moretto, E.; Pasquali, S.; Trivellato, B. Option pricing under deformed Gaussian distributions. *Phys. A Stat. Mech. Appl.* **2016**, *446*, 246–263. [CrossRef]
55. Moretto, E.; Pasquali, S.; Trivellato, B. A non-Gaussian option pricing model based on Kaniadakis exponential deformation. *Eur. Phys. J. B* **2017**, *90*, 179. [CrossRef]
56. Clementi, F.; Di Matteo, T.; Gallegati, M.; Kaniadakis, G. The κ-generalized distribution: A new descriptive model for the size distribution of incomes. *Phys. A Stat. Mech. Appl.* **2008**, *387*, 3201–3208. [CrossRef]
57. Clementi, F.; Gallegati, M.; Kaniadakis, G. A κ-generalized statistical mechanics approach to income analysis. *J. Stat. Mech. Theory Exp.* **2009**, *2009*, P02037. [CrossRef]
58. Clementi, F.; Gallegati, M.; Kaniadakis, G. A model of personal income distribution with application to Italian data. *Empir. Econ.* **2010**, *39*, 559–591. [CrossRef]
59. Clementi, F.; Gallegati, M.; Kaniadakis, G. A new model of income distribution: The κ-generalized distribution. *J. Econ.* **2012**, *105*, 63–91. [CrossRef]
60. Clementi, F.; Gallegati, M. *The Distribution of Income and Wealth: Parametric Modeling with the κ-Generalized Family*; Springer International Publishing AG: Cham, Switerland, 2016.
61. Clementi, F.; Gallegati, M.; Kaniadakis, G.; Landini, S. κ-generalized models of income and wealth distributions: A survey. *Eur. Phys. J. Spec. Top.* **2016**, *225*, 1959–1984. [CrossRef]
62. Clementi, F.; Gallegati, M. New economic windows on income and wealth: The κ-generalized family of distributions. *J. Soc. Econ. Stat.* **2017**, *6*, 1–15.
63. Landini, S. Mathematics of *strange* quantities: Why are κ-generalized models a good fit to income and wealth distributions? An explanation. In *The Distribution of Income and Wealth: Parametric Modeling with the κ-Generalized Family*; Clementi, F., Gallegati, M., Eds.; Springer International Publishing AG: Cham, Switerland, 2016; pp. 93–132.
64. Bartels, C.P.A.; van Metelen, H. *Alternative Probability Density Functions of Income: A Comparison of the Longnormal-, Gamma- and Weibull-Distribution with Dutch Data*; Research Memorandum 29; Department of Quantitative Studies, Faculty of Economics, Vrije Universiteit: Amsterdam, The Netherlands, 1975.
65. Bartels, C.P.A. *Economic Aspects of Regional Welfare: Income Distribution and Unemployment*; Martinus Nijhoff: Leiden, The Netherlands, 1977.
66. Espinguet, P.; Terraza, M. Essai d'extrapolation des distributions de salaires français. *Econ. Appliquée* **1983**, *36*, 535–561.
67. Bordley, R.F.; McDonald, J.B.; Mantrala, A. Something new, something old: Parametric models for the size distribution of income. *J. Income Distrib.* **1996**, *6*, 91–103. [CrossRef]
68. Brachmann, K.; Stich, A.; Trede, M. Evaluating parametric income distribution models. *Allg. Stat. Arch.* **1996**, *80*, 285–298.

69. Tachibanaki, T.; Suruga, T.; Atoda, N. Estimations of income distribution parameters for individual observations by maximum likelihood method. *J. Jpn. Stat. Soc.* **1997**, *27*, 191–203. [CrossRef]
70. Mandelbrot, B. The Pareto-Lévy law and the distribution of income. *Int. Econ. Rev.* **1960**, *1*, 79–106. [CrossRef]
71. Kakwani, N. *Income Inequality and Poverty: Methods of Estimation and Policy Applications*; Oxford University Press: New York, NY, USA, 1980.
72. Lorenz, M.O. Methods of measuring the concentration of wealth. *Publ. Am. Stat. Assoc.* **1905**, *9*, 209–219.
73. Gastwirth, J.L. A general definition of the Lorenz curve. *Econometrica* **1971**, *39*, 1037–1039. [CrossRef]
74. Okamoto, M. *Extension of the κ-Generalized Distribution: New Four-Parameter Models for the Size Distribution of Income and Consumption*; Working Paper 600; LIS Cross-National Data Center: Luxembourg, 2013. Available online: https://www.lisdatacenter.org/wps/liswps/600.pdf (accessed on 26 June 2023).
75. Gini, C. Sulla misura della concentrazione e della variabilità dei caratteri. *Atti Del R. Ist. Veneto Di Sci. Lett. Ed Arti* **1914**, *73*, 1201–1248.
76. Arnold, B.C.; Laguna, L. *On Generalized Pareto Distributions with Applications to Income Data*; Iowa State University Press: Ames, IA, USA, 1977.
77. Kleiber, C.; Kotz, S. *Statistical Size Distributions in Economics and Actuarial Sciences*; John Wiley & Sons: New York, NY, USA, 2003.
78. Allison, P.D. Measures of inequality. *Am. Sociol. Rev.* **1978**, *43*, 865–880. [CrossRef]
79. Cowell, F.A. Generalized entropy and the measurement of distributional change. *Eur. Econ. Rev.* **1980**, *13*, 147–159. [CrossRef]
80. Cowell, F.A. On the structure of additive inequality measures. *Rev. Econ. Stud.* **1980**, *47*, 521–531. [CrossRef]
81. Shorrocks, A.F. The class of additively decomposable inequality measures. *Econometrica* **1980**, *48*, 613–625. [CrossRef]
82. Cowell, F.A.; Kuga, K. Additivity and the entropy concept: An axiomatic approach to inequality measurement. *J. Econ. Theory* **1981**, *25*, 131–143. [CrossRef]
83. Cowell, F.A.; Kuga, K. Inequality measurement: An axiomatic approach. *Eur. Econ. Rev.* **1981**, *15*, 287–305. [CrossRef]
84. Theil, H. *Economics and Information Theory*; North-Holland: Amsterdam, The Netherlands, 1967.
85. Sarabia, J.M.; Jordá, V.; Remuzgo, L. The Theil indices in parametric families of income distributions—A short review. *Rev. Income Wealth* **2017**, *63*, 867–880. [CrossRef]
86. Atkinson, A.B. On the measurement of inequality. *J. Econ. Theory* **1970**, *2*, 244–263. [CrossRef]
87. Jenkins, S.P. Distributionally-sensitive inequality indices and the GB2 income distribution. *Rev. Income Wealth* **2009**, *55*, 392–398. [CrossRef]
88. Cowell, F.A. *Measuring Inequality*; Oxford University Press: New York, NY, USA, 2011.
89. Kleiber, C. The existence of population inequality measures. *Econ. Lett.* **1997**, *57*, 39–44. [CrossRef]
90. Rao, C. *Linear Statistical Inference and Its Applications*; John Wiley & Sons: New York, NY, USA, 1973.
91. Ghosh, J.K. *Higher Order Asymptotics*; Institute of Mathematical Statistics and American Statistical Association: Hayward, CA, USA, 1994.
92. Araar, A.; Duclos, J.Y. *User Manual for Stata Package DASP: Version 3.0*; PEP, World Bank, UNDP and Université Laval, 2022. Available online: http://dasp.ecn.ulaval.ca/dasp3/manual/DASP_MANUAL_V303.pdf (accessed on 26 June 2023).
93. Stacy, E.W. A generalization of the gamma distribution. *Ann. Math. Stat.* **1962**, *33*, 1187–1192. [CrossRef]
94. Chotikapanich, D.; Griffiths, W.; Hajargasht, G.; Karunarathne, W.; Rao, D. Using the GB2 income distribution. *Econometrics* **2018**, *6*, 21. [CrossRef]
95. Jordá, V.; Alonso, J.M. New estimates on educational attainment using a continuous approach (1970–2010). *World Dev.* **2017**, *90*, 281–293. [CrossRef]
96. Fellman, J. Estimation of Gini coefficients using Lorenz curves. *J. Stat. Econom. Methods* **2012**, *1*, 31–38.
97. Takayasu, H. *Fractals in Physical Science*; Manchester University Press: Manchester, UK, 1991.
98. Okamoto, M. Evaluation of the goodness of fit of new statistical size distributions with consideration of accurate income inequality estimation. *Econ. Bull.* **2012**, *32*, 2969–2982.
99. Vallejos, A.; Ormazábal, I.; Borotto, F.A.; Astudillo, H.F. A new κ-deformed parametric model for the size distribution of wealth. *Phys. Stat. Mech. Appl.* **2019**, *514*, 819–829. [CrossRef]
100. Clementi, F.; Gallegati, M.; Kaniadakis, G. A generalized statistical model for the size distribution of wealth. *J. Stat. Mech. Theory Exp.* **2012**, *2012*, P12006. [CrossRef]
101. StataCorp. *Stata Statistical Software: Release 17*; StataCorp: College Station, TX, USA, 2021. Available online: https://www.stata.com/ (accessed on 26 June 2023).
102. R Core Team. *R: A Language and Environment for Statistical Computing*; R Foundation for Statistical Computing: Vienna, Austria, 2023. Available online: https://www.R-project.org/ (accessed on 26 June 2023).

entropy

MDPI

Article

Nonlinear Fokker–Planck Equations, H-Theorem and Generalized Entropy of a Composed System

Luiz R. Evangelista [1,2,3,*] and Ervin K. Lenzi [4]

1 Dipartimento di Scienza Applicata e Tecnologia (DISAT) del Politecnico di Torino, Corso Duca degli Abruzzi 24, 10129 Torino, Italy
2 Istituto dei Sistemi Complessi del Consiglio Nazionale delle Ricerche (ISC-CNR) co Politecnico di Torino, Corso Duca degli Abruzzi 24, 10129 Torino, Italy
3 Departamento de Física, Universidade Estadual de Maringá, Avenida Colombo, 5790, Maringá 87020-900, Paraná, Brazil
4 Departamento de Física, Universidade Estadual de Ponta Grossa, Avenida General Carlos Cavalcanti, 4748, Ponta Grossa 84030-900, Paraná, Brazil; eklenzi@uepg.br
* Correspondence: luiz.evangelista@polito.it

Abstract: We investigate the dynamics of a system composed of two different subsystems when subjected to different nonlinear Fokker–Planck equations by considering the H–theorem. We use the H–theorem to obtain the conditions required to establish a suitable dependence for the system's interaction that agrees with the thermodynamics law when the nonlinearity in these equations is the same. In this framework, we also consider different dynamical aspects of each subsystem and investigate a possible expression for the entropy of the composite system.

Keywords: generalized entropy; H–theorem; entropy production; nonlinear Fokker–Planck equation

Citation: Evangelista, L.R.; Lenzi, E.K. Nonlinear Fokker–Planck Equations, H-Theorem and Generalized Entropy of a Composed System. *Entropy* **2023**, *25*, 1357. https://doi.org/10.3390/e25091357

Academic Editor: Jaume Masoliver

Received: 5 July 2023
Revised: 11 August 2023
Accepted: 13 August 2023
Published: 20 September 2023

1. Introduction

Thermodynamics and statistical mechanics have entropy as a fundamental tool connecting the properties of a system from the particles' microscopic dynamics with macroscopic quantities and, consequently, with thermodynamic quantities. The concept of entropy started with Clausius's studies of thermal machines [1]. Subsequently, the Boltzmann and Gibbs works incorporated the concept of probability, building up the fundamentals of statistical mechanics [2–4]. It has been successfully applied in many contexts, where the fundamental basis is the molecular chaos hypothesis, which assumes the close-range interaction of molecules and the absence of memory in the collision of particles [5,6]. However, for many physical systems (e.g., fractal and self-organizing structures), conditions for the fulfillment of the molecular chaos hypothesis are not observed as well as the range of the interactions, which are long-ranged [7–9]. These points have motivated the analysis of extensions for thermodynamics and statistical physics to cover these scenarios. As an example, Tsallis has proposed an extension of the entropy [10], which has been systematically applied in many contexts such as black holes [11], the electrocaloric effect in quantum dots [12], chemotaxis of biological populations [13], Bose–Einstein condensation [14,15], and stimulated the analysis of other entropies [16–20]. More applications can be found in Refs. [21–26]. These entropies verify the H–theorem [27–31], which represents an important result of nonequilibrium statistical mechanics by ensuring that a system will reach an equilibrium after a long time evolution. The H–theorem establishes a connection between the dynamics and entropy, which may be used to investigate the dynamics behind the law of additivity for the different entropies. In this framework, by considering a nonlinear Fokker–Planck equation, the H–theorem can show how the entropy additivity laws can be obtained when a system composed of many subsystems is taken into account. In addition, it can also allow us to obtain the equilibrium distributions.

Here, we investigate through the H–theorem the conditions on the dynamics equations, i.e., nonlinear Fokker–Planck equations [32–35], for each subsystem of a composed system to reach the equilibrium condition. The results show that generalized entropies imply a coupling between the nonlinear equations. The distributions that emerge from these dynamics equations have a power-law behavior, where each subsystem modifies the other. We also investigate the entropy production for this system. These developments are presented in Section 2. In Section 3, we present our discussions and conclusions.

2. The Problem

Let us start our analysis by establishing the nonlinear Fokker–Planck equations connected to the dynamics of each subsystem of a composed system. They are

$$\frac{\partial}{\partial t}\rho_1(x_1,t) = \Gamma\frac{\partial^2}{\partial x_1^2}P_1(\rho_1,t) - \frac{\partial}{\partial x_1}\left[F_1(x_1)\rho_1(x_1,t)\right] \tag{1}$$

and

$$\frac{\partial}{\partial t}\rho_2(x_2,t) = \Gamma\frac{\partial^2}{\partial x_2^2}P_2(\rho_2,t) - \frac{\partial}{\partial x_2}\left[F_2(x_2)\rho_2(x_2,t)\right], \tag{2}$$

where $F_i(x_i)$, with $i = 1$ or 2, represents the external force, i.e., $F_i = -\partial_{x_i}\phi_i(x_i)$ and $\phi_i(x_i)$ is a potential energy, while Γ stands for a generic diffusion coefficient. Notice that $P_1(\rho_1,t)$ and $P_2(\rho_2,t)$ present in the diffusive term may have the same form or a different form. Particular choices of $P_i(\rho_i,t)$ have been successfully analyzed in several problems such as in porous media [36], anomalous diffusion [37], overdamped systems [38], and the Boltzmann equation endowed with a correlation term [39]. In Equations (1) and (2), $P_i(\rho_i,t)$ will be determined by the H–theorem in connection with the entropic form used to describe the combination of subsystems 1 and 2. It is worth pointing out that the different possibilities may be considered by allowing us to obtain different results for the composite system of 1 and 2 subsystems, as discussed in Refs. [28,29]. However, the combination of these equations, which represent the subsystem 1 and 2, in connection with thermostatistics (e.g., the nonextensive statistics [40]) requires careful analysis with direct consequences on the entropic additivity and zeroth law [41–43]. To accomplish this task, we consider general scenarios with different dynamics to investigate possible conditions to Equations (1) and (2) to allow a thermostatistics context.

2.1. H-Theorem

We start our analysis in terms of the H–theorem first by considering $P_1(\rho_1)$ and $P_2(\rho_2)$ with the same functional form. Afterwards, we consider $P_1(\rho_1)$ and $P_2(\rho_2)$ with a different functional form. Each one of these cases has different implications for the entropy related to the composed system formed by the systems 1 and 2, with the dynamics given in terms of Equations (1) and (2). Following Ref. [28,29,31], we analyze the behavior of the time derivative of the Helmholtz free energy. This free energy is defined by $F = U - TS$, with the internal energy, U, given by

$$U = \int_{-\infty}^{\infty}dx_1\int_{-\infty}^{\infty}dx_2[\phi_1(x_1)+\phi_2(x_2)]\rho_1(x_1,t)\rho_2(x_2,t) \tag{3}$$

and the entropy, S, expressed in terms of an arbitrary function

$$\mathcal{S} = k\int_{-\infty}^{\infty}dx_1\int_{-\infty}^{\infty}dx_2 s(\rho_1,\rho_2). \tag{4}$$

Note that Equations (3) and (4) represent the total internal energy and the entropy of the system composed of two subsystems governed by Equations (1) and (2), respectively.

By using the previous equations, the total free energy of the system is given by

$$F = \int_{-\infty}^{\infty}\int_{-\infty}^{\infty} dx_1 dx_2 [\Psi(x_1,x_2)\rho_1(x_1,t)\rho_2(x_2,t) - kTs(\rho_1,\rho_2)], \quad (5)$$

with $\Psi(x_1,x_2) = \phi_1(x_1) + \phi_2(x_2)$. Before determining the time derivative of Equation (5), we assume that $P_1(\rho_1,t)$ and $P_2(\rho_2,t)$ have essentially the same functional forms and the entropy is a function of the product of the probability densities related to each subsystem, i.e., $s(\rho_1,\rho_2) = s(\rho_1\rho_2)$. It is then possible to show that

$$\frac{d}{dt}F = \int_{-\infty}^{\infty}\int_{-\infty}^{\infty} dx_1 dx_2 \left[\Psi(x_1,x_2) - kT\frac{\partial}{\partial\rho_{12}}s(\rho_{12})\right]\frac{\partial}{\partial t}[\rho_1(x_1,t)\rho_2(x_2,t)], \quad (6)$$

where $\rho_{12} = \rho_1\rho_2$, and

$$\begin{aligned} \frac{d}{dt}F &= \int_{-\infty}^{\infty}\int_{-\infty}^{\infty} dx_1 dx_2 \left\{\Psi(x_1,x_2)\rho_2 - kT\rho_2\frac{\partial}{\partial\rho_{12}}s(\rho_{12})\right\} \\ &\times \frac{\partial}{\partial x_1}\left\{\Gamma\frac{\partial}{\partial x_1}P_1(\rho_1,t) - F_1(x_1)\rho_1(x_1,t)\right\} \\ &+ \int_{-\infty}^{\infty}\int_{-\infty}^{\infty} dx_1 dx_2 \left\{\Psi(x_1,x_2)\rho_1 - kT\rho_1\frac{\partial}{\partial\rho_{12}}s(\rho_{12})\right\} \\ &\times \frac{\partial}{\partial x_2}\left\{\Gamma\frac{\partial}{\partial x_2}P_2(\rho_2,t) - F_2(x_2)\rho_2(x_2,t)\right\}. \end{aligned} \quad (7)$$

After integration by parts and applying the conditions $\rho_i(x \to \pm\infty, t) \to 0$ and $\partial_x\rho_i(x \to \pm\infty, t) \to 0$, we obtain

$$\begin{aligned} \frac{d}{dt}F &= -\int_{-\infty}^{\infty}\int_{-\infty}^{\infty} dx_1 dx_2 \left\{\frac{\partial}{\partial x_1}\phi_1(x_1)\rho_2 - kT\rho_2^2\frac{\partial\rho_1}{\partial x_1}\frac{\partial^2}{\partial\rho_{12}^2}s(\rho_{12})\right\} \\ &\times \left\{\Gamma\frac{\partial}{\partial x_1}P_1(\rho_1,t) - F_1(x_1)\rho_1(x_1,t)\right\} \\ &- \int_{-\infty}^{\infty}\int_{-\infty}^{\infty} dx_1 dx_2 \left\{\frac{\partial}{\partial x_2}\phi_2(x_2)\rho_1 - kT\rho_1^2\frac{\partial\rho_2}{\partial x_2}\frac{\partial^2}{\partial\rho_{12}^2}s(\rho_{12})\right\} \\ &\times \left\{\Gamma\frac{\partial}{\partial x_2}P_2(\rho_2,t) - F_2(x_2)\rho_2(x_2,t)\right\}. \end{aligned} \quad (8)$$

Now, let us focus on the term

$$\Gamma\frac{\partial}{\partial x_i}P_i(\rho_i,t) - F_i(x_i)\rho_i(x_i,t), \quad (9)$$

where $i = 1,2$, which will be directly connected with the properties of the entropy of the composite system. To proceed, we consider that

$$P_i(\rho_i,t) = \mathcal{D}_{j,\gamma}(t)\rho_i^{\gamma}(x_i,t) + \mathcal{D}_{j,\nu}(t)\rho_i^{\nu}(x_i,t), \quad (10)$$

with $j \neq i$, $j = 1,2$, and

$$\mathcal{D}_{j,\gamma}(t) = \alpha_\gamma \int_{-\infty}^{\infty} dx_j \rho_j^{\gamma}(x_j,t) \quad \text{and} \quad \mathcal{D}_{j,\nu}(t) = \alpha_\nu \int_{-\infty}^{\infty} dx_j \rho_j^{\nu}(x_j,t), \quad (11)$$

to be able to cover different scenarios, where α_γ and α_ν are constants. Note that the choice of the $\mathcal{D}_{j,\gamma}(t)$ and $\mathcal{D}_{j,\nu}(t)$ implies that each subsystem influences the other. This aspect of the problem can be associated to the feature that the nonlinearity present in Equations (1) and (2) introduces additional interactions between the subsystems during the thermalization process, where each subsystem works as an additional thermal bath to the other. By using the previous equations, we have

$$\begin{aligned} \frac{d}{dt}F &= -\int_{-\infty}^{\infty} dx_1 \frac{1}{\rho_1} \left\{ \int_{-\infty}^{\infty} dx_2 \left[\frac{\partial}{\partial x_1}\phi_1(x_1)\rho_2\rho_1 - kT\rho_2^2\rho_1 \frac{\partial \rho_1}{\partial x_1}\frac{\partial^2}{\partial \rho_{12}^2} s(\rho_{12}) \right] \right. \\ &\times \left. \int_{-\infty}^{\infty} dx_2 \left[\frac{\partial}{\partial x_1}\phi_1(x_1)\rho_1\rho_2 + \Gamma \frac{\partial \rho_1}{\partial x_1}\rho_2 \frac{\partial}{\partial \rho_{12}} \left(\alpha_\gamma \rho_2^\gamma \rho_1^\gamma + \alpha_\nu \rho_2^\nu \rho_1^\nu\right) \right] \right\} \\ &- \int_{-\infty}^{\infty} dx_2 \frac{1}{\rho_2} \left\{ \int_{-\infty}^{\infty} dx_1 \left[\frac{\partial}{\partial x_2}\phi_2(x_2)\rho_1\rho_2 - kT\rho_1^2\rho_2 \frac{\partial \rho_2}{\partial x_2}\frac{\partial^2}{\partial \rho_{12}^2} s(\rho_{12}) \right] \right. \\ &\times \left. \int_{-\infty}^{\infty} dx_1 \left[\frac{\partial}{\partial x_2}\phi_2(x_2)\rho_1\rho_2 + \Gamma \frac{\partial \rho_2}{\partial x_2}\rho_1 \frac{\partial}{\partial \rho_{12}} \left(\alpha_\gamma \rho_2^\gamma \rho_1^\gamma + \alpha_\nu \rho_2^\nu \rho_1^\nu\right) \right] \right\}. \end{aligned} \tag{12}$$

We verify that

$$\begin{aligned} \frac{d}{dt}F \leq 0 \quad \text{for} \quad -kT\rho_{12}\frac{\partial^2}{\partial \rho_{12}^2} s(\rho_{12}) &= \Gamma \frac{\partial}{\partial \rho_{12}} \left(\alpha_\gamma \rho_2^\gamma \rho_1^\gamma + \alpha_\nu \rho_2^\nu \rho_1^\nu\right) \\ &= \Gamma \frac{\partial}{\partial \rho_{12}} \left(\alpha_\gamma \rho_{12}^\gamma + \alpha_\nu \rho_{12}^\nu\right), \end{aligned} \tag{13}$$

which implies

$$\begin{aligned} \frac{d}{dt}F &= -\int_{-\infty}^{\infty} dx_1 \frac{1}{\rho_1} \left\{ \int_{-\infty}^{\infty} dx_2 \left[\frac{\partial}{\partial x_1}\phi_1(x_1)\rho_2\rho_1 - kT\rho_2^2\rho_1 \frac{\partial \rho_1}{\partial x_1}\frac{\partial^2}{\partial \rho_{12}^2} s(\rho_{12}) \right] \right\}^2 \\ &- \int_{-\infty}^{\infty} dx_2 \frac{1}{\rho_2} \left\{ \int_{-\infty}^{\infty} dx_1 \left[\frac{\partial}{\partial x_2}\phi_2(x_2)\rho_1\rho_2 - kT\rho_1^2\rho_2 \frac{\partial \rho_2}{\partial x_2}\frac{\partial^2}{\partial \rho_{12}^2} s(\rho_{12}) \right] \right\}^2. \end{aligned} \tag{14}$$

Consequently, by solving Equation (13) with $\Gamma = kT$ under the conditions defined in Refs. [28–31], we obtain

$$s(\rho_{12}) = \frac{\alpha_\gamma}{\gamma - 1}(\rho_{12} - \rho_{12}^\gamma) + \frac{\alpha_\nu}{\nu - 1}(\rho_{12} - \rho_{12}^\nu). \tag{15}$$

The entropy for the composite system is given by

$$\begin{aligned} \mathcal{S} &= \frac{\alpha_\gamma k}{\gamma - 1} \int_{-\infty}^{\infty} dx_1 \int_{-\infty}^{\infty} dx_2 (\rho_{12} - \rho_{12}^\gamma) \\ &+ \frac{\alpha_\nu k}{\nu - 1} \int_{-\infty}^{\infty} dx_1 \int_{-\infty}^{\infty} dx_2 (\rho_{12} - \rho_{12}^\nu), \end{aligned} \tag{16}$$

which can also be rewritten as

$$\begin{aligned} \mathcal{S} &= \frac{\alpha_\gamma k}{\gamma - 1} \int_{-\infty}^{\infty} dx_1 \int_{-\infty}^{\infty} dx_2 \left[\rho_1\rho_2 - (\rho_1\rho_2)^\gamma\right] \\ &+ \frac{\alpha_\nu k}{\nu - 1} \int_{-\infty}^{\infty} dx_1 \int_{-\infty}^{\infty} dx_2 \left[\rho_1\rho_2 - (\rho_1\rho_2)^\nu\right] \end{aligned} \tag{17}$$

and, consequently, as

$$\begin{aligned} \mathcal{S} &= \frac{\alpha_\gamma k}{\gamma - 1} \left[1 - \int_{-\infty}^{\infty} dx_1 \int_{-\infty}^{\infty} dx_2 (\rho_1\rho_2)^\gamma\right] \\ &+ \frac{\alpha_\nu k}{\nu - 1} \left[1 - \int_{-\infty}^{\infty} dx_1 \int_{-\infty}^{\infty} dx_2 (\rho_1\rho_2)^\nu\right]. \end{aligned} \tag{18}$$

Equation (18) has several particular cases, such as the Tsallis and Kaniadakis entropies, depending on the values of the parameters α_γ, α_ν, γ, and ν. It is noteworthy that this result preserves the additivity in the Penrose sense [3], i.e., $S(\rho_{12}) = S(\rho_1\rho_2)$ required for a system composed of independent subsystems when the standard entropy is employed.

In the previous context, Equations (1) and (2) can be written as follows:

$$\frac{\partial}{\partial t}\rho_1(x_1,t) = \overline{\mathcal{D}}_{2,\gamma}(t)\frac{\partial^2}{\partial x_1^2}\rho_1^\gamma(x_1,t) + \overline{\mathcal{D}}_{2,\nu}(t)\frac{\partial^2}{\partial x_1^2}\rho_1^\nu(x_1,t) - \frac{\partial}{\partial x_1}[F_1(x_1)\rho_1(x_1,t)] \tag{19}$$

and

$$\frac{\partial}{\partial t}\rho_2(x_2,t) = \overline{\mathcal{D}}_{1,\gamma}(t)\frac{\partial^2}{\partial x_2^2}\rho_2^\gamma(x_2,t) + \overline{\mathcal{D}}_{1,\nu}(t)\frac{\partial^2}{\partial x_2^2}\rho_2^\nu(x_2,t) - \frac{\partial}{\partial x_2}[F_2(x_2)\rho_2(x_2,t)], \tag{20}$$

with $\overline{\mathcal{D}}_{i,\gamma}(t) = \mathcal{D}_{i,\gamma}(t)\Gamma$ and $\overline{\mathcal{D}}_{i,\nu}(t) = \mathcal{D}_{i,\nu}(t)\Gamma$, by evidencing the influence of one of them on the other. In particular, the terms forming the diffusive part can also be connected with anomalous diffusion processes with different diffusion regimes. The stationary solutions obtained from Equations (19) and (20) are given by

$$\frac{\gamma}{\gamma-1}\overline{\mathcal{D}}_{2,\gamma}\rho_{1,st}^{\gamma-1}(x_1) + \frac{\nu}{\nu-1}\overline{\mathcal{D}}_{2,\nu}\rho_{1,st}^{\nu-1}(x_1) = \mathcal{C}_1 - \phi_1(x_1) \tag{21}$$

and

$$\frac{\gamma}{\gamma-1}\overline{\mathcal{D}}_{1,\gamma}\rho_{2,st}^{\gamma-1}(x_2) + \frac{\nu}{\nu-1}\overline{\mathcal{D}}_{1,\nu}\rho_{2,st}^{\nu-1}(x_2) = \mathcal{C}_2 - \phi_2(x_2), \tag{22}$$

where $\lim_{t\to\infty}\overline{\mathcal{D}}_{i,\gamma}(t) = \overline{\mathcal{D}}_{i,\gamma} =$ constant, $\phi_i(x)$ are potentials with a minimum, and $\mathcal{C}_i$ are constants. For the Tsallis entropy, by taking, for simplicity, $\overline{\mathcal{D}}_{i,\nu} = 0$, we have

$$\rho_{1,st}(x_1) = \frac{1}{\mathcal{Z}_1}\left[1-(\gamma-1)\frac{\mathcal{Z}_1^{\gamma-1}}{\gamma\overline{\mathcal{D}}_{2,\gamma}}\phi_1(x_1)\right]^{\frac{1}{\gamma-1}} = \frac{1}{\mathcal{Z}_1}\exp_\gamma\left[-\frac{\mathcal{Z}_1^{\gamma-1}}{\gamma\overline{\mathcal{D}}_{2,\gamma}}\phi_1(x_1)\right] \tag{23}$$

and

$$\rho_{2,st}(x_2) = \frac{1}{\mathcal{Z}_2}\left[1-(\gamma-1)\frac{\mathcal{Z}_2^{\gamma-1}}{\gamma\overline{\mathcal{D}}_{2,\gamma}}\phi_1(x_2)\right]^{\frac{1}{\gamma-1}} = \frac{1}{\mathcal{Z}_2}\exp_\gamma\left[-\frac{\mathcal{Z}_2^{\gamma-1}}{\gamma\overline{\mathcal{D}}_{1,\gamma}}\phi_2(x_2)\right], \tag{24}$$

where $\mathcal{Z}_i = 1/\{[(\gamma-1)/(\gamma\overline{\mathcal{D}}_{i,\gamma})]\mathcal{C}_i\}^{\frac{1}{\gamma-1}}$ is defined by the normalization condition and $\overline{\mathcal{D}}_{i,\gamma} = kT\int_{-\infty}^{\infty}dx_i\rho_{i,st}^\gamma(x_i)$. In the preceding equations, $\exp_q[x]$ is the $q-$exponential function, defined as follows [40]:

$$\exp_q[x] \equiv \begin{cases} (1+(q-1)x)^{1/(q-1)} & , \quad x > 1/(1-q), \\ 0 & , \quad x < 1/(1-q). \end{cases} \tag{25}$$

The presence of this function in the previous equations enables the identification of either a short- or a long-tailed behavior of the solution, depending on the value of the parameters γ and ν. Indeed, they may have a compact behavior for $\gamma > 1$ (or $\nu > 1$) due to the *cut-off* required by the q-exponential to retain the probabilistic interpretation of the distribution. On the other hand, for $\gamma < 1$ (or $\nu > 1$), the solutions may have the asymptotic limit governed by a power-law behavior, which may also be related to a Lévy distribution [44] and, consequently, asymptotically with the solutions of the fractional Fokker–Planck equations [45], which are asymptotically governed by power-laws.

From the stochastic point of view, Equations (19) and (20) are connected with the following Langevin equations:

$$\dot{x}_1 = F_1(x_1) + \sqrt{2\Gamma\Lambda_{1,2}(t)}\xi_1(t) \tag{26}$$

and

$$\dot{x}_2 = F_2(x_2) + \sqrt{2\Gamma\Lambda_{2,1}(t)}\xi_2(t), \tag{27}$$

where $\xi_1(t)$ and $\xi_2(t)$ are connected to the stochastic forces and $\Lambda_{i,j}(t) = \mathcal{D}_{j(i),\gamma(\nu)}(t)\,\rho_{i(j)}^{\gamma(\nu)}(x,t) + \mathcal{D}_{j(i),\nu(\gamma)}(t)\,\rho_{i(j)}^{\nu(\gamma)}(x,t)$. In particular, we have

$$\langle\xi_1\rangle = \langle\xi_2\rangle = 0\ , \quad \langle\xi_1\xi_2\rangle = \langle\xi_2\xi_1\rangle = 0 \tag{28}$$

and

$$\langle\xi_1(t)\xi_1(t')\rangle \propto \delta(t-t')\ , \quad \langle\xi_2(t)\xi_2(t')\rangle \propto \delta(t-t'). \tag{29}$$

The walkers related to this problem can be described, for simplicity, in the absence of external forces, in terms of the following equations [46,47]:

$$\rho_1(x_1, t+\tau) = \int_{-\infty}^{\infty} \Theta_{1,2}[x_1 - x_1', t; \rho(x_1 - x_1', t)]\rho_1(x_1 - x_1', t)\Phi(x_1')dx_1' \tag{30}$$

and

$$\rho_2(x_2, t+\tau) = \int_{-\infty}^{\infty} \Theta_{2,1}[x_2 - x_2', t; \rho(x_2 - x_2', t)]\rho_2(x_2 - x_2', t)\Phi(x_2')dx_2', \tag{31}$$

where

$$\Theta_{i,j}[x_i, t; \rho(x_i, t)] = \alpha_\gamma \int_{-\infty}^{\infty} dx_j \rho_j^\gamma(x_j, t)\rho_i^{\gamma-1}(x_i, t) + \alpha_\nu \int_{-\infty}^{\infty} dx_j \rho_j^\nu(x_j, t)\rho_i^{\nu-1}(x_i, t). \tag{32}$$

These equations, in the limit $\tau \to 0$ and $x_i' \to 0$, yield Equations (1) and (2) in the absence of external forces, respectively.

Let us now consider a general case, i.e., the one in which the diffusion terms have a different nonlinear dependence on the distributions. This means that the systems have different dynamical aspects governed by the nonlinear dependence on the distribution present in the diffusive term. By using the preceding equations and having in mind Equation (5), we may write

$$\begin{aligned} \frac{d}{dt}F &= \int_{-\infty}^{\infty}\int_{-\infty}^{\infty} dx_1 dx_2 \Big\{ \Psi(x_1, x_2)\frac{\partial}{\partial t}[\rho_1(x_1, t)\rho_2(x_2, t)] \\ &- kT\Big[\frac{\partial}{\partial \rho_1}s(\rho_1, \rho_2)\frac{\partial}{\partial t}\rho_1(x_1, t) + \frac{\partial}{\partial \rho_2}s(\rho_1, \rho_2)\frac{\partial}{\partial t}\rho_2(x_2, t)\Big]\Big\}, \end{aligned} \tag{33}$$

which implies

$$\begin{aligned} \frac{d}{dt}F &= \int_{-\infty}^{\infty}\int_{-\infty}^{\infty} dx_1 dx_2 \Big\{ \Psi(x_1, x_2)\rho_2 - kT\frac{\partial}{\partial \rho_1}s(\rho_1, \rho_2)\Big\} \\ &\times \frac{\partial}{\partial x_1}\Big\{\Gamma\frac{\partial}{\partial x_1}P_1(\rho_1, t) - F_1(x_1)\rho_1(x_1, t)\Big\} \\ &+ \int_{-\infty}^{\infty}\int_{-\infty}^{\infty} dx_1 dx_2 \Big\{ \Psi(x_1, x_2)\rho_1 - kT\frac{\partial}{\partial \rho_2}s(\rho_1, \rho_2)\Big\} \\ &\times \frac{\partial}{\partial x_2}\Big\{\Gamma\frac{\partial}{\partial x_2}P_2(\rho_2, t) - F_2(x_2)\rho_2(x_2, t)\Big\}. \end{aligned} \tag{34}$$

After some calculations, it is possible to show that

$$\begin{aligned}\frac{d}{dt}F &= -\int_{-\infty}^{\infty} dx_1 \frac{1}{\rho_1}\int_{-\infty}^{\infty} dx_2 \left\{ \frac{\partial}{\partial x_1}\phi(x_1)\rho_2\rho_1 - kT\left[\rho_1\frac{\partial^2}{\partial \rho_1^2}s(\rho_1,\rho_2)\right]\frac{\partial}{\partial x_1}\rho_1 \right\} \\ &\times \left\{ \left[\Gamma\frac{\partial}{\partial \rho_1}P_1(\rho_1,t)\right]\frac{\partial}{\partial x_1}\rho_1 - F_1(x_1)\rho_1 \right\} \\ &- \int_{-\infty}^{\infty} dx_1 \int_{-\infty}^{\infty} dx_2 \frac{1}{\rho_2}\left\{ \frac{\partial}{\partial x_2}\phi(x_2)\rho_1\rho_2 - kT\left[\rho_2\frac{\partial^2}{\partial \rho_2^2}s(\rho_1,\rho_2)\right]\frac{\partial}{\partial x_2}\rho_2 \right\} \\ &\times \frac{\partial}{\partial x_2}\left\{ \left[\Gamma\frac{\partial}{\partial \rho_2}P_2(\rho_2,t)\right]\frac{\partial}{\partial x_2}\rho_2 - F_2(x_2)\rho_2 \right\}. \end{aligned} \tag{35}$$

Let us analyze in particular the previous equation, for example, for the case

$$P_1(\rho_1,t) = \mathcal{D}_{2,\nu}(t)\rho_1^{\gamma}(x_1,t) \quad \text{and} \quad P_2(\rho_2,t) = \mathcal{D}_{1,\gamma}(t)\rho_2^{\nu}(x_2,t), \tag{36}$$

with

$$\mathcal{D}_{2,\nu}(t) = \frac{1}{\nu-1}\int_{-\infty}^{\infty} dx_2\rho_2^{\nu}(x_2,t) \quad \text{and} \quad \mathcal{D}_{1,\gamma}(t) = \frac{1}{\gamma-1}\int_{-\infty}^{\infty} dx_1\rho_1^{\gamma}(x_1,t), \tag{37}$$

which implies different dynamics for each subsystem. We notice that it is possible to take into account different aspects of the dynamics of each subsystem, and every choice has different implications for the total entropy of the composite system. Similar nonlinear Fokker–Planck equations were considered in Ref. [48] from the point of view of analyzing the interaction between the two subsystems. From Equation (37), we deduce that the entropy needs to satisfy the following equations:

$$-\rho_1\frac{\partial^2}{\partial \rho_1^2}s(\rho_1,\rho_2) = \frac{\gamma}{\nu-1}\rho_2^{\nu}\rho_1^{\gamma-1} \quad \text{and} \quad -\rho_2\frac{\partial^2}{\partial \rho_2^2}s(\rho_1,\rho_2) = \frac{\nu}{\gamma-1}\rho_2^{\nu-1}\rho_1^{\gamma} \tag{38}$$

in order to verify

$$\frac{d}{dt}F \leq 0, \tag{39}$$

and, consequently, to satisfy the H–theorem. A solution for the previous system of equations is

$$s(\rho_1,\rho_2) = \frac{1}{(\nu-1)(\gamma-1)}\left(\rho_1\rho_2 - \rho_2^{\nu}\rho_1^{\gamma}\right). \tag{40}$$

This result allows us to write the total entropy of this system as follows:

$$\mathcal{S} = \frac{k}{(\nu-1)(\gamma-1)}\left[1 - \int_{-\infty}^{\infty} dx_2\rho_2^{\nu}(x_2,t)\int_{-\infty}^{\infty} dx_1\rho_1^{\gamma}(x_1,t)\right]. \tag{41}$$

It is remarkable that this result for the entropy differs from the preceding one given by Equation (18), obtained from a different choice of nonlinear Fokker–Planck equations. Equation (41) results from a combination of different subsystems with different dynamics, which individually have different entropies associated with them. One of the consequences is that the entropy of the composite system, for this specific case, can not be written as $S(\rho_1\rho_2)$, only when $\gamma = \nu$. Another remarkable point is the connection of Equation (41) with the composition of Tsallis entropies of different q-indices [49,50]. The solution can be found in this framework using the q-exponential functions. In particular, it is possible to show that the solution for each nonlinear Fokker–Planck equation, in the absence of external force, is

$$\rho_1(x_1,t) = \exp_\gamma\left[-\beta_1(t)x_1^2\right]/\mathcal{Z}_1(t) \tag{42}$$

and

$$\rho_2(x_1,t) = \exp_\nu\left[-\beta_2(t)x_2^2\right]/\mathcal{Z}_2(t), \tag{43}$$

with $\beta_1(t)$, $\beta_2(t)$, $\mathcal{Z}_1(t)$, and $\mathcal{Z}_2(t)$ obtained from the following set of equations:

$$\frac{1}{2\beta_1}\frac{d}{dt}\beta_1 = -\frac{2\gamma}{\nu-1}\frac{\mathcal{I}_\nu}{\mathcal{Z}_2^\nu\sqrt{\beta_2}}\beta_1\mathcal{Z}_1^{1-\gamma} \quad , \quad -\frac{1}{\mathcal{Z}_1}\frac{d}{dt}\mathcal{Z}_1 = -\frac{2\gamma}{\nu-1}\frac{\mathcal{I}_\nu}{\mathcal{Z}_2^\nu\sqrt{\beta_2}}\beta_1\mathcal{Z}_1^{1-\gamma}, \tag{44}$$

$$\frac{1}{2\beta_2}\frac{d}{dt}\beta_2 = -\frac{2\nu}{\gamma-1}\frac{\mathcal{I}_\gamma}{\mathcal{Z}_1^\gamma\sqrt{\beta_1}}\beta_2\mathcal{Z}_2^{1-\nu} \quad , \quad -\frac{1}{\mathcal{Z}_2}\frac{d}{dt}\mathcal{Z}_2 = -\frac{2\nu}{\gamma-1}\frac{\mathcal{I}_\gamma}{\mathcal{Z}_1^\gamma\sqrt{\beta_1}}\beta_2\mathcal{Z}_2^{1-\nu}, \tag{45}$$

with

$$\mathcal{I}_\kappa = \begin{cases} \frac{\Gamma(\frac{1}{2})\Gamma(1+\frac{\kappa}{\kappa-1})}{\sqrt{\kappa-1}\Gamma(\frac{3}{2}+\frac{\kappa}{\kappa-1})} & 1 \le \kappa < 2 \\ \frac{\Gamma(\frac{1}{2})\Gamma(\frac{\kappa}{1-\kappa}-\frac{1}{2})}{\sqrt{1-\kappa}\Gamma(\frac{\kappa}{1-\kappa})} & 0 \le \kappa \le 1 \end{cases}, \tag{46}$$

where $\kappa = \gamma$ or ν.

Figure 1 shows the behavior of the mean square displacement for two different sets of γ and ν in the absence of external forces. The values chosen for the parameters γ and ν are responsible for different behaviors of the mean square displacement for each case, as pointed out in the inset of Figure 1. In particular, the diffusion present in this scenario is anomalous [51,52]. Figure 2 shows the behavior of Equation (41) for two different sets of γ and ν. Note that different values of $\beta_1(0)$ and $\beta_2(0)$ used to obtain Figures 1 and 2 are connected to different initial conditions for each subsystem. This is the reason why we initially verified different behaviors for each set of the parameters γ and ν, and, after some time, the mean square displacement has the same time dependence for both subsystems. The entropy production is shown in the inset in Figure 2, which corresponds to the behavior of Equation (61) for the entropy given by Equation (41). We underline that the system composed of these two systems reaches equilibrium in the limit of $t \to \infty$, since in this limit $\dot{S}(t) \to 0$. For general nonlinear Fokker–Planck equations, the entropy should simultaneously satisfy the following equations,

$$-\rho_1\frac{\partial^2}{\partial\rho_1^2}s(\rho_1,\rho_2) = \frac{\partial}{\partial\rho_1}P_1(\rho_1,t) \quad \text{and} \quad -\rho_2\frac{\partial^2}{\partial\rho_2^2}s(\rho_1,\rho_2) = \frac{\partial}{\partial\rho_2}P_2(\rho_2,t), \tag{47}$$

to verify $\frac{d}{dt}F \le 0$ and, consequently, satisfy the H–theorem. It is also significant to mention that, depending on the form of the nonlinear dependence in the Equations (1) and (2), which may not recover the standard form of the Fokker–Planck equation, the entropy associated with these equations will not recover the usual form.

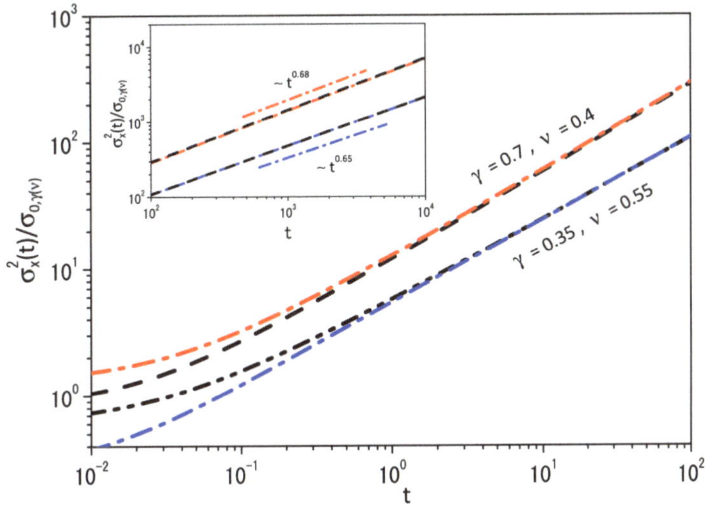

Figure 1. Behavior of $\sigma_x^2/\sigma_{0,\gamma(\nu)}$ versus t for two different sets of γ and ν, where $\sigma_{0,\gamma(\nu)} = \sigma_{\gamma(\nu)} \int_{-\infty}^{\infty} d\xi \xi^2 \exp_{\gamma(\nu)}(-\xi^2) / \int_{-\infty}^{\infty} d\xi \exp_{\gamma(\nu)}(-\xi^2)$, where $\sigma_{\gamma(\nu)}$ is chosen in order to collapse the curves for each set of values. We consider, for simplicity, $\beta_1(0) = 2$ and $\beta_2(0) = 1$. The red dashed-dotted and black dashed lines represent the case $\gamma = 0.4$ with $\nu = 0.7$. The blue dashed-dotted and black dashed-dotted-dotted lines represent the case $\gamma = 0.35$ with $\nu = 0.55$. Notice that the behavior for the cases worked out in this figure have different time dependence for the mean square displacement, as pointed out in the inset.

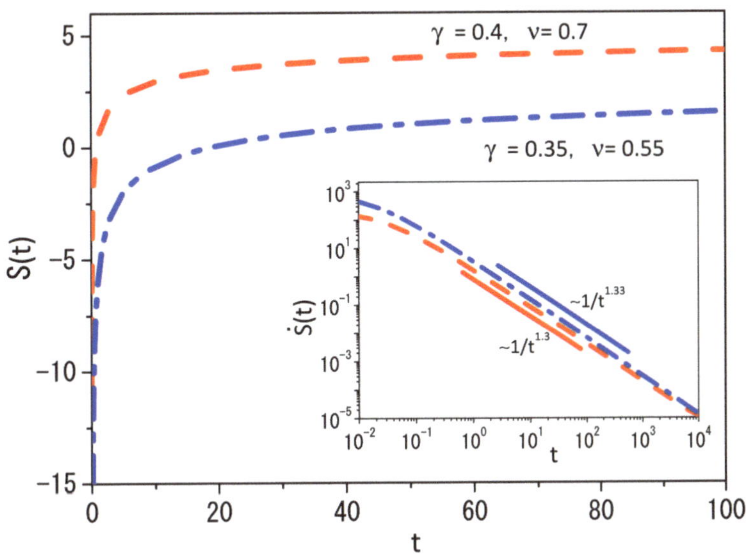

Figure 2. Behavior of Equation (41) versus t for two different sets of γ and ν. We consider, for simplicity, $\beta_1(0) = 2$ and $\beta_2(0) = 1$. The red dashed-dotted line represents the case $\gamma = 0.4$ with $\nu = 0.7$. The blue dashed-dotted line represents the case $\gamma = 0.35$ with $\nu = 0.55$. Notice that the behavior for the cases worked out in this figure have different time dependence for $\dot{S}(t)$, as pointed out in the inset.

2.2. Entropy Production

Let us analyze the entropy production related to Equation (17) with the dynamics of $\rho_1(x_1,t)$ and $\rho_2(x_2,t)$ given by Equations (19) and (20). By performing a time derivative of Equation (17), we obtain

$$\begin{aligned}\frac{d}{dt}\mathcal{S}(t) &= k\int_{-\infty}^{\infty}\int_{-\infty}^{\infty}dx_1dx_2\left[\frac{\partial}{\partial\rho_{12}}s(\rho_{12})\right]\frac{\partial}{\partial t}[\rho_1(x_1,t)\rho_2(x_2,t)] \\ &= -k\int_{-\infty}^{\infty}dx_1\int_{-\infty}^{\infty}dx_2\rho_2\frac{\partial}{\partial\rho_{12}}s(\rho_{12})\frac{\partial}{\partial x_1}\mathcal{J}_1(x_1,t) \\ &- k\int_{-\infty}^{\infty}dx_1\rho_1\int_{-\infty}^{\infty}dx_2\frac{\partial}{\partial\rho_{12}}s(\rho_{12})\frac{\partial}{\partial x_2}\mathcal{J}_2(x_2,t)\end{aligned} \tag{48}$$

and, consequently, performing integration by parts with the conditions $\mathcal{J}_1(x_1\to\pm\infty,t)\to 0$ and $\mathcal{J}_2(x_2\to\pm\infty,t)\to 0$, also

$$\begin{aligned}\frac{d}{dt}\mathcal{S}(t) &= k\int_{-\infty}^{\infty}dx_1\int_{-\infty}^{\infty}dx_2\left[\rho_2^2\frac{\partial^2}{\partial\rho_{12}^2}s(\rho_{12})\frac{\partial\rho_1}{\partial x_1}\right]\mathcal{J}_1(x_1,t) \\ &+ k\int_{-\infty}^{\infty}dx_1\int_{-\infty}^{\infty}dx_2\left[\rho_1^2\frac{\partial^2}{\partial\rho_{12}^2}s(\rho_{12})\frac{\partial\rho_2}{\partial x_2}\right]\mathcal{J}_2(x_2,t)\,.\end{aligned} \tag{49}$$

It is possible to simplify Equation (48) by using, from the H–theorem, the equations

$$-kT\rho_1\rho_2^2\frac{\partial\rho_1}{\partial x_1}\frac{\partial^2}{\partial\rho_{12}^2}s(\rho_{12}) = \Gamma\frac{\partial}{\partial x_1}P_1(\rho_1,t) \tag{50}$$

and

$$-kT\rho_2\rho_1^2\frac{\partial\rho_2}{\partial x_2}\frac{\partial^2}{\partial\rho_{12}^2}s(\rho_{12}) = \Gamma\frac{\partial}{\partial x_2}P_2(\rho_2,t), \tag{51}$$

in order to obtain

$$\begin{aligned}\frac{d}{dt}\mathcal{S}(t) &= -\frac{1}{T}\int_{-\infty}^{\infty}dx_1F_1(x_1)\mathcal{J}_1(x_1,t) - \frac{1}{T}\int_{-\infty}^{\infty}dx_2F_2(x_2)\mathcal{J}_2(x_2,t) \\ &+ \frac{1}{T}\int_{-\infty}^{\infty}dx_1\frac{\mathcal{J}_1^2(x_1,t)}{\rho_1(x_1,t)} + \frac{1}{T}\int_{-\infty}^{\infty}dx_2\frac{\mathcal{J}_2^2(x_2,t)}{\rho_2(x_2,t)},\end{aligned} \tag{52}$$

where

$$\mathcal{J}_1(x_1,t) = -\Gamma\frac{\partial}{\partial x_1}P_1(\rho_1,t) + F_1(x_1)\rho_1(x_1,t) \tag{53}$$

and

$$\mathcal{J}_2(x_2,t) = -\Gamma\frac{\partial}{\partial x_2}P_2(\rho_2,t) + F_2(x_2)\rho_2(x_2,t), \tag{54}$$

with $P_1(\rho_1,t)$ and $P_2(\rho_2,t)$ given by Equations (10) and (11). Equation (48) can be written as follows:

$$\frac{d}{dt}\mathcal{S} = \Pi - \Phi \tag{55}$$

where one identifies the entropy flux, representing the exchanges of entropy between the subsystems represented by ρ_1 and ρ_2 and their neighborhood,

$$\Phi = \frac{1}{T}\int_{-\infty}^{\infty}dx_1F_1(x_1)\mathcal{J}_1(x_1,t) + \frac{1}{T}\int_{-\infty}^{\infty}dx_2F_2(x_2)\mathcal{J}_2(x_2,t), \tag{56}$$

as well as the entropy-production contribution:

$$\Pi = \frac{1}{T}\int_{-\infty}^{\infty} dx_1 \frac{\mathcal{J}_1^2(x_1,t)}{\rho_1(x_1,t)} + \frac{1}{T}\int_{-\infty}^{\infty} dx_2 \frac{\mathcal{J}_2^2(x_2,t)}{\rho_2(x_2,t)}. \quad (57)$$

We underline that T and $\rho_i(x_i,t)$ are positive quantities, yielding the desirable result: $\Pi \geq 0$.

For the general case represented by Equation (4), we have

$$\begin{aligned} \frac{d}{dt}\mathcal{S}(t) &= k\int_{-\infty}^{\infty}\int_{-\infty}^{\infty} dx_1 dx_2 \left[\frac{\partial}{\partial \rho_1}s(\rho_1,\rho_2)\frac{\partial}{\partial t}\rho_1(x_1,t) + \frac{\partial}{\partial \rho_2}s(\rho_1,\rho_2)\frac{\partial}{\partial t}\rho_2(x_2,t)\right] \\ &= -k\int_{-\infty}^{\infty}\int_{-\infty}^{\infty} dx_1 dx_2 \frac{\partial}{\partial \rho_1}s(\rho_1,\rho_2)\frac{\partial}{\partial x_1}\mathcal{J}_1(x_1,t) \\ &- k\int_{-\infty}^{\infty}\int_{-\infty}^{\infty} dx_1 dx_2 \frac{\partial}{\partial \rho_2}s(\rho_1,\rho_2)\frac{\partial}{\partial x_2}\mathcal{J}_2(x_2,t). \end{aligned} \quad (58)$$

Performing integration by parts in Equation (58) and by taking into account the conditions $\mathcal{J}_1(x_1 \to \pm\infty, t) \to 0$ and $\mathcal{J}_2(x_2 \to \pm\infty, t) \to 0$, we obtain that

$$\begin{aligned} \frac{d}{dt}\mathcal{S}(t) &= k\int_{-\infty}^{\infty}\int_{-\infty}^{\infty} dx_1 dx_2 \left[\frac{\partial \rho_1}{\partial x_1}\frac{\partial^2}{\partial \rho_1^2}s(\rho_1,\rho_2)\right]\mathcal{J}_1(x_1,t) \\ &+ k\int_{-\infty}^{\infty}\int_{-\infty}^{\infty} dx_1 dx_2 \left[\frac{\partial \rho_2}{\partial x_2}\frac{\partial^2}{\partial \rho_2^2}s(\rho_1,\rho_2)\right]\mathcal{J}_2(x_2,t). \end{aligned} \quad (59)$$

By using the equations,

$$-\rho_1\frac{\partial \rho_1}{\partial x_1}\frac{\partial^2}{\partial \rho_1^2}s(\rho_1,\rho_2) = \frac{\partial}{\partial x_1}P_1(\rho_1,t) \quad \text{and} \quad -\rho_2\frac{\partial \rho_2}{\partial x_2}\frac{\partial^2}{\partial \rho_2^2}s(\rho_1,\rho_2) = \frac{\partial}{\partial x_2}P_2(\rho_2,t), \quad (60)$$

it is possible to simplify Equation (59) in order to obtain

$$\begin{aligned} \frac{d}{dt}\mathcal{S}(t) &= -\frac{1}{T}\int_{-\infty}^{\infty} dx_1 F_1(x_1)\mathcal{J}_1(x_1,t) - \frac{1}{T}\int_{-\infty}^{\infty} dx_2 F_2(x_2)\mathcal{J}_2(x_2,t) \\ &+ \frac{1}{T}\int_{-\infty}^{\infty} dx_1 \frac{\mathcal{J}_1^2(x_1,t)}{\rho_1(x_1,t)} + \frac{1}{T}\int_{-\infty}^{\infty} dx_2 \frac{\mathcal{J}_2^2(x_2,t)}{\rho_2(x_2,t)}, \end{aligned} \quad (61)$$

where

$$\mathcal{J}_1(x_1,t) = -\Gamma\frac{\partial}{\partial x_1}P_1(\rho_1,t) + F_1(x_1)\rho_1(x_1,t) \quad (62)$$

and

$$\mathcal{J}_2(x_2,t) = -\Gamma\frac{\partial}{\partial x_2}P_2(\rho_2,t) + F_2(x_2)\rho_2(x_2,t), \quad (63)$$

as before, with $P_1(\rho_1,t)$ and $P_2(\rho_2,t)$ arbitrary. Note that Equation (61) is formally equal to Equation (52), which evidences that the result obtained for the entropy production is invariant in form when the entropies are obtained from the H–theorem.

3. Discussion and Conclusions

We have investigated the entropy of a system composed of two subsystems governed by nonlinear Fokker–Planck equations. In this context, we have essentially analyzed two scenarios; in one of them, the subsystems have the same dynamics, and in the other one, they have different dynamics, i.e., the nonlinear Fokker–Planck equations are different. The first case allows the definition of entropy which can be connected to different cases and preserves the formal structure $S(\rho_1,\rho_2) = S(\rho_1\rho_2)$ also verified by the standard entropy of

Boltzmann–Gibbs. For the other case, we consider different dynamics for each subsystem, which allows the definition of an entropic form for which $S(\rho_1, \rho_2) \neq S(\rho_1\rho_2)$. In both cases, we have analyzed the entropy production and we have shown the effect of each subsystem on the composite system. In addition, we have shown that the time variation of the entropy (entropy production) for the total system is invariant in form for all the cases considered here.

Author Contributions: Conceptualization, L.R.E. and E.K.L.; methodology, L.R.E. and E.K.L.; validation, L.R.E. and E.K.L.; formal analysis, L.R.E. and E.K.L.; investigation, L.R.E. and E.K.L.; writing—original draft preparation, L.R.E. and E.K.L.; writing—review and editing, L.R.E. and E.K.L. All authors have read and agreed to the published version of the manuscript.

Funding: E.K.L. acknowledges the support of the CNPq (Grant No. 301715/2022-0).

Institutional Review Board Statement: Not applicable.

Data Availability Statement: Not applicable.

Acknowledgments: E.K.L. acknowledges the support of the CNPq and the National Institute of Science and Technology of Complex Systems-INCT-SC. LRE acknowledges the support of the Program of Visiting Professor of Politecnico di Torino.

Conflicts of Interest: The authors declare no conflict of interest.

References

1. Clausius, R. *The Mechanical Theory of Heat: With Its Appli- cations to the Steam-Engine and to the Physical Properties of Bodies*; Macmillan: New York, NY, USA, 1879.
2. Bryan, G. Elementary principles in statistical mechanics. *Nature* **1902**, *66*, 291–292. [CrossRef]
3. Penrose, O. *Foundations of Statistical Mechanics: A Deductive Treatment*; Courier Corporation: Washington, DC, USA, 2005.
4. Sandler, S.I.; Woodcock, L.V. Historical observations on laws of thermodynamics. *J. Chem. Eng. Data* **2010**, *55*, 4485–4490. [CrossRef]
5. Bogoliubov, N.N. *Problems of Dynamic Theory in Statistical Physics*; Technical Information Service, United States Atomic Energy Commission: Washington, DC, USA, 1960.
6. Hill, T.L. *Statistical Mechanics: Principles and Selected Applications*; Courier Corporation: Washington, DC, USA, 2013.
7. Abe, S.; Okamoto, Y. *Nonextensive Statistical Mechanics and Its Applications*; Springer Science & Business Media: Berlin/Heidelberg, Germany, 2001; Volume 560.
8. Mukamel, D. Statistical mechanics of systems with long range interactions. In *Proceedings of the AIP Conference Proceedings*; American Institute of Physics: Melville, NY, USA, 2008; Volume 970, pp. 22–38.
9. Singh, R.; Moessner, R.; Roy, D. Effect of long-range hopping and interactions on entanglement dynamics and many-body localization. *Phys. Rev. B* **2017**, *95*, 094205. [CrossRef]
10. Tsallis, C. Possible generalization of Boltzmann-Gibbs statistics. *J. Stat. Phys.* **1988**, *52*, 479–487. [CrossRef]
11. Nojiri, S.; Odintsov, S.D.; Faraoni, V. From nonextensive statistics and black hole entropy to the holographic dark universe. *Phys. Rev. D* **2022**, *105*, 044042. [CrossRef]
12. Khordad, R.; Sedehi, H. Electrocaloric effect in quantum dots using the non-extensive formalism. *Opt. Quantum Electron.* **2022**, *54*, 511. [CrossRef]
13. Chavanis, P.H. Nonlinear mean field Fokker-Planck equations. Application to the chemotaxis of biological populations. *Eur. Phys. J. B* **2008**, *62*, 179–208. [CrossRef]
14. Megías, E.; Timóteo, V.; Gammal, A.; Deppman, A. Bose–Einstein condensation and non-extensive statistics for finite systems. *Phys. A Stat. Mech. Its Appl.* **2022**, *585*, 126440. [CrossRef]
15. Rajagopal, A.; Mendes, R.; Lenzi, E. Quantum statistical mechanics for nonextensive systems: Prediction for possible experimental tests. *Phys. Rev. Lett.* **1998**, *80*, 3907. [CrossRef]
16. Lenzi, E.; Mendes, R.; Da Silva, L. Statistical mechanics based on Renyi entropy. *Phys. A Stat. Mech. Its Appl.* **2000**, *280*, 337–345. [CrossRef]
17. Lopes, A.M.; Machado, J.A.T. A review of fractional order entropies. *Entropy* **2020**, *22*, 1374. [CrossRef] [PubMed]
18. Lenzi, E.; Scarfone, A. Extensive-like and intensive-like thermodynamical variables in generalized thermostatistics. *Phys. A Stat. Mech. Its Appl.* **2012**, *391*, 2543–2555. [CrossRef]
19. da Silva, J.; da Silva, G.; Ramos, R. The Lambert-Kaniadakis Wκ function. *Phys. Lett. A* **2020**, *384*, 126175. [CrossRef]
20. Kaniadakis, G. Maximum entropy principle and power-law tailed distributions. *Eur. Phys. J. B* **2009**, *70*, 3–13. [CrossRef]
21. Oylukan, A.D.; Shizgal, B. Nonequilibrium distributions from the Fokker-Planck equation: Kappa distributions and Tsallis entropy. *Phys. Rev. E* **2023**, *108*, 014111. [CrossRef] [PubMed]
22. Lutz, E. Anomalous diffusion and Tsallis statistics in an optical lattice. *Phys. Rev. A* **2003**, *67*, 051402. [CrossRef]

23. Borland, L. Ito-Langevin equations within generalized thermostatistics. *Phys. Lett. A* **1998**, *245*, 67–72. [CrossRef]
24. Chavanis, P.H. Generalized Euler, Smoluchowski and Schrödinger equations admitting self-similar solutions with a Tsallis invariant profile. *Eur. Phys. J. Plus* **2019**, *134*, 353. [CrossRef]
25. Kaniadakis, G.; Lapenta, G. Microscopic dynamics underlying anomalous diffusion. *Phys. Rev. E* **2000**, *62*, 3246–3249. [CrossRef]
26. Frank, T. A Langevin approach for the microscopic dynamics of nonlinear Fokker–Planck equations. *Phys. A Stat. Mech. Its Appl.* **2001**, *301*, 52–62. [CrossRef]
27. Kaniadakis, G. H–theorem and generalized entropies within the framework of nonlinear kinetics. *Phys. Lett. A* **2001**, *288*, 283–291. [CrossRef]
28. Plastino, A.; Wedemann, R.; Nobre, F. H–theorems for systems of coupled nonlinear Fokker-Planck equations. *Europhys. Lett.* **2022**, *139*, 11002. [CrossRef]
29. Dos Santos, M.; Lenzi, E. Entropic nonadditivity, H theorem, and nonlinear Klein-Kramers equations. *Phys. Rev. E* **2017**, *96*, 052109. [CrossRef]
30. dos Santos, M.; Lenzi, M.; Lenzi, E. Nonlinear Fokker–Planck equations, H–theorem, and entropies. *Chin. J. Phys.* **2017**, *55*, 1294–1299. [CrossRef]
31. Casas, G.; Nobre, F.; Curado, E. Entropy production and nonlinear Fokker-Planck equations. *Phys. Rev. E* **2012**, *86*, 061136. [CrossRef] [PubMed]
32. Frank, T.D. *Nonlinear Fokker-Planck Equations: Fundamentals and Applications*; Springer Science & Business Media: Berlin/Heidelberg, Germany, 2005.
33. Schwämmle, V.; Nobre, F.D.; Curado, E.M. Consequences of the H theorem from nonlinear Fokker-Planck equations. *Phys. Rev. E* **2007**, *76*, 041123. [CrossRef] [PubMed]
34. Schwämmle, V.; Curado, E.M.; Nobre, F.D. A general nonlinear Fokker-Planck equation and its associated entropy. *Eur. Phys. J. B* **2007**, *58*, 159–165. [CrossRef]
35. Pedron, I.T.; Mendes, R.; Malacarne, L.C.; Lenzi, E.K. Nonlinear anomalous diffusion equation and fractal dimension: Exact generalized Gaussian solution. *Phys. Rev. E* **2002**, *65*, 041108. [CrossRef]
36. Vázquez, J.L. *The Porous Medium Equation: Mathematical Theory*; Oxford University Press: Oxford, UK, 2007.
37. Casas, G.A.; Nobre, F.D. Nonlinear Fokker-Planck equations in super-diffusive and sub-diffusive regimes. *J. Math. Phys.* **2019**, *60*, 053301. [CrossRef]
38. Plastino, A.R.; Wedemann, R.S.; Tsallis, C. Nonlinear fokker-planck equation for an overdamped system with drag depending on direction. *Symmetry* **2021**, *13*, 1621. [CrossRef]
39. Deppman, A.; Khalili Golmankhaneh, A.; Megías, E.; Pasechnik, R. From the Boltzmann equation with non-local correlations to a standard non-linear Fokker-Planck equation. *Phys. Lett. B* **2023**, *839*, 137752. [CrossRef]
40. Tsallis, C. *Introduction to Nonextensive Statistical Mechanics*; Springer: Berlin/Heidelberg, Germany, 2009; Volume 34.
41. Martınez, S.; Pennini, F.; Plastino, A. Thermodynamics' zeroth law in a nonextensive scenario. *Phys. A Stat. Mech. Its Appl.* **2001**, *295*, 416–424. [CrossRef]
42. Biró, T.; Ván, P. Zeroth law compatibility of nonadditive thermodynamics. *Phys. Rev. E* **2011**, *83*, 061147. [CrossRef]
43. Wang, Q.A.; Le Méhauté, A. Unnormalized nonextensive expectation value and zeroth law of thermodynamics. *Chaos Solitons Fractals* **2003**, *15*, 537–541. [CrossRef]
44. Tsallis, C.; Levy, S.V.; Souza, A.M.; Maynard, R. Statistical-mechanical foundation of the ubiquity of Lévy distributions in nature. *Phys. Rev. Lett.* **1995**, *75*, 3589–3592. [CrossRef] [PubMed]
45. Lenzi, E.K.; Neto, R.M.; Tateishi, A.A.; Lenzi, M.K.; Ribeiro, H. Fractional diffusion equations coupled by reaction terms. *Phys. A* **2016**, *458*, 9–16. [CrossRef]
46. Mendez, V.; Campos, D.; Bartumeus, F. *Stochastic Foundations in Movement Ecology*; Springer: Berlin/Heidelberg, Germany, 2014.
47. Lenzi, E.; Lenzi, M.; Ribeiro, H.; Evangelista, L. Extensions and solutions for nonlinear diffusion equations and random walks. *Proc. R. Soc. A* **2019**, *475*, 20190432. [CrossRef]
48. Marin, D.; Ribeiro, M.; Ribeiro, H.; Lenzi, E. A nonlinear Fokker–Planck equation approach for interacting systems: Anomalous diffusion and Tsallis statistics. *Phys. Lett. A* **2018**, *382*, 1903–1907. [CrossRef]
49. Wang, Q.A.; Nivanen, L.; Méhauté, A.L. A composition of different q nonextensive systems with the normalized expectation based on escort probability. *arXiv* **2006**, arXiv:cond-mat/0601255.
50. Nivanen, L.; Pezeril, M.; Wang, Q.; Le Méhauté, A. Applying incomplete statistics to nonextensive systems with different q indices. *Chaos Solitons Fractals* **2005**, *24*, 1337–1342. [CrossRef]
51. Evangelista, L.R.; Lenzi, E.K. *Fractional Diffusion Equations and Anomalous Diffusion*; Cambridge University Press: Cambridge, UK, 2018.
52. Evangelista, L.R.; Lenzi, E.K. *An Introduction to Anomalous Diffusion and Relaxation*; Springer: Berlin/Heidelberg, Germany, 2023.

entropy

MDPI

Article

Multi-Additivity in Kaniadakis Entropy

Antonio M. Scarfone [1,*] and Tatsuaki Wada [2]

1 Istituto dei Sistemi Complessi—Consiglio Nazionale delle Ricerche (ISC-CNR), c/o Dipartimento di Scienza Applicata e Tecnologia del Politecnico di Torino, Corso Duca degli Abruzzi 24, 10129 Torino, Italy

2 Region of Electrical and Electronic Systems Engineering, Ibaraki University, 4-12-1 Nakanarusawa-cho, Hitachi 316-8511, Ibaraki, Japan; tatsuaki.wada.to@vc.ibaraki.ac.jp

* Correspondence: antoniomaria.scarfone@cnr.it

Abstract: It is known that Kaniadakis entropy, a generalization of the Shannon–Boltzmann–Gibbs entropic form, is always super-additive for any bipartite statistically independent distributions. In this paper, we show that when imposing a suitable constraint, there exist classes of maximal entropy distributions labeled by a positive real number $\aleph > 0$ that makes Kaniadakis entropy multi-additive, i.e., $S_\kappa[p^{\rm A\cup B}] = (1+\aleph)\,(S_\kappa[p^{\rm A}] + S_\kappa[p^{\rm B}])$, under the composition of two statistically independent and identically distributed distributions $p^{\rm A\cup B}(x,\,y) = p^{\rm A}(x)\,p^{\rm B}(y)$, with reduced distributions $p^{\rm A}(x)$ and $p^{\rm B}(y)$ belonging to the same class.

Keywords: κ-entropy; pseudo-additivity; power-law distributions

1. Introduction

A possible generalization of conventional statistics, named κ-statistics, is founded on Kaniadakis entropy (κ-entropy) [1–3]. This is a continuous one-parameter deformation of the information functional, also known as the Shannon–Boltzmann–Gibbs (SBG) entropic form, defined in

$$S_\kappa[p] = -\int\limits_{\mathcal{D}} p(x)\,\ln_\kappa\left(p(x)\right)dx\,, \tag{1}$$

in the appropriate dimensionless unities, where $\mathcal{D}$ is a suitable integration domain and

$$\ln_\kappa(x) = \frac{x^\kappa - x^{-\kappa}}{2\,\kappa}\,, \tag{2}$$

is a deformed version of the standard logarithm that, in the $\kappa \to 0$ limit, reduces to the ordinary logarithm: $\ln_0(x) \equiv \ln(x)$. Is then clear that, in the same limit, entropy S_κ reproduces also the standard expression of SBG-entropy.

Over the last 15 years, the statistics theory based on the κ-entropy has attracted the interest of many researchers, who have studied its foundations on the physical ground [4–11] and its mathematical aspects [12–17]. Concurrently, κ-statistic has been employed in various fields. A non-exhaustive list of applications of κ-entropy and κ-distribution includes, to cite a few, those in thermodynamics [18,19]; plasma physics and astrophysics [20–26]; nuclear physics [27–30]; cosmological issues [31–33], including dark energy [34–36] and holographic theory [37–39]; information theory [40–46]; genomics [47,48]; complex networks [49,50]; the economy [51–53]; and finance [54–56].

As is known [57], for a joined statistical system described by the bipartite probability distribution $p^{\rm A\cup B}(x,\,y)$ of two statistically independent distributions $p^{\rm A}(x)$ and $p^{\rm B}(y)$, i.e., $p^{\rm A\cup B}(x,\,y) = p^{\rm A}(x)\,p^{\rm B}(y)$, the κ-entropy $S_\kappa\left[p^{\rm A\cup B}\right]$ is a super-additive quantity, being

$$S_\kappa\left[p^{\rm A\cup B}\right] > S_\kappa\left[p^{\rm A}\right] + S_\kappa\left[p^{\rm B}\right]\,. \tag{3}$$

Citation: Scarfone, A.M.; Wada, T. Multi-Additivity in Kaniadakis Entropy. *Entropy* **2024**, *26*, 77. https://doi.org/10.3390/e26010077

Academic Editor: Nikolai Leonenko

Received: 19 December 2023
Revised: 12 January 2024
Accepted: 16 January 2024
Published: 17 January 2024

The difference between the total entropy of a joined system A∪B, and the sum of the entropies of the single parts A and B is sometimes called the *entropic excess*. This is defined in

$$S^{\rm exc} = S\big[p^{\rm A\cup B}\big] - \Big(S\big[p^{\rm A}\big] + S\big[p^{\rm B}\big]\Big)\,, \tag{4}$$

where, depending on the entropy nature, it can be a positive or negative quantity. It quantifies the information gain or loss in a bipartite system, a property that may be related to the concept of super-stability or sub-stability in thermodynamics [58]. Entropy is super-additive, and the systems it describes are thermodynamically super-stable if the entropic excess is positive. On the other hand, entropy is sub-additive, and the systems it describes are thermodynamically sub-stable if the entropic excess is negative.

Following [59], in agreement with the second principle of thermodynamics, super-stable systems (with a positive entropic excess) tend to join together while sub-stable systems (with a negative entropic excess) tend to fragment.

The opposite of the entropic excess is called the *entropic defect*. Recently, the entropic defect has been investigated as a basic concept of thermodynamics which is able to characterize the entropic form describing a given physical system [60,61].

As discussed in [9], the entropic excess of the κ-entropy for any pair of statistically independent distributions is always positive (cf. Equation (3)), indicating that, in this case, it is a super-additive quantity useful for characterizing super-stable systems.

In general, the entropic excess depends on the bipartite distribution $p^{\rm A\cup B}$, and cannot be quantified in a precise manner. In this work, we show that, within κ-statistics, there are classes of maximal entropy probability distributions, labeled by a real positive parameter $\aleph > 0$, such that the entropic excess of a statistically independent bipartite system is proportional to the sum of the entropy of the single distributions. Thus, for any pair of probability distribution functions (pdf) belonging to the same $\aleph$-class, we have

$$S_\kappa^{\rm exc} = \aleph\,\Big(S_\kappa\big[p^{\rm A}\big] + S_\kappa\big[p^{\rm B}\big]\Big)\,, \tag{5}$$

so that the joint κ-entropy $S_\kappa\big[p^{\rm A\cup B}\big]$ turns out to be related directly to the sum of the κ-entropies of the single distributions, according to the relation

$$S_\kappa\big[p^{\rm A\cup B}\big] = (1+\aleph)\,\big(S_\kappa\big[p^{\rm A}\big] + S_\kappa\big[p^{\rm B}\big]\big)\,. \tag{6}$$

We call this propriety *multi-additivity* of κ-entropy.

The structure of this paper is as follows. In Section 2, we present the mathematical background related to κ-statistics and the proprieties of composability of κ-entropy for a bipartite statistically independent system. Section 3 contains our main results. There, we introduce the multi-additivity of κ-entropy, and investigate the variational problem concerning the maximization of the κ-entropy under the usual constraints given by the distribution momenta and the multi-additivity conditions. In Section 4, we show by a numerical evaluation that the problem admits solutions at least within the family of Gibbs-like distributions. Finally, Section 5 contains our conclusive comments.

2. Mathematical Background

To start with, let us consider the following functional-differential equation [1]:

$$\frac{d}{d\,x}\big[x\,\Lambda(x)\big] = \lambda\,\Lambda\Big(\frac{x}{\alpha}\Big)\,, \tag{7}$$

where λ and α are two scaling parameters to be determined. It is easy to verify that a solution of Equation (7), with the boundary conditions $\Lambda(1) = 0$ and $(d/d\,x)\,\Lambda(x)\big|_{x=1} = 1$, is given by the κ-logarithm (2), provide the scaling constants are set as

$$\lambda = \sqrt{1-\kappa^2}\,, \tag{8}$$

$$\alpha = \left(\frac{1-\kappa}{1+\kappa}\right)^{\frac{1}{2\kappa}}. \tag{9}$$

As will be clarified in a follow-up, the constant λ plays the role of the scaling factor in the argument of pdf, while the constant α is a κ-deformed version of the reciprocal Neperian number. In the $\kappa \to 0$ limit, $\lambda \to 1$ and $\alpha \to 1/e$.

The κ-logarithm, defined in $\Re^+ \to \Re$, is symmetric in $\kappa \to -\kappa$, with $\ln_\kappa(1) = 0$, $\lim_{x\to+\infty} \ln_\kappa(x) \to +\infty$, as well as $\lim_{x\to 0} \ln_\kappa(x) \to -\infty$. Furthermore, it is a continuous, strictly increasing $(d/dx)\ln_\kappa(x) > 0$ and concave $(d^2/dx^2)\ln_\kappa(x) > 0$ function for $|\kappa| < 1$. More importantly,

$$\ln_\kappa\left(\frac{1}{x}\right) = -\ln_\kappa(x)\,, \tag{10}$$

is a well-known propriety of standard logarithm that is preserved in its κ-deformed version. Finally, since it is in the $\kappa \to 0$ limit, κ-logarithm collapses to the standard logarithm function; this legitimates us considering the κ-logarithm a *faithful generalization of the logarithmic function*.

As the κ-logarithm is a monotonic function, its inverse, the κ-exponential, surely exists, and is given in

$$\exp_\kappa(x) = \left(\kappa\,x + \sqrt{1+\kappa^2\,x^2}\right)^{\frac{1}{\kappa}}. \tag{11}$$

It is a function defined in $\Re \to \Re^+$, symmetric in $\kappa \to -\kappa$, reduces to the standard exponential in the $\kappa \to 0$ limit, and, like the standard exponential, is a continuous, strictly increasing $(d/dx)\exp_\kappa(x) > 0$ and convex $(d^2/dx^2)\exp_\kappa(x) < 0$ function, with $\exp_\kappa(0) = 1$, $\lim_{x\to-\infty} \exp_\kappa(x) \to 0$ as well as $\lim_{x\to+\infty} \exp_\kappa(x) \to +\infty$.

Again, the well-known propriety of the standard exponential is also satisfied by its deformed version

$$\exp_\kappa(-x) = \frac{1}{\exp_\kappa(x)}\,, \tag{12}$$

and, therefore, the κ-exponential is a *faithful generalization of the exponential function*.

Another solution of Equation (7) with the same scaling constants (8) and (9), but with different boundary conditions $\Lambda(1) = 1$ and $(d/d\,x)\Lambda(x)\big|_{x=1} = 0$, is given by [62]

$$u_\kappa(x) = \frac{x^\kappa + x^{-\kappa}}{2}\,, \tag{13}$$

which is a function defined in $\Re^+ \to \Re^+$, and is symmetric in $\kappa \to -\kappa$, with $\lim_{x\to 0} u_\kappa(x) = \lim_{x\to+\infty} u_\kappa(x) \to +\infty$. Furthermore, $u_\kappa(x)$ is a continuous, concave $(d^2/dx^2)u_\kappa(x) > 0$ function for $|\kappa| < 1$, and obtains its minimum at $x = 1$, where $u_\kappa(1) = 1$. Finally, $u_\kappa(x)$ satisfies a dual relation of (10); that is,

$$u_\kappa\left(\frac{1}{x}\right) = u_\kappa(x)\,, \tag{14}$$

while, in the $\kappa \to 0$ limits, it becomes merely a constant, $u_0(x) = 1$. Therefore, there is not an equivalent function in the standard, undeformed formalism.

En passant, we observe that the two scaling constants α and λ are related by the relations

$$-\lambda\,\ln_\kappa(\alpha) = \lambda\,u_\kappa(\alpha) = 1\,. \tag{15}$$

One might easily persuade oneself that the functions $u_\kappa(x)$ and $\ln_\kappa(x)$ are strongly connected. In fact, two equivalent analytical expressions of these two functions are given by

$$\ln_\kappa(x) = \frac{1}{\kappa}\,\sinh\left(\kappa\,\ln(x)\right)\,, \tag{16}$$

$$u_\kappa(x) = \cosh\left(\kappa\,\ln(x)\right)\,, \tag{17}$$

which show their relationship with the trigonometric hyperbolic functions. Consequently, many properties of $\ln_\kappa(x)$ and $u_\kappa(x)$ follow from the corresponding properties of $\sinh(x)$ and $\cosh(x)$. In particular, it is ready to verify that

$$\ln_\kappa(x\,y) = \ln_\kappa(x)\,u_\kappa(y) + u_\kappa(x)\,\ln_\kappa(y)\,, \tag{18}$$

$$u_\kappa(x\,y) = u_\kappa(x)\,u_\kappa(y) + \kappa^2\,\ln_\kappa(x)\,\ln_\kappa(y)\,. \tag{19}$$

as a consequence of the additivity formulas of hyperbolic functions.

In addition, it is useful to recall the following relations relating these two functions

$$u_\kappa(x) = \sqrt{1+\kappa^2\,\ln_\kappa^2(x)} = \ln_\kappa(x) - \lambda\,\ln_\kappa(\alpha\,x)\,, \tag{20}$$

that become trivial relations $(1 = 1)$ in the $\kappa \to 0$ limit, since, in the same limit, $\lambda \to 1$ and $\alpha \to e^{-1}$.

By using the first of these relations, Equation (18) can be rewritten in

$$\ln_\kappa(x\,y) = \ln_\kappa(x)\,\sqrt{1+\kappa^2\,\ln_\kappa^2(y)} + \ln_\kappa(y)\,\sqrt{1+\kappa^2\,\ln_\kappa^2(x)}\,, \tag{21}$$

which implies the relevant inequality

$$\ln_\kappa(xy) < \ln_\kappa(x) + \ln_\kappa(y)\,, \tag{22}$$

holding in the statistically meaningful interval $0 < x,\,y < 1$.

The next step is to introduce the κ-entropy $S_\kappa[p]$. Like in the standard case, where SBG-entropy is defined as the negative of the linear average of the Hartley function (or surprise function) defined by $h(p) = \ln\left(p(x)\right)$, it is natural to introduce the κ-entropy as the negative of the linear average of the κ-deformed Hartley function, $h_\kappa(p) = \ln_\kappa\left(p(x)\right)$; that is,

$$S_\kappa[p] = -\langle h_\kappa(p)\rangle\,, \tag{23}$$

a relation that reproduces Equation (1), accounting for the usual definition of the linear average of a statistical observable $\mathcal{O}(x)$, given by

$$\langle\mathcal{O}\rangle = \int_{\mathcal{D}} \mathcal{O}(x)\,p(x)\,dx\,. \tag{24}$$

It is natural, in analogy with definition (23), to introduce the auxiliary function $\mathcal{I}_\kappa[p]$, as the linear average of $u_\kappa\left(p(x)\right)$, according to the relation

$$\mathcal{I}_\kappa[p] = \langle u_\kappa(p)\rangle\,. \tag{25}$$

This quantity is a positive definite, with $\mathcal{I}_\kappa[p] > 1$ for any normalized pdf and, in the $\kappa \to 0$ limit, it gives

$$\lim_{\kappa\to 0} \mathcal{I}_\kappa[p] = \int_{\mathcal{D}} p(x)\,dx = 1\,. \tag{26}$$

Therefore, there is no equivalent function in standard statistics. However, like $\ln_\kappa(x)$ and $u_\kappa(x)$, that are two strictly related functions, also $S_\kappa[p]$ and $\mathcal{I}_\kappa[p]$ turn out to be recurrent in the developing of the κ-statistics.

In particular, by taking the linear average of Equations (18) and (19) for a statistically independent bipartite distribution, with $p^{\rm A\cup B}(x,\,y) = p^{\rm A}(x)\,p^{\rm B}(y)$, we obtain

$$S_\kappa\big[p^{\rm A\cup B}\big] = \mathcal{I}_\kappa\big[p^{\rm A}\big]\,S_\kappa\big[p^{\rm B}\big] + S_\kappa\big[p^{\rm A}\big]\,\mathcal{I}_\kappa\big[p^{\rm B}\big]\,, \tag{27}$$

$$\mathcal{I}_\kappa\big[p^{\rm A\cup B}\big] = \mathcal{I}_\kappa\big[p^{\rm A}\big]\,\mathcal{I}_\kappa\big[p^{\rm B}\big] + \kappa^2\,S_\kappa\big[p^{\rm A}\big]\,S_\kappa\big[p^{\rm B}\big]\,, \tag{28}$$

stating the additivity rule of S_κ and $\mathcal{I}_\kappa$ for two statistically independent systems.

From Equation (27), we readily deduce the super-additive propriety of κ-entropy summarized in Equation (3). In particular, in the $\kappa \to 0$ limit, according to (26), we recover the usual additivity rule of the SBG entropy while, in the same limit, Equation (28) reduces to a trivial identity.

As discussed in [9], composition rule (27) can be rewritten also by means of κ-parentropy, a quantity defined as

$$S^*_\kappa[p] = -\lambda\,S_\kappa[\alpha\,p] - 1\,, \tag{29}$$

which is a scaled version of κ-entropy.

In fact, by taking the average of Equation (20), we can obtain the following relationship

$$\mathcal{I}_\kappa[p] = 1 + S^*_\kappa[p] - S_\kappa[p]\,, \tag{30}$$

that relates κ-entropy and κ-parentropy to the $\mathcal{I}_\kappa$ function. In this way, Equation (27) can be rewritten in

$$\begin{aligned} S_\kappa\big[p^{\rm A\cup B}\big] \;&=\; S_\kappa\big[p^{\rm A}\big] + S_\kappa\big[p^{\rm B}\big] - 2\,S_\kappa\big[p^{\rm A}\big]\,S_\kappa\big[p^{\rm B}\big] \\ &\quad + S_\kappa\big[p^{\rm A}\big]\,S^*_\kappa\big[p^{\rm B}\big] + S_\kappa\big[p^{\rm B}\big]\,S^*_\kappa\big[p^{\rm A}\big]\,, \end{aligned} \tag{31}$$

providing a composition rule for S_κ that formally only includes κ-entropy; however, it is important to remark that S_κ and S^*_κ are, actually, two independent quantities.

From Equation (31), the entropic excess of the κ-entropy is given by

$$S^{\rm exc}_\kappa = S_\kappa\big[p^{\rm A}\big]\,S^*_\kappa\big[p^{\rm B}\big] + S_\kappa\big[p^{\rm B}\big]\,S^*_\kappa\big[p^{\rm A}\big] - 2\,S_\kappa\big[p^{\rm A}\big]\,S_\kappa\big[p^{\rm B}\big] \ge 0\,, \tag{32}$$

that is a quantity defined only as a function of the κ-entropy.

Finally, let us note that Equations (27) and (28) can be combined to write the κ-entropy of a statistically independent multi-partite system in terms of S_κ and $\mathcal{I}_\kappa$ of a single distribution. For instance, given a statistically independent tri-partite system, we can obtain the relation

$$\begin{aligned} S_\kappa\big[p^{\rm A\cup B\cup C}\big] \;&=\; \kappa^2\,S_\kappa\big[p^{\rm A}\big]\,S_\kappa\big[p^{\rm B}\big]\,S_\kappa\big[p^{\rm C}\big] + S_\kappa\big[p^{\rm A}\big]\,\mathcal{I}_\kappa\big[p^{\rm B}\big]\,\mathcal{I}_\kappa\big[p^{\rm C}\big] \\ &\quad + \mathcal{I}_\kappa\big[p^{\rm A}\big]\,S_\kappa\big[p^{\rm B}\big]\,\mathcal{I}_\kappa\big[p^{\rm C}\big] + \mathcal{I}_\kappa\big[p^{\rm A}\big]\,\mathcal{I}_\kappa\big[p^{\rm B}\big]\,S_\kappa\big[p^{\rm C}\big]\,, \end{aligned} \tag{33}$$

and so on.

3. Multi-Additivity in Kaniadakis Entropy

Within κ-statistics, a maximal entropy pdf may be derived by maximizing the κ-entropy under certain appropriate boundary conditions. Quite often, they are given using linear averages of certain functions $\mathcal{O}_i(x)$ as

$$\langle \mathcal{O}_i \rangle = \int_{\mathcal{D}} \mathcal{O}_i(x)\, p(x)\, dx \,, \tag{34}$$

where $i = 0, 1, \ldots, M$, with $M+1$ being the number of given constraints. These relations fix the values of different quantities $\langle \mathcal{O}_i \rangle$, related to the system under inspection, whose spectra of possible outcomes are given by $\mathcal{O}_i(x)$. Many times, constraints are given by the momenta of a certain order n; that is, $\mathcal{O}_n(x) = x^n$. For instance, for $n = 0$, we pose $\mathcal{O}_0(x) = 1$ with $\langle \mathcal{O}_0 \rangle = 1$, which fixes the normalization of the distribution; for $n = 1$, we have $\mathcal{O}_1(x) = x$ with $\langle \mathcal{O}_1 \rangle \equiv \langle x \rangle$, which fixes the mean value of the distribution; for $n = 2$, we have $\mathcal{O}_2(x) = x^2$ with $\langle \mathcal{O}_2 \rangle \equiv \langle x^2 \rangle$, which is related to the variance of the distribution, etc.

In this case, the maximal entropy distribution can be derived from the following variational problem:

$$\frac{\delta}{\delta p(y)} \left(S_\kappa[p] - \sum_{i=0}^{M} \mu_i \int_{\mathcal{D}} \mathcal{O}_i(x)\, p(x)\, dx \right) = 0 \,, \tag{35}$$

where μ_i are Lagrange multipliers related to the $M+1$ constraints.

By accounting for Equation (7) and definition (1), we obtain the maximal entropy pdf in the form

$$p(x) = \alpha \, \exp_\kappa\left(-\frac{1}{\lambda} \sum_{i=0}^{M} \mu_i\, \mathcal{O}_i(x) \right) , \tag{36}$$

where the Lagrange multipliers $\mu_i(\langle \mathcal{O}_0 \rangle, \langle \mathcal{O}_1 \rangle, \ldots, \langle \mathcal{O}_M \rangle)$ are fixed throughout Equation (34) and are finally functions of the boundary conditions $\langle \mathcal{O}_i \rangle$.

It is worthwhile to observe that, given the analytical expression of $\exp_\kappa(x)$ given in (11), Distribution (36) has an asymptotic power-law behavior, i.e.,

$$p(x) \approx |\kappa\, x^n|^{1/\kappa} \,, \tag{37}$$

for large x, where n is the order of the maximal momenta. This fact justifies the κ-statistic in the study of those anomalous systems, often complex systems, characterized by pdfs with heavy tails.

In the following, let us generalize the optimal problem described above to the case in which, in addition to relations (34), we have further constraints that are functions of the pdf itself. In particular, we seek a class of distributions, labeled by a real constant $\aleph$, maximizing the κ-entropy under the further constraint given by

$$\mathcal{I}_\kappa[p] = 1 + \aleph \,, \qquad \forall p(x) \in \aleph - \text{class} \,. \tag{38}$$

As will be shown in the next section, this class always existed whenever $\aleph \geq 0$, at least for the Gibbs-like distributions, provided the constraint $\langle \mathcal{O}_1 \rangle \equiv \langle x \rangle$ falls in a given region fixed by $\aleph$.

Therefore, we can state the following: *for a bipartite probability distribution function $p^{A\cup B}(x, y) = p^A(x)\, p^B(y)$ of two statistically independent and identically distributed pdfs $p^A(x)$ and $p^B(y)$, belonging to the same $\aleph$-class that maximizing the κ entropy under the constraint (38), we have*

$$S_\kappa\left[p^{A\cup B}\right] = (1+\aleph) \left(S_\kappa\left[p^A\right] + S_\kappa\left[p^B\right] \right) . \tag{39}$$

We call this property *multi-additivity*, and we say that κ-entropy is $(1+\aleph)$-additive whenever relation (39) holds.

We observe that condition $\aleph > 0$ is fixed by the super-additive character of κ-entropy, while the condition $\aleph = 0$ admits only trivial solutions. In fact, as it is straightforward to verify, the two relations

$$\int_{\mathcal{D}} p(x)\,dx = 1\,, \tag{40}$$

$$\int_{\mathcal{D}} \frac{p(x)^{1+\kappa} + p(x)^{1-\kappa}}{2}\,dx = 1\,, \tag{41}$$

are consistent only in the trivial case of $\kappa = 0$ or for an exact distribution $p(x) = \delta(x)$.

To derive the pdf maximizing the κ-entropy under the constraints (34) and (38), we pose

$$\frac{\delta}{\delta p(x)}\left(S_\kappa[p] - \nu\,\mathcal{I}_\kappa[p] - \sum_{i=0}^{M}\mu_i\int_{\mathcal{D}} \mathcal{O}_i(x)\,p(x)\,dx\right) = 0\,, \tag{42}$$

where ν is the Lagrange multiplier related to Equation (38).

We obtain

$$-\frac{1}{2\,\kappa}\left[(1+\kappa)\,(1+\kappa\,\nu)\,p(x)^{\kappa} - (1-\kappa)\,(1-\kappa\,\nu)\,p(x)^{-\kappa}\right] - \sum_{i=0}^{M}\mu_i\,\mathcal{O}_i(x) = 0\,, \tag{43}$$

and pose

$$\begin{cases} (1+\kappa)\,(1+\kappa\,\nu) = \lambda(\nu)\,\alpha(\nu)^{-\kappa} \\ (1-\kappa)\,(1-\kappa\,\nu) = \lambda(\nu)\,\alpha(\nu)^{\kappa} \end{cases} \tag{44}$$

From Equation (43), we obtain

$$-\lambda(\nu)\ln_\kappa\left(\frac{p(x)}{\alpha(\nu)}\right) - \sum_{i=0}^{M}\mu_i\,\mathcal{O}_i(x) = 0\,, \tag{45}$$

that, solved for $p(x)$, gives the pdf in the form

$$p(x) = \alpha(\nu)\,\exp_\kappa\left(-\frac{1}{\lambda(\nu)}\sum_{i=0}^{M}\mu_i\,\mathcal{O}_i(x)\right)\,. \tag{46}$$

Although this distribution has the same structure as Equation (36), it differs from (36) in that the two functions $\lambda(\nu)$ and $\alpha(\nu)$, given by

$$\lambda(\nu) = \lambda\,\sqrt{1-\kappa^2\,\nu^2}\,, \tag{47}$$

$$\alpha(\nu) = \alpha\left(\frac{1-\kappa\,\nu}{1+\kappa\,\nu}\right)^{\frac{1}{2\kappa}}\,, \tag{48}$$

now depend on the Lagrange multiplier ν. They fulfill the relations

$$\lambda(\nu)\,\ln_\kappa(\alpha(\nu)) = -(1+\nu)\,, \tag{49}$$

$$\lambda(\nu)\,u_\kappa(\alpha(\nu)) = 1+\kappa\,\nu\,, \tag{50}$$

that reduce to (15) for $\nu = 0$.

The reality condition of distribution (46), which is still symmetric in $\kappa \to -\kappa$, requires

$$\begin{cases} \kappa \in (0,\, 1/\nu) \cup (1,\, +\infty) & \text{for} \quad \nu > 1\,, \\ \kappa \in (0,\, 1) \cup (1/\nu\, +\infty) & \text{for} \quad \nu < 1\,, \end{cases} \tag{51}$$

while functions $\lambda(\nu)$ and $\alpha(\nu)$ reduce to the constants (8) and (9), respectively, in the $\nu \to 0$ limit. Further, in the same limit, Problem (42) collapses into Problem (35).

Finally, plugging distribution (46) into Equations (34) and (38), they fix the Lagrange multipliers ν and μ_i as functions of boundary conditions $\langle \mathcal{O}_i \rangle$, that is, $\nu \equiv \nu(\aleph,\, \langle \mathcal{O}_1 \rangle,\, \ldots,\, \langle \mathcal{O}_M \rangle)$ and $\mu_i \equiv \mu_i(\aleph,\, \langle \mathcal{O}_{1,} \rangle\, \ldots,\, \langle \mathcal{O}_M \rangle)$, so that the problem is solved definitively.

It is remarkable to note that, when accounting for the normalization of pdf, from Equation (46), we have

$$\alpha(\nu)^{-1} = \int_{\mathcal{D}} \exp_\kappa \left(-\frac{1}{\lambda(\nu)} \sum_{i=0}^{M} \mu_i\, \mathcal{O}_i(x) \right) dx\,, \tag{52}$$

a relation that suggests the role of $\alpha(\nu)$ as a partition function, i.e., $Z \equiv \alpha(\nu)^{-1}$, in the present formalism. However, a word of caution is in order. As is well-known in standard statistics, the partition function accounted for the normalization. Thus, it is related to the corresponding Lagrange multiplier γ by the relation $\ln(Z) = 1 + \gamma$. This is not the case for pdf (46), since $\alpha(\nu)$ is related to the Lagrange multiplier of constraint (41), whereas Normalization (40) is controlled by the Lagrange multiplier μ_0.

To convince yourself of this, it is sufficient to consider the $\kappa \to 0$ limit. In this case, both Constraints (40) and (41) assume the same form, since $\mathcal{I}_0 \equiv \int p(x)\, dx$, and

$$\lim_{\kappa \to 0} \alpha(\nu) \to \exp(-1-\nu)\,, \tag{53}$$

is a constant. Therefore, in this limit, Distribution (46) becomes

$$\begin{aligned} \lim_{\kappa \to 0} p(x) &= \exp(-1-\nu)\, \exp\left(-\sum_{i=0}^{M} \mu_i\, \mathcal{O}_i(x) \right) \\ &= \exp(-1-\nu-\mu_0)\, \exp\left(-\sum_{i=1}^{M} \mu_i\, \mathcal{O}_i(x) \right) \\ &= \frac{1}{Z} \exp\left(-\sum_{i=1}^{M} \mu_i\, \mathcal{O}i_i(x) \right), \end{aligned} \tag{54}$$

where, with $\gamma = \nu + \mu_0$, we recover the usual definition of the partition function given above.

Finally, we remark that when the distribution has Expression (46), according to Constraint (38) and by using Equations (49) and (50), we obtain

$$(1+\kappa^2\,\nu) \left\langle \sqrt{\lambda(\nu)^2 + \kappa^2 \left(\sum_{i=0}^{M} \mu_i\, \mathcal{O}_i(x) \right)^2} \right\rangle - \kappa^2\,(1+\nu) \sum_{i=0}^{M} \mu_i\, \langle \mathcal{O}_i \rangle = (1+\aleph)\, \lambda(\nu)^2\,, \tag{55}$$

which is a consistent relationship between the Lagrange multipliers and the expectation values of the present statistical model.

4. A Numerical Example: The Gibbs-like Distribution

To show the existence of solutions to the problem under investigation, let us consider the simplest case of a problem with $M = 2$. Thus, we seek a family of pdf maximizing the κ-entropy under the following constraints:

$$\int_0^\infty p(x)\,dx = 1\,, \tag{56}$$

$$\int_0^\infty x\,p(x)\,dx = \langle x\rangle\,, \tag{57}$$

$$\int_0^\infty \frac{p(x)^{1+\kappa}+p(x)^{1-\kappa}}{2}\,dx = 1+\aleph\,, \tag{58}$$

corresponding, respectively, to the normalization, the linear average, and the multi-additivity constraints.

Solving the variational problem (43) in the present case, we obtain the optimizing pdf in the form

$$p(x) = \alpha(\nu)\,\exp_\kappa\left(-\frac{1}{\lambda(\nu)}(\mu_0+\mu_1\,x)\right). \tag{59}$$

This is a Gibbs-like distribution since, in the $\kappa \to 0$ limit, standard Gibbs-distribution $p(x) = \exp(-1-\nu-\mu_0-\mu_1\,x)$ is obtained. Otherwise, (59) is a pdf with an asymptotic power-law heavy tail, being $p(x) \approx (\kappa\,x)^{1/\kappa}$ for $\kappa\,x \gg 1$.

By plugging distribution (59) into Equations (56)–(58), we obtain the system of equations

$$I_1(z) = \frac{\mu_1}{\alpha(\nu)\,\lambda(\nu)}\,, \tag{60}$$

$$\frac{\mu_0}{\lambda(\nu)}\,I_1(z)+I_2(z) = \frac{\mu_1^2\,\langle x\rangle}{\alpha(\nu)\,\lambda(\nu)^2}\,, \tag{61}$$

$$\alpha(\nu)^\kappa\,I_3(z)+\alpha(\nu)^{-\kappa}\,I_4(z) = 2\,(1+\aleph)\,I_1(z)\,, \tag{62}$$

where $I_i(z), i = 1,\ldots,4$, are elementary integrals given by

$$I_1(z) = \int_0^z \frac{x^\kappa+x^{-\kappa}}{2}\,dx = \frac{z}{2}\left(\frac{z^\kappa}{1+\kappa}+\frac{z^{-\kappa}}{1-\kappa}\right), \tag{63}$$

$$I_2(z) = \int_0^z \frac{x^{2\kappa}+x^{-2\kappa}}{4\kappa}\,dx = \frac{z}{4\kappa}\left(\frac{z^{2\kappa}}{1+2\kappa}-\frac{z^{-2\kappa}}{1-2\kappa}\right), \tag{64}$$

$$I_3(z) = \int_0^z \frac{x^{2\kappa}+1}{2}\,dx = \frac{z}{2}\left(\frac{z^{2\kappa}}{1+2\kappa}+1\right), \tag{65}$$

$$I_4(z) = \int_0^z \frac{x^{-2\kappa}+1}{2}\,dx = \frac{z}{2}\left(\frac{z^{-2\kappa}}{1-2\kappa}+1\right), \tag{66}$$

as functions of the quantity

$$z = \exp_\kappa\left(-\frac{\gamma}{\lambda(\nu)}\right). \tag{67}$$

The system of Equations (60)–(62) can be solved numerically to obtain the Lagrange multipliers $\mu_0(\aleph, \langle x\rangle)$, $\mu_1(\aleph, \langle x\rangle)$ and $\nu(\aleph, \langle x\rangle)$ as functions of the constraints $\aleph$ and $\langle x\rangle$.

For any fixed value of $\aleph$, real solutions only exist in certain intervals of $\langle x \rangle$. This is shown in Figure 1, where the region of existence of real solutions (shaded areas) of the system of Equations (60)–(62) is depicted for several values of the deformation parameter κ. The case $\kappa = 0$ (not reported in the figure) corresponds to the horizontal line passing for $\aleph = 0$. In this case, κ-entropy becomes 1-additive for any pdf.

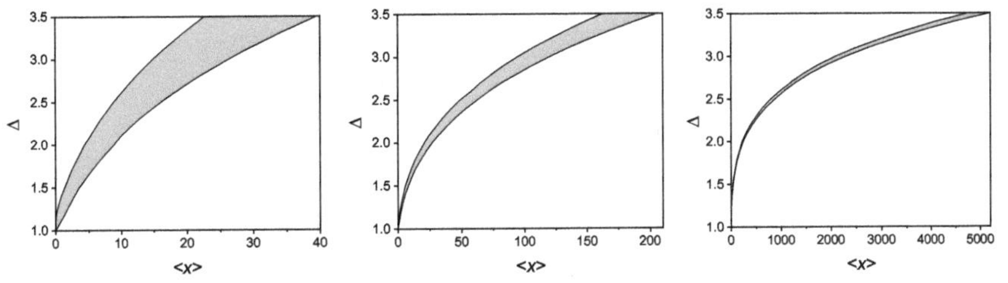

Figure 1. In the figure, we plotted the region of real solutions of system (60)–(62) in the plane of constraints, for several values of the deformation parameter κ. The shaded areas represent the admissible domine.

In other words, each value of $\aleph$ selects a class of distributions whose interval $(\langle x \rangle_{\min}, \langle x \rangle_{\max})$ determines the possible pdf for which the κ-entropy is $(1 + \aleph)$-additive.

In Table 1, we give some numerical values of the interval $(\langle x \rangle_{\min}, \langle x \rangle_{\max})$ for the $\aleph$-classes between 0.5 and 2.5, step 0.5, corresponding to the three values of the deformation parameter κ reported in the figure.

Table 1. Permitted interval $(\langle x \rangle_{\min}, \langle x \rangle_{\max})$ for the $\aleph$-classes between 0.5 and 2.5, step 0.5, corresponding to the three values of the deformation parameter κ reported in the figure.

	$\kappa = 0.2$		$\kappa = 0.3$		$\kappa = 0.4$	
$\aleph$	$\langle x \rangle_{\min}$	$\langle x \rangle_{\max}$	$\langle x \rangle_{\min}$	$\langle x \rangle_{\max}$	$\langle x \rangle_{\min}$	$\langle x \rangle_{\max}$
0.5	36.8	41.3	5.85	8.06	1.34	3.57
1.0	224.0	246.0	20.59	26.70	4.36	8.87
1.5	787.0	863.0	48.53	61.90	8.78	16.09
2.0	2107.0	2306.0	94.40	119.90	14.81	26.36
2.5	4752.0	5193.0	163.10	206.10	22.64	39.63

As an example, let us consider the case with $\kappa = 0.3$ and $\aleph = 1.0$. Any pair of κ-deformed Gibbs-like distributions with $20.59 < \langle x \rangle < 26.70$ is 2-additive; that is, $S_\kappa(p^{\rm A}\, p^{\rm B}) = 2\left(S_\kappa(p^{\rm A}) + S_\kappa(p^{\rm B})\right)$. For instance, take $\langle x_{\rm A} \rangle = 22.5$ and $\langle x_{\rm B} \rangle = 25.5$; we can evaluate the numerical values of the Lagrange multipliers corresponding to constraints (56)–(58).

These can be read from Table 2, where we show several numerical values of the Lagrange multipliers μ_0, μ_1 and ν, obtained from the system of Equations (60)–(62), for several values of constraints $\aleph$ and $\langle x \rangle$ belonging to the allowed region, and corresponding to the three values of the deformation parameter κ reported in the figure.

From this table, we can obtain the terna of multiplier values $(-2.1989, 0.01849, -0.8219)$, corresponding to the distribution $p^{\rm A}(22.5)$, and $(-7.3990, 0.009404, -1.0796)$, corresponding to the two distributions $p^{\rm B}(25.5)$. Then, the respective values of κ-entropy of these two distributions $p^{\rm A}$ and $p^{\rm B}$ are readily evaluated in $S_\kappa(p^{\rm A}) = 5.63627$ and $S_\kappa(p^{\rm B}) = 5.70258$, while the value of κ-entropy for the join system $S_\kappa(p^{\rm A\cup B}) = 22.6777$, which is exactly the attended result.

Table 2. Several numerical values of the Lagrange multipliers, μ_0, μ_1 and ν, obtained from the system (60)–(62), for some values of constraints $\aleph$ and $\langle x \rangle$ in the allowed region, corresponding to the three values of the deformation parameter κ reported in the figure.

	$\kappa = 0.2$				$\kappa = 0.3$				$\kappa = 0.4$			
$\aleph$	$\langle x \rangle$	μ_0	μ_1	ν	$\langle x \rangle$	μ_0	μ_1	ν	$\langle x \rangle$	μ_0	μ_1	ν
0.5	37	2.5761	0.03342	2.0607	6.0	0.8102	0.1980	0.5634	1.5	0.1690	0.8377	0.08488
	38	−1.3791	0.01671	−0.8298	6.5	−1.7817	0.09376	−0.9642	2.0	−2.1727	0.3281	−1.1016
	39	−3.6921	0.01204	−1.5989	7.0	−3.5685	0.06270	1.3475	2.5	−3.9538	0.1719	−1.2830
	40	−6.7617	0.009670	−1.9744	7.5	−6.1733	0.04751	−1.5050	3.0	−6835	0.1113	−1.2834
1	225	1.2720	0.003917	0.8586	21.0	0.3309	0.04680	0.2129	5.0	−1.4060	0.1438	−0.6946
	230	−1.8947	0.001918	−0.8003	22.5	−2.1989	0.01849	−0.8219	6.0	0.0608	−0.9088	−0.9088
	235	−4.1393	0.001391	−1.2201	24.0	−4.1330	0.01231	−1.0079	7.0	−5.0451	0.03775	−0.9106
	240	−7.2656	0.001124	−1.4248	25.5	−7.3990	0.009404	−1.0796	8.0	−8.6768	0.02712	−0.8879
1.5	790	2.2376	0.001219	1.4587	50.0	−0.5497	0.01190	−0.2549	9.0	0.1923	0.1678	0.1149
	810	−2.1479	0.0004171	−0.6953	52.5	−2.2195	0.006305	−0.6605	11.0	−2.7096	0.03009	−0.7113
	830	−4.9360	0.0002896	−1.0202	55.0	−3.6418	0.004508	−0.7731	13.0	−4.6906	0.01655	−0.7141
	850	−10.1219	0.0002300	−1.1654	57.5	−5.4875	0.003571	−0.8236	15.0	−8.3442	0.01136	−0.6919
2	2150	−1.3309	0.0001541	−0.4175	95	1.1453	0.01147	0.6736	16.0	−1.2338	0.03603	−0.4690
	2200	−3.9385	0.00009907	−0.7820	100	−1.7453	0.003185	−0.4927	19.0	−3.1742	0.01251	−0.5956
	2250	−7.4910	0.00007668	−0.9237	105	−3.2500	0.002083	−0.6232	22.0	−5.1612	0.007630	−0.5848
	2300	−26.3122	0.00006387	−1.0015	110	−5.0400	0.001591	−0.6727	25.0	−9.1792	0.005478	−0.5660
2.5	4800	0.08936	0.00008495	0.03164	170	−1.2742	0.001911	−0.3547	23.0	0.1565	0.05803	0.08391
	4900	−2.6290	0.00004530	−0.5566	180	−3.1060	0.001066	−0.5263	28.0	−2.9118	0.007853	−0.508
	5000	−5.0810	0.00003349	−0.7245	190	−5.2257	0.0007732	−0.5753	33.0	−5.0884	0.004362	−0.4981
	5100	−9.2667	0.00002734	−0.8089	200	−9.9583	0.0006167	−0.5964	38.0	−9.8852	0.003019	−0.4796

5. Conclusions

In this work, we showed that, within κ-statistics, there exist classes of pdf that maximize κ-entropy under the condition of constant $\mathcal{I}_\kappa[p]$, a problem that admits a solution at least for the family of Gibbs-like distributions. In this way, for any pair of distributions belonging to the same $\aleph$-class, fixed by the real number $\aleph > 0$, κ-entropy turns out to be $(1+\aleph)$-additive; that is, the value of κ-entropy of a bipartite statistically independent distribution, whose reduced belonging to the same $\aleph$-class is a multiple of the sum of the single κ-entropy according to Equation (39).

Equivalently, for any pair of distributions belonging to the same $\aleph$-class, the entropic excess is proportional to the sum of the κ-entropy of the single pdfs, according to (5).

On the physical ground, Distribution (46) describes a statistical ensemble constrained by condition (38). While the physical meaning of functional $\mathcal{I}_\kappa$ is still unclear, it seems to be related to the κ-partition function and, consequently, to the κ-free energy of the system, as discussed in [57] (see also [45]). This also applies to Distribution (59), which characterizes a canonical ensemble that is further constrained by (38). Moreover, given two independent physical systems, both members of the same $\aleph$-class but with different internal energy, the joined κ-entropy is an $\aleph$-multiple of the sum of their respective κ-entropies. This propriety could be useful for studying thermal and mechanical equilibrium, where the composability of entropy plays a role [62]. However, the potential impact that multi-additivity might have on this aspect of the κ-thermostatistic deserves further investigation.

Furthermore, looking at Equations (27) and (30), we see that the entropic excess in κ-statistics is related to the difference between the κ-entropy and κ-parentropy. In the $\kappa = 0$ case (standard statistics), such a difference is always null ($\aleph = 0$), i.e., parentropy and entropy have a constant gap equal to 1 for any pdf. Otherwise, when $\kappa > 0$, the difference between the κ-entropy and κ-parentropy depends on pdf. As shown in this paper, there exist classes of distributions that optimize the κ-entropy under constraint (38), such that the difference between the κ-entropy and κ-parentropy is fixed and equal to $\aleph$ for any pdf belonging to the same $\aleph$-class.

In other words, the difference between distribution (36) and distribution (46) can be stated as follows: the former assigns distinct values for S_κ and $\mathcal{I}_\kappa$, as these functionals both depend on the expectation values $\langle \mathcal{O}_i \rangle$. In contrast, the latter assigns distinct values for S_κ, but assumes a constant value for $\mathcal{I}_\kappa = 1 + \aleph$, fixed prior, for any distribution that falls within the same ℵ-class. In this way, the entropic excess turns out to be proportional to the sum of the κ-entropies of the two systems that are members of the same ℵ-class.

Author Contributions: Conceptualization, A.M.S. and T.W.; methodology, A.M.S. and T.W.; formal analysis, A.M.S. and T.W.; investigation, A.M.S. and T.W.; writing—original draft preparation, A.M.S. and T.W.; writing—review and editing, A.M.S. and T.W. All authors have read and agreed to the published version of the manuscript.

Funding: This research received no external funding.

Institutional Review Board Statement: Not applicable.

Data Availability Statement: No new data were created or analyzed in this study. Data sharing is not applicable to this article.

Conflicts of Interest: The authors declare no conflicts of interest.

References

1. Kaniadakis, G. Non-linear kinetics underlying generalized statistics. *Physical A* **2001**, *296*, 405–425. [CrossRef]
2. Kaniadakis, G. Statistical mechanics in the context of special relativity. *Phys. Rev. E* **2002**, *66*, 056125. [CrossRef] [PubMed]
3. Kaniadakis, G. Statistical mechanics in the context of special relativity. II. *Phys. Rev. E* **2005**, *72*, 036108. [CrossRef] [PubMed]
4. Silva, R. The H-theorem in κ-statistics: Influence on the molecular chaos hypothesis. *Phys. Lett. A* **2006**, *352*, 17–20. [CrossRef]
5. Kaniadakis, G. Towards a relativistic statistical theory. *Physical A* **2006**, *365*, 17–23. [CrossRef]
6. Oikonomou, T.; Baris Bagci, G. A completeness criterion for Kaniadakis, Abe and two-parameter generalized statistical theories. *Rep. Math. Phys.* **2010**, *66*, 137–146. [CrossRef]
7. Kaniadakis, G. Power-law tailed statistical distributions and Lorentz transformations. *Phys. Lett. A* **2011**, *375*, 356–359. [CrossRef]
8. Guo, L. Physical meaning of the parameters in the two-parameter (κ, ζ) generalized statistics. *Mod. Phys. Lett. B* **2012**, *26*, 1250064. [CrossRef]
9. Kaniadakis, G.; Scarfone, A.M.; Sparavigna, A.; Wada, T. Composition law of κ-entropy for statistically independent systems. *Phys. Rev. E* **2017**, *95*, 052112. [CrossRef]
10. Scarfone, A.M. Boltzmann configurational entropy revisited in the framework of generalized statistical mechanics. *Entropy* **2022**, *24*, 140. [CrossRef]
11. Alves, T.F.A.; Neto, J.F.D.S.; Lima, F.W.S.; Alves, G.A.; Carvalho, P.R.S. Is Kaniadakis κ-generalized statistical mechanics general? *Phys. Lett. B* **2023**, *843*, 138005. [CrossRef]
12. Naudts, J. Deformed exponentials and logarithms in generalized thermostatistics. *Physical A* **2002**, *316*, 323–334. [CrossRef]
13. Tempesta, P. Group entropies, correlation laws, and zeta functions. *Phys. Rev. E* **2011**, *84*, 021121. [CrossRef]
14. Scarfone, A.M. Entropic forms and related algebras. *Entropy* **2013**, *15*, 624–649. [CrossRef]
15. Souza, N.T.C.M.; Anselmo, D.H.A.L.; Silva, R.; Vasconcelos, M.S.; Mello, V.D. Analysis of fractal groups of the type d-(m, r)–Cantor within the framework of Kaniadakis statistics. *Phys. Lett. A* **2014**, *378*, 1691–1694. [CrossRef]
16. Scarfone, A.M. On the κ-deformed cyclic functions and the generalized Fourier series in the framework of the κ-algebra. *Entropy* **2015**, *17*, 2812–2833. [CrossRef]
17. Hirica, I.-E.; Pripoae, C.-L.; Pripoae, G.-T.; Preda, V. Lie Symmetries of the nonlinear Fokker-Planck equation based on weighted Kaniadakis entropy. *Mathematics* **2022**, *10*, 2776. [CrossRef]
18. Aliano, A.; Kaniadakis, G.; Miraldi, E. Bose-Einstein condensation in the framework of κ-statistics. *Physical A* **2003**, *325*, 35–40. [CrossRef]
19. Wada, T. Thermodynamic stabilities of the generalized Boltzmann entropies. *Physical A* **2004**, *340*, 126–130. [CrossRef]
20. Lourek, I.; Tribeche, M. Thermodynamic properties of the blackbody radiation: A Kaniadakis approach. *Phys. Lett. A* **2017**, *381*, 452–456. [CrossRef]
21. Raut, S.; Mondal, K.K.; Chatterjee, P.; Roy, S. Dust ion acoustic bi-soliton, soliton, and shock waves in unmagnetized plasma with Kaniadakis-distributed electrons in planar and nonplanar geometry. *Eur. Phys. J. D* **2023**, *77*, 100–115. [CrossRef]
22. Dubinov, A.E. Gas-dynamic approach to the theory of non-linear ion-acoustic waves in plasma with Kaniadakis' distributed species. *Adv. Space Res.* **2023**, *71*, 1108–1115. [CrossRef]
23. Khalid, M.; Khan, M.; Muddusir; Ata-Ur-Rahman; Irshad, M. Periodic and localized structures in dusty plasma with Kaniadakis distribution. *Z. Naturforsch. A* **2021**, *76*, 891–897. [CrossRef]
24. Curé, M.; Rial, D.F.; Christen, A.; Cassetti, J. A method to deconvolve stellar rotational velocities. *Astron. Astrophys.* **2014**, *565*, A85–A87. [CrossRef]
25. Bento, E.P.; Silva, J.R.P.; Silva, R. Non-Gaussian statistics, Maxwellian derivation and stellar polytropes. *Physical A* **2013**, *392*, 666–672. [CrossRef]
26. Carvalho, J.C.; Do Nascimento, J.D., Jr.; Silva, R.; De Medeiros, J.R. Non-gaussian statistics and stellar rotational velocities of main-sequence field stars. *Astrophys. J.* **2009**, *696*, L48–L51. [CrossRef]

27. de Abreu, W.V.; Martinez, A.S.; do Carmo, E.D.; Gonçalves, A.C. A novel analytical solution of the deformed Doppler broadening function using the Kaniadakis distribution and the comparison of computational efficiencies with the numerical solution. *Nucl. Eng. Technol.* **2022**, *54*, 1471–1481. [CrossRef]
28. Moradpour, H.; Javaherian, M.; Namvar, E.; Ziaie, A.H. Gamow temperature in Tsallis and Kaniadakis statistics. *Entropy* **2022**, *24*, 797. [CrossRef]
29. Guedes, G.; Palma, D.A.P. Quasi-Maxwellian interference term functions. *Ann. Nucl. Energy* **2021**, *151*, 107914. [CrossRef]
30. de Abreu, W.V.; Gonçalves, A.C.; Martinez, A.S. New analytical formulations for the Doppler broadening function and interference term based on Kaniadakis distributions. *Ann. Nucl. Energy* **2020**, *135*, 106960. [CrossRef]
31. Luciano, G.G. Modified Friedmann equations from Kaniadakis entropy and cosmological implications on baryogenesis and ^{7}Li-abundance. *Eur. Phys. J. C* **2022**, *82*, 314–318. [CrossRef]
32. Nojiri, S.; Odintsov, S.D.; Paul, T. Modified cosmology from the thermodynamics of apparent horizon. *Phys. Lett. B* **2022**, *835*, 137553. [CrossRef]
33. Luciano, G.G. Gravity and cosmology in Kaniadakis statistics: Current status and future challenges. *Entropy* **2022**, *24*, 1712–1717. [CrossRef] [PubMed]
34. Kumar, P.S.; Pandey, B.D.; Pankaj; Sharma, U.K. Kaniadakis agegraphic dark energy. *New Astron.* **2024**, *105*, 102085. [CrossRef]
35. Sania; Azhar, N.; Rani, S.; Jawad, A. Cosmic and thermodynamic consequences of Kaniadakis holographic dark energy in Brans–Dicke gravity. *Entropy* **2023**, *25*, 576. [CrossRef] [PubMed]
36. Singh, B.K.; Sharma, U.K.; Sharma, L.K.; Dubey, V.C. Statefinder hierarchy of Kaniadakis holographic dark energy with composite null diagnostic. *Int. J. Geom. Methods Mod. Phys.* **2023**, *20*, 2350074. [CrossRef]
37. Nojiri, S.; Odintsov, S.D.; Paul, T. Early and late universe holographic cosmology from a new generalized entropy. *Phys. Lett. B* **2022**, *831*, 137189. [CrossRef]
38. Ghaffari, S. Kaniadakis holographic dark energy in Brans-Dicke cosmology. *Mod. Phys. Lett. A* **2022**, *37*, 2250152. [CrossRef]
39. Rani, S.; Jawad, A.; Sultan, A.M.; Shad, M. Cosmographic and thermodynamic analysis of Kaniadakis holographic dark energy. *Int. J. Mod. Phys. D* **2022**, *31*, 2250078. [CrossRef]
40. Wada, T.; Matsuzoe, H.; Scarfone, A.M. Dualistic Hessian structures among the thermodynamic potentials in the κ-hermostatistics. *Entropy* **2015**, *17*, 7213–7229. [CrossRef]
41. Mehri-Dehnavi, H.; Mohammadzadeh, H. Thermodynamic geometry of Kaniadakis statistics. *J. Phys. A Math. Gen.* **2020**, *53*, 375009. [CrossRef]
42. Scarfone, A.M.; Matsuzoe, H.; Wada, T. Information geometry of κ-exponential families: Dually-flat, Hessian and Legendre structures. *Entropy* **2018**, *20*, 436. [CrossRef]
43. Wada, T.; Scarfone, A.M. Information geometry on the κ-thermostatistics. *Entropy* **2015**, *17*, 1204–1217. [CrossRef]
44. Quiceno Echavarría, H.R.; Arango Parra, J.C. A statistical manifold modeled on Orlicz spaces using Kaniadakis κ-exponential models. *J. Math. Anal.* **2015**, *431*, 1080–1098. [CrossRef]
45. Scarfone, A.M.; Wada, T. Legendre structure of kappa-thermostatistics revisited in the framework of information geometry. *J. Phys. A Math. Theor.* **2014**, *47*, 275002. [CrossRef]
46. Pistone, G. κ-Exponential models from the geometrical viewpoint. *Eur. Phys. J. B* **2009**, *70*, 29–37. [CrossRef]
47. Costa, M.O.; Silva, R.; Anselmo, D.H.A.L.; Silva, J.R.P. Analysis of human DNA through power-law statistics. *Phys. Rev. E* **2019**, **99**, 022112. [CrossRef] [PubMed]
48. Souza, N.T.C.M.; Anselmo, D.H.A.L.; Silva, R.; Vasconcelos, M.S.; Mello, V.D. A κ-statistical analysis of the Y-chromosome. *EPL* **2014**, *108*, 38004. [CrossRef]
49. Macedo-Filho, A.; Moreira, D.A.; Silva, R.; da Silva, L.R. Maximum entropy principle for Kaniadakis statistics and networks. *Phys. Lett. A* **2013**, *377*, 842–846. [CrossRef]
50. Stella, M.; Brede, M. A κ-deformed model of growing complex networks with fitness. *Physical A* **2014**, 407, 360–368. [CrossRef]
51. Clementi, F.; Gallegati, M.; Kaniadakis, G. A κ-generalized statistical mechanics approach to income analysis. *J. Stat. Mech.* **2009**, P02037. [CrossRef]
52. Bertotti, M.L.; Modenese, G. Exploiting the flexibility of a family of models for taxation and redistribution. *Eur. Phys. J. B* **2012**, *85*, 261–270. [CrossRef]
53. Modanese, G. Common origin of power-law tails in income distributions and relativistic gases. *Phys. Lett. A* **2016**, *380*, 29–32. [CrossRef]
54. Trivellato, B. The minimal κ-entropy martingale measure. *Int. J. Theor. Appl. Financ.* **2012**, *15*, 1250038. [CrossRef]
55. Trivellato, B. Deformed exponentials and applications to finance. *Entropy* **2013**, *15*, 3471–3489. [CrossRef]
56. Tapiero, O.J. A maximum (non-extensive) entropy approach to equity options bid-ask spread. *Physical A* **2013**, *392*, 3051–3060. [CrossRef]
57. Scarfone, A.M.; Wada, T. Canonical partition function for anomalous systems described by the κ-entropy. *Prog. Theor. Phys. Suppl.* **2006**, *162*, 45–52. [CrossRef]
58. Scarfone, A.M. Thermal and mechanical equilibrium among weakly interacting systems in generalized thermostatistics framework. *Phys. Lett. A* **2006**, *355*, 404–412. [CrossRef]
59. Landsberge, P.T.; Vedral, V. Distributions and channel capacities in generalized statistical mechanics. *Phys. Lett. A* **1998**, *247*, 211–217. [CrossRef]

60. Livadiotis, G.; McComas, D.J. Entropy defect: Algebra and thermodynamics. *Europhys. Lett.* **2023**, *144*, 21001. [CrossRef]
61. Livadiotis, G.; McComas, D.J. Entropy defect in thermodynamics. *Sci. Rep.* **2023**, *13*, 9033. [CrossRef] [PubMed]
62. Scarfone, A.M.; Wada, T. Thermodynamic equilibrium and its stability for microcanonical systems described by the Sharma-Taneja-Mittal entropy. *Phys. Rev. E* **2005**, *72*, 026123. [CrossRef] [PubMed]

entropy MDPI

Article

Relativistic Roots of κ-Entropy

Giorgio Kaniadakis

Dipartimento di Scienza Applicata e Tecnologia, Politecnico di Torino, Corso Duca degli Abruzzi 24, 10129 Torino, Italy; giorgio.kaniadakis@polito.it or giorgio.kaniadakis@gmail.com

Abstract: The axiomatic structure of the κ-statistcal theory is proven. In addition to the first three standard Khinchin–Shannon axioms of continuity, maximality, and expansibility, two further axioms are identified, namely the self-duality axiom and the scaling axiom. It is shown that both the κ-entropy and its special limiting case, the classical Boltzmann–Gibbs–Shannon entropy, follow unambiguously from the above new set of five axioms. It has been emphasized that the statistical theory that can be built from κ-entropy has a validity that goes beyond physics and can be used to treat physical, natural, or artificial complex systems. The physical origin of the self-duality and scaling axioms has been investigated and traced back to the first principles of relativistic physics, i.e., the Galileo relativity principle and the Einstein principle of the constancy of the speed of light. It has been shown that the κ-formalism, which emerges from the κ-entropy, can treat both simple (few-body) and complex (statistical) systems in a unified way. Relativistic statistical mechanics based on κ-entropy is shown that preserves the main features of classical statistical mechanics (kinetic theory, molecular chaos hypothesis, maximum entropy principle, thermodynamic stability, H-theorem, and Lesche stability). The answers that the κ-statistical theory gives to the more-than-a-century-old open problems of relativistic physics, such as how thermodynamic quantities like temperature and entropy vary with the speed of the reference frame, have been emphasized.

Keywords: relativistic statistical mechanics; power-law tailed distributions; κ-entropy; κ-distribution; κ-statistics; κ-exponential; κ-logarithm; κ-mathematics; κ-deformation; temperature of a moving body; relativistic temperature; relativistic thermodynamics

1. Introduction

The spread of the neologism κ-distribution within the astrophysical plasma community began after the publication of the seminal paper of Vasyliunas' [1,2] in 1968. The enormous existing literature on so-called κ-plasmas shows the undisputed success of the Vasyliunas κ-distribution, which still seems to be very relevant today. There have been a very high number of attempts to justify it, which shows that none of the proposals put forward are accepted by the whole community of physicists in the field and, therefore, the success of this distribution is mainly of an empirical nature. Curiously, there are no advanced proposals that consider the possibility of going beyond the Vasyliunas κ-distribution and investigating new possible distributions that have other analytical forms but describe the empirical data equally well.

This paper deals with the statistical theory [3,4] proposed in 2001, which goes beyond the Vasyliunas distribution. The underlying new distribution is also called a κ-distribution, which sometimes unintentionally causes some confusion for the reader. This choice was made because κ-plasmas represent one of the most natural fields of applications of the new κ-distribution. The promotion of the proposal of the new κ-distribution essentially arises from the purely theoretical need to have a statistical distribution that possesses the important property of self-duality, i.e., $f(-E)f(E) =$ constant, as in the case of the Boltzmann exponential factor $\exp(-\beta E)\exp(\beta E) = 1$ of ordinary Boltzmann–Gibbs statistical mechanics. This need was easily met thanks to the empirical evidence suggesting that the Pareto power law trend of the statistical distribution has purely asymptotic validity. The

Citation: Kaniadakis, G. Relativistic Roots of κ-Entropy. *Entropy* **2024**, *26*, 406. https://doi.org/10.3390/e26050406

Academic Editor: Yong Deng

Received: 5 April 2024
Revised: 24 April 2024
Accepted: 2 May 2024
Published: 7 May 2024

new κ-distribution asymptotically exhibits a power-law tail but gradually transforms in the intermediate range in its bulk range and takes on the features of the typical behavior of the standard exponential Boltzmann factor. In the two papers published in 2002 [5] and 2005 [6], it was shown that the new κ-distribution arises naturally in the context of Einstein's special relativity and generates a self-consistent κ-statistic, which turns out to be a relativistic extension of classical Boltzmann–Gibbs statistical mechanics. The entirety of κ-statistical mechanics can be traced back to κ-entropy

$$S_\kappa = \sum_{i=1}^{W} \frac{f_i^{1-\kappa} - f_i^{1+\kappa}}{2\kappa} \tag{1}$$

where $\{f_i\}$ is the statistical distribution. S_κ entropy is the relativistic generalization of classical Boltzmann–Gibbs–Shannon entropy, which recovers in the $\kappa \to 0$ limit. The corresponding κ-distribution behaves like the ordinary Boltzmann distribution at low energies, while it presents a power-law tail at high energies.

One of the greatest successes of κ-statistics is undoubtedly the explanation of the non-Boltzmannian spectrum of cosmic rays, which are relativistic particles. The persistent power-law tails of this spectrum, spanning 13 decades in terms of energy and 33 decades in terms of particle flux, turn out to be a purely relativistic effect correctly predicted by κ-statistics.

Remarkably, although the statistical theory based on S_κ can be traced back to the first principles of special relativity, it can also be introduced without reference to special relativity, as will be shown in Section II, since it also has applications outside relativistic physics. For this reason, statistical theory [7–13] based on the κ-distribution has attracted the interest of many researchers. In the last two decades, various authors have devoted themselves to the study of both the theoretical foundations of the theory and its applications not only in plasma physics but also in various other areas of the science of complex physical, natural, or artificial statistical systems. Some of these works deal with the H-theorem and the molecular chaos hypothesis [14,15], thermodynamic stability [16,17], Lesche stability [18–21], the Legendre structure of the resulting thermodynamics [22,23], the thermodynamics of non-equilibrium systems [24], quantum versions of the theory [25–28], the geometric structure of the theory [29,30], various mathematical aspects of the theory [31–44], etc. On the other hand, specific applications to physical systems have been considered, e.g., cosmic rays [5], relativistic [45] and classical [46] plasmas in presence of external electromagnetic fields, relaxation in relativistic plasmas under wave–particle interactions [47,48], electronic cooling [49], dark energy models [50–70], quantum gravity [71–78], quantum cosmology [79–83], gravitation and cosmology [84,85], anomalous diffusion [86,87], non-linear kinetics [88–94], the kinetics of interacting atoms and photons [95], particle kinetics in the presence of temperature gradients [96,97], particle systems in external conservative force fields [98], stellar distributions in astrophysics [99–102], quark–gluon plasma formation [103], quantum hadrodynamics models [104], fracture propagation [105], plasma physics [106–120], seismology [121–124], seismic imaging [125–129], nuclear physics [130–134], and quantum mechanics [135–137]. Other applications concern dynamical systems at the edge of chaos [138–140], fractal systems [141], field theories [142], genomic analysis [143–145], random matrix theory [146–148], robust statistical inference [149,150], error theory [151,152], game theory [153], the theory of complex networks [154], information theory [155], etc. Also, applications to economic systems have been considered, e.g., to study the personal income distribution [156–162], to model deterministic heterogeneity in tastes and product differentiation [163,164], in finance [165,166], in equity options [167], to construct taxation and redistribution models [168], etc.

In this paper, we present some new aspects of κ-statistical theory. Section 2 focuses on the axiomatic structure of the theory by proposing the five axioms from which the theory can be deduced without referring to the principles of special relativity. Section 3 focuses on the relativistic origin of the theory. Some peculiar aspects of the physical–mathematical formalism of the theory are emphasized and, in particular, it is shown how the axioms of the theory emerge in relativistic physics. Finally, in Section 4, a synthetic overview of the theory is given in the light of the results obtained in recent years.

2. An Axiomatic Approach to κ-Entropy

The concept of entropy was introduced in the second half of the nineteenth century in the context of thermodynamics by Clausius, who also gave it its name, and immediately afterward by Boltzmann in the context of statistical mechanics. This physical quantity, which emerged within the framework of classical physics, has retained its original form over time, even after the emergence of new branches of physics such as relativistic physics and quantum physics. This entropy, which is still used in physics, was also introduced towards the middle of the twentieth century in Shannon's information theory and subsequently in various fields of science to treat physical, natural, or artificial complex systems. Currently, this entropy is called Boltzmann–Gibbs–Shannon (BGS) entropy [169,170] and is a special case of the more general class of the trace form generalized entropic functional

$$S = \sum_{i=1}^{W} \sigma(f_i) = -\sum_{i=1}^{W} f_i \Lambda(f_i) = - < \Lambda(f_i) > \tag{2}$$

where $<>$ indicates the standard mean value, and in the distribution $f = \{f_1, f_2, \ldots, f_i, \ldots, f_W\}$, f_i represents the probability that the system is in the microstate i with $\sum_1^W f_i = 1$. The standard BGS entropy is obtained by setting $\Lambda(f_i) = \ln(f_i)$. In expression (2) of the generalized entropy [171,172], the function $\Lambda(f_i)$, called the generalized logarithm, is an arbitrary strictly increasing function that is negative on the interval $0 < f_i < 1$. The function $\sigma(f_i) = -f_i \Lambda(f_i)$ represents the contribution to entropy associated with the state i.

Some meaningful properties of the BGS entropy that are elevated to the rank of axioms [169,170], i.e., the Khinchin–Shannon (KS) axioms I, II, and III, can also apply to the generalized entropies. It is therefore assumed that the generalized entropy defined in Equation (2) obeys the following three KS axioms:

I. Continuity axiom: The entropy depends continuously on all the variables f_i. From this axiom follows the continuity of the function $\Lambda(p_i)$.
II. Maximality axiom: The entropy is maximized by the uniform distribution $f_W = \{f_1 = \frac{1}{W}, f_2 = \frac{1}{W}, \ldots, f_i = \frac{1}{W}, \ldots, f_W = \frac{1}{W}\}$, i.e., $S[f] \leq S[f_W]$. From this axiom follows the concavity property $\frac{d^2\sigma(f_i)}{df_i^2} < 0$.
III. Expansibility axiom: The $(W+1)$-component distribution g obtained after the expansion of the W-component distribution f by adding a component with probability equal to zero corresponds to the same entropy of the distribution g, i.e., $S[g] = S[f]$. From this axiom follows the property $0^+\Lambda(0^+) = 0$. We also recall that the particular probability distribution $f = \{\delta_{ia}, 1 \leq i \leq W\}$, where a is a given integer with $1 \leq a \leq W$, describes a state for which one has the maximum information. For this state, $S = 0$ must be set. This condition in turn states that $0^+\Lambda(0^+) = 0$ and also that $\Lambda(1) = 0$. Equivalently, we can set up $\sigma(0) = \sigma(1) = 1$.

It is noteworthy that although the above three KS axioms impose some properties on the function $\Lambda(f_i)$ and then on $\sigma(f_i)$, they do not uniquely determine its form. In the case of BGS entropy, the form of the function $\Lambda(f_i)$ is determined by the fourth KS axiom, i.e., the separability or strong additivity axiom, which implies the property $\Lambda(f_i\, g_j) = \Lambda(f_i) + \Lambda(g_j)$, from which $\Lambda(f_i) = \ln(f_i)$ follows. To go beyond the logarithmic BGS entropy and introduce new entropic functionals, it is necessary to abandon the fourth KS axiom, provided that the first three KS axioms are still equally valid. The fourth KS axiom is replaced by two meaningful properties of the BGS entropy that can equally define the BGS entropy form without invoking the additivity property of the ordinary logarithm function. These two properties are elevated to the status of new axioms and must also apply to the case of generalized entropies. The problem is, therefore, reduced to the search for new generalized entropies that, in addition to the ordinary BGS entropy, also obey the two new axioms.

Starting from the generalized logarithm $\Lambda(f_i)$, we introduce the function $\Lambda(1/f_i)$, which we will call generalized surprise or generalized unexpectedness in analogy to the terms surprise [173] or unexpectedness [174] used in the literature when the generalized logarithm is reduced to the ordinary logarithm. The generalized surprise/unexpectedness is a continuous, decreasing function that admits a unique zero at $f_i = 1$. The opposite of the generalized surprise/unexpectedness $\Lambda^*(f_i) = -\Lambda(1/f_i)$ is a continuous, increasing function and is referred to below as the dual generalized logarithm. The two generalized logarithms $\Lambda(f_i)$ and $\Lambda^*(f_i)$, are the duals of each other and are both increasing functions on the interval $0 < f_i < +\infty$ with a zero at $f_i = 1$. The two functions $\sigma(f_i) = -f_i\,\Lambda(f_i)$ and $\sigma^*(f_i) = -f_i\,\Lambda^*(f_i)$ can be employed to construct the two entropic functionals $S = \sum_i \sigma(f_i)$ and $S^* = \sum_i \sigma^*(f_i)$, respectively, both of which fulfill the first three KS axioms. In general, $S^* \neq S$ holds, and this leads to a theoretical dichotomy, since the two entropies S and S^* define two different statistical theories and, most worryingly, there is no criterion for choosing one of the two entropies. This dilemma does not exist in the case of ordinary BGS entropy, because the property $\ln(1/f_i) = -\ln(f_i)$ implies the self-duality of the logarithm $\ln^*(f_i) = \ln(f_i)$ and then the self-duality of the entropy, i.e., $S^* = S$. To guarantee the uniqueness of the entropy form when considering a generalized statistical theory, we must force the generalized logarithm to be self-dual, just as in the case of ordinary statistical theory. Then, we can introduce the following axiom:

IV. Self-duality axiom: The entropy defined in Equation (2) must be considered both as the standard mean value of the opposite of the generalized logarithm $-\Lambda(f_i)$ and as the standard mean value of the generalized surprise/unexpectedness $\Lambda(1/f_i)$, i.e.,

$$S = - < \Lambda(f_i) > = < \Lambda(1/f_i) > \tag{3}$$

or, equivalently, the generalized logarithm must possess the self-duality property

$$\Lambda(1/f_i) = -\Lambda(f_i) \tag{4}$$

It is noteworthy that axioms I, III, and IV concern some properties of the function $\sigma(f_i)$ or equivalently of $\Lambda(f_i)$, while axiom II concerns a precise property of the function $\frac{d^2\sigma(f_i)}{df_i^2}$. In the following, we will focus on a property of the function $\lambda(f_i) = -\frac{d\sigma(f_i)}{df_i} = \frac{d}{df_i} f_i\Lambda(f_i)$. First, recall that $\sigma(f_i)$ is a continuous and concave function with $\frac{d^2\sigma(f_i)}{df_i^2} < 0$, which has two zeros $\sigma(0) = \sigma(1) = 0$. Then, $\sigma(f_i)$ presents its maximum value at $f_i = 1/\epsilon$ with $\epsilon > 1$. This means that $\lambda(f_i)$ is a monotonically increasing function that has a zero at $f_i = 1/\epsilon$. These general features of the function $\lambda(f_i)$ are typical of a generalized logarithm, with the exception that the generalized logarithm has its zero at $f_i = 1$. Recall that in the case of the ordinary logarithm, the associated function $\lambda(f_i)$ is simply the ordinary logarithm after it has been properly scaled, i.e., $\lambda(f_i) = \frac{1}{\gamma}\ln(\epsilon f_i)$, with $\gamma = 1$ and $\epsilon = e$ (e is the Napier number). This scaling property of the ordinary logarithm must also apply to the generalized logarithm $\Lambda(f_i)$, so that the relationship $\lambda(f_i) = \frac{1}{\gamma}\Lambda(\epsilon f_i)$ must hold, with γ and ϵ being two scaling parameters that are connected by the Boltzmann limit $\lim_{\gamma\to 1}\epsilon = e$. The above scaling property of $\Lambda(f_i)$ is elevated to the status of the following axiom:

V. Scaling axiom: The generalized logarithm which appears in the definition of entropy (2) has the following property of scaling:

$$\frac{d}{df_i}\left(f_i\Lambda(f_i)\right) = \frac{1}{\gamma}\Lambda(\epsilon f_i) \tag{5}$$

where γ and ϵ are the scaling parameters.

The question naturally arises as to whether the BGS entropy is the only existing entropy that obeys the two axioms of self-duality and scaling or whether there is another generalized entropy that equally fulfills the two axioms mentioned. To answer this question, we start from Equation (5), which expresses the scaling axiom and is to be regarded as a differential–functional equation. We seek its general solution after we have correctly determined the free scaling parameters γ and ϵ. Equation (5) was solved in [5,32], and it was shown that besides the BGS entropy, there is a large class of entropies obeying the scaling axiom, some of which are already known in the literature [5,32]. This class of generalized entropies is drastically reduced if the generalized entropy must simultaneously satisfy the scaling and self-duality axioms. In this case, the above class of generalized entropies is reduced to only two entropies. Only the standard BGS entropy corresponding to $\Lambda(f_i) = \ln(f_i)$ and the so-called κ-entropy S_κ corresponding to the κ-logarithm $\Lambda(f_i) = \ln_\kappa(f_i)$ remain to obey the two scaling and self-duality axioms simultaneously. The κ-logarithm is defined by

$$\ln_\kappa(f_i) = \frac{f_i^\kappa - f_i^{-\kappa}}{2\kappa} = \frac{1}{\kappa}\sinh(\kappa \ln(f_i)) \tag{6}$$

The free parameter that appears in the expression of the κ-logarithm varies in the range of $0 < \kappa < 1$ and in the $\kappa \to 0$ limit, the κ-logarithm $\ln_\kappa(f_i)$ is reduced to the ordinary logarithm $\ln(f_i)$. The function $\ln_\kappa(f_i)$ can then be regarded as a one-parameter generalization of the ordinary logarithm. Remarkably, the meaning of the parameter κ emerges when the asymptotic behaviour of the κ-logarithm is considered. The asymptotic behaviour of the κ-logarithm results from Equation (6), i.e., for $f_i \to 0^+$, it obtains $\ln_\kappa(f_i) \propto -f_i^{-\kappa}$, while for $f_i \to +\infty$, according to self-duality axiom, it results in $\ln_\kappa(f_i) \propto f_i^\kappa$. The parameter κ turns out to be the Pareto index, which characterizes the power-law asymptotic behavior of the κ-logarithm. Finally, the constants γ and $\epsilon = \exp_\kappa(\gamma)$ are given by

$$\gamma = \frac{1}{\sqrt{1-\kappa^2}} \quad , \quad \epsilon = \left(\frac{1+\kappa}{1-\kappa}\right)^{\frac{1}{2\kappa}} \tag{7}$$

and in the $\kappa \to 0$ limit, they reduce to unity and Napier number e, respectively, reproducing the results of the standard logarithmic entropy. The connection between the parameters γ and ϵ follows directly from their expressions and is given by $\gamma = \ln_\kappa(\epsilon)$.

Besides the BGS entropy, the κ-entropy is the only one that simultaneously fulfills all five axioms presented above. Thanks to the self-duality property of the κ-logarithm, i.e., $\ln_\kappa(1/f_i) = -\ln_\kappa(f_i)$, κ-entropy can be written as follows:

$$S_\kappa = \sum_i^W \sigma_\kappa(f_i) = -\sum_i^W f_i \ln_\kappa(f_i) = \sum_i^W f_i \ln_\kappa(1/f_i) \tag{8}$$

and can be regarded as the standard mean of both the opposite of the κ-logarithm and its self-dual κ-surprise/unexpectedness i.e. $S_\kappa = - < \ln_\kappa(f_i) > = < \ln_\kappa(1/f_i) >$.

This axiomatic approach to the introduction of κ-entropy is typical of information theory. Remarkably, the two self-duality and scaling axioms that give rise to κ-entropy are also valid in the framework of BGS entropy, although they do not have the rank of axioms but rather express two important properties of standard entropy. The method of replacing the strong additivity axiom of BGS entropy with the new self-duality and scaling axioms that do not contradict any of the standard properties of BGS entropy, including its additivity, clearly leads to a new generalized entropy, namely κ-entropy, which can be employed to analyze physical or non-physical complex systems.

In the reference [175], about sixty different entropies are given, and the corresponding list is not complete. For each of these generalized entropies, it is in principle possible to identify the founding axioms that follow the standard lines of information theory [171]. In any case, it is important to emphasize that entropy is a physical concept that was first introduced in the context of classical thermodynamics and statistical physics. This means that a generalized entropy that claims to be physically meaningful should not only be introduced by postulating some mathematical axioms, as we have done here, but also that these founding axioms should emerge within the framework of a physical theory, starting from its first principles.

The task of the next section will be to show that the two axioms of self-duality and scaling, as well as the κ-logarithm form, follow naturally from the first principles of special relativity.

3. Special Relativity

3.1. Energy–Momentum Lorentz Transformations

Let us consider two identical particles A and B with rest mass m in the one-dimensional inertial frame $\mathcal{S}$, whose velocities, momenta, and total energies are given by v_A, $p_A = mv_A\gamma(v_A)$, and $E_A = mc^2\gamma(v_A)$ and v_A, $p_B = mv_B\gamma(v_B)$, and $E_B = mc^2\gamma(v_B)$, respectively, where $\gamma(v) = (1 - v^2/c^2)^{-1/2}$ is the Lorentz factor, and c is the light speed.

In the rest frame $\mathcal{S}'$ of particle B, the above variables in the case of particle B assume the values $v'_B = 0$, $p'_B = 0$, and $E'_B = mc^2$, respectively, while the velocity v'_A of particle A is given by the Einstein relativistic velocity additivity law $v'_A = (v_A - v_B)/(1 - v_Av_B/c^2)$. In the same frame $\mathcal{S}'$, the momentum p'_A and the energy E'_A of the particle A are given by the dynamic Lorentz transformations

$$p'_A = \gamma(v_B)p_A - c^{-2}v_B\gamma(v_B)E_A \tag{9}$$
$$E'_A = \gamma(v_B)E_A - v_B\gamma(v_B)p_A \tag{10}$$

After introducing the momentum $p_B = mv_B\gamma(v_B)$ and the energy $E_B = mc^2\gamma(v_B)$ of the particle B, the above transformations assume the form

$$p'_A = \frac{1}{mc^2}p_AE_B - \frac{1}{mc^2}E_Ap_B \tag{11}$$
$$E'_A = \frac{1}{mc^2}E_AE_B - \frac{1}{m}p_Ap_B \tag{12}$$

It will be more useful for our discussion hereafter to introduce the new dimensionless variables $(u, q, \mathcal{E})$ in place of the dimensional variables (v, p, E) through

$$\frac{v}{u} = \frac{p}{mq} = \sqrt{\frac{E}{m\mathcal{E}}} = \kappa c = v_* < c \tag{13}$$

where v_* is an arbitrary reference velocity. For a particle at rest, this results in $E(0) = m\,c^2$ and then $\mathcal{E}(0) = 1/\kappa^2$ so that $1/\kappa^2$ is the dimensionless rest energy of the particle. Finally, we note that the classical $c \to \infty$ limit is replaced now by the $\kappa \to 0$ limit.

The Lorentz transformations for the dimensionless momentum and energy variable q and $\mathcal{E}$ assume the form

$$q'_A = \kappa^2 q_A\mathcal{E}_B - \kappa^2 q_B\mathcal{E}_A \tag{14}$$
$$\mathcal{E}'_A = \kappa^2\mathcal{E}_A\mathcal{E}_B - q_Aq_B \tag{15}$$

3.2. Emergence of κ-Exponential Function in Special Relativity

By directly combining the Lorentz transformations, we obtain

$$\kappa^2\mathcal{E}'_A \pm \kappa q'_A = \left(\kappa^2\mathcal{E}_B \mp \kappa q_B\right)\left(\kappa^2\mathcal{E}_A \pm \kappa q_A\right) \tag{16}$$

The variables $\kappa^2\mathcal{E} \pm \kappa q \geq 0$ can be viewed as a dynamic light cone variable, and Equation (16) can be written in the form

$$\left(\kappa^2\mathcal{E}'_A \pm \kappa q'_A\right)^{1/\kappa} = \left(\kappa^2\mathcal{E}_B \mp \kappa q_B\right)^{1/\kappa}\left(\kappa^2\mathcal{E}_A \pm \kappa q_A\right)^{1/\kappa} \tag{17}$$

After taking into account that $\lim_{\kappa\to 0}\left(\kappa^2\mathcal{E} \pm \kappa q\right)^{1/\kappa} = \exp(\pm q)$, the latter relationship reduces to $\exp(\pm q'_A) = \exp(\mp q_B)\exp(\pm q_A)$, which implies the Galilei transformations for the momenta $q'_A = q_A - q_B$. This result suggests the introduction of the new variable

$$\exp_\kappa(q) = \left(\kappa^2\mathcal{E} + \kappa\, q\right)^{1/\kappa} \tag{18}$$

generalizing within the special relativity the ordinary exponential $\exp(q)$, which recovers in the classical limit, i.e., $\lim_{\kappa\to 0} \exp_\kappa(q) = \exp(q)$. The Lorentz transformations, as given by Equation (17), in terms of the κ-exponential function, assume the form

$$\exp_\kappa(\pm q'_A) = \exp_\kappa(\mp q_B)\, \exp_\kappa(\pm q_A) \tag{19}$$

Starting from the Lorentz transformations (14) and (15), the relationship expressing the Lorentz invariance, i.e., $\kappa^4\mathcal{E}'^2 - \kappa^2 q'^2 = \kappa^4\mathcal{E}^2 - \kappa^2 q^2$, can be obtained, and after identifying $\mathcal{S}'$ as the particle rest frame where $\mathcal{E}(0) = 1/\kappa^2$, the energy–momentum dispersion relation

$$\kappa^4\mathcal{E}^2 - \kappa^2 q^2 = 1 \tag{20}$$

follows. From the latter relationship, the expression of the dimensionless total energy $\mathcal{E}$ can be obtained in terms of the dimensionless momentum q

$$\mathcal{E} = \frac{1}{\kappa^2}\sqrt{1+\kappa^2\, q^2} \tag{21}$$

After inserting this expression of total energy in the definition (18) of the κ-exponential function, the explicit form is obtained as follows:

$$\exp_\kappa(q) = \left(\sqrt{1+\kappa^2 q^2} + \kappa q\right)^{1/\kappa} = \exp\left(\frac{1}{\kappa}\mathrm{arcsinh}(\kappa q)\right) \tag{22}$$

After the substitution of the expression of the dimensionless total energy given by Equation (21) into the first of the Lorentz transformations given by Equation (14), the relativistic additivity law of the dimensionless momenta assumes the form $q'_A = q_A \overset{\kappa}{\ominus} q_B = q_A \overset{\kappa}{\oplus} (-q_B)$, where the κ-sum $\overset{\kappa}{\oplus}$ is defined as

$$q_A \overset{\kappa}{\oplus} q_B = q_A\sqrt{1+\kappa^2\, q_B^2} + q_B\sqrt{1+\kappa^2\, q_A^2} \tag{23}$$

The following property of the κ-exponential holds:

$$\exp_\kappa(q_A \overset{\kappa}{\oplus} q_B) = \exp_\kappa(q_A)\exp_\kappa(q_B) \tag{24}$$

which is reminiscent of the analogous property of the classical exponential function $\exp(q_A + q_A) = \exp(q_A)\, \exp(q_B)$.

3.3. Emergence of κ-Logarithm Function in Special Relativity

The function $\ln_\kappa(w)$ is defined as the inverse function of $\exp_\kappa(w)$ through $\ln_\kappa(\exp_\kappa w) = \exp_\kappa(\ln_\kappa w) = w$. Its explicit expression is

$$\ln_\kappa(w) = \frac{w^\kappa - w^{-\kappa}}{2\kappa} = \frac{1}{\kappa}\,\sinh\left(\kappa \ln w\right) \tag{25}$$

and it reduces to the ordinary logarithm in the classical limit, i.e., $\lim_{\kappa\to 0}\ln_\kappa(w) = \ln(w)$.

The ordinary logarithm $\ln(w)$ is the only existing function unless a multiplicative constant is used, which results in the solution to the function equation $\ln(w_1 w_2) = \ln(w_1) + \ln(w_2)$. Let us now consider the relativistic generalization of this equation, which we obtain from the Lorentz transformation given by Equation (24) after posing $w = \exp_\kappa(q)$, i.e., $\ln_\kappa(w_1 w_2) = \ln_\kappa(w_1) \overset{\kappa}{\oplus} \ln_\kappa(w_2)$, which is written as

$$\ln_\kappa(w_1 w_2) = \ln_\kappa(w_1)\,\gamma_\kappa(\ln_\kappa(w_2)) + \ln_\kappa(w_2)\,\gamma_\kappa(\ln_\kappa(w_1)) \tag{26}$$

where $\gamma_\kappa(\ln_\kappa(w)) = \sqrt{1+\kappa^2\,\ln_\kappa^2(w)}$ is the Lorentz factor of argument $\ln_\kappa(w)$. By the direct substitution of the κ-logarithm in the expression of $\gamma_\kappa(\ln_\kappa(w))$, the further expression $\gamma_\kappa(\ln_\kappa(w)) = (w^\kappa + w^{-\kappa})/2$ is obtained. Starting from this latter relationship and after some tedious but straightforward calculation, a third expression is obtained of the function $\gamma_\kappa(\ln_\kappa(w))$, i.e.,

$$\gamma_\kappa(\ln_\kappa(w)) = \frac{1}{\gamma}\,\ln_\kappa(\epsilon w) - \ln_\kappa(w) \tag{27}$$

where the constant $\epsilon = ((1+\kappa)/(1-\kappa))^{1/2\kappa}$ represents the κ-generalization of the Napier number e, while the constant $\gamma = 1/\sqrt{1-\kappa^2}$ is the Lorentz factor corresponding to the reference velocity $v = v_*$. The two constants are linked through $\epsilon = \exp_\kappa(\gamma)$. Equation (27) expresses an important property of the κ-logarithm, which will be used in the following.

It is noteworthy that the introduction of the function $\ln_\kappa(w)$ allows us to write the additivity law of dimensionless relativistic moments defined in Equation (24) in the form

$$q_A \overset{\kappa}{\oplus} q_B = \ln_\kappa(\,\exp_\kappa(q_A)\exp_\kappa(q_B)) \tag{28}$$

3.4. Emergence of Self-Duality in Special Relativity

The dispersion relation (20) can be written in the factorized form $(\kappa^2\mathcal{E} + \kappa\,q)(\kappa^2\mathcal{E} - \kappa\,q) = 1$, and after noticing that $\kappa^2\mathcal{E} \pm \kappa\,q \geq 0$, the dispersion relation can be rewritten as follows:

$$\left(\kappa^2\mathcal{E} + \kappa\,q\right)^{1/\kappa}\left(\kappa^2\mathcal{E} - \kappa\,q\right)^{1/\kappa} = 1 \tag{29}$$

and finally, after involving the κ-exponential function, the relation can be rewritten as

$$\exp_\kappa(q)\,\exp_\kappa(-q) = 1 \tag{30}$$

The latter relationship expresses an important property of the κ-exponential function, which, in the classical limit, is reduced to the well-known property of the ordinary exponential function $\exp(q)\,\exp(-q) = 1$. As in the case of the ordinary exponential function, the values of the κ-exponential function for $q < 0$ are directly related to its values for $q > 0$, resulting in $\exp_\kappa(-q) = 1/\exp_\kappa(q)$. This self-duality property in terms of the κ-logarithm assumes the form

$$\ln_\kappa(1/w) = -\ln_\kappa(w) \tag{31}$$

and means that the values of the κ-logarithm function on the interval $w > 1$ are related to its values on the interval $0 < w < 1$. An important consequence of the relationship (30) is that the inverse transformations of the direct Lorentz transformations (19) assume the form

$$\exp_\kappa(\pm q_A) = \exp_\kappa(\pm q_B)\,\exp_\kappa(\pm q'_A) \tag{32}$$

A comparison of the direct (19) and inverse (32) Lorentz transformations shows that the inverse Lorentz transformations have the same structure as the direct transformations, except for the substitutions $q'_A \leftrightarrow q_A$ and $q_B \to -q_B$. This symmetry expresses the Galilean principle of relativity, which applies both in classical physics and in special relativity and prescribes the equivalence of all inertial frames. From this, we can conclude that the self-duality property $\exp(q)\exp(-q) = 1$ of the ordinary exponential function and the analogous property of the κ-exponential function, which is given by Equation (30), is enforced by the Galilean principle of relativity.

3.5. κ-Mathematics

The additivity law of dimensionless relativistic moments defined in Equation (23) with $q_A, q_B \in \mathbf{R}$ called κ-sum and denoted by $\overset{\kappa}{\oplus}$ is a generalized sum and can be viewed as a one-parameter, continuous deformation of the ordinary sum, which recovers in the classical limit $\kappa \to 0$, i.e., $q_A \overset{0}{\oplus} q_B = q_A + q_B$. The κ-sum has the following properties: (1) it is associative, where $(q_A \overset{\kappa}{\oplus} q_B) \overset{\kappa}{\oplus} q_C = q_A \overset{\kappa}{\oplus} (q_B \overset{\kappa}{\oplus} q_C)$; (2) it admits a neutral element, where $q_A \overset{\kappa}{\oplus} 0 = 0 \overset{\kappa}{\oplus} a_A = q_A$; (3) it admits an opposite element, where $q_A \overset{\kappa}{\oplus} (-q_A) = (-q_A) \overset{\kappa}{\oplus} q_A = 0$; (4) it is commutative, where $q_A \overset{\kappa}{\oplus} q_B = q_B \overset{\kappa}{\oplus} q_A$. Then, the algebraic structure $(\mathbf{R}, \overset{\kappa}{\oplus})$ forms an abelian group. The κ-difference $\overset{\kappa}{\ominus}$ is defined as $q_A \overset{\kappa}{\ominus} q_B = q_A \overset{\kappa}{\oplus} (-q_A)$.

Starting from the κ-sum, κ-mathematics can be introduced after defining the κ-exponential function as the solution to the functional Equation (24). The introduction of κ-functions can be performed starting from the κ-exponential and following the standard procedures of ordinary mathematics. For instance, κ-trigonometry (ordinary or hyperbolic) can be introduced by employing the κ-Euler formula, while the κ-inverse function follows after the inversion of their direct functions [3]. Also, κ-differential calculus can be introduced after defining the κ-derivative as the differential operator, which acts on the κ-exponential function, which subsequently produces the κ-exponential function itself.

Next, we revisit the κ-derivative and discuss how it arises within the special relativity. Let us consider two identical particles A and B in the one-dimension spatial frame $\mathcal{S}$ having dimensionless momenta $q_A = q$ and $q_B = \tilde{q}$, respectively. In the rest frame $\mathcal{S}'$ of particle B, which is an inertial frame that moves with velocity v_B with respect the inertial frame $\mathcal{S}$, the dimensionless moment of particle B is $q'_B = \tilde{q}' = 0$, while the dimensionless moment $q'_A = q'$ of particle A is given by $q' = q \overset{\kappa}{\ominus} \tilde{q}$. We suppose that $\tilde{q} \approx q$ and pose $dq \approx q - \tilde{q}$ and $dq' \approx q \overset{\kappa}{\ominus} \tilde{q}$. Starting from the limit

$$\lim_{\tilde{q} \to q} \frac{q \overset{\kappa}{\ominus} \tilde{q}}{q - \tilde{q}} = \frac{1}{\gamma_\kappa(q)} \tag{33}$$

with $\gamma_\kappa(q) = \sqrt{1 + \kappa^2 q^2}$ being the Lorentz factor, the differential dq' can be obtained as

$$dq' = \frac{dq}{\gamma_\kappa(q)} \tag{34}$$

The κ-differential $d_\kappa q = dq'$ has a very transparent physical meaning representing the infinitesimal variation in the momentum of a given particle, observed in the frame $\mathcal{S}'$. It is related to the infinitesimal variation in the momentum dq of the same particle, observed in the inertial frame $\mathcal{S}$ through the Lorentz factor. A further interesting property of the differentials $d_\kappa q$ is given by $d_\kappa q = d(\rho_\kappa(q))$ or simply $d_\kappa q = d\rho$, where $\rho = \rho_\kappa(q)$ is the κ-rapidity defined through

$$\rho_\kappa(q) = \frac{1}{\kappa} \mathrm{arcsinh}(\kappa\, q) \tag{35}$$

The variable $\phi_\kappa(u) = \mathrm{arctanh}(v/c) = \mathrm{arctanh}(\kappa u)$ was introduced into special relativity in 1910 by V. Varicak and E. T. Whittak and was named rapidity by A. Robb in 1911. The old rapidity is related to the κ-rapidity $\rho_\kappa(q)$ through $\phi_\kappa(u) = \kappa\, \rho_\kappa(q)$, which can be easily verified after taking into account that $u = q/\gamma_\kappa(q)$. The presence of the proportionality factor κ in the relation linking $\phi_\kappa(u)$ and $\rho_\kappa(q)$ is not trivial because, in the classical limit, the κ-rapidity reduces to the dimensionless momentum, i.e., $\rho_0(q) = q$, while in the same limit, the old rapidity does not reduce to the dimensionless velocity holding $\phi_0(u) = 0$. The relativistic composition law of κ-rapidity is given by

$$\rho_\kappa(q'_A) = \rho_\kappa(q_A) - \rho_\kappa(q_B) \tag{36}$$

and becomes identical to the ordinary difference $\rho'_A = \rho_A - \rho_B$. The expression of the κ-exponential function in terms of $\rho_\kappa(q)$ is given by

$$\exp_\kappa(q) = \exp(\rho_\kappa(q)) \tag{37}$$

The κ-derivative of the scalar function $f(q)$ is defined through

$$\frac{d\,f(q)}{d_\kappa\,q} = \gamma_\kappa(q)\,\frac{d\,f(q)}{d\,q} \tag{38}$$

It is important to note that $d\,f(q)$ is an ordinary differential, while $d_\kappa\,q$ is a κ-differential. It follows that the κ-derivative is proportional through the Lorentz factor $\gamma_\kappa(q)$ to the ordinary derivative and then obeys Leibniz's rules of the ordinary derivative.

3.6. The κ-Differential Equations

The dynamic variables of relativistic physics can be obtained as solutions of first-order differential equations involving the κ-derivative $df(q)/d_\kappa q$, which, in the classical limit, reduces to the corresponding differential equations of classical physics.

The solution to

$$\frac{d}{d_\kappa q}\,f(q) = 1 \tag{39}$$

with the condition $f(0) = 0$ is the rapidity function $f(q) = \rho_\kappa(q)$, i.e., $f(q) = \frac{1}{\kappa}\mathrm{arcsinh}(\kappa q)$.

The solution to

$$\frac{d}{d_\kappa q}\,f(q) = q \tag{40}$$

with the condition that $f(0) = 1/\kappa^2$ is the total energy $f(q) = \mathcal{E}_\kappa(q)$, i.e., $f(q) = \sqrt{1+\kappa^2 q^2}/\kappa^2$, while the solution to the same equation with the condition $f(0) = 0$ is the relativistic kinetic energy $f(q) = \mathcal{W}_\kappa(q)$ i.e., $f(q) = \left(\sqrt{1+\kappa^2 q^2} - 1\right)/\kappa^2$.

The solution to

$$\frac{d}{d_\kappa q}\,f(q) = \kappa^2 q \tag{41}$$

with the condition $f(0) = 1$ is the Lorentz factor $f(q) = \gamma_\kappa(q)$, i.e., $f(q) = \sqrt{1+\kappa^2 q^2}$.

The solution to

$$\frac{d}{d_\kappa q}\,f(q) = f(q) \tag{42}$$

with the condition $f(0) = 1$ is the κ-exponential function $f(q) = \left(\sqrt{1+\kappa^2 q^2} + \kappa q\right)^{1/\kappa}$.

Finally, the relativistic velocity $u_\kappa(q) = q/\sqrt{1+\kappa^2 q^2}$ is the solution $f(q) = u_\kappa(q)$ of the differential equation

$$\frac{d}{d_\kappa q}\,f(q) = \left(\frac{f(q)}{q}\right)^2 \tag{43}$$

with the condition $f(\pm\infty) = \pm 1/\kappa$.

3.7. The Scaling Property of κ-Logarithm

The differential equation

$$\sqrt{1+\kappa^2 q^2}\ \frac{d\ \exp_\kappa(q)}{dq} = \exp_\kappa(q) \tag{44}$$

obeyed by the $\exp_\kappa(q)$ can be easily inverted, obtaining

$$\frac{d\ \ln_\kappa(w)}{dw} = \frac{\gamma_\kappa(\ln_\kappa(w))}{w} \tag{45}$$

and after taking into account Equation (27), it follows that the κ-logarithm function obeys the first-order differential–functional equation

$$\frac{d}{dw}\left[w\ \ln_\kappa(w)\right] = \frac{1}{\gamma}\ \ln_\kappa(\epsilon w) \tag{46}$$

expressing the so-called scaling property of the κ-logarithm. In the $\kappa \to 0$ classical limit, the latter equation continues to hold and reduces to a well-known property of the ordinary logarithm, where scaling constants reduce to the values $\gamma = 1$ and $\epsilon = e$.

The two last equations, if combined, lead to the further property of κ-logarithm

$$\frac{d^2}{dw^2}\left[w\ \ln_\kappa(w)\right] = \frac{1}{\gamma\, w}\ \gamma_\kappa(\ln_\kappa(\epsilon w)) \geq 0 \tag{47}$$

4. κ-Statistical Physics

4.1. Maximum Entropy Principle and κ-Entropy

In proposing a relativistic statistical theory, the only guiding principle available is the metaphor of classical statistical physics, and the entropy form plays an important role in this context. The standard relativistic statistical theory is based on an entropic form identical to that of classical statistical physics, the BGS entropy. This is due to the great success of BGS entropy in classical many-body physics. In Einstein's special relativity, all microscopic physical quantities such as particle momentum or particle energy are properly generalized. Regarding macroscopic quantities such as temperature or pressure, there is still a debate about how they should be defined in a relativistic context. It is, therefore, an evident dichotomy that on one side, there is the BGS entropy, which dominates both classical and relativistic physics, and on the other side, there are all the other physical quantities, both microscopic and macroscopic, which are or could be modified in special relativity. We note that the BGS entropy in the relativistic context conducts to the Juttner distribution, which, when considered as a function of the relativistic particle energy, is exactly the Boltzmann exponential factor of classical physics. It has long been known that the Boltzmann factor does not correctly describe the spectrum of cosmic rays, which are relativistic particles.

In the following, we will present the relativistic statistical theory based on the κ-entropy S_κ, which is defined as the standard mean of the opposite of the κ-logarithm emerging in special relativity. The paradigm of classical statistical physics will be constantly present in our discussion, and the starting point will be the maximum entropy principle, the cornerstone of statistical theory. Let us consider the constrained entropy $\Phi(f) = S_\kappa(f) + \mathcal{C}(f)$, where the constraints functional $\mathcal{C}(f)$ is given in its simplest form by

$$\mathcal{C}(f) = \mathrm{a}_1\left[\sum_i f_i - 1\right] + a_2\left[I - \sum_i\ I_i f_i\right] \tag{48}$$

where a_1 and a_2 are the Lagrange multipliers, while $\{I_i\}$ is the generator function of the moment $I = \sum_i I_i f_i$. The variational equation $\frac{\delta\Phi(f)}{\delta f_i} = 0$ implies the maximization of S_κ under the constraints imposing the conservation of the norm of f_i and the a priori knowledge of the values of the moment $I = \sum_i I_i f_i$ generated by the generator function $\{I_i\}$. The solution to the above variational problem conducts to the equation $\frac{d}{df_i}(f_i \ln_\kappa(f_i)) = a_1 - a_2 I_i$, which, after taking into account the scaling axiom, assumes the form

$$\frac{1}{\gamma} \ln_\kappa(\epsilon f_i) = a_1 - a_2 I_i \tag{49}$$

In the case of a classical particle gas, the above equation reduces to $\ln(ef_i) = a_1 - a_2 I_i$, where the microscopic collisional invariant I_i is the classical particle energy in the state i, while Lagrange multipliers are related to the gas temperature T and chemical potential μ according to $a_2 = 1/k_B T$ and $a_1 = \mu/k_B T$. The same parameters T and μ will also occur in the case of a relativistic gas, while I_i will be the microscopic relativistic collisional invariant. Equation (49), after inversion, takes the form

$$f_i = \frac{1}{\epsilon} \exp_\kappa\left(-\frac{I_i - \mu}{k_B T_\kappa}\right) \tag{50}$$

with $T_\kappa = T/\gamma$.

Remarkably, thanks to the scaling property of the κ-logarithm, the expressions of the κ-entropy and the κ-distribution f_i are given in terms of the same function, which appears in its direct (κ-logarithm) or inverse (κ-exponential) form, just as in the classical case.

Let us pose $w_i = (I_i - \mu)/k_B T_\kappa$. When $w_i \to +\infty$, the asymptotic behavior of the function $\exp_\kappa(-w_i)$ is given by $\exp_\kappa(-w_i) \approx (2\kappa w_i)^{-1/\kappa}$. Consequently, the tail of the distribution (50) is described by a Pareto power law function, i.e., $f_i \approx \epsilon^{-1}(2\kappa w_i)^{-1/\kappa}$, instead of the exponential tails of the Juttner distribution $f_i = e^{-1}\exp(-w_i)$, originating from the BGS entropy. The power-law tail of the distribution (50) is one of its most interesting features and is consistent with the experimental evidence in relativistic particle physics, i.e., cosmic rays and the so-called κ-plasmas observed in laboratory or in astrophysics.

Let us introduce the κ-entropy $S_\kappa(g) = \sum_i \sigma(g_i)$, which refers to the arbitrary distribution $g = \{g_i\}$ and is subjected to the constraints described by the functional $C(g) = -\sum_i (\mathrm{a}_2 I_i - \mathrm{a}_1)\, g_i - \mathrm{a}_1 + \mathrm{a}_2 I$. Let us further denote by $f = \{f_i\}$ the optimal distribution defined according to the maximum entropy principle, defined in Equation (49), which takes the form $a_2\, I_i - a_1 = \frac{d\,\sigma(f_i)}{d\,f_i}$, so that the constraints functional can be written as follows:

$$C(g) = -\mathrm{a}_1 + \mathrm{a}_2 I - \sum_i \frac{d\,\sigma(f_i)}{d\,f_i}\, g_i \tag{51}$$

The difference in the constrained entropy $\Phi(g)$ from its maximum value $\Phi(f)$, i.e., $\Phi(f) - \Phi(g) = S_\kappa(f) - S_\kappa(g) + C(f) - C(g)$, finally assumes the form

$$\Phi(f) - \Phi(g) = \sum_i \left[\sigma(f_i) - \sigma(g_i) - \frac{d\,\sigma(f_i)}{d\,f_i}(f_i - g_i) \right] \tag{52}$$

When $g_i \approx f_i$, the Taylor expansion can be considered as

$$\sigma(g_i) \approx \sigma(f_i) + \frac{d\,\sigma(f_i)}{d\,f_i}(g_i - f_i) + \frac{1}{2}\frac{d^2\sigma(f_i)}{d\,f_i^2}(g_i - f_i)^2 \tag{53}$$

so that we obtain

$$\Phi(f) - \Phi(g) \approx -\sum_i \frac{1}{2}\frac{d^2\sigma(g_i)}{d\,g_i^2}(f_i - g_i)^2 \tag{54}$$

and after taking into account the expression of $\frac{d^2\sigma(g_i)}{d g_i^2}$ as given by Equation (47), we obtain

$$\Phi(f) - \Phi(g) \approx \sum_i \frac{\gamma_\kappa(\ln_\kappa(\epsilon g_i))}{2\gamma g_i}(f_i - g_i)^2 \geq 0 \tag{55}$$

The latter relationship tells us that $\Phi(f)$ represents the maximum value of $\Phi(g)$ and expresses the thermodynamic stability of the system.

Another stability that differs from thermodynamic stability is the Lesche stability condition, which prescribes that any physically meaningful entropy that depends on a probability distribution function g should exhibit a small relative error

$$R = \left| \frac{S(g) - S(h)}{S_{max}} \right| \tag{56}$$

with respect to small changes in the probability distributions $g \to h$

$$D = ||g - h|| \tag{57}$$

Mathematically, this means that for every $\varepsilon > 0$, there is a $\delta > 0$ so that $R \leq \varepsilon$ applies to all distribution functions that fulfill $D \leq \delta$. It is known that the Lesche stability condition holds for the Boltzmann–Shannon entropy, and in refs. [18,19], it was shown that the Lesche stability condition also holds for the κ-entropy. In addition, the κ-entropy is also Lesche-stable in the thermodynamic limit.

4.2. *κ-Kinetics*

Let us consider the first equation of the Bogoliubov–Born–Green–Kirkwood–Yvon hierarchy, which describes the evolution of a relativistic many-body system in the presence of an external force field and imposes particle conservation during collisions [6,176,177]:

$$p^\nu \partial_\nu f - mF^\nu \frac{\partial f}{\partial p^\nu} = \int \frac{d^3p'}{p'^0} \frac{d^3p_1}{p_1^0} \frac{d^3p'_1}{p'^0_1} \, G\left[C(f', f'_1) - C(f, f_1)\right] \tag{58}$$

The system is described by the one-particle correlation function or distribution function $f = f(x, p)$, where x and p are the four-vector position and momentum. In the above equation, both the streaming term and the Lorentz invariant integrations in the collision integral have the standard forms of relativistic kinetic theory. The two-particle correlation function $C(f, f_1)$, which is determined below, is postulated in the case of ordinary relativistic kinetics as $C(f, f_1) = f f_1$, which represents the molecular chaos hypothesis and reduces the above evolution equation to the relativistic Boltzmann equation.

Following standard lines of kinetic theory, we note that in stationary conditions, the collision integral vanishes and then $C(f, f_1) = C(f', f'_1)$. This relationship expresses a conservation law for the particle system and must have the form $L(f) + L(f_1) = L(f') + L(f'_1)$. In relativistic kinetics, the collision invariant $L(f)$, unless an additive constant, is proportional to the microscopic relativistic invariant $I(x, p)$, i.e.,

$$L(f) = -a_2 I(x, p) + a_1 \tag{59}$$

with a_1 and a_2 being two arbitrary constants. The more general microscopic relativistic invariant I, in the presence of an external electromagnetic field A^ν, has a form proportional to

$$I(x, p) = (p^\nu + qA^\nu/c)\, U_\nu - mc^2 \tag{60}$$

with U_ν being the hydrodynamic four-vector velocity with $U^\nu U_\nu = c^2$ [176].

The expression of the distribution function defined in Equation (50) holds in stationary conditions where the entropy of the particle system reaches its maximum value. According to the scaling axiom, after considering the correspondences $f_i \to f(x, p)$ and $I_i \to I(x, p)$

and after the identification of $a_2 = 1/k_B T$ and $a_1 = \mu/k_B T$, it follows that $L(f) = \lambda_\kappa(f)$, or more explicitly

$$L(f) = \frac{1}{\gamma}\,\ln_\kappa(\epsilon f) \tag{61}$$

while the stationary distribution assumes the form

$$f(x,p) = \frac{1}{\epsilon}\exp_\kappa\left(-\frac{I(x,p)-\mu}{k_B T_\kappa}\right) \tag{62}$$

with $T_\kappa = T/\gamma$.

4.3. *κ-Molecular Chaos Hypothesis*

In stationary conditions, $C(f,f_1) = C(f',f_1')$ applies. This relationship expresses a conservation law and can be written in the form $L(f) + L(f_1) = L(f') + L(f_1')$ after posing

$$C(f,f_1) = L^{-1}(\,L(f) + L(f_1)) \tag{63}$$

The function $L(w)$ increases monotonically on the interval $0 \le w < +\infty$, with $L(0) = -\infty$ and $L(+\infty) = +\infty$. These conditions imply that $C(0,f_1) = C(f,0) = 0$, just as in the case of the ordinary correlation function. After taking into account the expression of the function $L(f)$, the two-particle correlation function assumes the form

$$\epsilon\, C(f,f_1) = \exp_\kappa(\,\ln_\kappa(\epsilon f) + \ln_\kappa(\epsilon f_1)) \tag{64}$$

which can be written in a more compact form

$$\epsilon\, C(f,f_1) = (\epsilon f) \otimes (\epsilon f_1) \tag{65}$$

by involving the generalized product

$$g \otimes h = \exp_\kappa(\,\ln_\kappa g + \ln_\kappa h) \tag{66}$$

This κ-product between probabilities has the following properties:

(i) $(g \otimes h) \otimes l = g \otimes (h \otimes l)$, i.e., it is associative;
(ii) $g \otimes h = h \otimes g$, i.e., it is commutative;
(iii) $1 \otimes g = g$, i.e., it admits the unity as a neutral element;
(iv) $g \otimes (1/g) = 1$, i.e., the inverse element of g is $1/g$;
(v) It holds the property $g \otimes 0 = 0$;
(vi) $g \oslash h = g \otimes (1/h)$ defines the κ-division between probabilities.

The real, positive probability distribution functions form an abelian group. The properties of $\otimes$ are the same as those of the ordinary product, so the two products are isomorphic.

We can conclude that the relation given by Equation (64), which defines the two-particle correlation function by the κ-product, is the relativistic version of the molecular chaos hypothesis and reduces to its standard form $C(f,f_1) \propto f\,f_1$ in the classical limit $\kappa \to 0$.

4.4. *Four-Vector κ-Entropy and Relativistic H-Theorem*

In standard relativistic kinetics, it is known from the H-theorem that entropy production is never negative and that there is no entropy production under equilibrium conditions. In the following, we will demonstrate the H-theorem for the system governed by the kinetic Equation (58). We define the four-vector entropy $S^\nu = (S^0, \mathbf{S})$ as follows:

$$S^\nu = -\int \frac{d^3p}{p^0}\, p^\nu\, f\,\ln_\kappa(f) \tag{67}$$

and note that the scalar entropy $S^0 = S_\kappa$ coincides with the κ-entropy, while $\mathbf{S} = \mathbf{S}_\kappa$ is the κ-entropy flow. After considering the identity $d^3p/p^0 = d^4p\, 2\,\theta(p^0)\,\delta(p^\mu p_\mu - m^2c^2)$ and the observation that d^4p is a scalar because the Jacobian of the Lorentz transformation is equal to unity, we conclude that S^ν transforms as a four-vector, since p^ν transforms as a four-vector.

In order to calculate the entropy production $\partial_\nu S^\nu$, we start from the definition of S^ν and the relationship $\partial_\nu\,[f\ \ln_\kappa(f)] = [\,\partial\,[f\ \ln_\kappa(f)]/\,\partial f\,]\,\partial_\nu f = \lambda_\kappa(f)\,\partial_\nu f$ with $\lambda_\kappa(f) = \frac{1}{\gamma}\ln_\kappa(\epsilon f)$, obtaining

$$\partial_\nu S^\nu = -\int \frac{d^3p}{p^0}\,\lambda_\kappa(f)\ p^\nu\,\partial_\nu f \tag{68}$$

After taking into account the kinetic equation (58), the entropy production assumes the form

$$\partial_\nu S^\nu = -\int \frac{d^3p}{p^0}\frac{d^3p'}{p'^0}\frac{d^3p_1}{p_1^0}\frac{d^3p'_1}{p'^0_1}\ G\left[C(f',f'_1) - C(f,f_1)\right]\lambda_\kappa(f) - m\int \frac{d^3p}{p^0}\,\lambda_\kappa(f)\ F^\nu\frac{\partial f}{\partial p^\nu} \tag{69}$$

Since the Lorentz force F^ν has the properties $p^\nu F_\nu = 0$ and $\partial F^\nu/\partial p^\nu = 0$, the last term in the above equation involving F^ν is equal to zero [176]. Given the particular symmetry of the non-vanishing integral in Equation (69) we can write the entropy production as follows

$$\begin{aligned}\partial_\nu S^\nu = \frac{1}{4}\int &\frac{d^3p}{p^0}\frac{d^3p'}{p'^0}\frac{d^3p_1}{p_1^0}\frac{d^3p'_1}{p'^0_1}\ G\\ &\times\left[C(f',f'_1) - C(f,f_1)\right]\left[\lambda_\kappa(f') + \lambda_\kappa(f'_1) - \lambda_\kappa(f) - \lambda_\kappa(f_1)\right]\end{aligned} \tag{70}$$

From the definition of the two-particle correlation function, it follows that $\lambda_\kappa(f') + \lambda_\kappa(f'_1) - \lambda_\kappa(f) - \lambda_\kappa(f_1) = \lambda_\kappa(C(f',f'_1)) - \lambda_\kappa(C(f,f_1))$, and after posing $\alpha' = C(f',f'_1)$, $\alpha = C(f,f_1)$, finally, we write Equation () in the form

$$\partial_\nu S^\nu = \frac{1}{4}\int \frac{d^3p}{p^0}\frac{d^3p'}{p'^0}\frac{d^3p_1}{p_1^0}\frac{d^3p'_1}{p'^0_1}\ G\left[\alpha' - \alpha\right]\left[\lambda_\kappa(\alpha') - \lambda_\kappa(\alpha)\right] \tag{71}$$

With $\lambda_\kappa(\alpha)$ being an increasing function, it follows that $[\alpha' - \alpha][\lambda_\kappa(\alpha') - \lambda_\kappa(\alpha)] \geq 0$, $\forall \alpha', \alpha$, and then we can conclude that

$$\partial_\nu S^\nu \geq 0 \tag{72}$$

This last relation is the local formulation of the relativistic H-theorem, which represents the second law of thermodynamics for the system governed by the evolution Equation (58).

4.5. Relativistic Temperature

The construction of a thermodynamic theory compatible with the principles of special relativity is an old and still open problem, dating back to the first years immediately after the proposal of the relativistic theory. The proposals that have dealt with the question of how the thermodynamic quantities that characterize the physical system change when the inertial reference system changes are diverse and contradictory. Some of these proposals are still under consideration, and the problem is still highly topical. Let T denote the temperature of a body at rest and T' the temperature of the body when the body is observed from a reference frame moving at a speed characterized by the Lorentz factor γ. According to Planck and Einstein, the two temperatures are linked by $T' = T/\gamma$. According to Ott, $T' = \gamma T$. Finally, according to Landsberg, $T' = T$. In a series of subsequent articles, some of which have appeared recently, the scientific community has overwhelmingly sided with the Planck–Einstein proposal and accepted that a moving body is colder.

We do not intend to go into this important topic here. However, it is noteworthy that the present formalism proves to be consistent with the Planck–Einstein proposal. Let us consider the particle gas described by the distribution function (62), where the relevant temperature is given by

$$T_\kappa = \frac{1}{\gamma}\, T \tag{73}$$

with

$$\gamma = \frac{1}{\sqrt{1-\kappa^2}} = \frac{1}{\sqrt{1-(\frac{v^*}{c})^2}} \tag{74}$$

A possible interpretation of the above formula of temperature is the following. The temperature of the system at rest is $T_0 = T$ while $T_\kappa = T_0/\gamma < T_0$ is its temperature when it is moving at speed v^*; this is just the Planck–Einstein proposal. The two temperatures T_κ and T_0 are proportional to each other, and the proportionality factor is the Lorentz factor γ. Remarkably, entropy does not have this proportionality property. The entropy of the physical system moving at the speed v^* is $S_\kappa = -\sum_i f_i \ln_\kappa(f_i)$, while its entropy at rest reduces to the classical Boltzmann entropy $S_0 = -\sum_i f_i \ln(f_i)$.

5. Epilogue

Some of the results of the present theory, which were discussed in the previous sections, are emphasized below:

(i) Relativistic statistical theory: It is possible to construct a statistical theory within the framework of special relativity that preserves the main features of classical statistical theory (axiomatic structure, maximum entropy principle, thermodynamic stability, Lesche stability, molecular chaos hypothesis, local formulation of H-theorem, etc.).

(ii) Old problems of special relativity: Within the framework of the new relativistic statistical theory, answers naturally arise to questions that were formulated immediately after the proposal of special relativity as to how the temperature and entropy of a moving body change. In particular, it turns out that the temperature varies according to the law $T_\kappa = T_0/\gamma$ proposed by Planck and Einstein in 1906, where $\gamma = 1/\sqrt{1-\kappa^2}$ is the Lorentz factor.

(iii) Axiomatic structure of the theory: Although the statistical theory generated by the entropy S_κ was developed within the framework of Einstein's special relativity, it can also be introduced without reference to special relativity given its applications outside physics by following the guidelines of information theory, which emphasizes the axiomatic structure of the various theories. In the construction of κ-entropy, the first three Khinchin–Shannon axioms are taken into account, i.e., those of the continuity, maximality, and expansibility of the ordinary Boltzmann entropy. Subsequently, the fourth Khinchin–Shannon axiom of strong additivity is replaced by two new axioms, namely, those of self-duality and scaling, which express well-known properties of logarithmic Boltzmann entropy. In the final step, it is shown that these five axioms are not only able to generate the Boltzmann entropy but also a further and unique entropy, namely, κ-entropy, which turns out to be a one-parameter continuous generalization of the Boltzmann entropy. The axioms of self-duality and scaling can be seen as stemming from the first principles of special relativity. In any case, these two axioms can also be easily justified outside the special relativity, since they have general validity and can also generate the Boltzmann entropy.

(iv) κ-mathematical statistics: Statistical theory does not only include statistical mechanics, which is a physical theory. Mathematical statistics is another important tool for analyzing complex systems. Two important families of distributions dominate ordinary mathematical statistics. On the one hand, there is the family of distributions with exponential tails (generalized gamma distribution, Weibull distribution, logistic distribution, etc.), and on the other hand, the family of distributions with power-law tails (Pareto, Log-Logistic, Burr type XII or Singh-Maddala distribution, Dagum distribution, etc.). This dichotomy can be overcome in the framework of the present formalism by using the κ-exponential function instead of the ordinary exponential function in the construction of statistical distributions, obtaining a unique family of statistical distributions (κ-generalized gamma distribution, κ-Weibull distribution, κ-logistic distribution, etc.). The new unified class of κ-distributions [178] in the low spectral region reproduces the standard family of exponential distributions, while in the high spectral region, it exhibits Pareto power-law tails.

(v) κ-mathematics: In special relativity, the physical quantities such as momentum, kinetic energy, etc. are relativistically generalized and change their expressions relatively to the corresponding classical expressions. The composition laws of the various physical quantities are also properly generalized. The generalized sum of relativistic moments inevitably leads to the generalization of the entire mathematics. The resulting κ-calculus allows for the introduction of relativistic functions such as the κ-exponential, the κ-logarithm, the κ-trigonometry, and so on. κ-mathematics proves to be isomorphic to ordinary mathematics, which classically obtains the $\kappa \to 0$ limit.

(vi) The Gell-Mann plectic: κ-mathematics is based on a formalism that can handle both simple systems (relativistic one-particle physics) and complex systems (relativistic statistical physics). Furthermore, the same formalism makes it possible to treat physical and non-physical complex systems (statistical physics, information theory, and statistical mathematics) in a unified way. The above features of the κ-formalism give it the status of a candidate for the construction of the holistic theory of simple and complex systems, called *plectics* by Gell-Mann [179,180].

Funding: This research received no external funding.

Institutional Review Board Statement: Not applicable.

Data Availability Statement: Data sharing is not applicable.

Conflicts of Interest: The authors declare no conflicts of interest.

References

1. Vasyliunas, V.M. Low-Energy Electrons in the Magnetosphere as Observed by OGO-1 and OGO-3. *J. Geophys. Res.* **1968**, *73*, 2839. [CrossRef]
2. Kaniadakis, G.; Lissia, M. Editorial on News and expectations in thermostatistics. *Phys. A* **2004**, *340*, XV–XIX.
3. Kaniadakis, G. Non-linear kinetics underlying generalized statistics. *Phys. A* **2001**, *296*, 405–425. [CrossRef]
4. Kaniadakis, G. H-theorem and generalized entropies within the framework of nonlinear kinetics. *Phys. Lett. A* **2001**, *288*, 283–291. [CrossRef]
5. Kaniadakis, G. Statistical mechanics in the context of special relativity. *Phys. Rev. E* **2002**, *66*, 056125. [CrossRef] [PubMed]
6. Kaniadakis, G. Statistical mechanics in the context of special relativity II. *Phys. Rev. E* **2005**, *72*, 036108. [CrossRef] [PubMed]
7. Kaniadakis, G. Towards a relativistic statistical theory. *Phys. A* **2006**, *365*, 17–23. [CrossRef]
8. Kaniadakis, G. Relativistic Entropy and related Boltzmann kinetics. *Eur. Phys. J. A* **2009**, *40*, 275–287. [CrossRef]
9. Kaniadakis, G. Maximum Entropy Principle and power-law tailed distributions. *Eur. Phys. J. B* **2009**, *70*, 3–13. [CrossRef]
10. Kaniadakis, G. Relativistic kinetics and power-law-tailed distributions. *Europhys. Lett.* **2010**, *92*, 35002. [CrossRef]
11. Kaniadakis, G. Power-law tailed statistical distributions and Lorentz transformations. *Phys. Lett. A* **2011**, *375*, 356–359. [CrossRef]
12. Kaniadakis, G. Physical origin of the power-law tailed statistical distribution. *Mod. Phys. Lett. B* **2012**, *26*, 1250061. [CrossRef]
13. Kaniadakis, G. Theoretical Foundations and Mathematical Formalism of the Power-Law Tailed Statistical Distributions. *Entropy* **2013**, *15*, 3983–4010. [CrossRef]
14. Silva, R. The relativistic statistical theory and Kaniadakis entropy: An approach through a molecular chaos hypothesis. *Eur. Phys. J. B* **2006**, *54*, 499–502. [CrossRef]
15. Silva, R. The H-theorem in κ-statistics: Influence on the molecular chaos hypothesis. *Phys. Lett. A* **2006**, *352*, 17–20. [CrossRef]
16. Wada, T. Thermodynamic stabilities of the generalized Boltzmann entropies. *Phys. A* **2004**, *340*, 126–130. [CrossRef]
17. Wada, T. Thermodynamic stability conditions for nonadditive composable entropies. *Contin. Mechan. Thermod.* **2004**, *16*, 263–267. [CrossRef]
18. Kaniadakis, G.; Scarfone, A.M. Lesche stability of κ-entropy. *Phys. A* **2004**, *340*, 102–109. [CrossRef]
19. Abe, S.; Kaniadakis, G.; Scarfone, A.M. Stabilities of generalized entropy. *J. Phys. A Math. Gen.* **2004**, *37*, 10513. [CrossRef]
20. Naudts, J. Deformed exponentials and logarithms in generalized thermostatistics. *Phys. A* **2002**, *316*, 323–334. [CrossRef]
21. Naudts, J. Continuity of a class of entropies and relative entropies. *Rev. Math. Phys.* **2004**, *16*, 809–822. [CrossRef]
22. Scarfone, A.M.; Wada, T. Canonical partition function for anomalous systems described by the κ-entropy. *Prog. Theor. Phys. Suppl.* **2006**, *162*, 45–52. [CrossRef]

23. Yamano, T. On the laws of thermodynamics from the escort average and on the uniqueness of statistical factors. *Phys. Lett. A* **2003**, *308*, 364–368. [CrossRef]
24. Lucia, U. Maximum entropy generation and kappa-exponential mode. *Phys. A* **2010**, *389*, 4558–4563. [CrossRef]
25. Aliano, A.; Kaniadakis, G.; Miraldi, E. Bose-Einstein condensation in the framework of kappa-statistics. *Phys. B* **2003**, *325*, 35–40. [CrossRef]
26. Santos, A.P.; Silva, R.; Alcaniz, J.S.; Anselmo, D.H.A.L. Kaniadakis statistics and the quantum H-theorem. *Phys. Lett. A* **2011**, *375*, 352–355. [CrossRef]
27. Santos, A.P.; Silva, R.; Alcaniz, J.S.; Anselmo, D.H.A.L. Generalized quantum entropies. *Phys. Lett. A* **2011**, *375*, 3119–3123. [CrossRef]
28. Santos, A.P.; Silva, R.; Alcaniz, J.S.; Anselmo, D.H.A.L. Non-Gaussian effects on quantum entropies. *Phys. A* **2012**, *391*, 2182–2192. [CrossRef]
29. Pistone, G. $\kappa-$exponential models from the geometrical point of view. *Eur. Phys. J. B* **2009**, *70*, 29–37. [CrossRef]
30. Pistone, G.; Shoaib, M. Kaniadakis's Information Geometry of Compositional Data. *Entropy* **2023**, *25*, 1107. [CrossRef] [PubMed]
31. Kaniadakis, G.; Lissia, M.; Scarfone, A.M. Deformed logarithms and entropies. *Phys. A* **2004**, *40*, 41–49. [CrossRef]
32. Kaniadakis, G.; Lissia, M.; Scarfone, A.M. Two-parameter deformations of logarithm, exponential, and entropy: A consistent framework for generalized statistical mechanics. *Phys. Rev. E* **2005**, *71*, 046128. [CrossRef]
33. Kaniadakis, G.; Scarfone, A.M. A new one-parameter deformation of the exponential function. *Phys. A* **2002**, *305*, 69–75. [CrossRef]
34. Oikonomou, T.; Bagci, G.B. A completness criterion for Kaniadakis, Abe, and two-parameter generalized statistical theories. *Rep. Math. Phys.* **2010**, *66*, 137–146. [CrossRef]
35. Stankovic, M.S.; Marinkovic, S.D.; Rajkovic, P.M. The deformed exponential functions of two variables in the context of various statistical mechanics. *Appl. Math. Comput.* **2011**, *218*, 2439–2448. [CrossRef]
36. Tempesta, P. Group entropies, correlation laws, and zeta functions. *Phys. Rev. E* **2011**, *84*, 021121. [CrossRef] [PubMed]
37. Deossa Casas, D.E. Sobre Funciones Exponenciales y Logaritmicas Deformadas Segun Kaniadakis. Master's Thesis, Universidad EAFIT, Medellin, Colombia, 2011. Available online: http://hdl.handle.net/10784/156 (accessed on 1 May 2024).
38. Vigelis, R.F.; Cavalcante, C.C. On φ-Families of probability distributions. *J. Theor. Probab.* **2013**, *26*, 870–884. [CrossRef]
39. Scarfone, A.M. Entropic Forms and Related Algebras. *Entropy* **2013**, *15*, 624–649. [CrossRef]
40. Biró, T.S. Kaniadakis Entropy Leads to Particle–Hole Symmetric Distribution. *Entropy* **2022**, *24*, 1217. [CrossRef] [PubMed]
41. Sfetcu, R.-C.; Sfetcu, S.-C.; Preda, V. Some Properties of Weighted Tsallis and Kaniadakis Divergences. *Entropy* **2022**, *24*, 1616. [CrossRef] [PubMed]
42. Sfetcu, R.-C.; Sfetcu, S.-C.; Preda, V. On Tsallis and Kaniadakis Divergences. *Math. Phys. An. Geom.* **2022**, *25*, 7. [CrossRef]
43. Wada, T.; Scarfone, A.M. On the Kaniadakis Distributions Applied in Statistical Physics and Natural Sciences. *Entropy* **2023**, *25*, 292. [CrossRef]
44. Scarfone, A.M.; Wada, T. Multi-Additivity in Kaniadakis Entropy. *Entropy* **2024**, *26*, 77. [CrossRef] [PubMed]
45. Guo, L.; Du, J.; Liu, Z. The property of κ-deformed statistics for a relativistic gas in an electromagnetic field: κ parameter and κ-distribution. *Phys. Lett. A* **2007**, *367*, 431–435. [CrossRef]
46. Guo, L.; Du, J. The κ parameter and κ-distribution in κ-deformed statistics for the sysstems in an external field. *Phys. Lett. A* **2007**, *362*, 368–370. [CrossRef]
47. Lapenta, G.; Markidis, S.; Marocchino, A.; Kaniadakis, G. Relaxation of relativistic plasmas under the effect of wave-particle interactions. *Astrophys. J.* **2007**, *666*, 949–954. [CrossRef]
48. Lapenta, G.; Markidis, S.; Kaniadakis, G. Computer experiments on the relaxation of collisionless plasmas. *J. Stat. Mech.* **2009**, *9*, P02024. [CrossRef]
49. Da Silva, S.L.E.F. Newton's cooling law in generalised statistical mechanics. *Phys. A* **2021**, *565*, 125539. [CrossRef]
50. Rani, S.; Jawad, A.; Sultan, A.M.; Shad, M. Cosmographic and thermodynamic analysis of Kaniadakis holographic dark energy. *Int. J. Mod. Phys. D* **2022**, *31*, 2250078. [CrossRef]
51. Ghaffari, S. Kaniadakis holographic dark energy in Brans-Dicke cosmology. *Mod. Phys. Lett. A* **2022**, *37*, 2250152. [CrossRef]
52. Sharma, U.K.; Dubey, V.C.; Ziaie, A.H.; Moradpour, H. Kaniadakis holographic dark energy in nonflat universe. *Int. J. Mod. Phys. D* **2022**, *31*, 2250013. [CrossRef]
53. Drepanou, N.; Lymperis, A.; Saridakis, E.N.; Yesmakhanova, K. Kaniadakis holographic dark energy and cosmology. *Eur. Phys. J. C* **2022**, *82*, 449. [CrossRef]
54. Korunur, S. Kaniadakis holographic dark energy with scalar field in Bianchi type-V universe. *Int. J. Mod. Phys. A* **2022**, *37*, 2250214. [CrossRef]
55. Hernandez-Almada, A.; Leon, G.; Magana, J.; Garcia-Aspeitia, M.A.; Motta, V.; Saridakis, E.N.; Yesmakhanova, K.; Millano, A.D. Observational constraints and dynamical analysis of Kaniadakis horizon-entropy cosmology. *Mon. Not. R. Astron. Soc.* **2022**, *512*, 5122–5134. [CrossRef]
56. Hernandez-Almada, A.; Leon, G.; Magana, J.; Garcia-Aspeitia, M.A.; Motta, V.; Saridakis, E.N.; Yesmakhanova, K. Kaniadakis-holographic dark energy: Observational constraints and global dynamics. *Mon. Not. R. Astron. Soc.* **2022**, *511*, 4147–4158. [CrossRef]
57. Blasone, M.; Lambiase, G.; Luciano, G.G. Kaniadakis entropy-based characterization of IceCube PeV neutrino signals. *Phys. Dark Universe* **2023**, *42*, 101342. [CrossRef]

58. Sania, A.N.; Rani, S.; Jawad, A. Cosmic and Thermodynamic Consequences of Kaniadakis Holographic Dark Energy in Brans-Dicke Gravity. *Entropy* **2023**, *25*, 576. [CrossRef] [PubMed]
59. Dubey, V.C.; Kumar, M.; Sharma, L.K.; Sharma, U.K. Some features of Kaniadakis holographic dark energy model. *Int. J. Geom. Meth. Mod. Phys.* **2023**, *20*, 2350036. [CrossRef]
60. Jawad, A.; Ul Abideen, Z.; Rani, S. Study of cosmic acceleration in modified theories of gravity through Kaniadakis holographic dark energy. *Mod. Phys. Lett. A* **2023**, *38*, 2350037. [CrossRef]
61. Singh, B.K.; Sharma, U.K.; Sharma, L.K.; Dubey, V.C. Statefinder hierarchy of Kaniadakis holographic dark energy with composite null diagnostic. *Int. J. Geom. Meth. Mod. Phys.* **2023**, *20*, 2350074. [CrossRef]
62. Kumar, P.S.; Pandey, B.D.; Sharma, U.K.; Pankaj. Holographic dark energy through Kaniadakis entropy in non flat universe. *Eur. Phys. J. C* **2023**, *83*, 143. [CrossRef]
63. Sharma, U.K.; Kumar, P.S.; Pankaj. Quintessence scalar field of Kaniadakis holographic dark energy model with statefinder analysis. *Int. J. Geom. Meth. Mod. Phys.* **2023**, *20*, 2450004. [CrossRef]
64. Sadeghi, J.; Gashti, S.N.; Azizi, T. Complex quintessence theory, Tsallis and Kaniadakis holographic dark energy and Brans-Dicke cosmology. *Mod. Phys. Lett. A* **2023**, *38*, 2350076. [CrossRef]
65. Kumar, P.S.; Pandey, B.D.; Pankaj; Sharma, U.K. Kaniadakis agegraphic dark energy. *New Astr.* **2024**, *105*, 102085. [CrossRef]
66. Sultana, S.; Chattopadhyay, S. Intermediate inflation through Nojiri–Odintsov holographic dark fluid with the cosmological settings of Kaniadakis. *Int. J. Geom. Meth. Mod. Phys.* **2024**, *21*, 2450133. [CrossRef]
67. Chokyi, K.K.; Chattopadhyay, S. Cosmology of Tsallis and Kaniadakis holographic dark energy in Saez-Ballester theory and consideration of viscous van der Waals fluid. *Ann. Phys.* **2024**, *463*, 169611. [CrossRef]
68. Sarfraz, A.; Muhammad, K.; Umber, S. Cosmic implications of Kaniadakis HDE model in Chern-Simons modified gravity. *New Astron.* **2024**, *110*, 102226.
69. Ganeswara Raoa, B.; Mohanty, D.J.; Adityac, Y.; Divya Prasanthid, U.Y. Cosmological evolution of Bianchi type-V I_0 Kaniadakis Holographic dark energy model. *East Eur. J. Phys.* **2024**, *1*, 43–54.
70. Yarahmadi, M.; Salehi, A. Using the Kaniadakis horizon entropy in the presence of neutrinos to alleviate the Hubble and S_8 tensions. *Eur. Phys. J. C* **2024**, *84*, 443.
71. Abreu, E.M.C.; Neto, J.A.; Barboza, E.M.; Nunes, R.C. Jeans instability criterion from the viewpoint of Kaniadakis statistics. *EPL* **2016**, *114*, 55001. [CrossRef]
72. Abreu, E.M.C.; Neto, J.A.; Barboza, E.M.; Nunes, R.C. Tsallis and Kaniadakis statistics from the viewpoint of entropic gravity formalism. *Int. J. Mod. Phys.* **2017**, *32*, 1750028. [CrossRef]
73. Chen, H.; Zhang, S.X.; Liu, S.Q. Jeans gravitational instability with kappa-deformed Kaniadakis distribution. *Chin. Phys. Lett.* **2017**, *34*, 075101. [CrossRef]
74. Abreu, E.M.C.; Neto, J.A.; Mendes, A.C.R.; Bonilla, A. Tsallis and Kaniadakis statistics from a point of view of the holographic equipartition law. *EPL* **2018**, *121*, 45002. [CrossRef]
75. Abreu, E.M.C.; Neto, J.A.; Mendes, A.C.R.; Bonilla, A.; de Paula, R.M. Cosmological considerations in Kaniadakis statistics. *EPL* **2018**, *124*, 30003. [CrossRef]
76. Abreu, E.M.C.; Neto, J.A.; Mendes, A.C.R.; de Paula, R.M. Loop quantum gravity Immirzi parameter and the Kaniadakis statistics. *Chaos Sol. Fractals* **2019**, *118*, 307–310. [CrossRef]
77. Yang, W.; Xiong, Y.; Chen, H.; Liu, S. Jeans instability of dark-baryonic matter model in the context of Kaniadakis' statistic distribution. *J. Taibah Univ. Sci.* **2022**, *16*, 337–343. [CrossRef]
78. He K.-R. Jeans analysis with κ-deformed Kaniadakis distribution in f (R) gravity. *Phys. Scr.* **2022**, *97*, 025601. [CrossRef]
79. Moradpour, H.; Javaherian, M.; Namvar, E.; Ziaie, A.H. Gamow Temperature in Tsallis and Kaniadakis Statistics. *Entropy* **2022**, *24*, 797. [CrossRef] [PubMed]
80. Luciano, G.G. Modified Friedmann equations from Kaniadakis entropy and cosmological implications on baryogenesis and 7Li -abundance. *Eur. Phys. J. C* **2022**, *82*, 314. [CrossRef]
81. Luciano, G.G.; Saridakis, E.N. P-v criticalities, phase transitions and geometrothermodynamics of charged AdS black holes from Kaniadakis statistics. *J. High Energy Phys.* **2023**, *2023*, 114. [CrossRef]
82. Lambiase, G.; Luciano, G.G.; Sheykhi, A. Slow-roll inflation and growth of perturbations in Kaniadakis modification of Friedmann cosmology. *Eur. Phys. J. C* **2023**, *83*, 936. [CrossRef]
83. Sheykhi, A. Corrections to Friedmann equations inspired by Kaniadakis entropy. *Phys. Lett.* **2024**, *850*, 138495. [CrossRef]
84. Luciano, G.G. Gravity and Cosmology in Kaniadakis Statistics: Current Status and Future Challenges. *Entropy* **2022**, *24*, 1712. [CrossRef] [PubMed]
85. Sadeghnezhad, N. Entropic gravity and cosmology in Kaniadakis statistics. *Int. J. Mod. Phys. D* **2023**, *32*, 2350002. [CrossRef]
86. Wada, T.; Scarfone, A.M. Asymptotic solutions of a nonlinear diffusive equation in the framework of κ-generalized statistical mechanics. *Eur. Phys. J. B* **2009**, *70*, 65–71. [CrossRef]
87. Wada, T. A nonlinear drift which leads to kappa-generalized distributions. *Eur. Phys. J. B* **2010**, *73*, 287–291. [CrossRef]
88. Kaniadakis, G.; Quarati, P.; Scarfone, A.M. Kinetical foundations of non-conventional statistics. *Phys. A* **2002**, *305*, 76–83. [CrossRef]
89. Biro, T.S.; Kaniadakis, G. Two generalizations of the Boltzmann equation. *Eur. Phys. J. B* **2006**, *50*, 3–6. [CrossRef]

90. Casas, G.A.; Nobre, F.D.; Curado, E.M.F. Entropy production and nonlinear Fokker-Planck equations. *Phys. Rev. E* **2012**, *86*, 061136. [CrossRef]
91. Hirica, I.-E.; Pripoae, C.-L.; Pripoae, G.-T.; Preda, V. Lie Symmetries of the Nonlinear Fokker-Planck Equation Based on Weighted Kaniadakis Entropy. *Mathematics* **2022**, *10*, 2776. [CrossRef]
92. Gomez, I.S.; da Costa, B.G.; dos Santos, M.A.F. Inhomogeneous Fokker–Planck equation from framework of Kaniadakis statistics. *Commun. Nonlin. Sci. Num. Sim.* **2023**, *119*, 107131. [CrossRef]
93. Evangelista, L.R.; Lenzi, E.K. Nonlinear Fokker–Planck Equations, H-Theorem and Generalized Entropy of a Composed System. *Entropy* **2023**, *25*, 1357. [CrossRef] [PubMed]
94. Guha, P. The κ-Deformed Calogero-Leyvraz Lagrangians and Applications to Integrable Dynamical Systems. *Entropy* **2023**, *24*, 1673. [CrossRef]
95. Rossani, A.; Scarfone, A.M. Generalized kinetic equations for a system of interacting atoms and photons: Theory and Simulations. *J. Phys. A* **2004**, *37*, 4955–4975. [CrossRef]
96. Guo, L.N.; Du, J.L. The two parameters (κ, r) in the generalized statistics. *Phys. A* **2010**, *389*, 47–51. [CrossRef]
97. Guo, L.N. Physical meaning of the parametres in the two-parameter (κ, ζ generalized theory. *Modern Phys. Lett. B* **2012**, *26*, 1250064. [CrossRef]
98. Silva, J.M.; Silva, R.; Lima, J.A.S. Conservative force fields in non-Gaussian statistics. *Phys. Lett. A* **2008**, *372*, 5754–5757. [CrossRef]
99. Carvalho, J.C.; Silva, R.; do Nascimento, J.D., Jr.; De Medeiros, J.R. Power law statistics and stellar rotational velocities in the Pleiades. *Europhys. Lett.* **2008**, *84*, 59001. [CrossRef]
100. Carvalho, J.C.; do Nascimento, J.D., Jr.; Silva, R.; De Medeiros, J.R. Non-gaussian statistics and stellar rotational velocities of main sequence field stars. *Astrophys. J. Lett.* **2009**, *696*, L48–L51. [CrossRef]
101. Carvalho, J.C.; Silva, R.; do Nascimento, J.D., Jr.; Soares, B.B.; De Medeiros, J.R. Observational measurement of open stellar clusters: A test of Kaniadakis and Tsallis statistics. *Europhys. Lett.* **2010**, *91*, 69002. [CrossRef]
102. Bento, E.P.; Silva, J.R.P.; Silva, R. Non-Gaussian statistics, Maxwellian derivation and stellar polytropes. *Phys. A* **2013**, *392*, 666–672. [CrossRef]
103. Teweldeberhan, A.M.; Miller, H.G.; Tegen, R. κ-deformed Statistics and the formation of a quark-gluon plasma. *Int. J. Mod. Phys. E* **2003**, *12*, 669–673. [CrossRef]
104. Pereira, F.I.M.; Silva, R.; Alcaniz, J.S. Non-gaussian statistics and the relativistic nuclear equation of state. *Nucl. Phys. A* **2009**, *828*, 136–148. [CrossRef]
105. Cravero, M.; Iabichino, G.; Kaniadakis, G.; Miraldi, E.; Scarfone, A.M. A κ-entropic approach to the analysis of the fracture problem. *Phys. A* **2004**, *340*, 410–417. [CrossRef]
106. Gougam, L.A.; Tribeche, M. Electron-acoustic waves in a plasma with a kappa-deformed Kaniadakis electron distribution. *Phys. Plasmas* **2016**, *23*, 014501. [CrossRef]
107. Lourek, I.; Tribeche, M. On the role of the κ-deformed Kaniadakis distribution in nonlinear plasma waves. *Phys. A* **2016**, *441*, 215–220. [CrossRef]
108. Lopez, R.A.; Navarro, R.E.; Pons, S.I.; Araneda, J.A. Landau damping in Kaniadakis and Tsallis distributed electron plasmas. *Phys. Plasmas* **2017**, *24*, 102119. [CrossRef]
109. Chen, H.; Zhang, S.X.; Liu, S.Q. Te longitudinal plasmas modes of κ-deformed kaniadakis distributed plasmas. *Phys. Plasmas* **2017**, *24*, 022125. [CrossRef]
110. Saha, A.; Tamang, J. Qualitative analysis of the positron-acoustic waves in electron-positron-ion plasmas with kappa deformed Kaniadakis distributed electrons and hot positrons. *Phys. Plasmas* **2017**, *24*, 082101. [CrossRef]
111. Lourek, I.; Tribeche, M. Dust charging current in non equilibrium dusty plasma in the context of Kaniadakis generalization. *Phys. A* **2019**, *517*, 522–529. [CrossRef]
112. Khalid, M.; Rahman, A.-U. Oblique ion acoustic excitations in a magnetoplasma having κ-deformed Kaniadakis distributed electrons. *Astr. Space Sc.* **2020**, *365*, 75. [CrossRef]
113. Tan, L.; Yang, Q.; Chen, H.; Liu, S. The Longitudinal Plasma Modes of κ-Deformed Kaniadakis Distributed Plasmas Carrying Orbital Angular Momentum. *Entropy* **2022**, *24*, 1211. [CrossRef] [PubMed]
114. Irshad, M.; Khalid, M.; Ata-ur-Rahman. Modulational instability of ion acoustic excitations in a plasma with a κ-deformed Kaniadakis electron distribution. *Eur. Phys. J. Plus* **2022**, *137*, 893. [CrossRef]
115. Bellahsene, Z.; Bacha, M.; Zerguini, T.H. The role of κ-deformed Kaniadakis distributed electrons on the dust ion-acoustic waves in charge-varying dusty plasma. *Contrib. Plasma Phys.* **2023**, *64*, e202300155. [CrossRef]
116. Dubinov, A.E. Gas-dynamic approach to the theory of non-linear ion-acoustic waves in plasma with Kaniadakis' distributed species. *Adv. Space Res.* **2023**, *71*, 1108–1115. [CrossRef]
117. Raut, S.; Mondal, K.K.; Chatterjee, P.; Roy, S. Dust ion acoustic bi-soliton, soliton, and shock waves in unmagnetized plasma with Kaniadakis-distributed electrons in planar and nonplanar geometry. *Eur. Phys. J. D* **2023**, *77*, 100. [CrossRef]
118. Irshad, M.; Ata-Ur-Rahman; Khalid, M.; Khan, S.; Alotaibi, B.M.; El-Sherif, L.S.; El-Tantawy, S.A. Effect of κ-deformed Kaniadakis distribution on the modulational instability of electron-acoustic waves in a non-Maxwellian plasma. *Phys. Fluids* **2023**, *35*, 105116. [CrossRef]
119. Khalid, M.; Kabir, A.; Jan, S.U.; Eldin, S.M. Coexistence of Compressive and Rarefactive Positron-Acoustic Electrostatic Excitations in Unmagnetized Plasma with Kaniadakis Distributed Electrons and Hot Positrons. *Braz. J. Phys.* **2023**, *53*, 66. [CrossRef]

120. Bala, P.; Kaur, G. Modulational instability of ion-acoustic waves in multicomponent plasma using κ-deformed Kaniadakis distribution. *Pramana J. Phys.* **2024**, *98*, 7. [CrossRef]
121. Hristopulos, D.T.; Petrakis, M.P.; Kaniadakis, G. Finite-size effects on return interval distributions for weakest-link-scaling systems. *Phys. Rev. E* **2014**, *89*, 052142. [CrossRef] [PubMed]
122. Hristopulos, D.T.; Petrakis, M.P.; Kaniadakis, G. Weakest-Link Scaling and Extreme Events in Finite-Sized Systems. *Entropy* **2015**, *17*, 1103–1122. [CrossRef]
123. Da Silva, S.L.E.F. κ-generalised Gutenberg–Richter law and the self-similarity of earthquakes. *Chaos Solit. Fractals* **2021**, *143*, 110622. [CrossRef]
124. Hristopulos, D.T.; Baxevani, A. Kaniadakis Functions beyond Statistical Mechanics: Weakest-Link Scaling, Power-Law Tails, and Modified Lognormal Distribution. *Entropy* **2022**, *24*, 1362. [CrossRef] [PubMed]
125. Da Silva, S.L.E.F.; Carvalho, P.T.; de Araújo, J.M.; Corso, G. Full-waveform inversion based on Kaniadakis statistics. *Phys. Rev. E* **2020**, *101*, 053311. [CrossRef] [PubMed]
126. Da Silva, S.L.E.F.; dos Santos Lima, G.Z.; de Araújo, J.M.; Corso, G. Extensive and nonextensive statistics in seismic inversion. *Phys. A* **2021**, *563*, 125496. [CrossRef]
127. Da Silva, S.L.E.F.; Kaniadakis, G. κ–statistics approach to optimal transport waveform inversion. *Phys. Rev. E* **2022**, *106*, 034113. [CrossRef] [PubMed]
128. Da Silva, S.L.E.F.; de Araújo, J.M.; de la Barra, E.; Corso, G. A Graph-Space Optimal Transport Approach Based on Kaniadakis κ-Gaussian Distribution for Inverse Problems Related to Wave. *Entropy* **2023**, *25*, 990. [CrossRef]
129. Da Silva, S.L.E.F.; Kaniadakis, G. A graph-space optimal transport FWI approach based on κ-generalized Gaussian distribution. In *Third International Meeting for Applied Geoscience & Energy*; Society of Exploration Geophysicists: Houston, TX, USA, 2023; pp. 670–674.
130. De Abreu, W.V.; Goncalves, A.C.; Martinez, A.S. Analytical solution for the Doppler broadening function using the Kaniadakis distribution. *Ann. Nucl. Energy* **2019**, *126*, 262–268. [CrossRef]
131. De Abreu, W.V.; Martinez, A.S. New analytical formulations for the Doppler broadening function and interference term based on Kaniadakis distributions. *Ann. Nucl. Energy* **2020**, *135*, 106960. [CrossRef]
132. De Abreu, W.V.; Maciel, J.M.; Martinez, A.S.; Gonçalves, A.D.C.; Schmidt, L. Doppler Broadening of Neutron Cross-Sections Using Kaniadakis Entropy. *Entropy* **2022**, *24*, 1437. [CrossRef] [PubMed]
133. De Abreu, W.V.; Martinez, A.S.; do Carmo, E.D.; Gonçalves, A.C. A novel analytical solution of the deformed Doppler broadening function using the Kaniadakis distribution and the comparison of computational efficiencies with the numerical solution. *Nucl. Eng. Techn.* **2022**, *54*, 1471–1481. [CrossRef]
134. Martinez, A.S.; de Abreu, W.V. The Scientific Contribution of the Kaniadakis Entropy to Nuclear Reactor Physics: A Brief Review. *Entropy* **2023**, *25*, 478. [CrossRef] [PubMed]
135. Da Costa, B.G.; Gomez, I.S.; Portesi, M. κ-Deformed quantum and classical mechanics for a system with position-dependent effective mass. *J. Math. Phys.* **2020**, *61*, 082105. [CrossRef]
136. Chung, W.S.; Hassanabadi, H. Investigation of Some Quantum Mechanics Problemswith κ-Translation Symmetry. *Int. J. Theor. Phys.* **2022**, *61*, 110. [CrossRef]
137. Santos, F.F.; Boschi-Filho, H. Black branes in asymptotically Lifshitz spacetimes with arbitrary exponents in κ-Horndeski gravity. *Phys. Rev. D* **2024**, *109*, 064035. [CrossRef]
138. Coraddu, M.; Lissia, M.; Tonelli, R. Statistical descriptions of nonlinear systems at the onset of chaos. *Phys. A* **2006**, *365*, 252–257. [CrossRef]
139. Tonelli, R.; Mezzorani, G.; Meloni, F.; Lissia, M.; Coraddu, M. Entropy production and Pesin identity at the onset of chaos. *Prog. Theor. Phys.* **2006**, *115*, 23–29. [CrossRef]
140. Celikoglu, A.; Tirnakli, U. Sensitivity function and entropy increase rates for z-logistic map family at the edge of chaos. *Phys. A* **2006**, *372*, 238–242. [CrossRef]
141. Olemskoi, A.I.; Kharchenko, V.O.; Borisyuk, V.N. Multifractal spectrum of phase space related to generalized thermostatistics. *Phys. A* **2008**, *387*, 1895–1906. [CrossRef]
142. Olemskoi, A.I.; Borisyuk, V.N.; Shuda, I.A. Statistical field theories deformed within different calculi. *Eur. Phys. J. B* **2010**, *77*, 219–231. [CrossRef]
143. Souza, N.T.C.M.; Anselmo, D.H.A.L.; Silva, R.; Vasconcelos, M.S.; Mello, V.D. A kappa-statistical analysis of the Y-chromosome. *EPL* **2014**, *108*, 28004. [CrossRef]
144. Costa, M.O.; Silva, R.; Anselmo, D.H.A.L.; Silva, J.R.P. Analysis of human DNA through power-law statistics. *Phys. Rev. E* **2019**, *99*, 022112. [CrossRef] [PubMed]
145. De Lima, M.M.F.; Anselmo, D.H.A.L.; Silva, R.; Nunes, G.H.S.; Fulco, U.L.; Vasconcelos, M.S.; Mello, V.D. A Bayesian Analysis of Plant DNA Length Distribution via κ-Statistics. *Entropy* **2022**, *24*, 1225. [CrossRef] [PubMed]
146. Abul-Magd, A.Y. Nonextensive random-matrix theory based on Kaniadakis entropy. *Phys. Lett. A* **2007**, *361*, 450–454. [CrossRef]
147. Abul-Magd, A.Y. Nonextensive and superstatistical generalizations of random-matrix theory. *Eur. Phys. J. B* **2009**, *70*, 39–48. [CrossRef]
148. Abul-Magd, A.Y.; Abdel-Mageed, M. Kappa-deformed random-matrix theory based on Kaniadakis statistics. *Modern Phys. Lett. B* **2012**, *26*, 1250059. [CrossRef]

149. Da Silva, S.L.E.F.; dos Santos Lima, G.Z.; Volpe, E.V.; de Araújo, J.M.; Corso, G. Robust approaches for inverse problems based on Tsallis and Kaniadakis generalised statistics. *Eur. Phys. J. Plus* **2021**, *136*, 518. [CrossRef]
150. Dos Santos Lima, G.Z.; de Lima, J.V.T.; de Araújo, J.M.; Corso, G.; da Silva, S.L.E.F. Generalized statistics: Applications to data inverse problems with outlier-resistance. *PLoS ONE* **2023**, *18*, e0282578. [CrossRef] [PubMed]
151. Wada, T.; Suyari, H. κ-generalization of Gauss' law of error. *Phys. Lett. A* **2006**, *348*, 89–93. [CrossRef]
152. Da Silva, S.L.E.F.; Silva, R.; dos Santos Lima, G.Z.; de Araújo, J.M.; Corso, G. An outlier-resistant κ-generalized approach for robust physical parameter estimation. *Phys. A* **2022**, *600*, 127554. [CrossRef]
153. Topsoe, F. Entropy and equilibrium via games of complexity. *Phys. A* **2004**, *340*, 11–31. [CrossRef]
154. Macedo-Filho, A.; Moreira, D.A.; Silva, R.; da Silva, L.R. Maximum entropy principle for Kaniadakis statistics and networks. *Phys. Lett. A* **2013**, *377*, 842–846. [CrossRef]
155. Wada, T.; Suyari, H. A two-parameter generalization of Shannon-Khinchin axioms and the uniqueness teorem. *Phys. Lett. A* **2007**, *368*, 199–205. [CrossRef]
156. Clementi, F.; Gallegati, M.; Kaniadakis, G. κ-generalized statistics in personal income distribution. *Eur. Phys. J. B* **2007**, *57*, 187–193. [CrossRef]
157. Clementi, F.; Di Matteo, T.; Gallegati, M.; Kaniadakis, G. The κ-generalized distribution: A new descriptive model for the size distribution of incomes. *Phys. A* **2008**, *387*, 3201–3208. [CrossRef]
158. Clementi, F.; Gallegati, M.; Kaniadakis, G. A κ-generalized statistical mechanics approach to income analysis. *J. Stat. Mech.* **2009**, P02037. [CrossRef]
159. Clementi, F.; Gallegati, M.; Kaniadakis, G. A model of personal income distribution with application to Italian data. *Empir. Econ.* **2011**, *39*, 559–591. [CrossRef]
160. Clementi, F.; Gallegati, M.; Kaniadakis, G. A new model of income distribution: The kappa-generalized distribution. *J. Econ.* **2012**, *105*, 63–91. [CrossRef]
161. Clementi, F.; Gallegati, M.; Kaniadakis, G. A generalized statistical model for the size distribution of wealth. *J. Stat. Mech.* **2012**, P12006. [CrossRef]
162. Clementi, F. The Kaniadakis Distribution for the Analysis of Income and Wealth Data. *Entropy* **2023**, *25*, 1141. [CrossRef] [PubMed]
163. Rajaonarison, D.; Bolduc, D.; Jayet, H. The K-deformed multinomial logit model. *Econ. Lett.* **2005**, *86*, 13–20. [CrossRef]
164. Rajaonarison, D. Deterministic heterogeneity in tastes and product differentiation in the K-logit model. *Econ. Lett.* **2008**, *100*, 396–399. [CrossRef]
165. Trivellato, B. The minimal κ-entropy martingale measure. *Int. J. Theor. Appl. Financ* **2012**, *15*, 1250038. [CrossRef]
166. Trivellato, B. Deformed exponentials and applications to finance. *Entropy* **2013**, *15*, 3471–3489. [CrossRef]
167. Tapiero, O.J. A maximum (non-extensive) entropy approach to equity options bid-ask spread. *Phys. A* **2013**, *392*, 3051–3060. [CrossRef]
168. Bertotti, M.L.; Modenese, G. Exploiting the flexibility of a family of models for taxation and redistribution. *Eur. Phys. J. B* **2012**, *85*, 261–270. [CrossRef]
169. Shannon, C.E. A mathematical theory of communication. *Bell Syst. Tech. J.* **1948**, *27*, 379. [CrossRef]
170. Khinchin, A.Y. *Mathematical Foundations of Information Theory*; Dover Publications: Mineola, NY, USA, 1957.
171. Csiszár, I. Information-Type Measures of Difference of Probability Distributions and Indirect Observations. *Stud. Sci. Math. Hung.* **1967**, *2*, 299–318.
172. Zografos, K. On reconsidering entropies and divergences and their cumulative counterparts: Csiszár's, DPD's and Fisher's type cumulative and survival measures. *Probab. Eng. Informational Sci.* **2023**, *37*, 294–321. [CrossRef]
173. Watanabe, S. *Knowing and Guessing*; Wiley: New York, NY, USA, 1969.
174. Barlow, H. Conditions for versatile learning, Helmholtz's unconscious inference, and the task of perception. *Vis. Res.* **1990**, *30*, 1561–1571. [CrossRef]
175. Ilic, V.M.; Korbel, J.; Gupta, S.; Scarfone, A.M. An overview of generalized entropic forms. *EuroPhysics Lett.* **2021**, *133*, 5. [CrossRef]
176. Groot, S.R.; van Leeuwen, W.A.; van Weert, C.G. *Relativistic Kinetic Theory: Principles and Applications*; North-Holand: Amsterdam, The Netherlands, 1990.
177. Cercignani, C.; Medeiros Kremer, G. *The Relativistic Boltzmann Equation: Theory and Applications*; Birkhauser: Basel, Switzerland, 2002.
178. Kaniadakis, G. New power-law tailed distributions emerging in kstatistics. *EPL* **2021**, *133*, 10002. [CrossRef]
179. Gell-Mann, M. Let's Call It Plectics. *Complexity* **1995**, *1*, 96.
180. Gell-Mann, M. Plectics: The study of simplicity and complexity. *Europhys. New* **2002**, *33*, 17–20. [CrossRef]

MDPI AG
Grosspeteranlage 5
4052 Basel
Switzerland
Tel.: +41 61 683 77 34

Entropy Editorial Office
E-mail: entropy@mdpi.com
www.mdpi.com/journal/entropy

www.ingramcontent.com/pod-product-compliance
Lightning Source LLC
Chambersburg PA
CBHW041416210926
51850CB00013B/3284
* 9 7 8 3 7 2 5 8 3 6 4 9 9 *